Symplectic Geometry and Mirror Symmetry

Symplectic Geometry and Mirror Symmetry

Proceedings of the 4th KIAS Annual International Conference

Korea Institute for Advanced Study

Seoul, South Korea   14 – 18 August 2000

Editors

K. Fukaya
Kyoto University, Japan

Y.-G. Oh
University of Wisconsin-Madison, USA &
Korea Institute for Advanced Study, South Korea

K. Ono
Hokkaido University, Japan

G. Tian
Massachusetts Institute of Technology, USA

World Scientific
New Jersey • London • Singapore • Hong Kong

Published by

World Scientific Publishing Co. Pte. Ltd.
5 Toh Tuck Link, Singapore 596224
USA office: 27 Warren Street, Suite 401-402, Hackensack, NJ 07601
UK office: 57 Shelton Street, Covent Garden, London WC2H 9HE

British Library Cataloguing-in-Publication Data
A catalogue record for this book is available from the British Library.

SYMPLECTIC GEOMETRY AND MIRROR SYMMETRY
Proceedings of the 4th KIAS Annual International Conference

ISBN-13 978-981-02-4714-0
ISBN-10 981-02-4714-1

Speakers

D. Auroux (Ecole Polytechnique), T. Eguchi (University of Tokyo), E. Getzler (Northwestern University), A. Givental (University of California-Berkeley), M. Gross (Warwick University), K. Hori (Harvard University) E. Ionel (University of Wisconsin-Madison), Bumsig Kim (Postech), Xiaobo Liu (University of Notre Dame), D. McDuff (SUNY-Stony Brook), H. Nakajima (Kyoto University), R. Pandharipande (Caltech), Jaesuk Park (Columbia University), Wei-Dong Ruan (University of Illinois-Chicago), P. Seidel (Ecole Polytechnique), R. Thomas (Imperial College)

Preface

In 1993, M. Kontsevich proposed a conceptual framework for explaining the phenomenon of mirror symmetry. Mirror symmetry had been discovered by physicists in string theory as a duality between families of three-dimensional Calabi–Yau manifolds. Kontsevich's proposal uses Fukaya's construction of the A_∞-category of Lagrangian submanifolds on the symplectic side and the derived category of coherent sheaves on the complex side. The theory of mirror symmetry was further enhanced by physicists in the language of D-branes and also by Strominger–Yau–Zaslow in the geometric set-up of (special) Lagrangian torus fibrations. It rapidly expanded its scope across from geometry, topology, algebra to physics.

In this volume, leading experts in the field explore recent developments in relation to homological mirror symmetry, Floer theory, D-branes and Gromov–Witten invariants. Kontsevich–Soibelman describe their solution to the mirror conjecture on the abelian variety based on the deformation theory of A_∞-categories, and Ohta describes recent work on the Lagrangian intersection Floer theory by Fukaya–Oh–Ohta–Ono which takes an important step towards a rigorous construction of the A_∞-category. There follow a number of contributions on the homological mirror symmetry, D-branes and the Gromov–Witten invariants, e.g. Getzler shows how the Toda conjecture follows from recent work of Givental, Okounkov and Pandharipande. This volume provides a timely presentation of the important developments of recent years in this rapidly growing field.

We would like to take the opportunity to thank the participants of the conference and in particular the speakers for their contribution to make the conference a success and the contributors to this volume for making this proceedings successful. We also like to thank the Korea Institute for Advanced Study for its financial support, and the KIAS supporting staffs for their organizational help during the conference.

May 2001

K. Fukaya
Y.-G. Oh
K. Ono
G. Tian

Contents

ESTIMATED TRANSVERSALITY IN SYMPLECTIC GEOMETRY AND PROJECTIVE MAPS

DENIS AUROUX

Centre de Mathématiques, Ecole Polytechnique, F-91128 Palaiseau, France
E-mail: auroux@math.polytechnique.fr

1 Introduction

Since Donaldson's original work,[7] approximately holomorphic techniques have proven themselves most useful in symplectic geometry and topology, and various classical constructions from algebraic geometry have been extended to the case of symplectic manifolds.[3,4,8,11] All these results rely on an estimated transversality statement for approximately holomorphic sections of very positive bundles, obtained by Donaldson.[7,8] However, the arguments require transversality not only for sections but also for their covariant derivatives, which makes it necessary to painstakingly imitate the arguments underlying Thom's classical ***strong transversality theorem*** for jets.

It is our aim in this paper to formulate and prove a general result of estimated transversality with respect to finite stratifications in jet bundles. The transversality properties obtained in the various above-mentioned papers then follow as direct corollaries of this result, thus allowing some of the arguments to be greatly simplified. The result can be formulated as follows (see §2 and §3 for definitions) :

Theorem 1.1 *Let $(E_k)_{k\gg 0}$ be an asymptotically very ample sequence of locally splittable complex vector bundles over a compact almost-complex manifold (X, J). Let $\mathcal{S}_k$ be asymptotically holomorphic finite Whitney quasi-stratifications of the holomorphic jet bundles $\mathcal{J}^r E_k$. Finally, let $\delta > 0$ be a fixed constant. Then there exist constants K and η such that, given any asymptotically holomorphic sections s_k of E_k over X, there exist asymptotically holomorphic sections σ_k of E_k with the following properties for all $k \geq K$:*

(1) $|\sigma_k - s_k|_{C^{r+1},g_k} < \delta$;

(2) *the jet $j^r\sigma_k$ of σ_k is η-transverse to the quasi-stratification $\mathcal{S}_k$.*

We start by introducing in §2 a general notion of ampleness over an almost-complex manifold. Then, in §3 we define the notion of approximately holomorphic quasi-stratification of a jet bundle. Theorem 1.1 and its one-parameter version are proved in §4. Finally, we discuss applications in §5.

2 Ample bundles over almost-complex manifolds

The most general setup in which one can try to define a notion of ampleness is the following. Let X be a compact $2n$-dimensional manifold (possibly with boundary), endowed with an almost-complex structure J. In order to make estimates, we also endow X with a Riemannian metric g compatible with J (i.e. J is g-antisymmetric).

Definition 2.1 *Given positive constants c and δ, a complex line bundle L over X endowed with a Hermitian metric and a connection ∇^L is* (c,δ)-ample *if its curvature 2-form F_L satisfies the inequalities $iF_L(v,Jv) \geq c\,g(v,v)$ for every tangent vector $v \in TX$, and $\sup|F_L^{0,2}| \leq \delta$.*

A sequence of complex line bundles L_k with metrics and connections is asymptotically very ample *if there exist fixed constants δ and $(C_r)_{r\geq 0}$, and a sequence $c_k \to +\infty$, such that the curvature F_k of L_k satisfies the following properties :* (1) $iF_k(v,Jv) \geq c_k\,g(v,v)$ *for every tangent vector* $v \in TX$; (2) $\sup|F_k^{0,2}| \leq \delta\, c_k^{1/2}$; (3) $\sup|\nabla^r F_k| \leq C_r\, c_k\ \forall r \geq 0$.

Most of this definition is a natural extension to the almost-complex setup of the classical notion of ampleness on a complex manifold. Because the notion of holomorphic bundle is not relevant in the case of a non-integrable complex structure, one should allow the curvature to contain a non-trivial $(0,2)$-part. However, because $F_L^{0,2}$ is an obstruction to the existence of holomorphic sections, we need uniform bounds on this quantity in order to hope for the existence of approximately holomorphic sections.

The last condition in the definition seems less natural and should largely be considered as a technical assumption needed to obtain some control over the behavior of sections; it is likely that a suitable argument, possibly involving plurisubharmonic techniques, could allow the bounds to be significantly weakened.

Observe that the curvature of a (c,δ)-ample line bundle over X defines, after multiplication by $\frac{i}{2\pi}$, a J-tame symplectic structure on X with integral cohomology class (J is compatible with this symplectic structure if and only if the curvature is of type $(1,1)$).

Conversely, assume that X carries a J-compatible symplectic form ω with integral cohomology class, and choose g to be the Riemannian metric induced by J and ω. Then there exists a line bundle L with first Chern class $c_1(L) = [\omega]$ and a connection with curvature $-2\pi i\omega$ on L. By construction the line bundle L is $(2\pi, 0)$-ample; moreover, the line bundles $L^{\otimes k}$ with the induced connections are $(2\pi k, 0)$-ample and define an asymptotically very ample sequence of line bundles. This example is by far the most interesting one for applications, but many other situations can be considered as well.

In the rest of this section, we consider an asymptotically very ample sequence of line bundles over X, and study the properties of L_k for large values of k. In order to make the estimates below easier to understand, we rescale the metric by setting $g_k = c_k\, g$, which amounts to dividing by $c_k^{r/2}$ the norm of all r-tensors; the Levi-Civita connection is not affected by this rescaling. The bounds of Definition 2.1 imply that : $iF_k(v, Jv) \geq g_k(v,v)$; $|F_k^{0,2}|_{g_k} = O(c_k^{-1/2})$; $|F_k|_{g_k} = O(1)$; $|\nabla^r F_k|_{g_k} = O(c_k^{-1/2})\ \forall r \geq 1$. Also observe that $|\nabla^r J|_{g_k} = O(c_k^{-1/2})\ \forall r \geq 1$. Better bounds on higher-order derivatives are trivially available but we won't need them.

Lemma 2.1 *Let L_k be a sequence of asymptotically very ample line bundles L_k over X, and denote by F_k the curvature of L_k. Let $\omega_k = iF_k$, and let c_k be the constants appearing in Definition 2.1. Denote by ∇ the Levi-Civita connection associated to g. Then, for large enough k there exist ω_k-compatible almost-complex structures $\tilde{J}_k$ such that $|\nabla^r(\tilde{J}_k - J)|_{g_k} = O(c_k^{-1/2})\ \forall r \geq 0$.*

Proof. We construct $\tilde{J}_k$ locally; patching together the local constructions in order to obtain a globally defined almost-complex structure still satisfying the same type of bounds is an easy task left to the reader (recall that the space of ω_k-compatible almost-complex structures is pointwise contractible).

Let e_1 be a local tangent vector field of unit g_k-length and with $|\nabla^r e_1|_{g_k} = O(c_k^{-1/2})\ \forall r \geq 1$ (observe that, because of the rescaling process, (X, g_k) is almost flat for large k). We define $e_1' = Je_1$, and observe that e_1' has unit g_k-length (J is g-unitary and hence g_k-unitary) and $\omega_k(e_1, e_1') \geq 1$. Next, we proceed inductively, assuming that we have defined local vector fields $e_1, e_1', \dots, e_m, e_m'$ with the following properties for all $i, j \leq m$: $e_1, \dots, e_m$ have unit g_k-length; $\omega_k(e_i, e_j) = \omega_k(e_i', e_j') = 0$; $\omega_k(e_i, e_j') = 0$ if $i \neq j$; $\omega_k(e_i, e_i') \geq 1$; $e_i' - Je_i \in \mathrm{span}(e_1, e_1', \dots, e_{i-1}, e_{i-1}')$; $|e_i' - Je_i|_{g_k} = O(c_k^{-1/2})$; $|\nabla^r e_i|_{g_k} = O(c_k^{-1/2})$ and $|\nabla^r e_i'|_{g_k} = O(c_k^{-1/2})\ \forall r \geq 1$.

We choose e_{m+1} to be a g_k-unit vector field which is ω_k-orthogonal to $e_1, e_1', \dots, e_m, e_m'$. The bound on $|\nabla^r \omega_k|_{g_k}$ implies that we can choose e_{m+1} in such a way that $|\nabla^r e_{m+1}|_{g_k} = O(c_k^{-1/2})$. Next, we define

$$e_{m+1}' = Je_{m+1} + \sum_{i=1}^{m} \frac{\omega_k(e_i', Je_{m+1})e_i - \omega_k(e_i, Je_{m+1})e_i'}{\omega_k(e_i, e_i')}.$$

By construction, $\omega_k(e_i, e_{m+1}') = \omega_k(e_i', e_{m+1}') = 0$ for all $i \leq m$. Moreover, $\omega_k(e_{m+1}, e_{m+1}') = \omega_k(e_{m+1}, Je_{m+1}) \geq 1$.

Since $\omega_k(Je_i, e_{m+1}) = 0$, and because $\omega_k(Je_i, e_{m+1}) - \omega_k(e_i, Je_{m+1})$ is a component of $\omega_k^{0,2}$, we have $\omega_k(e_i, Je_{m+1}) = O(c_k^{-1/2})$. Similarly, $\omega_k(e'_i, Je_{m+1}) = \omega_k(Je_i, Je_{m+1}) + \omega_k(e'_i - Je_i, Je_{m+1})$; the first term differs from $\omega_k(e_i, e_{m+1}) = 0$ by a $(0,2)$-term and is therefore bounded by $O(c_k^{-1/2})$, while the bound on $e'_i - Je_i$ implies that the second term is also bounded by $O(c_k^{-1/2})$. Therefore we have $\omega_k(e'_i, Je_{m+1}) = O(c_k^{-1/2})$. Finally, using the lower bound on $\omega_k(e_i, e'_i)$ we obtain that $|e'_{m+1} - Je_{m+1}|_{g_k} = O(c_k^{-1/2})$. Finally, it is trivial that $|\nabla^r e'_{m+1}|_{g_k} = O(c_k^{-1/2})$; therefore we can proceed with the induction process.

We now define the almost-complex structure $\tilde{J}_k$ by the identities $\tilde{J}_k(e_i) = e'_i$ and $\tilde{J}_k(e'_i) = -e_i$. By construction, $\tilde{J}_k$ is compatible with ω_k, and the corresponding Riemannian metric $\tilde{g}_k$ admits $e_1, e'_1, \ldots, e_n, e'_n$ as an orthonormal frame. The required bounds on $\tilde{J}_k$ immediately follow from the available estimates. □

Remark. As suggested by the referee, Lemma 2.1 can also be proved more efficiently using the following argument : $\tilde{J}_k$ is characterized by a linear map $\mu_k : TX_J^{(1,0)} \to TX_J^{(0,1)}$, such that $TX_{\tilde{J}_k}^{(1,0)} = \{v + \mu_k\, v,\ v \in TX_J^{(1,0)}\}$ and $TX_{\tilde{J}_k}^{(0,1)} = \{v + \bar{\mu}_k\, v,\ v \in TX_J^{(0,1)}\}$. The compatibility of $\tilde{J}_k$ with ω_k is expressed by the condition $\omega_k(u + \mu_k\, u, v + \mu_k\, v) = 0\ \forall u, v \in TX_J^{(1,0)}$, i.e.

$$\omega_k^{1,1}(\mu_k\, \cdot, \cdot) + \omega_k^{1,1}(\cdot, \mu_k\, \cdot) + O(|\mu_k|^2) = -\omega_k^{2,0}.$$

Since $\omega_k^{1,1}$ defines a non-degenerate pairing between $TX_J^{(1,0)}$ and $TX_J^{(0,1)}$, and because the first two terms of the left-hand side are the antisymmetric part of $\omega_k^{1,1}(\mu_k\, \cdot, \cdot)$, the existence of a small solution μ_k to this equation follows directly from the smallness of $\omega_k^{2,0}$ and the implicit function theorem. □

Lemma 2.1 makes it possible to recover the main ingredients of Donaldson theory in the more general setting described here. We now introduce some basic definitions and results, imitating Donaldson's original work and subsequent papers.[7,3]

In what follows, L_k is an asymptotically very ample sequence of line bundles over X, c_k are the same constants as in Definition 2.1, and $g_k = c_k\, g$.

Lemma 2.2 *Near any point $x \in X$, and for any value of k, there exist local complex Darboux coordinates $(z_k^1, \ldots, z_k^n) : (X, x) \to (C^n, 0)$ for the symplectic structure $\omega_k = iF_k$, such that, denoting by ψ_k the inverse of the coordinate map, the following bounds hold uniformly in x and k over a ball of fixed g-radius around x : $|z_k^i(y)| = O(dist_{g_k}(x, y))$; $|\nabla^r \psi_k|_{g_k} = O(1)$ $\forall r \geq 1$; and, with respect to the almost-complex structure J on X and the* *canonical complex structure on* $\mathbb{C}^n$, $|\bar\partial\psi_k(z)|_{g_k} = O(c_k^{-1/2} + c_k^{-1/2}|z|)$, *and* $|\nabla^r\bar\partial\psi_k(z)|_{g_k} = O(c_k^{-1/2})$ *for all* $r \geq 1$.

Proof. The argument is very similar to that used by Donaldson,[7] except that one needs to be slightly more careful in showing that the various bounds hold uniformly in k. Fix a point $x \in X$: then we can find a neighborhood U of x and a local coordinate map $\phi : U \to \mathbb{C}^n$, such that U contains a ball of fixed uniform g-radius around x, and such that the expressions of g and J in these local coordinates satisfy uniform bounds independently of x (these uniformity properties follow from the compactness of X). A linear transformation can be used to ensure that the differential of ϕ at the origin is $\mathbb{C}$-linear with respect to J. Next, we rescale the coordinates by $c_k^{1/2}$ to obtain a new coordinate map $\phi_k : U \to \mathbb{C}^n$, in which J coincides with the standard almost-complex structure at the origin and has derivatives bounded by $O(c_k^{-1/2})$, while the expression of g_k is bounded between fixed constants and has derivatives bounded by $O(c_k^{-1/2})$.

Next, we observe that the bound on $|\omega_k|$ and the lower bound on $\omega_k(v, Jv)$ imply that the expression of $\omega_k^{(1,1)}$ at the origin of the coordinate chart is bounded from above and below by uniform constants. Therefore, after composing ϕ_k with a suitable element of $GL(n, \mathbb{C})$, we can assume without affecting the bounds on J and g_k that $(\phi_k^{-1})^*(\omega_k^{(1,1)})$ coincides with the standard Kähler form ω_0 of $\mathbb{C}^n$ at the origin.

Define over $\phi_k(U) \subset \mathbb{C}^n$ the symplectic form $\omega_1 = (\phi_k^{-1})^*\omega_k$. By construction, $\omega_1(0) - \omega_0(0) = O(c_k^{-1/2})$. Observe that, in the chosen coordinates, the Levi-Civita connection of g_k differs from the trivial connection by $O(c_k^{-1/2})$; therefore, the bounds on $|\nabla^r\omega_k|_{g_k}$ imply that the derivatives of ω_1 are also bounded by $O(c_k^{-1/2})$, and that $|\omega_1(z) - \omega_0(z)| = O(c_k^{-1/2} + c_k^{-1/2}|z|)$.

In particular, decreasing the size of U by at most a fixed factor if necessary, we obtain that the closed 2-forms $\omega_t = t\omega_1 + (1-t)\omega_0$ over $\phi_k(U)$ are all symplectic, and we can apply Moser's argument to construct in a controlled way a symplectomorphism between a subset of $(\phi_k(U), \omega_1)$ and a subset of $(\mathbb{C}^n, \omega_0)$. More precisely, it follows immediately from Poincaré's lemma that we can choose a 1-form α such that $\omega_1 - \omega_0 = d\alpha$, and such that $\alpha(0) = 0$, $|\alpha(z)| = O(c_k^{-1/2}|z| + c_k^{-1/2}|z|^2)$, $|\nabla\alpha(z)| = O(c_k^{-1/2} + c_k^{-1/2}|z|)$ and $|\nabla^r\alpha(z)| = O(c_k^{-1/2})$ $\forall r \geq 2$. Next, we define vector fields X_t by the identity $i_{X_t}\omega_t = \alpha$; clearly X_t and its derivatives satisfy the same bounds as α.

Integrating the flow of the vector fields X_t we obtain diffeomorphisms ρ_t, and it is a classical fact that the map $\tilde\phi_k = \rho_1 \circ \phi_k$ is a local symplectomorphism between (X,ω_k) and $(\mathbb{C}^n,\omega_0)$ and therefore defines Darboux coordinates. Because $|z| = O(c_k^{1/2})$ over a ball of fixed g-radius around x, the vector fields X_t satisfy a uniform bound of the type $|X_t(z)| \le \lambda|z|$ for some constant λ, so that $|\rho_t(z)| \le e^{\lambda t}|z|$, and therefore $\tilde{\phi}_k$ is well-defined over a ball of fixed g-radius around x. Moreover, the bounds $|\nabla(\rho_1 - \mathrm{Id})| = O(c_k^{-1/2} + c_k^{-1/2}|z|)$, obtained by integrating the bounds on $\nabla\alpha$, and $|\bar\partial(\phi_k^{-1})| = O(c_k^{-1/2}|z|)$, obtained from the bounds on the expression on J in the local coordinates, imply that $|\bar\partial(\tilde\phi_k^{-1})| = O(c_k^{-1/2} + c_k^{-1/2}|z|)$. Similarly, the bounds $|\nabla^{r+1}\rho_1| = O(c_k^{-1/2})$ and $|\nabla^r\bar\partial(\phi_k^{-1})| = O(c_k^{-1/2})$ for all $r \ge 1$ imply that $|\nabla^r\bar\partial(\tilde\phi_k^{-1})| = O(c_k^{-1/2})$. This completes the proof of Lemma 2.2. □

Definition 2.2 *A family of sections of L_k is* asymptotically J-holomorphic *for $k \to \infty$ if there exist constants $(C_r)_{r\ge 0}$ such that every section $s \in \Gamma(L_k)$ in the family satisfies at every point of X the bounds $|\nabla^r s|_{g_k} \le C_r$ and $|\nabla^r\bar\partial_J s|_{g_k} \le C_r\, c_k^{-1/2}$ for all $r \ge 0$, where $\bar\partial_J$ is the $(0,1)$-part of the connection on L_k.*

A family of sections of L_k has uniform Gaussian decay *properties if there exist a constant $\lambda > 0$ and polynomials $(P_r)_{r\ge 0}$ with the following property : for every section s of L_k in the family, there exists a point $x \in X$ such that for all $y \in X$ and for all $r \ge 0$, $|\nabla^r s(y)|_{g_k} \le P_r(d_k(x,y))\exp(-\lambda\, d_k(x,y)^2)$, where $d(.,.)$ is the distance induced by g_k.*

Lemma 2.3 *For all large enough values of k and for every point $x \in X$, there exists a section $s_{k,x}^{\mathrm{ref}}$ of L_k with the following properties : (1) the family of sections $(s_{k,x}^{\mathrm{ref}})_{x\in X,k\gg 0}$ is asymptotically J-holomorphic; (2) the family $(s_{k,x}^{\mathrm{ref}})_{x\in X,k\gg 0}$ has uniform Gaussian decay properties, each section $s_{k,x}^{\mathrm{ref}}$ being concentrated near the point x; (3) there exists a constant $\kappa > 0$ independent of x and k such that $|s_{k,x}^{\mathrm{ref}}| \ge \kappa$ at every point of the ball of g_k-radius 1 centered at x.*

Proof. The argument is a direct adaptation of the proof of Proposition 11 in Donaldson's paper.[7] Pick a value of k and a point $x \in X$. We work in the approximately J-holomorphic Darboux coordinates given by Lemma 2.2, and use a trivialization of L_k in which the connection 1-form becomes $\frac{1}{4}\sum(z_j d\bar z_j - \bar z_j dz_j)$. Then, we define a local section of L_k by $s(z) = \exp(-\frac{1}{4}|z|^2)$ and observe that s is holomorphic with respect to the standard complex structure of $\mathbb{C}^n$. Multiplying s by a cut-off function which equals 1 over the ball of radius $c_k^{1/6}$ around the origin, we obtain a globally defined section of L_k; because of the estimates on the Darboux coordinates one easily checks that the families of sections constructed in this way are asymptotically holomorphic and have uniform Gaussian decay properties.[7] □

We are also interested in working with higher rank bundles. The definition of ampleness becomes the following :

Definition 2.3 *A sequence of complex vector bundles E_k with metrics and connections is* asymptotically very ample *if there exist constants δ, $(C_r)_{r\geq 0}$, and $c_k \to +\infty$, such that the curvature F_k of E_k satisfies the following properties :* (1) $\langle iF_k(v,Jv).u,u\rangle \geq c_k\, g(v,v)\,|u|^2$, $\forall v \in TX$, $\forall u \in E_k$; (2) $\sup |F_k^{0,2}|_g \leq \delta_r\, c_k^{1/2}$; (3) $\sup |\nabla^r F_k|_g \leq C_r\, c_k\ \forall r \geq 0$.

A sequence of asymptotically very ample complex vector bundles E_k with metrics $|.|_k$ and connections ∇_k is locally splittable *if, given any point $x \in X$, there exists over a ball of fixed g-radius around x a decomposition of E_k as a direct sum $L_{k,1} \oplus \cdots \oplus L_{k,m}$ of line bundles, such that the following properties hold :* (1) *the $|.|_k$-determinant of a local frame consisting of unit length local sections of $L_{k,1}, \dots, L_{k,m}$ is bounded from below by a fixed constant independently of x and k;* (2) *denoting by $\nabla_{k,i}$ the connection on $L_{k,i}$ obtained by projecting $\nabla_{k|L_{k,i}}$ to $L_{k,i}$, and by ∇'_k the direct sum of the $\nabla_{k,i}$, the* 1*-form $\alpha_k = \nabla_k - \nabla'_k$ (the non-diagonal part of ∇_k) satisfies the uniform bounds $|\nabla^r \alpha_k|_g = O(c_k^{r/2})\ \forall r \geq 0$ independently of x.*

For example, if E is a fixed complex vector bundle and L_k are asymptotically very ample line bundles, then the vector bundles $E \otimes L_k$ are locally splittable and asymptotically very ample; so are direct sums of vector bundles of this type.

Observe that, if E_k is an asymptotically very ample sequence of locally splittable vector bundles, then near any given point $x \in X$ the summands $L_{k,1}, \dots, L_{k,m}$ are asymptotically very ample line bundles. Therefore, by Lemma 2.3 they admit asymptotically holomorphic sections $s^{\mathrm{ref}}_{k,x,i}$ with uniform Gaussian decay away from x. Moreover, these sections, which define a local frame for E_k, are easily checked to be asymptotically J-holomorphic not only as sections of $L_{k,i}$ but also as sections of E_k.

3 Estimated transversality in jet bundles

3.1 Asymptotically holomorphic stratifications

Throughout this section, we will denote by F_k be a sequence of complex vector bundles over X, or more generally fiber bundles with almost-complex manifolds as fibers. We also fix, in a manner compatible with the almost-complex structures J^v of the fibers, metrics g^v on the fibers of F_k and a connection on F_k. Finally, we fix a sequence of constants $c_k \to +\infty$.

The connection on F_k induces a splitting $TF_k = T^vF_k \oplus T^hF_k$ between horizontal and vertical tangent spaces ; this splitting makes it possible to define a metric $\hat{g}_k$ and an almost-complex structure $\hat{J}_k$ on the total space of F_k, obtained by orthogonal direct sum of g^v and J^v on T^vF_k together with the pullbacks of J and $g_k = c_k\, g$ on $T^hF_k \simeq \pi^*TX$.

We want to consider approximately holomorphic stratifications of the fibers of F_k, depending in an approximately holomorphic way on the point in the base manifold X. For simplicity, we assume that the topological picture is the same in every fiber of F_k, i.e. we restrict ourselves to stratifications which are everywhere transverse to the fibers. We will denote the strata by $(S_k^a)_{a\in A_k}$; we assume that the number of strata is finite. Each S_k^a is a possibly non-closed submanifold in F_k, whose closure is obtained by adding other lower dimensional strata : writing $b \prec a$ iff S_k^b is contained in $\overline{S_k^a}$, we have

$$\partial S_k^a \stackrel{\text{def}}{=} \overline{S_k^a} - S_k^a = \bigcup_{b\prec a} S_k^b.$$

We only consider *Whitney stratifications* ; in particular, transversality to a given stratum S_k^a implies transversality over a neighborhood of S_k^a to all the strata whose closure contains S_k^a, i.e. all the S_k^b for $b \succ a$. Also note that we discard any open strata, as they are irrelevant for transversality purposes ; so each S_k^a has codimension at least 1.

Definition 3.1 *Let (M, J) be an almost-complex manifold, with a Riemannian metric, and let s be a complex-valued function over M or a section of an almost-complex bundle with metrics and connection. Given two constants C and c, we say that s is C^2-*approximately holomorphic with bounds (C, c), *or C^2-*AH(C, c)*, if it satisfies the following estimates :*

$$|s| + |\nabla s| + |\nabla\nabla s| \le C, \qquad |\bar\partial s| + |\nabla\bar\partial s| \le C\, c^{-1/2}.$$

*Moreover, given constants $c_k \to +\infty$, we say that a sequence $(s_k)_{k\gg 0}$ of functions or sections is C^2-*asymptotically holomorphic, *or C^2-*AH, *if there exists a fixed constant C such that each section s_k is C^2-AH(C, c_k).*

Definition 3.2 *Let F_k be a sequence of almost-complex bundles over X, endowed with metrics and connections as above. For all values of k, let $(S_k^a)_{a\in A_k}$ be finite Whitney stratifications of F_k ; assume that the total number of strata is bounded by a fixed constant independently of k, and that all strata are transverse to the fibers of F_k.*

We say that this sequence of stratifications is asymptotically holomorphic *if, given any bounded subset $U_k \subset F_k$, and for every $\epsilon > 0$, there exist positive constants C_ϵ and ρ_ϵ depending only on ϵ and on the size of the subset U_k but not on k, with the following property. For every point $x \in U_k$ lying in a certain stratum S_k^a and at $\hat{g}_k$-distance greater than ϵ from $\partial S_k^a = \overline{S_k^a} - S_k^a$, there exist complex-valued functions $f_1, \dots, f_p$ over the ball $B = B_{\hat{g}_k}(x, \rho_\epsilon)$ with the following properties :*

(1) *a local equation of S_k^a over B is $f_1 = \cdots = f_p = 0$;*

(2) *$|df_1 \wedge \cdots \wedge df_p|_{\hat{g}_k}$ is bounded from below by ρ_ϵ at every point of B ;*

(3) *the restrictions of f_i to each fiber of F_k near x are C^2-AH(C_ϵ, c_k) ;*

(4) *for any constant $\lambda > 0$, and for any local section s of F_k which is C^2-AH(λ, c_k) with respect to the metric g_k on X and which intersects non-trivially the ball B, the function $f_i \circ s$ is C^2-AH$(\lambda C_\epsilon, c_k)$; moreover, given a local C^2-AH(λ, c_k) section θ of $s^* T^v F_k$, the functions $df_i \circ \theta$ are C^2-AH$(\lambda C_\epsilon, c_k)$;*

(5) *at every point $y \in B$ belonging to a stratum S_k^b such that $S_k^a \subset \partial S_k^b$, the norm of the orthogonal projection onto the normal space $N_y S_k^b$ of any unit length vector $v \in T_y F_k$ such that $df_1(v) = \cdots = df_p(v) = 0$ is bounded by $C_\epsilon \operatorname{dist}_{\hat{g}_k}(y, S_k^a)$.*

These conditions on the stratification can be reformulated more geometrically as follows. First, the strata must be uniformly transverse to the fibers of F_k, i.e. one requires the minimum angle[11] between TS_k^a and $T^v F_k$ to be bounded from below. Second, the submanifolds $S_k^a \subset F_k$ must be asymptotically $\hat{J}_k$-holomorphic, i.e. $\hat{J}_k(TS_k^a)$ and TS_k^a lie within $O(c_k^{-1/2})$ of each other. Third, the curvature of S_k^a as a submanifold of F_k must be uniformly bounded. Finally, the quantity measuring the lack of $\hat{J}_k$-holomorphicity of S_k^a must similarly vary in a controlled way.

We finish this section by introducing the notion of estimated transversality between a section and a stratification. Observe that, given any submanifold $N \subset M$, we can define over a neighborhood of N a "parallel" distribution $D_N \subset TM$ by parallel transport of TN in the normal direction to N. Also recall that the *minimum angle* between two linear subspaces U and V is defined as the minimum angle between a vector orthogonal to U and a vector orthogonal to V.[11] The minimum angle between U and V is non-zero if and only if they are transverse to each other, and in that case it can also be defined as the minimum angle between non-zero vectors orthogonal to $U \cap V$ in U and V.

Definition 3.3 *Given a constant $\eta > 0$, we say that a section s of a vector bundle carrying a metric and a connection is η-transverse to 0 if, at every* *point x such that $|s(x)| \le \eta$, the covariant derivative $\nabla s(x)$ is surjective and admits a right inverse of norm less than η^{-1}.*

Fix a constant $\eta > 0$, and a section s of a bundle carrying a metric and a finite Whitney stratification $\mathcal{S} = (S^a)_{a \in A}$ everywhere transverse to the fibers. We say that s is η-transverse to the stratification $\mathcal{S}$ if, at every point where s lies at distance less than η from some stratum S^a, the graph of s is transverse to the parallel distribution D_{S^a}, with a minimum angle greater than η.

Finally, we say that a sequence of sections is uniformly tranverse to 0 (resp. to a sequence of stratifications) if there exists a fixed constant $\eta > 0$ such that all sections in the sequence are η-transverse to 0 (resp. the stratifications).

Note that the above condition of transversality of the section s to each stratum S^a is only well-defined outside of a small neighborhood of the lower-dimensional strata contained in ∂S^a ; however, near these strata the assumption that $\mathcal{S}$ is Whitney makes transversality to S^a a direct consequence of the η-transversality to the lower-dimensional strata.

Another way in which uniform transversality to a stratification can be formulated is to use local equations of the strata, as in Definition 3.2. One can then define η-transversality as follows : at every point where s lies at distance less than η from S^a, and considering local equations $f_1 = \cdots = f_p = 0$ of S^a such that each $|df_i|$ is bounded by a fixed constant and $|df_1 \wedge \cdots \wedge df_p|$ is bounded from below by a fixed constant, the function $(f_1 \circ s, \dots, f_p \circ s)$ with values in $\mathbb{C}^p$ must be η-transverse to 0. The two definitions are equivalent up to changing the constant η by at most a bounded factor.

3.2 Quasi-stratifications in jet bundles

Let E_k be an asymptotically very ample sequence of locally splittable rank m vector bundles over the compact almost-complex manifold (X, J). We can introduce the *holomorphic jet bundles*

$$\mathcal{J}^r E_k = \bigoplus_{j=0}^{r} (T^*X^{(1,0)})^{\otimes j}_{\mathrm{sym}} \otimes E_k.$$

More precisely, the holomorphic part of the r-jet of a section s of E_k is defined inductively as follows : $T^*X^{(1,0)}$ and E_k, as complex vector bundles carrying a connection over an almost-complex manifold, are endowed with ∂ operators (the $(1,0)$ part of the connection) ; the r-jet of s is $j^r s = (s, \partial_{E_k} s, \partial_{T^*X^{(1,0)} \otimes E_k}(\partial_{E_k} s)_{\mathrm{sym}}, \dots)$.

Observe that, because the almost-complex structure J is not integrable and because the curvature of E_k is not necessarily of type $(1,1)$, the derivatives of order ≥ 2 are not symmetric tensors, but rather satisfy equality relations involving curvature terms and lower-order derivatives. However, we will only consider the symmetric part of the jet ; for example, the 2-tensor component of $j^r s$ is defined by $(\partial\partial s)_{\mathrm{sym}}(u,v) = \frac{1}{2}(\langle\partial(\partial s), u\otimes v\rangle + \langle\partial(\partial s), v\otimes u\rangle)$. Note that, anyway, in the case of asymptotically holomorphic sections, the antisymmetric terms are bounded by $O(c_k^{-1/2})$, because the $(2,0)$ curvature terms and Nijenhuis tensor are bounded by $O(c_k^{-1/2})$.

The metrics and connections on TX and on E_k naturally induce Hermitian metrics and connections on $\mathcal{J}^r E_k$ (to define the metric we use the rescaled metric g_k on X). In fact, it is easy to see that the vector bundles $\mathcal{J}^r E_k$ are asymptotically very ample.

Recall that, near any given point $x \in X$, there exist local approximately holomorphic coordinates ; besides a local identification of X with $\mathbb{C}^n$, these coordinates also provide an identification of $T^*X^{(1,0)}$ with $T^*\mathbb{C}^{n\,(1,0)}$. Moreover, by Lemma 2.3 there exist asymptotically holomorphic sections $s^{\mathrm{ref}}_{k,x,i}$ of E_k with Gaussian decay away from x and defining a local frame in E_k. Using these sections to trivialize E_k, we can locally identify $\mathcal{J}^r E_k$ with a space of jets of holomorphic $\mathbb{C}^m$-valued maps over $\mathbb{C}^n$. Observe however that, when we consider the holomorphic parts of jets of approximately holomorphic sections of E_k, the integrability conditions normally satisfied by jets of holomorphic functions only hold in an approximate sense.

In general, the various possible choices of trivializations of $\mathcal{J}^r E_k$ differ by approximately holomorphic diffeomorphisms of $\mathbb{C}^n$ and also by the action of approximately holomorphic local sections of the automorphism bundle $\mathrm{GL}(E_k)$. However, when E_k is of the form $\mathbb{C}^m \otimes L_k$ where L_k is a line bundle, the only automorphisms of E_k which we need to consider are multiplications by complex-valued functions.

Denote by $\mathcal{J}^r_{n,m}$ the space of r-jets of holomorphic maps from $\mathbb{C}^n$ to $\mathbb{C}^m$: pointwise, the identifications of the fibers of $\mathcal{J}^r E_k$ with $\mathcal{J}^r_{n,m}$ given by local trivializations differ from each other by the action of $GL_n(\mathbb{C}) \times GL_m(\mathbb{C})$ (or $GL_n(\mathbb{C}) \times \mathbb{C}^*$ when $E_k = \mathbb{C}^m \otimes L_k$), where $GL_n(\mathbb{C})$ corresponds to changes in the identification of $T^*X^{(1,0)}$ with $T^*\mathbb{C}^{n\,(1,0)}$ and $GL_m(\mathbb{C})$ or $\mathbb{C}^*$ corresponds to changes in the trivialization of E_k. Some stratifications of $\mathcal{J}^r_{n,m}$ are invariant under the actions of $GL_n(\mathbb{C})$ and $GL_m(\mathbb{C})$ (resp. $\mathbb{C}^*$). Given such a stratification it becomes easy to construct an asymptotically holomorphic sequence of finite Whitney stratifications of $\mathcal{J}^r E_k$, modelled in each fiber on the given stratification of $\mathcal{J}^r_{n,m}$. Many important examples of asymptotically holomorphic stratifications, and in a certain sense all the geometrically relevant ones, are obtained by this construction (see Proposition 3.1 below).

We also wish to consider cases where the available structure is not exactly a Whitney stratification but behaves in a similar manner with respect to transversality. We call such a structure a "Whitney quasi-stratification". Given a submanifold $S \subset \mathcal{J}^r_{n,m}$, one can introduce the subset Θ_S of all points $\sigma \in S$ such that there exists a holomorphic $(r+1)$-jet whose r-jet component is σ and which, considered as a 1-jet of r-jets, intersects S transversely at σ. For example, if S is the subset of all jets $(\sigma_0, \dots, \sigma_r)$ such that $\sigma_0 = 0$, the subset Θ_S consists of those jets such that $\sigma_0 = 0$ and σ_1 is surjective.

Similarly, when S is a submanifold in $\mathcal{J}^r E_k$, we can view an element of $\mathcal{J}^{r+1} E_k$ as the holomorphic 1-jet of a section of $\mathcal{J}^r E_k$. More precisely, for any point $x \in X$, we can associate to any $\sigma = (\sigma_0, \dots, \sigma_{r+1}) \in (\mathcal{J}^{r+1} E_k)_x$ the 1-jet at x of a local section $\tilde{\sigma}$ of $\mathcal{J}^r E_k$, such that $\tilde{\sigma}(x) = (\sigma_0, \dots, \sigma_r)$, $(\partial\tilde{\sigma}(x))^{\text{sym}} = (\sigma_1, \dots, \sigma_{r+1})$, $(\partial\tilde{\sigma}(x))^{\text{antisym}} = 0$, and $\bar{\partial}\tilde{\sigma}(x) = 0$ (in this definition, $\partial\tilde{\sigma}(x) \in T^*X^{1,0} \otimes (\bigoplus (T^*X^{1,0})^{\otimes j}_{\text{sym}} \otimes E)$ is decomposed into a symmetric part and an antisymmetric part). Then, we define Θ_S as the set of points of S for which there exists an element $\sigma \in \mathcal{J}^{r+1} E_k$ such that the corresponding 1-jet $\tilde{\sigma}$ in $\mathcal{J}^r E_k$ intersects S transversely at the given point. For example, if S is the set of r-jets $(\sigma_0, \dots, \sigma_r)$ such that $\sigma_0 = 0$, then Θ_S is the set of r-jets such that $\sigma_0 = 0$ and σ_1 is surjective. Also observe that Θ_S is always empty when the codimension of S is greater than n.

Definition 3.4 *Given a finite set $(A, \prec)$ carrying a binary relation without cycles (i.e., $a_1 \prec \cdots \prec a_p \Rightarrow a_p \not\prec a_1$), a finite Whitney quasi-stratification of $\mathcal{J}^r_{n,m}$ indexed by A is a collection $(S^a)_{a \in A}$ of smooth submanifolds of $\mathcal{J}^r_{m,n}$, not necessarily mutually disjoint, with the following properties :* (1) $\partial S^a = \overline{S^a} - S^a \subseteq \bigcup_{b \prec a} S^b$; (2) *given any point $p \in \partial S^a$, there exists $b \prec a$ such that $p \in S^b$ and such that either $p \notin \Theta_{S^b}$ or $S^b \subset \partial S^a$ and the Whitney regularity condition is satisfied at all points of S^b.*

Similarly, we can define the notion of asymptotically holomorphic finite Whitney quasi-stratifications of $\mathcal{J}^r E_k$. This is similar to Definition 3.2, except that the collections (S^a_k) are quasi-stratifications rather than stratifications, i.e. $\partial S^a_k \subseteq \bigcup_{b \prec a} S^b_k$, and for every $p \in \partial S^a_k$ there exists $b \prec a$ such that either $p \in S^b_k - \Theta_{S^b_k}$ or $p \in S^b_k \subset \partial S^a_k$; in the latter case the Whitney condition is required. Also, observe that condition (5) in Definition 3.2 is only required in the second case, and not for all b such that $a \prec b$.

It is important to understand that the notion of quasi-stratification is merely an attempt at simplifying the framework for applications of Theorem 1.1. In fact, most quasi-stratifications can be refined into genuine stratifications by suitably subdividing the strata into smaller pieces. However, by definition these modifications occur at points of $\mathcal{J}^r E_k$ that no generic jet can hit, thus making them utterly irrelevant to transversality.

Proposition 3.1 *Let $\mathcal{S} = (S^a)_{a\in A}$ be a finite Whitney quasi-stratification of $\mathcal{J}^r_{n,m}$ by complex submanifolds, invariant under the action of $GL_n(\mathbb{C}) \times GL_m(\mathbb{C})$ or $GL_n(\mathbb{C}) \times \mathbb{C}^*$. Let E_k be an asymptotically very ample sequence of rank m complex vector bundles over X, trivialized near every point by suitable choices of local asymptotically holomorphic coordinates and sections. Assume that $\mathcal{S}_k = (S^a_k)_{a\in A}$ are quasi-stratifications of $\mathcal{J}^r E_k$ such that, in each local trivialization, the intersection of S^a_k with every fiber becomes identified with S^a. Then the sequence of quasi-stratifications $\mathcal{S}_k$ is asymptotically holomorphic.*

The proof of this result is easy and left to the reader ; the independence on k of the model holomorphic quasi-stratification of $\mathcal{J}^r_{n,m}$ and the availability of asymptotically holomorphic local trivializations of $\mathcal{J}^r E_k$ (Lemma 2.2 and Lemma 2.3) immediately yield the necessary estimates on the strata of $\mathcal{S}_k$. The only important point to observe is that, because the strata of $\mathcal{S}$ are $GL_m(\mathbb{C})$-invariant (resp. $\mathbb{C}^*$-invariant), the local trivializations identifying S^a_k with S^a also identify $\Theta_{S^a_k}$ with Θ_{S^a}. This is e.g. due to the fact that, up to a suitable change in the choice of the local coordinates on X and local reference sections of L_k, i.e. up to a local gauge transformation, we can assume that the connection on $\mathcal{J}^r E_k$ agrees at a given point $x \in X$ with the trivial connection on $\mathcal{J}^r_{n,m}$; above x, the identification of $(r+1)$-jets with 1-jets of r-jets then becomes the same in $\mathcal{J}^r E_k$ as in $\mathcal{J}^r_{n,m}$, so that the definitions of Θ_S in $\mathcal{J}^r_{n,m}$ and in $\mathcal{J}^r E_k$ agree with each other.

Various examples of applications of Proposition 3.1 will be given in §5.

Finally, we state a one-parameter version of Theorem 1.1. Consider a continuous one-parameter family $(J_t)_{t\in[0,1]}$ of almost-complex structures on X, and a one-parameter family of asymptotically holomorphic finite Whitney (quasi)-stratifications $(\mathcal{S}_{k,t})_{k\gg 0, t\in[0,1]}$ of almost-complex bundles $F_{k,t}$ over (X, J_t). We say that the (quasi)-stratifications $\mathcal{S}_{k,t}$ depend continuously on t if the following one-parameter version of Definition 3.2 is true : for every $\epsilon > 0$, there exist constants ρ_ϵ and C_ϵ with the following property. Given any continuous path $(x_t)_{t\in[t_1,t_2]}$ of points all belonging to the fibers of $F_{k,t}$ above a same point in X, and assuming that all the points x_t belong to certain strata $S^a_{k,t}$ while lying at distance more than ϵ from $\partial S^a_{k,t}$, there exist for all $t \in [t_1, t_2]$ complex-valued functions $f_{1,t}, \dots, f_{p,t}$ defined over the ball $B_{\hat{g}_{k,t}}(x_t, \rho_\epsilon)$ and depending continuously on t, satisfying the various properties of Definition 3.2 for all values of t.

With this understood, the result is the following :

Theorem 3.2 *Let $(J_t)_{t\in[0,1]}$ be a continuous one-parameter family of almost-complex structures on the compact manifold X, and let $(E_{k,t})_{k\gg 0,t\in[0,1]}$ be a family of complex vector bundles over X endowed with metrics and connections depending continuously on t and such that the sequence $E_{k,t}$ is asymptotically very ample and locally splittable over (X, J_t) for all t. Let $\mathcal{S}_{k,t}$ be asymptotically holomorphic finite Whitney quasi-stratifications of $\mathcal{J}^r E_{k,t}$ depending continuously on t. Finally, let $\delta > 0$ be a fixed constant. Then there exist constants K and η such that, given any one-parameter family of asymptotically holomorphic sections $s_{k,t}$ of $E_{k,t}$ over X depending continuously on t, there exist asymptotically holomorphic sections $\sigma_{k,t}$ of $E_{k,t}$, depending continuously on t, with the following properties for all $k \geq K$ and for all $t \in [0,1]$:*

(1) $|\sigma_{k,t} - s_{k,t}|_{C^{r+1},g_k} < \delta$;

(2) *the jet $j^r\sigma_{k,t}$ of $\sigma_{k,t}$ is η-transverse to $\mathcal{S}_{k,t}$.*

4 Proof of the main result

The proof of Theorem 1.1 is quite similar to the arguments in previous papers.[3,4,7,8,11] It relies heavily on the fact that the estimated transversality of the r-jet of a section to a given submanifold of the jet bundle is a local and C^{r+1}-open property in the following sense.[3] Given a submanifold S of $\mathcal{J}^r E_k$, a constant $\eta > 0$ and a point $x \in X$, say that a section s of E_k satisfies the property $\mathcal{P}(S, \eta, x)$ if either the r-jet $j^r s(x)$ lies at distance more than η from S or $j^r s$ is η-transverse to S at x (in the sense of Definition 3.3, i.e. the minimum angle between the graph of $j^r s$ and the parallel distribution to S is at least η). The property $\mathcal{P}(S, \eta, x)$ depends only on the $(r+1)$-jet of s at x ("locality"). Moreover, if s satisfies $\mathcal{P}(S, \eta, x)$, then any section σ such that $|j^{r+1}\sigma(x) - j^{r+1}s(x)| < \epsilon$ satisfies $\mathcal{P}(S, \eta - C\epsilon, x)$, where C is some fixed constant involving only the curvature bounds of S ("openness").

A first consequence is that Theorem 1.1 can be proved by successively perturbing the given sections s_k in order to ensure transversality to the various strata. To show this, we first remark that, given any index b, the uniform transversality of $j^r s_k$ to all the strata S_k^a with $a \prec b$ implies its uniform transversality to S_k^b over a neighborhood of ∂S_k^b.

Indeed, first consider a pair of indices $a \prec b$ such that $S_k^a \subset \partial S_k^b$. By condition (5) of Definition 3.2, near a point of S_k^a the tangent space to S_k^b almost contains the parallel distribution to TS_k^a ; therefore, there exists a constant κ (independent of a and b) such that, for any small $\alpha > 0$, the α-transversality of $j^r s_k$ to S_k^a implies its $\frac{\alpha}{4}$-transversality to S_k^b over the $\kappa\alpha$-neighborhood of S_k^a. Next, consider a pair of indices $a \prec b$ and a point $p \in \partial S_k^b \cap (S_k^a - \Theta_{S_k^a})$: in this case, if the graph of $j^r s_k$ is α-transverse to S_k^a but intersects the ball of radius $\frac{\alpha}{2}$ around p, we can find an approximately holomorphic section σ_k of E_k differing from s_k by less than $\frac{3\alpha}{4}$ and whose jet goes through p. By definition of $\Theta_{S_k^a}$, all lifts of p in $\mathcal{J}^{r+1}E_k$, including $j^{r+1}\sigma_k$, correspond to local sections which intersect S_k^a non-transversely; because the antisymmetric and antiholomorphic terms in $\nabla(j^r\sigma_k)$ are smaller than $O(c_k^{-1/2})$, the minimum angle between $j^r\sigma_k$ and S_k^a at p is bounded by $O(c_k^{-1/2})$. However, since σ_k is close to s_k, its r-jet should be $\frac{\alpha}{4}$-transverse to S_k^a, which gives a contradiction. Therefore, $j^r s_k$ remains at distance more than $\frac{\alpha}{2}$ from p; this implies the $\frac{\alpha}{4}$-transversality to S_k^b of $j^r s_k$ over the $\frac{\alpha}{4}$-neighborhood of every point of $(S_k^a - \Theta_{S_k^a}) \cap \partial S_k^b$. Since these are the only two possible cases near the boundary of S_k^b, the uniform transversality of $j^r s_k$ to S_k^a for all $a \prec b$ implies its uniform transversality to S_k^b near ∂S_k^b.

Now, extend the binary relation $\prec$ on the set of strata of each $\mathcal{S}_k$ into a total order relation $<$, so that the indices in A_k can be identified with integers and the closure of a given stratum consists only of strata appearing before it. Assume that a first perturbation by less than $\delta_0 = \frac{\delta}{2}$ makes it possible to obtain for large k the η_1-transversality of $j^r s_k$ to the first stratum S_k^1, for some constant η_1 independent of k. Next, let δ_1 be a constant sufficiently smaller than δ and η_1 (but independent of k), and assume that a perturbation by at most δ_1 allows us to obtain the η_2-transversality of $j^r s_k$ to the second stratum S_k^2 outside of the $\frac{1}{4}\kappa\eta_1$-neighborhood of ∂S_k^2, for some constant η_2. Because this new perturbation is small enough, the resulting sections remain $\frac{\eta_1}{2}$-transverse to S_k^1 ; also, by the above observation this automatically implies the estimated transversality to S_k^2 of $j^r s_k$ near the points of $\partial S_k^2 \subseteq S_k^1$.

We can continue in this way until all strata have been considered ; each perturbation added to ensure estimated transversality to a new stratum outside of a small fixed size neighborhood of its boundary is chosen small enough in order not to affect the previously obtained transversality properties.

The fact that estimated transversality is local and open also makes it possible to reduce to a purely local setup, using a globalization principle due to Donaldson[7] and which can be formulated as follows (Proposition 3 of [3]) :

Proposition 4.1 *Let $\mathcal{P}_k(\eta, x)_{x\in X, \eta>0, k\gg 0}$ be local and C^{r+1}-open properties of sections of E_k over X. Assume that there exist constants c, c' and ν such that, given any $x \in X$, any small enough $\delta > 0$, and asymptotically holomorphic sections s_k of E_k, there exist, for all large enough k, asymptotically holomorphic sections $\tau_{k,x}$ of E_k with the following properties :* (a) $|\tau_{k,x}|_{C^{r+1},g_k} < \delta$, (b) *the sections $\frac{1}{\delta}\tau_{k,x}$ have uniform Gaussian decay away from x, and* (c) *the sections $s_k + \tau_{k,x}$ satisfy the property $\mathcal{P}_k(\eta, y)$ for all $y \in B_{g_k}(x, c)$, with $\eta = c'\delta\log(\delta^{-1})^{-\nu}$.*

Then, given any $\alpha > 0$ and asymptotically holomorphic sections s_k of E_k, there exist, for all large enough k, asymptotically holomorphic sections σ_k of E_k such that $|s_k - \sigma_k|_{C^{r+1},g_k} < \alpha$ and the sections σ_k satisfy $\mathcal{P}_k(\epsilon, x)$ $\forall x \in X$ for some $\epsilon > 0$ independent of k.

Proposition 4.1 is in fact slightly stronger than the previous results, as the notion of asymptotic holomorphicity has been extended to a more general framework in §2, but the argument remains strictly the same.

With this result, we are reduced to the problem of finding a localized perturbation of s_k near a given point x in order to ensure transversality to a given stratum. More precisely, fix an index $a \in A_k$ in each stratification, and remember that, from the previous steps of the inductive argument, we can restrict ourselves to considering only asymptotically holomorphic sections whose jet is γ-transverse to the strata S_k^b for $b < a$, for some fixed constant γ (this constant γ is half of the transversality estimate obtained in the previous step ; by assumption we only consider perturbations which are small enough to preserve γ-transversality to the previous strata). With this understood, say that a section s_k satisfies $\mathcal{P}_k(\eta, x)$ if either $j^r s_k(x)$ lies at distance more than η from S_k^a, or $j^r s_k(x)$ lies at distance less than $\frac{1}{4}\kappa\gamma - \eta$ from ∂S_k^a, or $j^r s_k$ is η-transverse to S_k^a at x. We want to show that the assumptions of Proposition 4.1 are satisfied by these properties.

Fix a point $x \in X$ and a constant $0 < \delta < \frac{1}{20}\kappa\gamma$, and consider asymptotically holomorphic sections s_k of E_k. First, if $j^r s_k(x)$ lies at distance less than $\frac{3}{20}\kappa\gamma$ from a point of $\partial S_k^a \cap S_k^b$ for some $b \prec a$, then the uniform bounds on covariant derivatives of s_k imply that the graph of $j^r s_k$ remains within distance less than $\frac{1}{5}\kappa\gamma$ of this point over a ball of fixed radius c_1 (independent of k, x or δ) around x. So, the property $\mathcal{P}_k(\frac{1}{20}\kappa\gamma, y)$ holds at every point $y \in B_{g_k}(x, c_1)$, and no perturbation is needed. In the rest of the argument, we can therefore assume that $j^r s_k(x)$ lies at distance at least $\frac{3}{20}\kappa\gamma$ from ∂S_k^a.

Let $\epsilon = \frac{1}{10}\kappa\gamma$, and let ρ_ϵ be the radius appearing in Definition 3.2. Without loss of generality we can assume that $\rho_\epsilon < \epsilon$. Assume that $j^r s_k(x)$ lies at distance more than $\frac{1}{2}\rho_\epsilon$ from S_k^a. Then, the bounds on covariant derivatives of s_k imply that the graph of $j^r s_k$ remains at distance more than $\frac{1}{4}\rho_\epsilon$ from S_k^a over a ball of fixed radius c_2 around x, and therefore that s_k satisfies $\mathcal{P}_k(\frac{1}{4}\rho_\epsilon, y)$ at every point $y \in B_{g_k}(x, c_2)$. No perturbation is needed.

Therefore, we may assume that $j^r s_k(x)$ lies at distance less than $\frac{1}{2}\rho_\epsilon$ from a certain point $u_0 \in S_k^a$. We may also safely assume that $\delta < \frac{1}{4}\rho_\epsilon$. One easily checks that u_0 lies at distance more than ϵ from ∂S_k^a. So we can find complex-valued functions $f_1, \dots, f_p$ over the ball $B_{\hat{g}_k}(u_0, \rho_\epsilon)$ such that a local equation of S_k^a is $f_1 = \dots = f_p = 0$ and satisfying the various properties listed in Definition 3.2. Let c_3 be a fixed positive constant (independent of k, x and δ) such that the graph of $j^r s_k$ over $B_{g_k}(x, c_3)$ is contained in $B_{\hat{g}_k}(u_0, \frac{3}{4}\rho_c)$, and define the $\mathbb{C}^p$-valued function $h = (f_1 \circ j^r s_k, \ldots, f_p \circ j^r s_k)$ over $B_{g_k}(x, c_3)$. By property (4) of Definition 3.2, the function h is C^2-approximately holomorphic.

Recall from Lemma 2.2 that there exist local approximately holomorphic ω_k-Darboux coordinates $z_1, \ldots, z_n$ over a neighborhood of x in X. Also recall from Lemma 2.3 that there exist approximately holomorphic sections $s^{\text{ref}}_{k,x,i}$ of E_k with Gaussian decay away from x and defining a local frame in E_k. For any $(n+1)$-tuple $I = (i_0, i_1, \ldots, i_n)$ with $1 \le i_0 \le m$, $i_1, \ldots, i_n \ge 0$, and $i_1 + \cdots + i_n \le r$, we define $s^{\text{ref}}_{k,x,I} = z_1^{i_1} \ldots z_n^{i_n} s^{\text{ref}}_{k,x,i_0}$. Clearly, these sections of E_k are asymptotically holomorphic and have uniform Gaussian decay away from x ; moreover it is easy to check that their r-jets define a local frame in $\mathcal{J}^r E_k$ near x. After multiplication by a suitable fixed constant factor, we can also assume that $|s^{\text{ref}}_{k,x,I}|_{C^{r+1}, g_k} \le \frac{1}{p}$. For each tuple I, define a $\mathbb{C}^p$-valued function Θ_I by $\Theta_I(y) = (df_1(j^r s_k(y)).j^r s^{\text{ref}}_{k,x,I}(y), \ldots, df_p(j^r s_k(y)).j^r s^{\text{ref}}_{k,x,I}(y))$. The functions Θ_I measure the variations of the function h when small multiples of the localized perturbations $s^{\text{ref}}_{k,x,I}$ are added to s_k ; by condition (4) of Definition 3.2, they are C^2-asymptotically holomorphic.

The fact that the jets of $s^{\text{ref}}_{k,x,I}$ define a frame of $\mathcal{J}^r E_k$ near x implies, by condition (2) of Definition 3.2, that the values $\Theta_I(x)$ generate all of $\mathbb{C}^p$. Moreover, for $1 \le i \le p$ there exist complex constants $\lambda_{I,i}$ with $\sum_I |\lambda_{I,i}| \le 1$ such that, defining the linear combinations $\sigma_{k,x,i} = \sum_I \lambda_{I,i} s^{\text{ref}}_{k,x,I}$ and $\Theta_i = \sum_I \lambda_{I,i} \Theta_I$, the quantity $|\Theta_1(x) \wedge \cdots \wedge \Theta_p(x)|$ is larger than some fixed positive constant $\beta > 0$ depending only on ϵ (and not on k, x or δ). The uniform bounds on derivatives imply that, for some fixed constant $0 < c_4 < c_3$, the norm of $\Theta_1 \wedge \cdots \wedge \Theta_p$ remains larger than $\frac{1}{2}\beta$ at every point of $B_{g_k}(x, c_4)$. Therefore, over this ball we can express h in the form $h = \mu_1 \Theta_1 + \cdots + \mu_p \Theta_p$, and the $\mathbb{C}^p$-valued function $\mu = (\mu_1, \ldots, \mu_p)$ is easily checked to be C^2-AH as well.

Finally, use once more the local approximately holomorphic coordinates to identify $B_{g_k}(x, c_4)$ with a neighborhood of the origin in $\mathbb{C}^n$. After rescaling the coordinates by a fixed constant factor, we can assume that this neighborhood contains the ball B^+ of radius $\frac{11}{10}$ around the origin in $\mathbb{C}^n$, and that there exists a fixed constant $0 < c_5 < c_4$ such that the inverse image of the unit ball B in $\mathbb{C}^n$ contains $B_{g_k}(x, c_5)$. Composing μ with the coordinate map, we obtain a $\mathbb{C}^p$-valued function $\tilde{\mu}$ over B^+ ; by construction $\tilde{\mu}$ is C^2-AH.

We may now use the following local result, due to Donaldson[8] (the case $p = 1$ is an earlier result of Donaldson;[7] the comparatively much easier case

$p > n$ is handled in [3]) :

Proposition 4.2 (Donaldson [8]) *Let f be a function with values in $\mathbb{C}^p$ defined over the ball B^+ of radius $\frac{11}{10}$ in $\mathbb{C}^n$. Let δ be a constant with $0 < \delta < \frac{1}{2}$, and let $\eta = \delta \log(\delta^{-1})^{-\nu}$ where ν is a suitable fixed integer depending only on n and p. Assume that f satisfies the following bounds over B^+:*

$$|f| \leq 1, \qquad |\bar{\partial} f| \leq \eta, \qquad |\nabla \bar{\partial} f| \leq \eta.$$

Then, there exists $w \in \mathbb{C}^p$, with $|w| \leq \delta$, such that $f - w$ is η-transverse to 0 over the interior ball B of radius 1.

Let $\eta = \delta \log(\delta^{-1})^{-\nu}$ as in the statement of the proposition, and observe that, if k is large enough, the antiholomorphic derives of $\tilde{\mu}$, which are bounded by a fixed multiple of $c_k^{-1/2}$, are smaller than η. Therefore, if k is large enough we can apply Proposition 4.2 (after a suitable rescaling to ensure that $\tilde{\mu}$ is bouded by 1) and find a constant $w = (w_1, \dots, w_p) \in \mathbb{C}^p$, smaller than δ, such that $\tilde{\mu} - w$ is η-transverse to 0 over the unit ball B. Going back through the coordinate map, this implies that $\mu - w$ is $c'_1\eta$-transverse to 0 over $B_{g_k}(x, c_5)$ for some fixed constant c'_1. Multiplying by the functions $\Theta_1, \dots, \Theta_p$, we obtain that $h - (w_1\Theta_1 + \dots + w_p\Theta_p)$ is $c'_2\eta$-transverse to 0 over $B_{g_k}(x, c_5)$ for some fixed constant c'_2.

Let $\tau_{k,x} = -(w_1\sigma_{k,x,1} + \dots + w_p\sigma_{k,x,p})$: by construction, the sections $\tau_{k,x}$ of E_k are asymptotically holomorphic, their norm is bounded by δ, and they have uniform Gaussian decay properties. Let $\tilde{s}_k = s_k + \tau_{k,x}$, and observe that by construction the graph of $j^r\tilde{s}_k$ over $B_{g_k}(x, c_3)$ is contained in $B_{\hat{g}_k}(u_0, \rho_\epsilon)$. Define $\tilde{h} = (f_1 \circ j^r\tilde{s}_k, \dots, f_p \circ j^r\tilde{s}_k)$; by construction, and because of the bounds on second derivatives of $f_1, \dots, f_p$, we have the equality $\tilde{h} = h - (w_1\Theta_1 + \dots + w_p\Theta_p) + O(\delta^2)$. If δ is assumed to be small enough, the quadratic term in this expression is much smaller than η ; therefore, under this assumption we get that $\tilde{h}$ is $c'_3\eta$-transverse to 0 over $B_{g_k}(x, c_5)$ for some fixed constant c'_3. Finally, recalling the characterization of estimated transversality to a submanifold defined by local equations given at the end of §3.1, we conclude that the graph of $j^r\tilde{s}_k$ is $c'_4\eta$-transverse to S_k^a over $B_{g_k}(x, c_5)$ for some fixed constant c'_4, i.e. $\tilde{s}_k$ satisfies the property $\mathcal{P}_k(c'_4\delta\log(\delta^{-1})^{-\nu}, y)$ at every point $y \in B_{g_k}(x, c_5)$.

Putting together the various possible cases (according to the distance between $j^r s_k(x)$ and S_k^a or its boundary), we obtain that the properties $\mathcal{P}_k$ satisfy the assumptions of Proposition 4.1. Therefore, for all large values of k a small perturbation can be added to s_k in order to achieve uniform transversality to S_k^a away from ∂S_k^a. The inductive argument described at the beginning of this section then makes it possible to complete the proof of Theorem 1.1.

The proof of Theorem 3.2 follows the same argument, but for one-parameter families of sections. One easily checks that the various results of §2 (Lemma 2.1, 2.2, 2.3) remain valid for families of objects depending continuously on a parameter $t \in [0,1]$. Moreover, Propositions 4.1 and 4.2 also extend to the one-parameter case.[8,3] So we only need to check that the argument used above to verify that the properties $\mathcal{P}_k$ satisfy the assumptions of Proposition 4.1 extends to the case of one-parameter families.

As before, fix a stratum $S^a_{k,t}$ in each stratification, a constant $\delta > 0$, a point $x \in X$, and asymptotically holomorphic sections $s_{k,t}$ of $E_{k,t}$. With the same notations as above, let $\Omega_k \subset [0,1]$ be the set of values of t such that $j^r s_{k,t}(x)$ lies at distance more than $\frac{3}{20}\kappa\gamma$ from $\partial S^a_{k,t}$, and within distance $\frac{1}{2}\rho_\epsilon$ from $S^a_{k,t}$. Let $\Omega_k^- \subset \Omega_k$ be the set of values of t such that $j^r s_{k,t}(x)$ lies at distance more than $\frac{1}{5}\kappa\gamma$ from $\partial S^a_{k,t}$ and within distance $\frac{1}{4}\rho_\epsilon$ from $S^a_{k,t}$. Observe that, if $t \notin \Omega_k^-$, a certain uniform transversality property with respect to $S^a_{k,t}$ is already satisfied by $j^r s_{k,t}$ over a small ball centered at x, and therefore no specific perturbation is needed : if x lies within distance $\frac{1}{5}\kappa\gamma$ from $\partial S^a_{k,t}$, then $\mathcal{P}_k(\frac{1}{40}\kappa\gamma, y)$ is satisfied at every point of a ball of fixed radius, while if x lies at distance more than $\frac{1}{4}\rho_\epsilon$ from $S^a_{k,t}$ then $\mathcal{P}_k(\frac{1}{8}\rho_\epsilon, y)$ holds over a ball of fixed radius around x. Even better, if δ is small enough compared to γ and ρ_ϵ, then any perturbation of $s_{k,t}$ by less than δ still satisfies a similar transversality property (with decreased estimates).

For t in Ω_k, the proximity of $j^r s_{k,t}(x)$ to $S^a_{k,t}$ makes it possible to locally define complex-valued functions $f_{1,t}, \dots, f_{p,t}$ depending continuously on t and such that a local equation of $S^a_{k,t}$ is $f_{1,t} = \cdots = f_{p,t} = 0$ (recall the definition of the continuous dependence of the stratifications $\mathcal{S}_{k,t}$ upon the parameter t given in §3.2). This lets us define as above the function $h_t = (f_{1,t} \circ j^r s_{k,t}, \dots, f_{p,t} \circ j^r s_{k,t})$, depending continuously on t. As in the non-parametric case, we can construct asymptotically holomorphic sections $s^{\text{ref}}_{k,x,t,I}$ of $E_{k,t}$, with Gaussian decay away from x and defining local frames in $\mathcal{J}^r E_{k,t}$, simply by multiplying the sections of Lemma 2.3 by polynomials of degree at most r in the local coordinates (all these sections depend continuously on t). We can then find linear combinations $\sigma_{k,x,t,1}, \dots, \sigma_{k,x,t,p}$ of the sections $s^{\text{ref}}_{k,x,t,I}$, with constant coefficients depending continuously on t, such that, denoting by $\Theta_{t,i}$ the $\mathbb{C}^p$-valued functions expressing the variations of h_t upon adding small multiples of $\sigma_{k,x,t,i}$ to $s_{k,t}$, the norm of $\Theta_{t,1} \wedge \cdots \wedge \Theta_{t,p}$ is bounded from below at x and over a small ball surrounding it.

Constructing the functions $\tilde{\mu}_t$ as in the proof of Theorem 1.1 and applying the one-parameter version of Proposition 4.2, we obtain, if k is large enough, a continuous one-parameter family of constants $w_t \in \mathbb{C}^p$, depending continuously on $t \in \Omega_k$ and bounded by δ for all t, such that $\tilde{\mu}_t - w_t$ is η-transverse to 0 over the unit ball in $\mathbb{C}^n$. It follows that, denoting by $\tau_{k,x,t}$ the asymptotically holomorphic perturbations $-(w_{t,1}\sigma_{k,x,t,1} + \cdots + w_{t,p}\sigma_{k,x,t,p})$, bounded by δ, with Gaussian decay away from x, and depending continuously on $t \in \Omega_k$, the sections $s_{k,t}+\tau_{k,x,t}$ satisfy the desired transversality property over a small ball centered at x. However these perturbations are only well-defined for $t \in \Omega_k$. In order to extend their definition to all values of t, let $\chi_k : [0,1] \to [0,1]$ be a continuous cut-off function such that $\chi_k(t) = 1$ for all $t \in \Omega_k^-$ and $\chi_k(t) = 0$ for all $t \notin \Omega_k$, and let $\tilde{\tau}_{k,x,t} = \chi_k(t)\tau_{k,x,t}$ (for $t \notin \Omega_k$ we set $\tilde{\tau}_{k,x,t} = 0$). For $t \in \Omega_k^-$ we have $\tilde{\tau}_{k,x,t} = \tau_{k,x,t}$, so the sections $s_{k,t} + \tilde{\tau}_{k,x,t}$ satisfy the required transversality property ; for $t \notin \Omega_k^-$, the sections $s_{k,t}$ already satisfy such a property and, because we have assumed δ to be small enough, transversality is not affected by adding $\tilde{\tau}_{k,x,t}$. Therefore, the assumptions of Proposition 4.1 are satisfied even in the one-parameter setting, and we can conclude the argument in the same way as in the non-parametric case.

Remark. In many cases, Theorems 1.1 and 3.2 can be proved without using Proposition 4.2 (estimated Sard lemma) in its full generality. Indeed, given suitable asymptotically holomorphic quasi-stratifications $\mathcal{S}_k$ of $\mathcal{J}^r E_k$, we can define quasi-stratifications $\tilde{\mathcal{S}}_k$ of $\mathcal{J}^{r+1}E_k$ in the following way. View each element of $\mathcal{J}^{r+1}E_k$ as a 1-jet in $\mathcal{J}^r E_k$, as in §3.2; for each stratum S_k^a of $\mathcal{S}_k$ with codimension greater than n in $\mathcal{J}^r E_k$, let $\tilde{S}_k^a$ be the set of points in $\mathcal{J}^{r+1}E_k$ whose r-jet component belongs to S_k^a. For each stratum S_k^a of $\mathcal{S}_k$ with codimension $p \leq n$, and for each value $0 \leq i \leq p-1$, let $\tilde{S}_k^{a,i}$ be the set of points in $\mathcal{J}^{r+1}E_k$ whose r-jet component belongs to S_k^a and such that the corresponding element in $T^*X^{(1,0)} \otimes \mathcal{J}^r E_k$, after projection to the normal space to TS_k^a, has rank equal to i. In other terms, the union of $\tilde{S}_k^{a,i}$ is the set of $(r+1)$-jets which intersect S_k^a non-transversely.

In a large number of examples, those of the $\tilde{S}_k^{a,i}$ which are not empty are approximately holomorphic submanifolds of $\mathcal{J}^{r+1}E_k$, transverse to the fibers and of codimension at least $n+1$. These submanifolds determine finite Whitney quasi-stratifications $\tilde{\mathcal{S}}_k$ of $\mathcal{J}^{r+1}E_k$, satisfying properties similar to those of Definition 3.2 but with C^1 estimates only instead of C^2 bounds. Still, the same argument as in the proof of Theorem 1.1 shows that, given asymptotically holomorphic sections s_k of E_k, small perturbations can be added for large enough k in order to ensure the uniform transversality of $j^{r+1}s_k$ to $\tilde{\mathcal{S}}_k$; the argument only uses Proposition 4.2 in the case $p > n$, where the proof becomes much easier [3] and C^1 bounds are sufficient. Because all the strata are of codimension greater than n, the η-transversality of $j^{r+1}s_k$ to $\tilde{\mathcal{S}}_k$ simply means that the graph of $j^{r+1}s_k$ remains at distance more than η from the strata of $\tilde{\mathcal{S}}_k$. By definition of $\tilde{\mathcal{S}}_k$, this is equivalent to the uniform transversality of $j^r s_k$ to $\mathcal{S}_k$, which was the desired result.

5 Examples and applications

We now consider various examples of (quasi)-stratifications to which we can apply Theorems 1.1 and 3.2. The fact that they are asymptotically holomorphic is in all cases a direct consequence of Proposition 3.1.

To make things more topological, we place ourselves in the case where the almost-complex structure J on X is tamed by a given symplectic form ω. In this context, the various approximately J-holomorphic submanifolds of X appearing in the constructions are automatically symplectic with respect to ω. Moreover, remember that the space of ω-tame or ω-compatible almost-complex structures on X is contractible. In most applications, asymptotically very ample bundles are constructed from line bundles with first Chern class proportional to $[\omega]$; in that situation, the ampleness properties of these bundles do not depend on the choice of an ω-compatible almost-complex structure J. Theorem 3.2 then implies that all the constructions described below are, for large enough values of k, canonical up to isotopy, independently of the choice of J. In the general case, the constructions are still canonical up to isotopy, but the space of possible choices for J is constrained by the necessity for the bundles E_k to be ample.

The first application is the construction of symplectic submanifolds as zero sets of asymptotically holomorphic sections of vector bundles over X, as initially obtained by Donaldson [7] and later extended to a slightly more general setting.[2]

Corollary 5.1 *Let (X, ω) be a compact symplectic manifold endowed with an ω-tame almost-complex structure J, and let E_k be an asymptotically very ample sequence of locally splittable vector bundles over (X, J). Then, for all large enough values of k there exist asymptotically holomorphic sections s_k of E_k which are uniformly transverse to 0 and whose zero sets are smooth symplectic manifolds in X. Moreover these sections and submanifolds are, for large k, canonical up to isotopy, indepedently of the chosen almost-complex structure on X.*

Proof. Let $\mathcal{S}_k$ be the stratification of $\mathcal{J}^0 E_k = E_k$ in which the only stratum is the zero section of E_k (these stratifications are obviously asymptotically holomorphic). By Theorem 1.1, starting from any asymptotically holomorphic sections of E_k (e.g. the zero sections) we can obtain for large k asymptotically holomorphic sections of E_k which are uniformly transverse to $\mathcal{S}_k$, i.e. uniformly transverse to 0. It is then a simple observation that the zero sets of these sections are, for large k, smooth approximately J-holomorphic (and therefore symplectic) submanifolds of X.[7] Finally, the uniqueness of the construction up to isotopy is a direct consequence of the one-parameter result Theorem 3.2.[2] □

The next example is that of determinantal submanifolds as constructed by Muñoz, Presas and Sols.[11]

Corollary 5.2 *Let (X, ω) be a compact symplectic manifold endowed with an ω-tame almost-complex structure J, let L_k be an asymptotically very ample sequence of line bundles over (X, J), and let E and F be complex vector bundles over X. Then, for all large enough values of k there exist asymptotically holomorphic sections s_k of $E^* \otimes F \otimes L_k$ such that the determinantal loci $\Sigma_i(s_k) = \{x \in X,\ \mathrm{rk}(s_k(x)) = i\}$ are stratified symplectic submanifolds in X. Moreover these sections and submanifolds are, for large k, canonical up to isotopy, indepedently of the chosen almost-complex structure on X.*

Proof. Let $E_k = E^* \otimes F \otimes L_k$, and let $\mathcal{S}_k$ be the stratification of $\mathcal{J}^0 E_k = E_k$ consisting of strata S_k^i, $0 \leq i < \min(\mathrm{rk}\, E, \mathrm{rk}\, F)$, defined as follows : viewing the points of E_k as elements of $\mathrm{Hom}(E, F)$ with coefficients in L_k, each S_k^i is the set of all elements in E_k whose rank is equal to i. By Proposition 3.1, the stratifications $\mathcal{S}_k$ are asymptotically holomorphic. Applying Theorem 1.1 to these stratifications and starting from the zero sections, we obtain asymptotically holomorphic sections of E_k which are uniformly transverse to $\mathcal{S}_k$. The determinantal locus $\Sigma_i(s_k)$ is precisely the set of points where the graph of s_k intersects the stratum S_k^i. The result of uniqueness up to isotopy is obtained by applying Theorem 3.2. □

However, our main application is that of maps to projective spaces. Observe that, given a section $s = (s_1, \ldots, s_{m+1})$ of a vector bundle of the form $\mathbb{C}^{m+1} \otimes L$, where L is a line bundle over X, we can construct away from its zero set a projective map $\mathbb{P}s = (s_1 : \ldots : s_{m+1}) : X - s^{-1}(0) \to \mathbb{CP}^m$.

Recall that the space of jets of holomorphic maps from $\mathbb{C}^n$ to $\mathbb{C}^m$ carries a natural partition into submanifolds, the Boardman "stratification".[1,6] Restricting oneself to *generic* r-jets, the strata Σ_I, labelled by r-tuples $I = (i_1, \ldots, i_r)$ with $i_1 \geq \cdots \geq i_r \geq 0$, are defined in the following way. Given a generic holomorphic map f, call $\Sigma_i(f)$ the set of points where $\dim \mathrm{Ker}\, df = i$, and denote by Σ_i the set of holomorphic 1-jets corresponding to such points (i.e., Σ_i is the set of 1-jets (σ_0, σ_1) such that $\dim \mathrm{Ker}\, \sigma_1 = i$). The submanifolds Σ_i determine a stratification of $\mathcal{J}^1_{n,m}$. For a generic holomorphic map f the critical loci $\Sigma_I(f)$ are smooth submanifolds defining a partition of $\mathbb{C}^n$. Therefore, we can define inductively $\Sigma_{i_1,\dots,i_r}(f)$ as the set of points of $\Sigma_{i_1,\dots,i_{r-1}}(f)$ where the kernel of the restriction of df to $T\Sigma_{i_1,\dots,i_{r-1}}(f)$ has dimension i_r (in particular, $\Sigma_{i_1,\dots,i_{r-1},0}(f)$ is open in $\Sigma_{i_1,\dots,i_{r-1}}(f)$ and corresponds to the set of points where f restricts to $\Sigma_{i_1,\dots,i_{r-1}}(f)$ as an immersion).

It is easy to check that the r-jet of f at a given point of $\mathbb{C}^n$ completely determines in which $\Sigma_I(f)$ it lies ; therefore, one can define $\Sigma_I \subset \mathcal{J}^r_{n,m}$ as the set of r-jets $j^r f(x)$ of generic holomorphic maps $f : \mathbb{C}^n \to \mathbb{C}^m$ at points $x \in \Sigma_I(f)$. In other terms, $\Sigma_I(f) = \{x \in \mathbb{C}^n,\ j^r f(x) \in \Sigma_I\}$. It is a classical result[6] that the Σ_I's are smooth submanifolds and define a partition of the space of generic holomorphic r-jets (an open subset in $\mathcal{J}^r_{n,m}$ whose complement has codimension $\geq n+1$), which can be extended into a partition of $\mathcal{J}^r_{n,m}$ by smooth submanifolds.

The Boardman classes Σ_I play a fundamental role in singularity theory, and they completely determine the classification of singularities in certain dimensions. For low enough values of r, m or n, the submanifolds Σ_I define a genuine stratification of the jet space $\mathcal{J}^r_{n,m}$. However, as observed by Boardman, things become more complicated as the dimension increases, and the boundary of Σ_I is in general not a union of entire strata ; in high dimensions Boardman classes do not even define a quasi-stratification.

Still, there exist well-known methods that allow Boardman's partitions to be refined into finite Whitney stratifications of $\mathcal{J}^r_{n,m}$. An example of such a construction can be found in the work of Mather[10] (the constructed object is tautologically a finite Whitney stratification, and one easy checks that each Boardman class is a union of several of its strata).

We now consider the case of maps to projective spaces defined by asymptotically holomorphic sections of $E_k = \mathbb{C}^{m+1} \otimes L_k$ over X. We want to construct a natural approximately holomorphic analogue of the Thom-Boardman stratifications, by defining certain submanifolds in $\mathcal{J}^r E_k$. In order to make things easier by avoiding a lengthy analysis of the boundary structure at the points where the vanishing of the section prevents the definition of a projective map, our aim will only be to construct quasi-stratifications of $\mathcal{J}^r E_k$ rather than genuine stratifications.

We first define $Z = \{(\sigma_0, \dots, \sigma_r) \in \mathcal{J}^r E_k,\ \sigma_0 = 0\}$, i.e. Z is the set of r-jets of sections which vanish at the considered point. As observed in §3.2, Θ_Z consists of all points of Z such that σ_1 is surjective. Next, observe that any point $(\sigma_0, \dots, \sigma_r) \in \mathcal{J}^r E_k$ which does not belong to Z determines the (symmetric) holomorphic r-jet $(\phi_0, \dots, \phi_r)$ of a map to $\mathbb{CP}^m$: $\phi_0 \in \mathbb{CP}^m$,

$\phi_1 \in T_x^* X^{1,0} \otimes T_{\phi_0}\mathbb{CP}^m, \ldots, \phi_r \in (T_x^* X^{1,0})^{\otimes r}_{\text{sym}} \otimes T_{\phi_0}\mathbb{CP}^m$ are defined in terms of $\sigma_0, \ldots, \sigma_r$ by expressions involving the projection map from $\mathbb{C}^{m+1} - \{0\}$ to $\mathbb{CP}^m$ and its derivatives. In fact, one easily checks that, if $(\sigma_0, \ldots, \sigma_r) = j^r s$ is the symmetric holomorphic part of the r-jet of a section of E_k, then $(\phi_0, \ldots, \phi_r) = j^r f$ is the symmetric holomorphic part of the r-jet of the corresponding projective map. Using this notation, define

$$\Sigma_i = \{(\sigma_0, \ldots, \sigma_r) \in \mathcal{J}^r E_k,\ \sigma_0 \neq 0,\ \dim \operatorname{Ker} \phi_1 = i\}.$$

For $\max(0, n-m) < i \leq n$, one easily checks that Σ_i is a smooth submanifold of $\mathcal{J}^r E_k$, and that $\partial \Sigma_i$ is the union of $\bigcup_{j>i} \Sigma_j$ and a subset of $Z - \Theta_Z$: indeed, observe that if $n \geq m$, then for any $(\sigma_0, \ldots, \sigma_r) \in \bar{\Sigma}_i \cap Z$ we have $\dim \operatorname{Ker} \sigma_1 \geq i - 1 > n - (m+1)$ and therefore σ_1 is not surjective, while in the case $n < m$ dimensional reasons prevent σ_1 from being surjective.

Next, we assume that $r \geq 2$, and observe that Θ_{Σ_i} is the set of points $(\sigma_0, \ldots, \sigma_r) \in \Sigma_i$ such that

$$\Xi_{i;(\sigma_0,\ldots,\sigma_r)} = \{u \in T_x X^{1,0},\ (\iota_u \sigma_1, \ldots, \iota_u \sigma_r, 0) \in T_{(\sigma_0,\ldots,\sigma_r)} \Sigma_i\}$$

has the expected codimension in $T_x X^{1,0}$ (i.e., the same codimension as Σ_i in $\mathcal{J}^r E_k$). Indeed, by definition $(\sigma_0, \ldots, \sigma_r)$ belongs to Θ_{Σ_i} if and only if the $(r+1)$-jet $(\sigma_0, \ldots, \sigma_r, 0)$, viewed as a 1-jet in $\mathcal{J}^r E_k$, intersects Σ_i transversly (because the definition of Σ_i involves only σ_0 and σ_1, the choice of a lift in $\mathcal{J}^{r+1} E_k$ does not matter, so we can choose the $(r+1)$-tensor component to be zero). By convention (see §3.2), this element of $\mathcal{J}^{r+1} E_k$ corresponds to the 1-jet of a local section σ of $\mathcal{J}^r E_k$ satisfying, at the given point $x \in X$, $\sigma(x) = (\sigma_0, \ldots, \sigma_r)$ and $\nabla \sigma(x) = (\sigma_1, \ldots, \sigma_{r+1})$: the covariant derivative contains no antiholomorphic or antisymmetric terms. The graph of σ intersects Σ_i transversely if and only if $\{u \in TX,\ \nabla \sigma(x).u \in T_{\sigma(x)} \Sigma_i\}$ has the expected dimension, hence the above criterion.

With this understood, we can define inductively, for $p + 1 \leq r$,

$$\Sigma_{i_1,\ldots,i_{p+1}} = \{\sigma \in \Theta_{\Sigma_{i_1,\ldots,i_p}},\ \dim(\operatorname{Ker} \phi_1 \cap \Xi_{(i_1,\ldots,i_p);\sigma}) = i_{p+1}\},$$

where $\Xi_{I;\sigma} = \{u \in T_x X^{1,0},\ (\iota_u \sigma_1, \ldots, \iota_u \sigma_r, 0) \in T_{(\sigma_0,\ldots,\sigma_r)} \Sigma_I\}$ as above, and Θ_{Σ_I} again consists of all points $\sigma \in \Sigma_I$ such that $\Xi_{I;\sigma}$ has the same codimension in $T_x X^{1,0}$ as Σ_I in $\mathcal{J}^r E_k$.

For $i_1 \geq \cdots \geq i_{p+1} \geq 1$, $\Sigma_{i_1,\ldots,i_{p+1}}$ is a smooth submanifold in $\mathcal{J}^r E_k$, and its closure inside $\Sigma_{i_1,\ldots,i_p}$ is obtained by adding $\bigcup_{j>i_{p+1}} \Sigma_{i_1,\ldots,i_p,j}$ and a subset of $\Sigma_{i_1,\ldots,i_p} - \Theta_{\Sigma_{i_1,\ldots,i_p}}$. However, it is quite difficult to fully understand the boundary structure of $\bar{\Sigma}_{i_1,\ldots,i_{p+1}}$; the situation is exactly the same as in standard Boardman theory for holomorphic jets, except that, besides pieces of $\Sigma_{j_1,\ldots,j_q}$ where $q \le p+1$ and $(j_1,\ldots,j_q) \ge (i_1,\ldots,i_q)$ for the lexicographic order, the boundary of $\Sigma_{i_1,\ldots,i_{p+1}}$ also contains a subset of $Z - \Theta_Z$.

In low dimensions and/or for low values of r, it can be checked that the submanifolds $Z, \Sigma_i, \Sigma_{i_1,i_2},\ldots,\Sigma_{i_1,\ldots,i_r}$ determine a finite Whitney quasi-stratification of $\mathcal{J}^r E_k$; for example when $r = 1$ this is an immediate consequence of the above discussion.

However, in larger dimensions it is necessary to refine Boardman's construction as in the holomorphic case. The important observation is that, when $\mathcal{J}^r E_k$ is trivialized by choosing local asymptotically holomorphic coordinates and sections, the partition of $\mathcal{J}^r E_k - Z$ described above corresponds exactly to the partition of the space of r-jets of maps to $\mathbb{CP}^m$ given by Boardman classes. Therefore, we can circumvent the problem by refining Boardman's partition of $\mathcal{J}^r_{n,m}$ into a genuine stratification as explained above, lifting it by the projectivization map to a stratification of the space of non-vanishing jets in $\mathcal{J}^r_{n,m+1}$, and finally pull it back to obtain a stratification of $\mathcal{J}^r E_k - Z$. As in the holomorphic case, the Σ_I classes are realized as unions of strata ; therefore, transversality to this stratification implies transversality to the Σ_I's. Moreover, all strata (except for the open one which we discard anyway) are contained in the closure of Σ_1, so that the boundary structures near Z are entirely contained in $Z - \Theta_Z$; therefore adding Z to this stratification yields a quasi-stratification of $\mathcal{J}^r E_k$.

Definition 5.1 *Given asymptotically very ample line bundles L_k over (X^{2n}, J), and setting $E_k = \mathbb{C}^{m+1} \otimes L_k$, the* Boardman stratification *of $\mathcal{J}^r E_k$ is the quasi-stratification given by the submanifold Z and by a refined Thom-Boardman stratification of $\mathcal{J}^r E_k - Z$.*

Corollary 5.3 *Let (X, ω) be a compact symplectic manifold endowed with an ω-tame almost-complex structure J, let L_k be an asymptotically very ample sequence of line bundles over (X, J), and let $E_k = \mathbb{C}^{m+1} \otimes L_k$. Then, for all large enough values of k there exist asymptotically holomorphic sections s_k of E_k such that the r-jets $j^r s_k$ are uniformly transverse to the Boardman stratifications of $\mathcal{J}^r E_k$.*

In particular, the zero sets $Z_k = s_k^{-1}(0)$ are smooth symplectic codimension $2m$ submanifolds in X, and the holomorphic r-jets of the projective maps $f_k = \mathbb{P}s_k : X - Z_k \to \mathbb{CP}^m$ behave at every point in a manner similar to those of generic holomorphic maps from a complex n-fold to $\mathbb{CP}^m$. Moreover, the singular loci $\Sigma_I(f_k) = \{x \in X - Z_k,\ j^r f_k(x) \in \Sigma_I\}$ are smooth symplectic submanifolds of the expected codimension and define a partition of $X - Z_k$. Finally, the sections s_k and the maps f_k are, for large k, canonical up to isotopy, independently of the chosen almost-complex structure on X.

Proof. By construction the Boardman stratifications of $\mathcal{J}^r E_k$ satisfy the assumptions of Proposition 3.1, as in every fiber of $\mathcal{J}^r E_k$ they can be identified with the same holomorphic quasi-stratification of $\mathcal{J}^r_{n,m+1}$. As a consequence, they are asymptotically holomorphic, and the existence of asymptotically holomorphic sections of E_k with the desired transversality properties is an immediate consequence of Theorem 1.1. The properties of Z_k follow immediately from the uniform transversality to the stratum Z of vanishing sections, while the properties of f_k are direct consequences of the uniform transversality to the Boardman strata (recall that each Σ_I is smooth and is a union of strata). Finally, the uniqueness result is obtained by applying Theorem 3.2. □

Corollary 5.3 is, in a certain sense, a fundamental result of asymptotically holomorphic singularity theory. Still, it falls short of the natural goal that one may have in mind at this point, namely the construction of approximately holomorphic projective maps which are near every point of X topologically conjugate in approximately holomorphic coordinates to generic holomorphic maps between complex manifolds.

Indeed, in order to achieve such a result, one needs to obtain some control on the antiholomorphic part of the jet of f_k at the points of the singular loci $\Sigma_I(f_k)$: roughly speaking, $\bar{\partial} f_k$ must be much smaller than ∂f_k in every direction and at every point, and when ∂f_k is singular this is no longer an immediate consequence of asymptotic holomorphicity and transversality. Note however that the behavior of f_k near the set of base points Z_k is always the expected one.

In many cases, it is possible to perturb slightly the sections s_k (by less than a fixed multiple of $c_k^{-1/2}$, which affects neither holomorphicity nor transversality properties) along the singular loci in order to obtain the proper topological picture for f_k.

The easiest case is $m \geq 2n$, where it is enough to consider 1-jets, and all the strata turn out to be of codimension greater than n ; the uniform transversality of s_k to the Boardman stratification then implies that the maps f_k are approximately holomorphic immersions. Moreover, when $m \geq 2n+1$ an arbitrarily small perturbation is enough to get rid of multiple points, thus giving approximately holomorphic embeddings into projective spaces, a result already obtained by Muñoz, Presas and Sols.[11]

Next, we can consider the case $m = 1$, where 1-jets are again sufficient, and the only interesting Boardman stratum is Σ_n, of complex codimension n, corresponding to critical points of $\mathbb{CP}^1$-valued maps. The sections of $\mathbb{C}^2 \otimes L_k$ given by Corollary 5.3 vanish along smooth codimension 4 base loci ; moreover, the differential ∂f_k of the $\mathbb{CP}^1$-valued map f_k only vanishes at isolated points, and does so in a non-degenerate way. These transversality properties are precisely those imposed by Donaldson in his construction of symplectic Lefschetz pencils;[8] the only missing ingredient is an extra perturbation near the zeroes of ∂f_k in order to get rid of the antiholomorphic terms and therefore ensure that they are genuine non-degenerate critical points, thus making f_k a complex Morse function.

The last case we will consider is when $m = 2$. In this case, we need to consider 2-jets, and the relevant Boardman strata are Σ_{n-1}, of complex codimension $n-1$, and $\Sigma_{n-1,1}$, of complex codimension n (the other strata have codimension greater than n). The sections of $\mathbb{C}^3 \otimes L_k$ constructed by Corollary 5.3 vanish along smooth codimension 6 base loci. The $\mathbb{CP}^2$-valued maps f_k are submersions outside of the smooth symplectic curves $R_k = \Sigma_{n-1}(f_k)$, and the restriction of f_k to R_k is an immersion except at the points of $C_k = \Sigma_{n-1,1}(f_k)$. After a suitable perturbation in order to ensure the vanishing of some antiholomorphic derivatives of f_k along R_k, one obtains a situation similar to that described in previous papers:[3,4] at every point of $R_k - C_k$, a local model for f_k in approximately holomorphic coordinates is $(z_1, \dots, z_n) \mapsto (z_1^2 + \cdots + z_{n-1}^2, z_n)$, while at the points of C_k the local model becomes $(z_1, \dots, z_n) \mapsto (z_1^3 + z_1 z_n + z_2^2 + \cdots + z_{n-1}^2, z_n)$ and the symplectic curve $f_k(R_k) \subset \mathbb{CP}^2$ presents an isolated cusp singularity.

In the general case, the most promising strategy to achieve topological conjugacy to generic holomorphic local models is to perturb the sections s_k in order to make sure that, along each stratum $\Sigma_I(f_k)$, the germ of f_k is holomorphic along the normal directions to $\Sigma_I(f_k)$. Such perturbations should be relatively easy to construct by methods similar to those in the above-mentioned papers,[3,4] provided that one starts from the strata of lowest dimension. This approach will be developped in a forthcoming paper.

Finally, let us formulate some natural extensions of Corollary 5.3 to more general situations. First, we mention the case when the asymptotically very ample line bundles L_k are replaced by vector bundles of rank $\nu \geq 2$. In that case, and provided that $m \geq \nu$, the projective maps defined by sections s_k of $\mathbb{C}^{m+1} \otimes L_k$ are replaced by maps $\mathrm{Gr}(s_k)$ taking values in the Grassmannian $\mathrm{Gr}(\nu, m+1)$ of ν-planes in $\mathbb{C}^{m+1}$, defined at every point of X where the $m+1$ chosen sections generate the whole fiber of L_k. More precisely, at every such point there exist $m+1-\nu$ independent linear relations between the $m+1$ components $s_k^1, \dots, s_k^{m+1}$, and these $m+1-\nu$ linear equations in $m+1$ variables determine a ν-dimensional complex subspace $\mathrm{Gr}(s_k)$ in $\mathbb{C}^{m+1}$. By adapting Corollary 5.3 to this situation, it is for example possible to recover the Grassmannian embedding result of Muñoz, Presas and Sols.[11]

Another direction in which Corollary 5.3 can be improved is by adding extra transversality requirements to the projective maps f_k. For example, given a stratified holomorphic submanifold $\mathcal{D} = (D_a)_{a\in A}$ in $\mathbb{CP}^m$, we can require the transversality of the map f_k to $\mathcal{D}$. Indeed, $\mathcal{D}$ induces a stratification $\tilde{\mathcal{D}}_k$ of $\mathcal{J}^r E_k$, in which each stratum consists of the jets $(\sigma_0, \dots, \sigma_r)$ such that $\mathbb{P}\sigma_0$ belongs to a certain stratum D_a of $\mathcal{D}$ (in fact, this stratification only involves the 0-jet part). Starting from sections s_k of E_k given by Corollary 5.3, we can apply Theorem 1.1 to the stratifications $\tilde{\mathcal{D}}_k$ (which one easily shows to to be asymptotically holomorphic by Proposition 3.1) ; this yields asymptotically holomorphic sections $\tilde{s}_k$ which are uniformly transverse to $\tilde{\mathcal{D}}_k$ but differ from s_k by an amount small enough to ensure that the transversality of the jets to the Boardman stratification is preserved. In this way, one obtains projective maps which have the same properties as in Corollary 5.3 and additionally are uniformly transverse to the stratified submanifold $\mathcal{D}$. This extends a result of Muñoz, Presas and Sols [11] where asymptotically holomorphic embeddings are made transverse to a given submanifold of $\mathbb{CP}^m$.

Another class of stratifications of $\mathcal{J}^r E_k$ that we can consider are those obtained from lower-dimensional Boardman stratifications by linear projections. Namely, fix $q < m$, and let $\pi : \mathbb{C}^{m+1} \to \mathbb{C}^{q+1}$ be a linear projection ; π induces maps $\tilde{\pi} : \mathcal{J}^r(\mathbb{C}^{m+1} \otimes L_k) \to \mathcal{J}^r(\mathbb{C}^{q+1} \otimes L_k)$, and the inverse images by $\tilde{\pi}$ of the Boardman stratifications of $\mathcal{J}^r(\mathbb{C}^{q+1} \otimes L_k)$ are asymptotically holomorphic quasi-stratifications of $\mathcal{J}^r(\mathbb{C}^{m+1} \otimes L_k)$. The transversality of $j^r s_k$ to these quasi-stratifications is equivalent to that of $j^r(\pi(s_k))$ to the Boardman stratifications ; denoting by $\bar{\pi}$ the map from $\mathbb{CP}^m$ to $\mathbb{CP}^q$ induced by π, this is also equivalent to the genericity of the holomorphic jets of the projective maps $\bar{\pi} \circ f_k$. Therefore, by applying Theorem 1.1 as in the previous example, we can obtain projective maps f_k with the same genericity properties as in Corollary 5.3 and such that the maps $\bar{\pi} \circ f_k$ also enjoy similar properties. Even better, by iteratedly applying Theorem 1.1 we can obtain the same property for any given finite family of linear projections. For example, when $m = 2$ and considering projections of $\mathbb{C}^3$ to $\mathbb{C}^2$ along coordinate axes, one obtains exactly the transversality properties which are needed in order to extend Moishezon-Teicher braid group techniques to the study of symplectic manifolds.[5,4]

To conclude, let us mention a different class of potential applications of Theorem 1.1, following the ideas of Donaldson and Smith. As shown by Donaldson,[8] any compact symplectic 4-manifold carries structures of symplectic Lefschetz pencils obtained from pairs of sections of asymptotically very ample line bundles L_k ; after blowing up the base points, we obtain Lefschetz fibrations over $\mathbb{CP}^1$, which may also be thought of as maps from $\mathbb{CP}^1$ to the moduli space $\bar{M}_g$ of stable curves of a certain genus g. These maps become asymptotically holomorphic as one considers pencils given by sections of L_k for $k \to +\infty$. In a largely unexplored class of constructions, one considers certain vector bundles over $\mathbb{CP}^1$ naturally arising from the Lefschetz fibrations : for example, spaces of holomorphic sections of certain bundles over each fiber, or pull-backs by the maps from $\mathbb{CP}^1$ to $\bar{M}_g$ of vector bundles over $\bar{M}_g$. It often turns out that these bundles over $\mathbb{CP}^1$ either are naturally asymptotically very ample or become so after tensor product by the line bundles $O(k)$. Theorem 1.1 can then used in order to obtain sections with suitable genericity properties, which in turn give rise to interesting geometric or topological structures. In some cases the objects naturally arising are sheaves rather than bundles, but the same type of argument should remain valid. It is to be expected that some interesting results about symplectic 4-manifolds and Lefschetz pencils can be obtained in this way, as similar considerations (but at a much more sophisticated level) have for example led to Donaldson and Smith's proof of the existence of a pseudo-holomorphic curve realizing the canonical class via Lefschetz fibrations.[9]

References

1. V.I. Arnold, S.M. Gusein-Zadeh, A.N. Varchenko, *Singularities of smooth mappings*, Monographs in Mathematics, Birkhäuser (1982).
2. D. Auroux, *Asymptotically holomorphic families of symplectic submanifolds*, Geom. Funct. Anal. **7** (1997), 971–995.
3. D. Auroux, *Symplectic 4-manifolds as branched coverings of* $\mathbb{CP}^2$, Invent. Math. **139** (2000), 551–602.
4. D. Auroux, *Symplectic maps to projective spaces and symplectic invariants*, to appear in Proc. 7th Gökova Geometry-Topology Conference (2000), International Press.
5. D. Auroux, L. Katzarkov, *Branched coverings of* $\mathbb{CP}^2$ *and invariants of symplectic 4-manifolds*, to appear in Invent. Math.
6. J.M. Boardman, *Singularities of differentiable maps*, Publ. Math. IHES, **33** (1967), 21–57.
7. S.K. Donaldson, *Symplectic submanifolds and almost-complex geometry*, J. Differential Geom. **44** (1996), 666–705.
8. S.K. Donaldson, *Lefschetz pencils on symplectic manifolds*, preprint.
9. S.K. Donaldson, I. Smith, in preparation.
10. J.N. Mather, *How to stratify mappings and jet spaces*, Singularités d'Applications Différentiables (Plans-sur-Bex 1975), Lecture Notes in Math. **535**, Springer (1976), 128–176.

11. V. Muñoz, F. Presas, I. Sols, *Almost holomorphic embeddings in Grassmannians with applications to singular symplectic submanifolds*, preprint.

LOCAL MIRROR SYMMETRY AND FIVE-DIMENSIONAL GAUGE THEORY

TOHRU EGUCHI

Department of Physics, Faculty of Science, University of Tokyo, Tokyo, Japan 113

We propose an exact solution of 5-dimensional $SU(2)$ gauge theory on $M_4 \times S^1$ for arbitrary values of the radius R of S^1. Our amplitude is constructed by compactifying type II string theory on a degenerate Calabi-Yau manifold which is the total space of the canonical bundle of the Hirzebruch surface $\mathbf{F}_2$. It is shown that our result reproduces the Seiberg-Witten theory at the 4-dimensional limit $R \to 0$ and as well as the result of the uncompactified 5-dimensional theory at $R \to \infty$. We also discuss $SU(2)$ gauge theory with $1 \leq N_f \leq 4$ matter and show that they are described by the geometry of the canonical bundle of $\mathbf{F}_2$ blown up at N_f points.

1 Introduction

In this talk I would like to discuss my recent work with Kanno [1] on the dynamics of 5-dimensional $SU(2)$ gauge theories on $M_4 \times S^1$. In ref.[1] we proposed an exact solution for the low-energy effective Lagrangian of the 5-dimensional $SU(2)$ theory for arbitrary values of the radius R of S^1. In the limit $R \to 0$ our effective action reduces to the Seiberg-Witten solution [2] of the $\mathcal{N} = 2$ 4-dimensional gauge theory and also in the opposite limit $R \to \infty$ it reproduces the well-known result on the uncompactified 5-dimensional gauge theory [3,4,5].

Our result is obtained directly from the amplitude of the type II string theory compactified on singular Calabi-Yau (CY) manifolds using the method of local mirror symmetry[6,7,8,9,10]. Local mirror symmetry is an application of mirror symmetry to the case of non-compact CY manifolds. Typical non-compact CY's are obtained by taking the limit of large size $t_E \to \infty$ of the fiber in the elliptically fibered CY over the Hirzebruch surface $\mathbf{F}_n$. The resulting non-compact manifold is modeled by the local $\mathbf{F}_n$, i.e. $\mathbf{F}_n$ inside a CY with its normal bundle being given by the canonical bundle of $\mathbf{F}_n$. The limit $t_E \to \infty$ may also be regarded as the limit of shrinking $\mathbf{F}_n$ with the size of the fiber being kept fixed.

Previously it was pointed out that the amplitudes of the string theory compactified on local models $\mathbf{F}_n$ with $n = 0, 1, 2$ all give the same results in the $R \to 0$ limit and reduce to the Seiberg-Witten solution[7]. What we would like to emphasize in this paper is that only the $\mathbf{F}_2$ model has the correct behavior in the opposite limit $R \to \infty$ and reproduce the known result. We thus propose the low energy effective action based on the local $\mathbb{F}_2$ as the exact solution of 5-dimensional $SU(2)$ gauge theory on $R^4 \times S^1$ for any radius R. Note that $\mathbf{F}_2$ has an anti-canonical class $-K$ which is positive semi-definite and may be considered as a degenerate limit of Del Pezzo surfaces. The degeneracy seems to be the feature necessary to reproduce the $R \to \infty$ limit of gauge theory.

Thus our solution is obtained from the type II theory compactified on the local $\mathbf{F}_2$ in the case of pure $SU(2)$ gauge theory without matter. Similarly, $SU(2)$ gauge theories with N_f matter in vector representations are also obtained from the type II theory compactified on the local $\mathbb{F}_2$ blown up at N_f points $(0 \le N_f \le 4)$.

It turns out that our models at $R = \infty$ has an infinite bare coupling constant and yield non-trivial interacting field theories in the infra-red limit.

2 $SU(2)$ Gauge Theory without Matter

Let us start from the case of 5-dimensional gauge theory without matter. We consider the local $\mathbf{F}_n$ $n = 0, 1, 2$ models which are described by the toric data given in the Appendix. Following the standard procedure [7,8,9,10] one obtains elliptic curves in the B-model given by

$$\mathbb{F}_0 : y^2 = (a_1 x^2 + a_0 x + a_3)^2 - 4a_2 a_4 x^2 \ , \tag{1}$$
$$\mathbb{F}_1 : y^2 = (a_1 x^2 + a_0 x + a_3)^2 - 4a_2 a_4 x \ , \tag{2}$$
$$\mathbb{F}_2 : y^2 = (a_1 x^2 + a_0 x + a_3)^2 - 4a_2 a_4. \tag{3}$$

If we choose $a_1 = a_3 = 1, a_0 = s, 4a_2 a_4 = K^4$, we find

$$\mathbf{F}_0 : y^2 = (x^2 + sx + 1)^2 - K^4 x^2, \ z_F = \frac{1}{s^2}, \ z_B = \frac{K^4}{4s^2}, \tag{4}$$

$$\mathbf{F}_1 : y^2 = (x^2 + sx + 1)^2 - K^4 x, \ z_F = \frac{1}{s^2}, \ z_B = \frac{K^4}{-4s}, \tag{5}$$

$$\mathbf{F}_2 : y^2 = (x^2 + sx + 1)^2 - K^4, \ z_F = \frac{1}{s^2}, \ z_B = \frac{K^4}{4}, \tag{6}$$

where z_F, z_B are parameters of complex structure deformation. $\mathbb{F}_2$ curve (6) is in fact the curve proposed by Nekrasov for the description of 5-dimensional gauge theory [11]. Its properties have been studied in [12] in some detail.

If we introduce a variable U which is the analogue of u of the Seiberg-Witten solution, the parameter s is written as

$$s = 2R^2U \tag{7}$$

where R is the radius of the circle S^1. If one shifts $x \to x - s/2$ and uses variables U and R, $\mathbf{F}_2$ curve reads as

$$y^2 = (x^2 - R^4(U^2 - \frac{1}{R^4}))^2 - K^4. \tag{8}$$

By comparing (8) with the Seiberg-Witten curve

$$y^2 = (x^2 - u)^2 - \Lambda^4, \tag{9}$$

we find that the parameters U and u are related as

$$u \approx R^4(U^2 - \frac{1}{R^4}). \tag{10}$$

Relation (10) holds for small values of R. We also note $K \approx \Lambda$. Strong coupling region $u \approx 0$ mapps to $U \approx \pm 1/R^2$ and the weak coupling region $u \approx \infty$ corresponds to $U \approx R^{-2}exp(2RA)$. A is the analogue of a (VEV of the scalar field of vector multiplet) of the Seiberg-Witten solution.

Note that in the weak coupling region $s \approx 2$ (when R is small). After the shift $x \to x - s/2 \approx x - 1$ both of the $\mathbf{F}_0, \mathbf{F}_1$ curves (4), (5) also reduce essentially to the Seiberg-Witten curve (9). This is the mechanism by which the local models $\mathbf{F}_n$ $n = 0, 1, 2$ reproduce the results of Seiberg-Witten theory.

In the brane-probe interpretation of Seiberg-Witten solution [13,14], u-plane is identified as a local region around one of the four fixed points (orientifold 7-planes) in type I' theory compactified on T^2 to 8-dimensions. $u = 0$ is identified as the location of the one of the orientifold planes. Since the mapping between U and u is 2:1 (10) and $u = 0$ is mapped to $U = \pm 1/R^2$, U-variable describes a pair of u-plane regions with their centers separated by a distance $\approx 1/R^2$. Thus the curve (8) describes a theory whose moduli space contains two of the orientifold planes. Since each fixed plane acts like a reflecting mirror for the external probe, D3-brane probe will possess an infinite number of mirror images when inserted into such a background. These mirror images are separated by distances n/R^2, $n \in \mathbf{Z}$ and open strings connecting them generate Kaluza-Klein modes of supersymmetric gauge fields. Thus the curve (8) effectively describes a theory on a 5-dimensional manifold $M_4 \times S^1$ with R being the radius of S^1. In the limit of $R \to 0$ one of the u-planes moves off to ∞ and the model (8) reduces to the Seiberg-Witten theory.

Periods of the B-model of local $\mathbf{F}_2$ (6) is determined by solving the differential equations (Gelfand-Kapranov-Zelevinskij (GKZ) system) associated with the toric data. Differential equations are given by

$$\mathcal{L}_1 = z_F(4\theta_{z_F}^2 + 2\theta_{z_F}) + \theta_{z_F}(2\theta_{z_B} - \theta_{z_F}), \tag{11}$$

$$\mathcal{L}_2 = z_B(2\theta_{z_B} - \theta_{z_F} + 1)(2\theta_{z_B} - \theta_{z_F}) - \theta_{z_B}^2, \tag{12}$$

$$\theta_{z_F} \equiv z_F \frac{\partial}{\partial z_F}, \quad \theta_{z_B} \equiv z_F \frac{\partial}{\partial z_F} \tag{13}$$

$$\mathcal{L}_1 \omega = 0, \qquad \mathcal{L}_2 \omega = 0. \tag{14}$$

These equations have a regular singular point at $z_F = z_B = 0$ and possess two "single-log" solutions ω_F, ω_B and one "double-log" solution Ω behaving as

$$\omega_F = \log(z_F) + \text{power series in } z_F, z_B \tag{15}$$

$$\omega_B = \log(z_B) + \text{power series in } z_F, z_B \tag{16}$$

$$\Omega = \log(z_f)^2 + \log(z_F)\log(z_B) + \text{power series in } z_F, z_B. \tag{17}$$

We identify the two single-log solutions as the Kähler parameters t_F, t_B of the A-model: t_F represents the size of the P^1 fiber of $\mathbf{F}_2$ and t_B the size of its base P^1.

t_F is given by

$$\begin{aligned}
-t_F &\equiv \omega_F \\
&= \log z_F + \left[\sum_{n\geq 1, m\geq 0} \frac{2(2n-1)!}{(n-2m)!n!m!^2} z_F^n z_B^m - \sum_{m\geq 1} \frac{(2m-1)!}{m!^2} z_B^m \right], \\
&= -2\log\left(\frac{1}{\sqrt{4z_F}} + \sqrt{\frac{1}{4z_F} - 1}\right) \\
&\quad + \left[\sum_{n\geq 1, m\geq 1} \frac{2(2n-1)!}{(n-2m)!n!m!^2} z_F^n z_B^m - \sum_{m\geq 1} \frac{(2m-1)!}{m!^2} z_B^m \right].
\end{aligned} \tag{18}$$

Similarly, t_B is given by

$$-t_B \equiv \omega_B = \log z_B + \sum_{n\geq 1} \frac{2(2n-1)!}{n!^2} z_B^n$$
$$= -2\log\left(\frac{1}{\sqrt{4z_B}} + \sqrt{\frac{1}{4z_B} - 1}\right). \tag{19}$$

In interpreting the string theory amplitude as that of the 5-dimensional gauge theory, the size of the fiber t_F is identified as the vacuum value A of the scalar field in the vector-multiplet

$$t_F = 4RA. \tag{20}$$

On the other hand, the size of the base t_B is associated with the dynamical mass parameter Λ as

$$e^{-t_B} = 4R^4\Lambda^4 . \tag{21}$$

Then the mirror transformation (19) becomes

$$K = \frac{2R\Lambda}{\sqrt{1+4R^4\Lambda^4}} . \tag{22}$$

Note that Katz, Klemm and Vafa suggested the identification of variables (20),(21) with R replaced by ϵ, a small parameter going to zero. In [15] ϵ was identified as the radius of a circle S^1, however, the full mirror transformation (18),(19) was not obtained.

Now we note that one of the single-log solutions, ω_B depends only on z_B and not on z_F. Hence the parameter K of the elliptic curve depends only on Λ and not on U. This is the special feature of the local mirror transformation associated with $\mathbf{F}_2$. In the case of $\mathbf{F}_0$ and $\mathbf{F}_1$, K depends both on U and Λ. When we consider blow ups of $\mathbf{F}_2$ in order to incorporate matter fields, still the parameter K depends only on Λ and not on U or the matter mass parameters. This seems to be the key feature of the local $\mathbf{F}_2$ so that the model exhibits a proper behavior in $R \to \infty$ limit.

One can invert the relations (18),(19) perturbatively and express the B-model parameters z_F,z_B in terms of t_F, t_B. In the case of local mirror symmetry the holomorphic solution of GKZ system is a constant and the mirror transformation is simpler than in the compact CY case. One can expand the double-log solution in terms of the Kähler parameters as

$$\Omega = t_F^2 + t_F t_B + 4\sum_{n=1} \frac{1}{n^2} q_F^n + 4q_B\left(\sum_{n=1} n^2 q_F^n\right)$$
$$+ q_B^2\left(q_F^2 + 36q_F^3 + 260q_F^4 + 1100q_F^5 + \cdots\right) + \mathcal{O}(q_B^3) \tag{23}$$

where

$$q_F = e^{-t_F}, \qquad q_B = e^{-t_B}. \tag{24}$$

The double-log solution is identified as A_D, the dual of the variable A

$$\Omega = -8\pi i R A_D. \tag{25}$$

First two terms of (23) represent the classical intersection numbers of the CY manifold and the remaining terms represent the contribution of world-sheet instantons. According to ref[10] the double-log solution has a generic form

$$\Omega = \sum_{i,j=1}^{2} t_i t_j \langle J^i J^j \rangle + \sum_{k=1} \sum_{n,m \geq 0} (\sum_i x^i \partial_{t_i}) d_{n,m} \frac{q_1^{kn} q_2^{km}}{k^3} \tag{26}$$

for a local model of a surface S with two Kähler parameters. J_i denotes the Kähler classes of S and $\langle J_i J_j \rangle$ their intersection numbers. Numerical coefficients x^i are defined by

$$c_1(S) = \sum_i x^i J_i \tag{27}$$

where $c_1(S)$ is the first Chern class of S. $d_{n,m}$ denotes the number of rational holomorphic curves in the homology class $nJ_1 + mJ_2$. Sum over k in (26) represents the multiple-cover factor.

Substituting geometrical data for $\mathbf{F}_2$ one has

$$\Omega = t_F t_B + t_F^2 - 2\frac{\partial}{\partial t_F} \sum_{k,n,m} d_{n,m} \frac{q_1^{kn} q_2^{km}}{k^3} \tag{28}$$

Comparing (23) and (26) (we set $q_1 = q_F, q_2 = q_B$) we find [7,10]

$$\begin{aligned} d_{1,0} &= -2, \;\; d_{n,0} = 0, \\ d_{n,1} &= -2n, \\ d_{1,2} &= d_{2,2} = 0, \;\; d_{3,2} = -6, \;\; d_{4,2} = -32, \cdots \end{aligned} \tag{29}$$

Integrating A_D over A we obtain the prepotential F for the local $\mathbf{F}_2$ model

$$F = \frac{1}{32\pi i R^2} \left[-\frac{t_F^3}{3} - \frac{t_F^2 t_B}{2} + 4 \sum_{n=1} \frac{1}{n^3} q_F^n \right.$$

$$+4q_B \left(\sum_{n=1} n q_F^n \right) + q_B^2 \left(\frac{1}{2} q_F^2 + 12 q_F^3 + 65 q_F^4 \right. \tag{30}$$

$$\left. +220 q_F^5 + \cdots \right) + \mathcal{O}(q_B^3) \Big] . \tag{31}$$

3 Small and Large Radius Limits

Let us next examine the small and large radius limits $R \to 0, \infty$ of (30). We first consider the 4-dimensional limit $R \to 0$ following [7]. Due to the relations (20), (21) small R corresponds to the base of $\mathbf{F}_2$ becoming large ($t_B \to \infty$) and the fiber becoming small ($t_F \to 0$) while the ratio e^{-t_B}/t_F^4 is kept fixed. As explained by Katz, Klemm and Vafa [7], quantum parts of F are suppressed because of the powers of $q_B \approx R^4$ and the surviving contributions come from the divergence of the series $\sum d_{n,m} q_F^n$ as $q_F = e^{-4RA} \to 1$. If fact when $d_{n,m}$ behaves at large n as

$$d_{n,m} \approx m^{4n-3} \tag{1}$$

for each fixed m, we have

$$\sum_{n=1} d_{n,m} q_F^n q_B^m \approx A^2 \left(\frac{\Lambda}{A}\right)^{4m} \tag{2}$$

and one obtains a contribution which is identified as the m space-time instanton amplitude in gauge theory.

Using (31) we find the behavior of the gauge coupling $\tau = \partial^2 F/\partial A^2$ at small R as

$$\begin{aligned} \tau = &\frac{4i}{\pi} RA - \frac{i}{2\pi}\log(4R^4\Lambda^4) - \frac{2i}{\pi}\sum_{n=1}\frac{q_F^n}{n} \\ &-\frac{8i}{\pi}R^4\Lambda^4\sum_{n=1} n^3 q_F^n + \cdots \\ \approx &\frac{2i}{\pi}\log(2\sqrt{2}\frac{A}{\Lambda}) - \frac{3i}{16\pi}\frac{\Lambda^4}{A^4} + \cdots \end{aligned} \tag{3}$$

(3) reproduces the 1-loop beta function and the 1-instanton contribution to the Seiberg-Witten solution [2]. We can check the agreement with Seiberg-Witten theory for higher instanton terms.

In (3) the world-sheet instanton expansion of type II theory is converted into the space-time instanton expansion of gauge theory in the $R \to 0$ limit. Coefficients of the m-instanton amplitudes of gauge theory are determined by the asymptotic behavior of the number of holomorphic curves $d_{n,m}$ as $n \to \infty$.

Let us next examine the $R \to \infty$ limit of the uncompactified 5-dimensional gauge theory. It is known [16,17] that the gauge theory on M_5 has no quantum corrections and its gauge coupling is simply expressed in terms of classical intersection numbers of CY manifold. In fact by taking $R \to \infty$ world-sheet instanton terms disappear and we have

$$\lim_{R\to\infty} \frac{\tau}{2\pi i R} \equiv \tau_5 = \frac{2}{\pi^2} A \tag{4}$$

(we have rescaled τ so that τ_5 corresponds to the gauge coupling of 5-dimensional theory, $\tau_5 = 1/g_5^2$). In the next section we will discuss local $\mathbf{F}_2$ model blown up at N_f points. We then find that the above formula is generalized as

$$\tau_5 = \frac{2}{\pi^2}\left(A - \sum_{i=1}^{N_f} \frac{1}{16}|A - M_i| - \sum_{i=1}^{N_f} \frac{1}{16}|A + M_i|\right). \tag{5}$$

(5) is exactly the behavior of gauge coupling of $SU(2)$ theory with N_f matter in vector representations (at infinite bare coupling) [3,5,4]. Thus we reproduce correct results also in the uncompactified limit $R \to \infty$.

4 $SU(2)$ theory with matter

Let us next consider the case of $SU(2)$ gauge theory coupled to N_f matter ($1 \leq N_f \leq 4$) in vector representations. We first discuss the $N_f = 1$ case. Relevant geometry is given by the local $\mathbf{F}_2$ with 1-point blown up. Corresponding curve is given by (see Appendix)

$$y^2 = (x^2 + sx + 1)^2 - K^3(xw + \frac{1}{w}). \tag{1}$$

Comparison with the $N_f = 1$ Seiberg-Witten curve suggests the identification

$$w = e^{RM} \tag{2}$$

where M denotes the bare mass of the matter multiplet. Complex moduli of the B-model are given by (we denote z_{F-E} as z_F and z_E as $z_{F'}$ for notational simplicity)

$$z_F = \frac{w^2}{s}, \quad z_{F'} = \frac{1}{sw^2}, \quad z_B = \frac{K^3}{4w}. \tag{3}$$

In this case the GKZ system is given by 5 partial differential equations in 3 variables and is of considerable complexity. Here we content ourselves with the analysis of prepotential at the tree and 1-loop level ignoring the space-time instantons. This suffices for our purpose of extracting the $R \to \infty$ behavior of the theory. The small R behavior has already been studied in [8] and argued to reproduce Seiberg-Witten theory (we have also verified that the 1-instanton term is correctly reproduced).

When we ignore instantons, single log solutions are given by

$$-t_F \equiv \omega_F = \log(z_F) + \sum_{n=1} \frac{(2n-1)!}{n!^2}(z_F z_{F'})^n, \tag{4}$$

$$-t_{F'} \equiv \omega_{F'} = \log(z_{F'}) + \sum_{n=1} \frac{(2n-1)!}{n!^2}(z_F z_{F'})^n, \tag{5}$$

$$-t_B \equiv \omega_B = \log(z_B). \tag{6}$$

Identification with the variables of the gauge theory is given by

$$t_F = 2(A-M), \quad t_{F'} = 2(A+M), \quad e^{-t_B} = \frac{2R^3\Lambda^3}{w}. \tag{7}$$

Then by inverting relations (4),(5) we find

$$z_F = \frac{e^{-2R(A-M)}}{1+e^{-4RA}}, \quad z_{F'} = \frac{e^{-2R(A+M)}}{1+e^{-4RA}}. \tag{8}$$

The double log solution

$$\begin{aligned} \Omega &= t_F^2 + \frac{1}{2}t_{F'}^2 + 2t_F t_{F'} + t_B(t_F + t_{F'}) \\ &+ \sum_{n=1} \frac{4(2n-1)!}{n!^2} \sum_{j=1}^{n} \frac{1}{j} (z_F z_{F'})^n \\ &+ \left(-\sum_{n>\ell} + \sum_{n<\ell}\right) \frac{(n+\ell-1)!}{n!\ell!} \frac{1}{n-\ell} z_F^n z_{F'}^\ell \end{aligned} \tag{9}$$

can then be re-expressed as

$$\begin{aligned} \Omega = t_F^2 + \frac{1}{2}t_{F'}^2 + 2t_F t_{F'} + t_B(t_F + t_{F'}) + 4\sum_{n=1} \frac{1}{n^2} e^{-4nRA} \\ -\sum_{n=1} \frac{1}{n^2} e^{-2nR(A+M)} - \sum_{n=1} \frac{1}{n^2} e^{-2nR(A-M)} \end{aligned} \tag{10}$$

using (8).

When we take the derivative in A, (10) gives

$$\begin{aligned} \tau = \frac{i}{\pi}[2\log\sinh(2RA) \\ -\frac{1}{4}\log\sinh(R(A+M))\sinh(R(A-M))]. \end{aligned} \tag{11}$$

In the 4-dimensional limit (11) becomes

$$\tau \approx \frac{i}{\pi}(2-\frac{1}{2})\log A \tag{12}$$

which gives the 1-loop beta function of Seiberg-Witten theory. On the other hand in the 5-dimensional limit we find

$$\tau_5 = \lim_{R\to\infty} \frac{\tau}{2\pi i R} = \frac{2}{\pi^2}\left[A - \frac{1}{16}|A-M| - \frac{1}{16}|A+M|\right] \tag{13}$$

as we have claimed before. We note that in formulas (4) and (13) τ_5 does not contain an additive constant which is identified as the bare coupling constant $\tau_{5,B} = 1/g^2_{5,B}$. Thus the theory sits at the infinite bare coupling constant limit $g^2_{5,B} = \infty$ which yields non-trivial 5-dimensional theory in the infra-red regime [3].

We can similarly study the system with more matter up to $N_f = 4$ using the local mirror symmetry. Curves of the B-model are given by (see Appendix)

$$y^2 = (x^2 + sx + 1)^2 - K^{4-N_f} \prod_{i=1}^{N_f} (w_i x + \frac{1}{w_i}) \tag{14}$$

where parameters w_i correspond to the bare masses of the matter multiplets

$$w_i = e^{M_i R}, \quad i = 1, \cdots, 4 \tag{15}$$

(14) agrees with the curve suggested by [11] and [18] (at $N_f = 4$ the factor K^{4-N_f} should be replaced by a dimensionless parameter q).

In the cases $N_f \geq 2$ the analysis of GKZ system becomes further involved: we will instead use a simpler method based on the Picard-Fuchs (PF) equation derived for the elliptic curves (14). It turns out that the PF equation of the elliptic curve is an ordinary differential equation of 3rd order in the variable A and can be studied relatively easily. This is the method used in ref [12]. This system, however, is not complete unlike the GKZ case. The periods are determined only up to integration constants and one can not completely fix the mirror transformation. This ambiguity, however, affects only the quantum part of the computation and one still obtain exact results for the prepotential at the tree and 1-loop level.

First we note that the quadratic curve $y^2 = ax^4 + 4bx^3 + 6cx^2 + 4dx + e$ is transformed into the Weierstrass form

$$y^2 = 4x^3 - g_2 x - g_3 \tag{16}$$

by the relation

$$g_2 = ae - 4bd + 3c^2, \quad g_3 = ace + 2bcd - ad^2 - b^2 e - c^3. \tag{17}$$

We regard g_2, g_3 as functions of the parameter s. Periods ω of the elliptic curve (16) obey the PF equation [19,20]

$$\frac{d^2\omega}{ds^2} + c_1(s)\frac{d\omega}{ds} + c_0(s)\omega = 0 \ , \tag{18}$$

where

$$c_1 = -\frac{d}{ds}\log\left(\frac{3}{2\Delta}(2g_2\frac{dg_3}{ds} - 3\frac{dg_2}{ds}g_3)\right) \ , \tag{19}$$

$$c_0 = \frac{1}{12}c_1\frac{d}{ds}\log\Delta + \frac{1}{12}\frac{d^2}{ds^2}\log\Delta$$
$$-\frac{1}{16\Delta}\left(g_2(\frac{dg_2}{ds})^2 - 12(\frac{dg_3}{ds})^2\right) \tag{20}$$

and $\Delta = (g_2)^3 - 27(g_3)^2$ denotes the discriminant of the curve. Since dA/ds is one of the periods of the curve, it satisfies the PF equation. Regarding s as a function of A we obtain

$$\frac{ds}{dA}\frac{d^3s}{dA^3} - 3(\frac{d^2s}{dA^2})^2 + c_1(\frac{ds}{dA})^2\frac{d^2s}{dA^2} - c_0(\frac{ds}{dA})^4 = 0 \ . \tag{21}$$

This determines s in terms of A.

Similarly $dA_D/ds = dA/ds \cdot d^2F/dA^2$ satisfies the PF equation and we obtain

$$(\frac{ds}{dA})^{-1}\frac{d^4F}{dA^4} - 3(\frac{ds}{dA})^{-2}\frac{d^2s}{dA^2}\frac{d^3F}{dA^3} + c_1\frac{d^3F}{dA^3} = 0 \ . \tag{22}$$

This equation can be integrated once and we find

$$\frac{d^3F}{dA^3} = \frac{const.}{\Delta}\left(2g_2\frac{dg_3}{ds} - 3\frac{dg_2}{ds}g_3\right)\left(\frac{ds}{dA}\right)^3 . \tag{23}$$

Solving (23) determines the prepotential (const. in the right-hand-side is fixed by a suitable normalization of F).

In the case $N_f = 2$, we find the gauge coupling at the classical and 1-loop level as

$$\tau = \frac{i}{\pi}[2\log\sinh(2RA)$$
$$-\frac{1}{4}\prod_{i=1}^{2}\log\sinh(R(A+M_i))\sinh(R(A-M_i))]. \tag{24}$$

In the 4-dimensional limit (24) reads as

$$\tau \approx \frac{2i}{\pi}(2 - \frac{1}{2}\cdot 2)\log A \tag{25}$$

which gives the beta function of $N_f = 2$ Seiberg-Witten theory. On the other hand, in the five-dimensional limit we have

$$\tau_5 = \frac{2}{\pi^2}\left(A - \sum_{i=1}^{2}\frac{1}{16}|A - M_i| - \sum_{i=1}^{2}\frac{1}{16}|A + M_i|\right) \tag{26}$$

in agreement with (5).

We have checked that the local model of $\mathbf{F}_2$ blown up at 3 and 4 points (see Appendix) also reproduce (5). Thus we have obtained a model which appears to describe correctly the physics of 5-dimensional theory up to the number of flavors $N_f = 4$ making use of the mirror symmetry.

5 Discussions

Since $\mathbf{F}_2$ is closely related to $\mathbf{F}_1$ which is a 1-point blow up of $\mathbf{P}^2$, we effectively have up to 5-point blow ups of $\mathbf{P}^2$ describing the gauge theory. Unfortunately, beyond the 5-point blow up mirror symmetry of del Pezzo surfaces can not be described by toric geometry and we can not apply our analysis for these cases. In fact in the range $N_f \geq 5$ something drastic must happen since in this asymptotically non-free region Seiberg-Witten solution does not exist and we should not have a smooth 4-dimensional limit. On the other hand, in this range the 5-dimensional gauge theory is expected to possess E_n, $n = 6, 7, 8$ global symmetries and are of particular interests. It is a challenging problem to clarify the physics of gauge theory with E_n symmetry. It may shed some light on the nature of asymptotically non-free field theories in 4-dimensions.

Our result shows that from the point of view of the type II/M theory compactified on CY manifold, low energy 5-dimensional theory on $M_4 \times S^1$ emerges most naturally with its prepotential being exactly the same as that of the string theory. On the other hand, 4-dimensional Seiberg-Witten theory is obtained only in the fine-tuned limit of the Kähler parameters. It seems possible that our 5-dimensional model is an example of an "M-theoretic" lift of various 4-dimensional quantum field theories. We may imagine most of the 4-dimensional SUSY field theories in fact have a lift to 5-dimensions where the quantum effects of the loops and instantons are replaced by purely geometrical effects of world-sheet instantons. It will be also quite interesting to see if there is a further lift of quantum field theory to 6-dimensions as suggested by the duality between F- and M-theory.

I would like to thank H. Kanno for his collaborations. Research of T.E. is supported in part by the fund for the Special Priority Area No.707, Japan Ministry of Education.

Appendix

The elliptic curve for five dimensional supersymmetric gauge theory on S^1 arises as a curve of the B model of the local mirror symmetry.

A.1 $N_f = 0$

For pure Yang-Mills case we take the Hirzebruch surface $\mathbf{F}_2$, whose toric diagram has five vertices

$$\nu_0 = (0,0) , \quad \nu_1 = (1,0) , \quad \nu_2 = (0,1) , \quad \nu_3 = (-1,0) , \\ \nu_4 = (-2,-1) . \tag{A.1}$$

This is a two dimensional reflexive polyhedra. The charge vectors of linear relation

$$\sum_i \ell_i^{(k)} \nu_i = 0 \tag{A.2}$$

are

$$\ell^{(F)} = (-2; 1,0,1,0) , \quad \ell^{(B)} = (0; 0,1,-2,1) . \tag{A.3}$$

The constraint

$$\prod_{\ell_i^{(k)}>0} Y_i^{\ell_i^{(k)}} = \prod_{\ell_i^{(k)}<0} Y_i^{-\ell_i^{(k)}} \tag{A.4}$$

gives

$$Y_0^2 = Y_1 Y_3 , \qquad Y_2 Y_4 = Y_3^2 , \tag{A.5}$$

and we have a solution

$$Y = (sx; x^2, z, s^2, \frac{s^4}{z}) . \tag{A.6}$$

Set $s = 1$. Then we obtain the following curve in the B model side

$$P = \sum a_i Y_i , \tag{A.7}$$

$$= a_0 x + a_1 x^2 + a_2 z + a_3 + a_4 \frac{1}{z} = 0 . \tag{A.8}$$

Introducing a new variable $y = a_2 z - (a_4/z)$, we can rewrite the curve as

$$y^2 = (a_1 x^2 + a_0 x + a_3)^2 - 4 a_2 a_4 . \tag{A.9}$$

For each charge vector $\ell^{(k)}$, an intrinsic complex structure moduli is given by

$$z_k = \prod_i a_i^{\ell_i^{(k)}} . \tag{A.10}$$

In the present case we have

$$z_F = \frac{a_1 a_3}{a_0^2} , \qquad z_B = \frac{a_2 a_4}{a_3^2} . \tag{A.11}$$

The comparison of the curve (A.9) with our $5D$ curve gives

$$a_1 = a_3 = 1 , \quad a_0 = -s , \quad 4a_2 a_4 = K^4 \tag{A.12}$$

Hence we have

$$z_F = \frac{1}{s^2} , \qquad z_B = \frac{K^4}{4} . \tag{A.13}$$

A.2 $\mathbf{F}_0$ *and* $\mathbf{F}_1$

Let us compare other Hirzebruch surfaces $\mathbf{F}_0$ and $\mathbf{F}_1$ with $\mathbf{F}_2$. Firstly we note that the position of the 4-th vertex ν_4 is the only difference in toric data of these surfaces. Consequently, the charge vector $\ell^{(F)}$ for the fiber is the same and there is a simple change in the charge vector $\ell^{(B)}$ for the base. The difference of the curve of the B-model appears only in the last term with coefficient a_4. Due to the change in the charge vector $\ell^{(B)}$, the identification of the moduli z_B is also changed.

The toric data of $\mathbf{F}_1$ is

$$\begin{aligned} &\nu_0 = (0,0) , \quad \nu_1 = (1,0) , \quad \nu_2 = (0,1) , \quad \nu_3 = (-1,0) , \\ &\nu_4 = (-1,-1) \end{aligned} \tag{A.14}$$

and the charge vectors are

$$\ell^{(F)} = (-2; 1,0,1,0) , \quad \ell^{(B)} = (-1; 0,1,-1,1) . \tag{A.15}$$

The constraint (A.4) gives

$$Y = (sx; x^2, z, s^2, \frac{xs^3}{z}) . \tag{A.16}$$

Setting $s = 1$, we obtain the following curve

$$y^2 = (a_1 x^2 + a_0 x + a_3)^2 - 4a_2 a_4 x . \tag{A.17}$$

Similarly, the toric data of $\mathbf{F}_0 = \mathbf{P}^1 \times \mathbf{P}^1$ is

$$\begin{aligned} &\nu_0 = (0,0) , \quad \nu_1 = (1,0) , \quad \nu_2 = (0,1) , \quad \nu_3 = (-1,0) , \\ &\nu_4 = (0,-1) \end{aligned} \tag{A.18}$$

and the charge vectors are

$$\ell^{(F)} = (-2;1,0,1,0) \ , \quad \ell^{(B)} = (-2;0,1,0,1) \ . \tag{A.19}$$

The constraint (A.4) is solved by

$$Y = (sx; x^2, z, s^2, \frac{x^2 s^2}{z}) \ . \tag{A.20}$$

and the curve is

$$y^2 = (a_1 x^2 + a_0 x + a_3)^2 - 4a_2 a_4 x^2 \ . \tag{A.21}$$

If we assume the following relation to physical moduli parameters as is the case of $\mathbf{F}_2$,

$$a_1 = a_3 = 1 \ , \quad a_0 = -s \ , \quad 4a_2 a_4 = K^4 \tag{A.22}$$

then we have

$$z_F = \frac{a_1 a_3}{a_0^2} = \frac{1}{s^2} \ , \qquad z_B = \frac{a_2 a_4}{a_0 a_3} = \frac{4K^4}{-4s} \ . \tag{A.23}$$

for $\mathbf{F}_1$ and

$$z_F = \frac{a_1 a_3}{a_0^2} = \frac{1}{s^2} \ , \qquad z_B = \frac{a_2 a_4}{a_0^2} = \frac{K^4}{4s^2} \ . \tag{A.24}$$

for $\mathbf{F}_0$.

Finally we make a comparison of differential operators of GKZ. They are

$$\mathcal{L}_B = \theta_B^2 - z_B(2\theta_B - \theta_F)(2\theta_B - \theta_F + 1) \ , \tag{A.25}$$

$$\mathcal{L}_F = \theta_F(\theta_F - 2\theta_B) - z_F(2\theta_F)(2\theta_F + 1) \ . \tag{A.26}$$

for $\mathbf{F}_2$,

$$\mathcal{L}_B = \theta_B^2 - z_B(\theta_B + 2\theta_F)(\theta_B - \theta_F) \ , \tag{A.27}$$

$$\mathcal{L}_F = \theta_F(\theta_F - \theta_B) - z_F(2\theta_F + \theta_B)(2\theta_F + \theta_B + 1) \ . \tag{A.28}$$

for $\mathbf{F}_1$ and

$$\mathcal{L}_B = \theta_B^2 - z_B(2\theta_B + 2\theta_F)(2\theta_B + 2\theta_F + 1) \ , \tag{A.29}$$

$$\mathcal{L}_F = \theta_F^2 - z_F(2\theta_F + 2\theta_B)(2\theta_F + 2\theta_B + 1) \ . \tag{A.30}$$

for $\mathbf{F}_0$.

A.3 $N_f = 1$

Matter in the fundamental representation is obtained by a point blow up (see, for instance[21]) and we expect a blow up of $\mathbf{F}_2$ provides a correct description of $N_f = 1$ theory. We note that this is also obtained as a blow up of $\mathbf{F}_1$, since we can shift n of the Hirzebruch surface $\mathbf{F}_n$ by blow ups and downs. The corresponding reflexive polyhedra has the vertices

$$\nu_0 = (0,0)\ ,\nu_1 = (1,0)\ ,\nu_2 = (0,1)\ , \tag{A.31}$$

$$\nu_3 = (-1,1)\ ,\nu_4 = (-1,0)\ ,\nu_5 = (-1,-1)\ . \tag{A.32}$$

We see that the charge vectors are

$$\ell^{(B)} = (0;0,0,1,-2,1)\ , \tag{A.33}$$

$$\ell^{(F-E)} = (-1;0,1,-1,1,0)\ , \tag{A.34}$$

$$\ell^{(E)} = (-1;1,-1,1,0,0)\ . \tag{A.35}$$

The constraint

$$Y_3Y_5 = Y_4^2\ , \quad Y_0Y_3 = Y_2Y_4\ , \quad Y_0Y_2 = Y_1Y_3\ , \tag{A.36}$$

is solved by

$$Y = (sx; sx^2, \frac{tx}{z}, \frac{s}{z}, s, zs)\ . \tag{A.37}$$

Setting $s = 1$ gives the curve

$$P = a_0x + a_1x^2 + a_2\frac{x}{z} + a_3\frac{1}{z} + a_4 + a_5z\ , \tag{A.38}$$

or

$$y^2 = (a_1x^2 + a_0x + a_4)^2 - 4a_5(a_2x + a_3)\ . \tag{A.39}$$

If we set

$$a_1 = a_4 = 1\ ,\ a_0 = -s\ ,\ 4a_5 = K^3\ ,\ a_2 = w\ ,\ a_3 = w^{-1}\ , \tag{A.40}$$

we obtain the curve

$$y^2 = (x^2 + sx + 1)^2 - K^3(wx + \frac{1}{w}). \tag{A.41}$$

Complex moduli are given by

$$z_B = \frac{a_3a_5}{a_4^2} = \frac{K^3}{4w}\ , \quad z_{F-E} = \frac{a_2a_4}{a_0a_3} = \frac{w^2}{s}\ ,$$

$$z_E = \frac{a_1a_3}{a_0a_2} = \frac{1}{sw^2}\ . \tag{A.42}$$

A.4 $N_f = 2$

We choose the following vertices

$$\begin{aligned} \nu_0 &= (0,0)\ ,\ \nu_1 = (1,0)\ ,\ \nu_2 = (0,1)\ , \\ \nu_3 &= (-1,1)\ ,\ \nu_4 = (-1,0)\ ,\ \nu_5 = (-1,-1)\ , \\ \nu_6 &= (1,1)\ . \end{aligned} \tag{A.43}$$

The charge vectors

$$\ell^{(B)} = (0;0,0,1,-2,1,0)\ , \tag{A.44}$$
$$\ell^{(F-E)} = (-1;0,1,-1,1,0,0)\ , \tag{A.45}$$
$$\ell^{(E)} = (-1;1,-1,1,0,0,0)\ , \tag{A.46}$$
$$\ell^{(E')} = (-1;1,1,0,0,0,-1)\ , \tag{A.47}$$

gives a solution

$$Y = (sx; sx^2, \frac{sx}{z}, \frac{s}{z}, s, zs, \frac{sx^2}{z})\ . \tag{A.48}$$

The curve is

$$P = a_0 x + a_1 x^2 + a_2 \frac{x}{z} + a_3 \frac{1}{z} + a_4 + a_5 z + a_6 \frac{x^2}{z} = 0\ , \tag{A.49}$$

or

$$y^2 = (a_1 x^2 + a_0 x + a_4)^2 - 4a_5(a_6 x^2 + a_2 x + a_3)\ . \tag{A.50}$$

Substituting the relation

$$a_1 = a_4 = 1,\ a_0 = -s,\ 4a_5 = K^2, \tag{A.51}$$
$$a_6 = w_1 w_2,\ a_2 = -\left(\frac{w_2}{w_1} + \frac{w_1}{w_2}\right),\ a_3 = (w_1 w_2)^{-1}, \tag{A.52}$$

we obtain the complex structure moduli

$$z_B = \frac{a_3 a_5}{a_4^2} = \frac{K^2}{4w_1 w_2},\ z_{F-E} = \frac{a_2 a_4}{a_0 a_3} = \frac{w_1^2 + w_2^2}{s}, \tag{A.53}$$
$$z_E = \frac{a_1 a_3}{a_0 a_2} = \frac{1}{s(w_1^2 + w_2^2)},\ z_{E'} = \frac{a_1 a_2}{a_0 a_6} = \frac{w_1^2 + w_2^2}{s w_1^2 w_2^2}. \tag{A.54}$$

Cases of $N_f = 3$ and $N_f = 4$ are similar. Readers are referred to [1].

References

1. T. Eguchi and H. Kanno, "Five-dimensional gauge theories and local mirror symmetry", Nucl. Phys. B586 (2000) 331, hep-th/0005008.
2. N. Seiberg and E. Witten, "Electro-magnetic duality, Monopole condensation and confinement in $N = 2$ supersymmetric Yang-Mills theory", Nucl. Phys. B426 (1994) 19, hep-th/9407087; "Monopoles, duality and chiral symmetry breaking in $N = 2$ supersymmetric QCD", Nucl. Phys. B431 (1994) 484, hep-th/9408099.
3. N. Seiberg, "Five dimensional SUSY field theories, non-trivial fixed points and string dynamics", Phys. Lett. B 388 (1996) 753, hep-th/9608111.
4. D.R. Morrison and N. Seiberg, "Extremal transitions and five-dimensional supersymmetric theories", Nucl. Phys. B 483 (1997) 229, hep-th/9609070.
5. M.R. Douglas, S. Katz and C. Vafa, "Small instantons, del Pezzo surfaces and type I' theory", Nucl. Phys. B 497 (1997) 155, hep-th/9609071.
6. A. Klemm, W. Lerche, P. Mayr, C. Vafa and N. Warner "Self-Dual Strings and $\mathcal{N} = 2$ Supersymmetric Field Theory", Nucl. Phys. **B483** (1997) 209, hep-th/9604034.
7. S. Katz, A. Klemm and C. Vafa, "Geometric engineering of quantum field theory", Nucl. Phys. B497 (1997) 173, hep-th/9609239.
8. A. Klemm, "On the geometry behind $N = 2$ supersymmetric effective actions in four dimensions", hep-th/9705131.
9. S. Katz, P. Mayr and C. Vafa, "Mirror symmetry and Exact Solution of 4D N=2 Gauge Theories I", Adv.Theor.Math.Phys. 1 (1998) 53, hep-th/9706110.
10. T.-M. Chiang, A. Klemm, S.-T. Yau and E. Zaslow, "Local mirror symmetry: calculations and interpretation", hep-th/9903053.
11. N. Nekrasov, "Five dimensional gauge theories and relativistic integrable systems", Nucl. Phys. B531 (1998) 323, hep-th/9609219.
12. H. Kanno and Y. Ohta, "Picard-Fuchs equation and prepotential of five-dimensional SUSY gauge theory compactified on a circle", Nucl. Phys. B530 (1998) 73, hep-th/9801036.
13. A. Sen, "F-theory and orientifolds", Nucl. Phys. B475 (1996) 526, hep-th/9605150.
14. T. Banks, M.R. Douglas and N. Seiberg, "Probing F-theory with branes", Phys.Lett. B387 (1996) 278, hep-th/9605199.
15. A. Lawrence and N. Nekrasov, "Instanton sums and five dimensional gauge theories", Nucl. Phys. B513 (1998) 239, hep-th/9706025.
16. A.C. Cadavid, A. Ceresole, R. D'Auria and S. Ferrara, "11-dimensional

supergravity compactified on Calabi-Yau threefolds", Phys.Lett. B357 (1995) 76, hep-th/9506144.
17. E. Witten "Phase transitions in M-Theory and F-Theory", Nucl.Phys. B471 (1996) 195, hep-th/9603150.
18. A. Brandhuber, N. Itzhaki, J. Sonnenschein, S. Theisen and S. Yankielowizc, "On the M-Theory approach to (compactified) 5D field theories", Phys.Lett. B415 (1997) 127, hep-th/9709010.
19. A. Klemm, S. Theisen and M.G. Schmidt, "Correlation functions for topological Landau-Ginzburg models with $c \leq 3$", Int. J. Mod. Phys. **A 7** (1992) 6215.
20. A. Klemm, B.H. Lian, S.S. Roan and S.T. Yau, "A Note on ODEs from mirror Symmetry", hep-th/9407192.
21. S. Katz and C. Vafa, "Matter from geometry", Nucl.Phys. **B497** (1997) 146, hep-th/9606086.

THE TODA CONJECTURE

EZRA GETZLER

RIMS, Kyoto University, Kitashirakawa Oiwake-cho, Sakyo-ku, Kyoto 602, Japan

Northwestern University, Evanston, IL 60208, USA

Consider the **Gromov-Witten potential**

$$F = \sum_{g=0}^{\infty} \varepsilon^{2g} F_g$$

of $\mathbb{CP}^1$. Eguchi and Yang [4] have conjectured that

$$Z = \exp(\varepsilon^{-2} F)$$

is a τ-function of the Toda hierarchy. (Similar ideas were also proposed by Dubrovin, cf. [2].) In this paper, we will explore this conjecture using the bihamiltonian method in the theory of integrable systems.

Let P be the puncture operator, with descendents $\tau_{k,P}$, $k \geq 0$, and let Q be the operator Poincaré dual to a point, with descendents $\tau_{k,Q}$; denote the corresponding coordinates on the large phase space by s_k and t_k respectively.

Let ∂ and ∂_Q be the vector fields of differentiation with respect to s_0 and t_0, let $\mathsf{E} = e^{\varepsilon \partial}$, and introduce the operators $\nabla = \varepsilon^{-1}(\mathsf{E}^{1/2} - \mathsf{E}^{-1/2})$ and $[2] = \mathsf{E}^{1/2} + \mathsf{E}^{-1/2}$. Let u and v be the functions $\nabla^2 F$ and $\nabla \partial_Q F$.

The **Toda conjecture** consists of the Toda equation

$$\partial_Q^2 F = q e^u, \tag{1}$$

whose implications have been studied by Pandharipande [25], and the recursion

$$(v\nabla + [2]\,\partial_Q)\langle\langle \tau_{k-1,Q} \rangle\rangle = (k+1)\nabla \langle\langle \tau_{k,Q} \rangle\rangle. \tag{2}$$

The large phase space of $\mathbb{CP}^1$ may be identified with the jet-space of the space with coordinates u and v, that is, it has coordinates $\{\partial^n u, \partial^n v\}_{n \geq 0}$. The Toda conjecture implies that, in these coordinates, the flows $\partial_{k,Q}$ are the flows of the Toda lattice hierarchy*.

Eguchi and Yang also give a matrix integral representation of the Gromov-Witten potential of $\mathbb{CP}^1$. Studying this representation, Eguchi, Hori and Yang [5] were led to conjecture that Z satisfies a sequence of constraints

*The original paper of Eguchi and Yang [4] had the incorrect identifications $u = \partial \nabla F$ and $v = \partial \partial_Q F$; the corrected form of the conjecture is found in Eguchi, Hori and Yang [5].

$z_n = 0$, $n \geq -1$, where $z_{-1} = 0$ is the string equation and $z_0 = 0$ is Hori's equation. This conjecture is analogous to the formulation of Witten's KdV conjecture for topological gravity in terms of an action of the Virasoro algebra, and is now called the **Virasoro conjecture** for $\mathbb{CP}^1$; it has recently been proved by Givental [13] (along with its generalization to higher-dimensional projective spaces).

In Section 5, we discuss the relationship between the Toda conjecture and the Virasoro conjecture for $\mathbb{CP}^1$; the main result is that, if (2) holds, then the Virasoro conjecture is equivalent of the following recursion:

$$(v\nabla + [2]\,\partial_Q)\langle\langle\tau_{k-1,P}\rangle\rangle = k\nabla\langle\langle\tau_{k,P}\rangle\rangle + 2\nabla\langle\langle\tau_{k-1,Q}\rangle\rangle. \tag{3}$$

We show that, when written in terms of the coordinates $\{\partial^n u, \partial^n v\}_{n\geq 0}$, the commuting flows $\partial_{k,P}$ associated to the descendents of the puncture operator P are Hamiltonian flows; in this way, we obtain a new hierarchy of Hamiltonians in involution with each other and with the flows of the Toda lattice.

In Section 6, we show that, in the presence of the Virasoro conjecture, the Toda conjecture follows once it is known to hold along the submanifold $\{s_k = 0\}_{k>1}$ of the large phase space. Since Okounkov and Pandharipande have recently proved the Toda conjecture on this submanifold [23], the Toda conjecture is established.

Dubrovin and Zhang [3] have proved that for homogenous spaces, and in particular for $\mathbb{CP}^1$, the Virasoro conjecture determines the Gromov-Witten potential. (They actually prove the analogous result in the more general context of semisimple Frobenius manifolds.) This poses the interesting problem of understanding how the Toda conjecture might follow directly from the Virasoro conjecture.

In this paper, we work over the field $\mathbb{Q}_\varepsilon = \mathbb{Q}((\varepsilon))$ of Laurent polynomials with rational coefficients. The parameter ε is known in physics as the loop expansion parameter: this simply means that integrals over moduli spaces of genus g are weighted by a factor of ε^{2g}. In the theory of the Toda lattice, ε is the lattice spacing — it is the identification of the lattice spacing with the genus expansion parameter that lies at the heart of the Toda conjecture.

1 Witten's conjecture

The Toda conjecture is the analogue for $\mathbb{CP}^1$ of a famous conjecture of Witten [27], proved by Kontsevich [18], that the Gromov-Witten potential of a point is a τ-function of the KdV hierarchy. (See also Itzykson and Zuber [15] and Looijenga [19] for illuminating discussions of the proof, and Okounkov and Pandharipande [22] for an enumerative proof.) In this section, we recall Witten's conjecture.

Let $\overline{\mathcal{M}}_{g,n}$ be the moduli space of n-pointed stable curves of arithmetic genus g, introduced by Deligne, Mumford and Knudsen; it is an orbifold of dimension $3(g-1)+n$.

Let $\mathcal{L}_i$ be the line bundle on $\overline{\mathcal{M}}_{g,n}$ whose fibre at a stable curve $(C, z_1, \dots, z_n)$ is the cotangent line $T^*_{z_i}C$, and let $\psi_i = c_1(\mathcal{L}_i) \in H^2(\overline{\mathcal{M}}_{g,n}, \mathbb{Z})$ be its first Chern class. Witten's conjecture is a formula for the values of the intersection numbers

$$\langle \tau_{k_1} \dots \tau_{k_n} \rangle_g = \int_{\overline{\mathcal{M}}_{g,n}} \psi_1^{k_1} \dots \psi_n^{k_n}.$$

It is convenient to assemble these numbers into generating functions on the large phase space; this is a space with coordinates t_k, $k \geq 0$. Denote by ∂_k the vector field $\partial/\partial t_k$ on the large phase space. The vector field $\partial = \partial_0$ plays a special role.

Introduce generating functions

$$F_g = \sum_{n=0}^{\infty} \frac{1}{n!} \sum_{k_1,\dots,k_n} t_{k_1} \dots t_{k_n} \langle \tau_{k_1} \dots \tau_{k_n} \rangle_g,$$

with partial derivatives $\langle\langle \tau_{k_1} \dots \tau_{k_n} \rangle\rangle_g = \partial_{k_1} \dots \partial_{k_n} F_g$, and let F be the total potential

$$F = \sum_{g=0}^{\infty} \varepsilon^{2g} F_g,$$

with partial derivatives $\langle\langle \tau_{k_1} \dots \tau_{k_n} \rangle\rangle = \partial_{k_1} \dots \partial_{k_n} F$.

The total potential F satisfies the **string equation** $\mathcal{L}_{-1}F + \frac{1}{2}t_0^2 = 0$ and the equation $\mathcal{L}_0 F + \frac{1}{8}\varepsilon^2 = 0$, where $\mathcal{L}_{-1}$ and $\mathcal{L}_0$ are the vector fields

$$\mathcal{L}_{-1} = \sum_{k=0}^{\infty} t_{k+1}\partial_k - \partial_0, \quad \mathcal{L}_0 = \sum_{k=0}^{\infty} (k+\tfrac{1}{2}) t_k \partial_k - \tfrac{3}{2}\partial_1.$$

Proposition 1.1. *If f is a function on the large phase space such that ∂f and $\mathcal{L}_{-1} f$ are constant, then f is constant. If in addition, $\lambda \mathcal{L}_0 f = f$ for some constant λ, then $f = 0$.*

Proof. We give an outline of the proof (see Section 3 of Getzler [10] for more details): if

$$(\partial + \mathcal{L}_{-1}) f = \sum_{k=0}^{\infty} t_{k+1} \partial_k f$$

is a constant, it follows that f is a constant. Hence $\mathcal{L}_0 f = 0$; if in addition $\lambda\mathcal{L}_0 f = f$, then $f = 0$. □

Theorem 1.2. *Let $u = \partial^2 F$. The functions $\partial^k u$ $(= \partial^{k+2}F)$, $k \geq 0$, form a coordinate system on the large phase space.*

Proof. The string equation, in conjunction with the genus 0 topological recursion relation

$$\langle\langle\tau_k\tau_\ell\tau_m\rangle\rangle_0 = \langle\langle\tau_{k-1}\tau_0\rangle\rangle_0\langle\langle\tau_0\tau_\ell\tau_m\rangle\rangle_0,$$

implies that $\partial_k(\partial^\ell u)|_{t_*=0,\varepsilon=0} = \delta_{k\ell}$. □

In the coordinate system $\{\partial^k u\}_{k\geq 0}$, the vector fields $\mathcal{L}_{-1}$ and $\mathcal{L}_0$ have the formulas

$$\mathcal{L}_{-1} = -\frac{\partial}{\partial u}, \quad \mathcal{L}_0 = -\sum_{k=0}^{\infty}(\tfrac{1}{2}k+1)\,\partial^k u\frac{\partial}{\partial(\partial^k u)}. \tag{1.1}$$

We now recall the definition of the Kortweg-deVries (KdV) hierarchy; this is a sequence of commuting vector fields on the jet-space of the affine line. Let $\mathcal{A}$ be a differential ring (a commutative ring with differential ∂), and let $\Psi(\mathcal{A})$ be the algebra of pseudodifferential operators defined over $\mathcal{A}$: this is the algebra

$$\Psi(\mathcal{A}) = \bigcup_{N=0}^{\infty}\Big\{\sum_{i=-\infty}^{N} a_i\partial^i \;\Big|\; a_i \in \mathcal{A}\Big\},$$

with product determined by the relations $\partial^i \cdot \partial^j = \partial^{i+j}$ and

$$\partial^i \cdot a = \sum_{j=0}^{\infty}\binom{i}{j}\partial^j a \cdot \partial^{i-j}.$$

Let $A \mapsto A_+$ be the projection on the space of pseudodifferential operators

$$\Big(\sum_{i=-\infty}^{N} a_i\partial^i\Big)_+ = \sum_{i=0}^{N} a_i\partial^i,$$

and let $A_- = A - A_+$.

We are interested in the case where $\mathcal{A}$ is the algebra of differential polynomials

$$\mathcal{A} = \mathbb{Q}_\varepsilon[\![u, \partial u, \partial^2 u, \dots]\!];$$

the differential ∂ acts on the generators as $\partial(\partial^k u) = \partial^{k+1}u$.

The Lax operator is the differential operator $L = \frac{1}{2}\varepsilon^2\partial^2 + u \in \Psi(\mathcal{A})$. There is a unique square root $D \in \Psi(\mathcal{A})$ of $2\varepsilon^{-2}L = \partial^2 + 2\varepsilon^{-2}u$, which commutes with L and has the form

$$D = \partial + \varepsilon^{-2}u\partial^{-1} + \dots.$$

Let k be a natural number. The **Lax equation** is the equation

$$\delta_k L = [(L^k D)_+, L],$$

or equivalently, $\delta_k L = -[(L^k D)_-, L]$. From these two equations, we see that $\delta_k L$ is an element of $\mathcal{A}$. Write

$$L^k D = \sum_{i=-\infty}^{2k+1} a_i(k)\partial^i,$$

where $a_i(k)$ is an element of $\mathcal{A}$. The differential polynomial $f_k = \varepsilon^2 a_{-1}(k)$is called the kth **Gelfand-Dickii polynomial**. Since $\delta_k L$ is the constant term in the commutator $-[(L^k D)_-, L]$, we see that

$$\delta_k L = \partial f_k \in \mathcal{A}.$$

The Lax equation determines a derivation of $\mathcal{A}$, defined on generators by

$$\delta_k(\partial^n u) = \partial^n(\delta_k L) = \partial^{n+1} f_k.$$

The essential property of the Lax equation is that these flows commute: since $\delta_k D = [(L^k D)_+, D]$, it follows that $\delta_m (L^n D)_+ = [(L^m D)_+, L^n D]_+$, and we see that

$$\begin{aligned}
&[\delta_m, \delta_n]L = \delta_m[(L^n D)_+, L] - \delta_n[(L^m D)_+, L] \\
&= [\delta_m(L^n D)_+, L] + [(L^n D)_+, \delta_m L] - [\delta_n(L^m D)_+, L] - [(L^m D)_+, \delta_n L] \\
&= [[(L^m D)_+, L^n D]_+, L] - [[(L^n D)_+, L^m D]_+, L] - [[(L^m D)_+, (L^n D)_+], L] \\
&= [[L^m D, L^n D]_+, L] = 0.
\end{aligned}$$

Theorem 1.3. *Let $\mathsf{K} = \frac{1}{8}\varepsilon^2\partial^3 + u\partial + \frac{1}{2}\partial u$. The differential polynomials f_k are characterized by two properties: the recursion $\mathsf{K} f_{k-1} = \partial f_k$ holds, and f_k has vanishing constant term.*

Proof. It is clear that the recursion $\mathsf{K} f_{k-1} = \partial f_k$ determines f_k up to a constant, since the kernel of the linear map $\partial : \mathcal{A} \to \mathcal{A}$ consists of multiples of the identity. Furthermore, f_k has vanishing constant term, since $D|_{u=0} = \partial$, hence $(L^k D)_-|_{u=0}$ vanishes. It remains to prove the recursion.

The vanishing of the coefficient of ∂^i in the equation $[L, L^k D] = 0$ gives

$$\partial a_{i-1}(k) = -\tfrac{1}{2}\partial^2 a_i(k) + \varepsilon^{-2} \sum_{j=1}^{\infty} \tbinom{i+j}{j} a_{i+j}(k)\, \partial^j u.$$

Taking $i = -2$ and $i = -1$, we see that

$$\partial a_{-3}(k) = -\tfrac{1}{2}\partial^2 a_{-2}(k) - \varepsilon^{-2}\partial u\, a_{-1}(k) = \left(\tfrac{1}{4}\partial^3 - \varepsilon^{-2}\partial u\right) a_{-1}(k).$$

By considering the coefficient of ∂^i in the equations $L^{k+1}D = (L^k D)L$, we see that

$$a_i(k+1) = \tfrac{1}{2}\varepsilon^2 a_{i-2}(k) + \sum_{j=0}^{\infty} \tbinom{i+j}{j} \partial^j u\, a_{i+j}(k),$$

and in particular, that $a_{-1}(k+1) = \frac{1}{2}\varepsilon^2 a_{-3}(k) + u\, a_{-1}(k)$. Taking a derivative of this equation gives

$$\partial a_{-1}(k+1) = \tfrac{1}{2}\varepsilon^2 \partial a_{-3}(k) + \partial u\, a_{-1}(k) + u\, \partial a_{-1}(k),$$

and the recursion follows. $\square$

Since $f_0 = u$, we see from Theorem 1.3 that $f_1 = \frac{1}{8}\varepsilon^2\partial^2 u + \frac{3}{4}u^2$. In particular, $\delta_0 = \partial$, while $\delta_1 u = \frac{1}{8}(\varepsilon^2\partial^3 u + 12u\partial u)$ is the KdV equation.

Let

$$\alpha_k = \langle\langle \tau_0 \tau_k \rangle\rangle - \frac{2^k}{(2k+1)!!} f_k;$$

in particular, $\alpha_0 = 0$. Witten's conjecture has a number of equivalent formulations:

$$\boxed{\text{for all } k \geq 0,\ \mathsf{K}\langle\langle \tau_0\tau_k \rangle\rangle = (k + \tfrac{1}{2})\, \partial\langle\langle \tau_0\tau_{k+1} \rangle\rangle} \tag{i}$$

$$\boxed{\text{for all } k \geq 0,\ \alpha_k = 0} \tag{ii}$$

$$\boxed{\text{for all } k \geq 0, \text{ the vector field } \partial_k \text{ equals } \tfrac{2^k}{(2k+1)!!}\delta_k} \tag{iii}$$

It is obvious that (ii) implies (i); let us show that (i) implies (ii). Since $\mathcal{L}_{-1}u = -1$, we see that $\mathcal{L}_{-1}(L^k D) = -(k+\frac{1}{2})L^{k-1}D$, hence $\mathcal{L}_{-1} f_k = -(k+\frac{1}{2})\, f_{k-1}$; it follows that $\mathcal{L}_{-1}\alpha_k = -\alpha_{k-1}$. Likewise, we may prove by induction, using the recursion $\mathsf{K} f_{k-1} = \partial f_k$, that

$$\mathcal{L}_0 f_k = -\sum_{n=0}^{\infty} (\tfrac{1}{2}n + 1)\, \partial^n u \frac{\partial}{\partial(\partial^n u)} f_k = -(k+1) f_k,$$

hence that $\mathcal{L}_0\alpha_k = -(k+1)\alpha_k$. Together, (i) and Theorem 1.3 imply that $\mathsf{K}\alpha_{k-1} = (k+\frac{1}{2})\partial\alpha_k$. Using Proposition 1.1, we may now argue by induction that $\alpha_k = 0$.

Likewise, it is obvious that (ii) implies (iii), since $\partial\alpha_k = \big(\partial_k - \frac{2^k}{(2k+1)!!}\delta_k\big)u$. The proof of the converse is similar to the proof that (i) implies (ii).

The involutivity of the vector fields δ_k is essential to the formulation of Witten's conjecture: it is seen to be an integrability condition for the existence of the coordinates t_k on the large phase space. Clearly, this conjecture determines ∂F; in combination with the equations $\mathcal{L}_{-1}F + \frac{1}{2}t_0^2 = 0$ and $\mathcal{L}_0 F + \frac{1}{8}\varepsilon^2 = 0$, it follows from Proposition 1.1 that it determines F.

2 The Toda lattice

In this section, we introduce the Toda lattice, in the form in which it enters into the Toda conjecture: the limit in which the lattice spacing ε is infinitesimal (Takasaki and Takebe [26]). We follow the approach of Kupershmidt [18].

If $(\mathcal{A},\partial)$ is a commutative algebra with derivation ∂ over $\mathbb{Q}_{\varepsilon,q} = \mathbb{Q}_{\varepsilon,q}$, let $\mathsf{E} : \mathcal{A} \to \mathcal{A}$ be the automorphism $e^{\varepsilon\partial}$. Let $\Phi(\mathcal{A})$ be the algebra of twisted Laurent series with coefficients $\mathcal{A}$; as a vector space, $\Phi(\mathcal{A})$ is the space of Laurent series $\mathcal{A}((\Lambda^{-1}))$, and the product is given by the formula

$$\sum_i a_i\Lambda^i \cdot \sum_j b_j\Lambda^j = \sum_{i,j}(\mathsf{E}^{-j/2}a_i)(\mathsf{E}^{i/2}b_j)\Lambda^{i+j}.$$

Extend the automorphism E to $\Phi(\mathcal{A})$ by letting E act trivially on Λ. For example, if $\mathcal{A} = C_c^\infty(\mathbb{R})$ and $\partial = d/dt$, then $\Phi(\mathcal{A})$ is a continuum limit of the algebra of infinite matrices $(M_{ij})_{i,j\in\mathbb{Z}}$ such that $M_{ij} = 0$ for $|i-j| \gg 0$.

Let $A \mapsto A_+$ be the projection on $\Phi(\mathcal{A})$ defined the formula

$$\Big(\sum_{i=-\infty}^{N} a_i\Lambda^i\Big)_+ = \sum_{i=0}^{N} a_i\Lambda^i,$$

and let $A_- = A - A_+$.

The commutative algebra with derivation which we will use is

$$\mathcal{A} = \mathbb{Q}_{\varepsilon,q}[\![e^u, \partial^n u, \partial^n v \mid n \geq 0]\!].$$

The derivation ∂ acts on the generators in the evident way:

$$\partial e^u = e^u\,\partial u, \qquad \partial(\partial^n u) = \partial^{n+1}u, \qquad \partial(\partial^n v) = \partial^{n+1}v.$$

The kernel of the operators ∂ and ∇ on $\mathcal{A}$ equals $\mathbb{Q}_{\varepsilon,q}$. Let P be the infinite-order differential operator

$$\mathsf{P} = \frac{\partial}{\nabla} = \sum_{g=0}^{\infty} \frac{\varepsilon^{2g}(2^{1-2g}-1)B_{2g}}{(2g)!}\,\partial^{2g} = 1 - \tfrac{1}{24}\,\varepsilon^2\,\partial^2 + O(\varepsilon^4).$$

Obviously, $\partial = \mathsf{P} \circ \nabla$.

Definition 2.1. *The* ***Lax operator*** *of the Toda lattice is*

$$L = \Lambda + v + qe^u\Lambda^{-1} \in \Phi(\mathcal{A}).$$

Define elements $p_k(n) \in \mathcal{A}$, $n \geq 0$, $k \in \mathbb{Z}$, as follows:

$$L^n = \sum_{k=-\infty}^{n} p_k(n)\Lambda^k.$$

Lemma 2.1. $p_{-1}(n) = qe^u p_1(n)$

Proof. Taking the coefficient of Λ^0 in the equation $[L, L^n] = 0$, we see that

$$\nabla p_{-1}(n) = \nabla(qe^u p_1(n)).$$

Thus $p_{-1}(n) - qe^u p_1(n) \in \mathbb{Q}_{\varepsilon,q}$. But when $u = v = 0$, the Lax operator equals $\Lambda + q\Lambda^{-1}$, so that $p_{-k}(n)|_{u=v=0} = q^k p_k(n)|_{u=v=0}$. □

The following proposition shows that the functions $p_k(n)$ have certain homogeneity properties.

Proposition 2.2. *Let e and $\mathcal{E}$ be the vector fields*

$$e = \frac{\partial}{\partial v} \text{ and } \mathcal{E} = \sum_{n=0}^{\infty} \partial^n v\,\frac{\partial}{\partial(\partial^n v)} + 2\,\frac{\partial}{\partial u}. \tag{2.1}$$

Then $e(p_k(n)) = np_k(n-1)$ and $\mathcal{E}(p_k(n)) = (n-k)p_k(n)$.

Proof. The vector fields e and $\mathcal{E}$ both commute with ∂, hence with the operator E. It follows that they induce derivations of the algebra $\Phi(\mathcal{A})$. Since $e(L) = 1$, we see that $e(L^n) = nL^{n-1}$, and expanding in powers of Λ, that $e(p_k(n)) = np_k(n-1)$.

Likewise, since $\mathcal{E}(L) = v + 2qe^u\Lambda^{-1} = (1 - \Lambda\partial_\Lambda)L$ and $\Lambda\partial_\Lambda$ is a derivation of $\Phi(\mathcal{A})$, we see that

$$\mathcal{E}(L^n) = (n - \Lambda\partial_\Lambda)L^n;$$

expanding in powers of Λ, it follows that $\mathcal{E}(p_k(n)) = (n-k)p_k(n)$. □

The nth **Toda flow** is determined by the Lax equation

$$\delta_n L = \varepsilon^{-1}[L^n_+, L] = -\varepsilon^{-1}[L^n_-, L].$$

Since L^n_+ involves only positive powers of Λ and L^n_- involves only negative powers of Λ, the coefficient of Λ^i in $\delta_n L$ vanishes unless i equals 0 or -1, and

$$\delta_n L = \nabla p_{-1}(n) + qe^u \nabla p_0(n)\Lambda^{-1}.$$

There is a unique derivation δ_n of the algebra $\mathcal{A}$, which commutes with ∂ and is characterized by the formulas $\delta_n u = \nabla p_0(n)$ and $\delta_n v = \nabla p_{-1}(n)$. In particular, the derivation δ_1 is the original Toda flow:

$$\delta_1 u = \nabla v, \quad \delta_1 v = q\nabla e^u.$$

Eliminating v, we obtain the Toda equation $\delta_1^2 u = q\nabla^2 e^u$.

Being defined by Lax equations, the Toda flows commute: by the formula $\delta_m L^n_+ = [L^m_+, L^n]_+$, we see that

$$\begin{aligned}
\varepsilon^2[\delta_m, \delta_n]L &= \delta_m[L^n_+, L] - \delta_n[L^m_+, L] \\
&= [\delta_m(L^n)_+, L] + [L^n_+, \delta_m L] - [\delta_n(L^m)_+, L] - [L^m_+, \delta_n L] \\
&= [[L^m_+, L^n]_+, L] - [[L^n_+, L^m]_+, L] - [[L^m_+, L^n_+], L] \\
&= [[L^m, L^n]_+, L] = 0.
\end{aligned}$$

Let $\Omega_{\mathcal{A}}$ be the algebra of differential forms

$$\Omega^\bullet_{\mathcal{A}} = \mathcal{A}[d(\partial^n u), d(\partial^n v) \mid n \geq 0],$$

generated over $\mathcal{A}$ by Grassmann variables $\{d(\partial^n u), d(\partial^n v)\}_{n\geq 0}$ of degree 1. There is a unique derivation ∂ on $\Omega_{\mathcal{A}}$ which agrees with the derivation ∂ on $\mathcal{A}$ and commutes with the exterior differential d.

The space of **functional differential forms** is the cokernel of ∂:

$$\mathcal{R}^\bullet = \Omega^\bullet_{\mathcal{A}}/\partial\Omega^\bullet_{\mathcal{A}}.$$

The image of an element $\alpha \in \Omega^\bullet_{\mathcal{A}}$ in $\mathcal{R}^\bullet$ is denoted $\int \alpha\, dt$; this notation is intended to indicate that integration by parts is permitted under the integral sign:

$$\int \partial\alpha \wedge \beta\, dt = -\int \alpha \wedge \partial\beta\, dt.$$

The exterior differential d on $\Omega^\bullet_{\mathcal{A}}$ induces a differential on $\mathcal{R}^\bullet$. Elements of $\mathcal{R}^0$ are called **functionals**.

There is a natural identification between $\mathcal{R}^1$ and $\mathcal{A}\,du \oplus \mathcal{A}\,dv$, since

$$\int f_k\, d(\partial^k u)\, dt = \int (-\partial)^k f_k\, du\, dt \text{ and } \int g_k\, d(\partial^k v)\, dt = \int (-\partial)^k g_k\, dv\, dt.$$

Under this identification, the exterior differential $d : \mathcal{R}^0 \to \mathcal{R}^1$ may be written

$$d = \delta_u \, du + \delta_v \, dv,$$

where δ_u and $\delta_v : \mathcal{R}^0 \to \mathcal{A}$ are the **variational derivatives** (also known as **Euler-Lagrange operators**)

$$\delta_u = \sum_{k=0}^{\infty} (-\partial)^k \frac{\partial}{\partial(\partial^k u)}, \quad \delta_v = \sum_{k=0}^{\infty} (-\partial)^k \frac{\partial}{\partial(\partial^k v)}. \tag{2.2}$$

Lemma 2.3. *The* **residue** $\mathrm{Res} : \Phi(\Omega_{\mathcal{A}}) \to \mathcal{R}$, *defined by the formula*

$$\mathrm{Res}(a_n \Lambda^n) = \begin{cases} \int a_0 \, dt, & n = 0, \\ 0, & \textit{otherwise}, \end{cases}$$

vanishes on graded commutators and satisfies $d\,\mathrm{Res}(f) = \mathrm{Res}(df)$.

Proof. It is clear that $d\,\mathrm{Res}(f) = \mathrm{Res}(df)$, and that $\mathrm{Res}[a\Lambda^k, b\Lambda^\ell]$ vanishes unless $k + \ell = 0$, while $\mathrm{Res}[a\Lambda^k, b\Lambda^{-k}] = (\mathsf{E}^{k/2} - \mathsf{E}^{-k/2})(ab) \in \partial\Omega^\bullet_{\mathcal{A}}$. □

The functionals $h_n = \frac{1}{n+1}\mathrm{Res}(L^{n+1}) = \frac{1}{n+1}\int p_0(n+1)\,dt \in \mathcal{R}^0$ are the **Hamiltonians** of the Toda lattice hierarchy.
Proposition 2.4. $\delta_u h_n = p_{-1}(n)$ *and* $\delta_v h_n = p_0(n)$

Proof. It follows from Lemma 2.3 that $dh_n = \mathrm{Res}(L^n dL)$. Since

$$dL = dv + qe^u du \Lambda^{-1},$$

we see that $\mathrm{Res}(L^n dL) = \int \big(p_0(n)dv + qe^u p_1(n)du\big)dt$. □

In the dispersionless limit $\varepsilon \to 0$, the algebra $\Phi(\mathcal{A})$ becomes the commutative algebra of Laurent series $\mathcal{A}((\Lambda^{-1}))$, and it is straightforward to calculate a generating function for these Hamiltonians (Fairlie and Strachan [7]).
Proposition 2.5.

$$\lim_{\varepsilon \to \infty} \sum_{n=0}^{\infty} t^n p_0(n) = \big(1 - 2tv + t^2(v^2 - 4qe^u)\big)^{-1/2}$$

Proof. Let $w = v - t^{-1}$. Let γ be a small circular contour around the origin of the complex plane. Using the residue formula, we may write the generating function which we wish to calculate as

$$\lim_{\varepsilon\to\infty}\sum_{n=0}^{\infty} t^n p_0(n) = \frac{1}{2\pi i}\int_\gamma \frac{1}{1-t(\Lambda+v+qe^u\Lambda^{-1})}\frac{d\Lambda}{\Lambda}$$
$$= -\frac{1}{2\pi i t}\int_\gamma \frac{d\Lambda}{(\Lambda+\frac{1}{2}w+\frac{1}{2}(w^2-4qe^u)^{1/2})(\Lambda+\frac{1}{2}w-\frac{1}{2}(w^2-4qe^u)^{1/2})}$$
$$= \frac{1}{2\pi i t(w^2-4qe^u)^{1/2}}\int_\gamma\Big(\frac{d\Lambda}{\Lambda+\frac{1}{2}w+\frac{1}{2}(w^2-4qe^u)^{1/2}} - \frac{d\Lambda}{\Lambda+\frac{1}{2}w-\frac{1}{2}(w^2-4qe^u)^{1/2}}\Big).$$

The two poles of the integrand are at $\frac{1}{2}w+\frac{1}{2}(w^2-4qe^u)^{1/2} = O(t)$ and $\frac{1}{2}w-\frac{1}{2}(w^2-4qe^u)^{1/2} = -t^{-1}+O(1)$ respectively; thus, for sufficiently small values of t, only the first contributes, with residue 1, and the generating function equals $t^{-1}(w^2-4qe^u)^{-1/2} = (1-2tv+t^2(v^2-4qe^u))^{-1/2}$. □

Corollary 2.6. *Let $P_n(x)$ be the nth Legendre polynomial. Then*

$$p_0(n) = (v^2-4qe^u)^{n/2}P_n\big(v/(v^2-4qe^u)^{1/2}\big).$$

Proof. The Legendre polynomials have generating function

$$\sum_{n=0}^{\infty} t^n P_n(x) = (1-2xt+t^2)^{-1/2}.$$

Setting $x = v/(v^2-4qe^u)^{1/2}$ and $t = t(v^2-4qe^u)^{1/2}$, the result follows. □

The following theorem of Kupershmidt [18] gives a pair of formulas for the derivations δ_n, in terms of h_n, and h_{n-1} respectively. (He also proves a third formula for δ_n, in terms of h_{n-2}.)

Theorem 2.7. *Let $\mathsf{C}_u = \varepsilon^{-1}q(\mathsf{E}^{1/2}\cdot e^u\cdot\mathsf{E}^{1/2} - \mathsf{E}^{-1/2}\cdot e^u\cdot\mathsf{E}^{-1/2})$. Then*

$$\delta_n\begin{bmatrix} v\\ u\end{bmatrix} = \begin{bmatrix} 0 & \nabla\\ \nabla & 0\end{bmatrix}\begin{bmatrix}\delta_v h_n\\ \delta_u h_n\end{bmatrix} = \begin{bmatrix}\mathsf{C}_u & v\nabla\\ \nabla v & \varepsilon^{-1}(\mathsf{E}-\mathsf{E}^{-1})\end{bmatrix}\begin{bmatrix}\delta_v h_{n-1}\\ \delta_u h_{n-1}\end{bmatrix}.$$

Proof. We have already proved the first of these formulas. To prove the second formula, observe that by the equations $L^n = L^{n-1}L = LL^{n-1}$, we have

$$p_k(n) = \mathsf{E}^{-1/2}p_{k-1}(n-1) + (\mathsf{E}^{k/2}v)\,p_k(n-1) + q(\mathsf{E}^{(k+1)/2}e^u)\,\mathsf{E}^{1/2}p_{k+1}(n-1)$$
$$= \mathsf{E}^{1/2}p_{k-1}(n-1) + (\mathsf{E}^{-k/2}v)\,p_k(n-1) + q(\mathsf{E}^{-(k+1)/2}e^u)\,\mathsf{E}^{-1/2}p_{k+1}(n-1).$$

In particular, the equality of these two formulas for $p_0(n)$ shows that

$$\nabla p_{-1}(n) = q\nabla(e^u p_1(n)). \tag{2.3}$$

Further, taking $\mathsf{E}^{1/2}$ times the first formula minus $\mathsf{E}^{-1/2}$ times the second, with k respectively -1 and 0, gives

$$\begin{aligned}\nabla p_{-1}(n) &= v\nabla p_{-1}(n-1) + \mathsf{C}_u p_0(n-1)\\ \nabla p_0(n) &= \nabla(v p_0(n-1)) + \varepsilon^{-1} q(\mathsf{E}-\mathsf{E}^{-1})(e^u p_1(n-1)).\end{aligned}$$

Since $q(\mathsf{E}-\mathsf{E}^{-1})(e^u p_1(n-1)) = (\mathsf{E}-\mathsf{E}^{-1})p_{-1}(n-1)$ by Lemma 2.3, the result follows. □

Corollary 2.8. $p_0(n) = vp_0(n-1) + [2]\, p_{-1}(n-1)$

Proof. Since $\nabla p_0(n) = \nabla\big(vp_0(n-1) + [2]\, p_{-1}(n-1)\big)$, we see that

$$p_0(n) - vp_0(n-1) - [2]\, p_{-1}(n-1) \in \mathbb{Q}_{\varepsilon,q}.$$

To show that this vanishes, we evaluate it at the point $u = v = 0$: since $L|_{u=v=0} = \Lambda + q\Lambda^{-1}$, we see that

$$p_0(n)|_{u=v=0} = (qp_1(n-1) + p_{-1}(n-1))|_{u=v=0} = 2\, p_{-1}(n-1)|_{u=v=0},$$

as required. □

3 Hamiltonian operators and the Toda lattice

In this section, we introduce the variational Schouten Lie algebra; this is an infinite dimensional analogue of the usual Schouten Lie algebra, developed by Dorfman and Gelfand [9]. We explain how it may be used to give an alternative approach to the Toda lattice. For more details, together with applications to other hierarchies, see Dorfman [1], Getzler [11], Manin [20] and Olver [24].

We start by introducing the free graded commutative algebra Λ_∞ over the algebra $\mathcal{A}$, with generators $\{\partial^n\theta_v, \partial^n\theta_u\}_{n\geq 0}$ of degree 1. The derivation ∂ is extended to Λ_∞ by the formulas

$$\partial(\partial^n\theta_v) = \partial^{n+1}\theta_v, \qquad \partial(\partial^n\theta_u) = \partial^{n+1}\theta_u.$$

The kernel of ∂ is spanned by $1 \in \Lambda_\infty$, and the cokernel of ∂ is denoted $\mathcal{L}$. Denote the image of an element $F \in \Lambda_\infty$ in $\mathcal{L}$ by $\int F\, dt$.

The variational derivatives δ_u and δ_v on Λ_∞ are defined by the same formulas (2.2) as on $\mathcal{A}$, while the associated Grassmann variational derivatives, which we denote by δ^u and δ^v, are defined by the formulas

$$\delta^u = \sum_{n=0}^{\infty}(-\partial)^n \frac{\partial}{\partial(\partial^n\theta_u)} \qquad \delta^u = \sum_{n=0}^{\infty}(-\partial)^n \frac{\partial}{\partial(\partial^n\theta_v)}.$$

Since all of these operators vanish on the image of ∂, they descend to linear operators from $\mathcal{L}$ to Λ_∞. The **Schouten bracket** is the bilinear operation on $\mathcal{L}$ defined by the formula

$$\left[\textstyle\int F\,dt, \int G\,dt\right] = \textstyle\int \big(\delta^u F\,\delta_u G + \delta^v F\,\delta_v G + (-1)^{|F|}(\delta_u F\,\delta^u G + \delta_v F\,\delta^v G)\big)\,dt.$$

With this bracket, the graded vector space $\mathcal{L}$ becomes a graded Lie algebra: $[f,g] = (-1)^{|f|\,|g|}[g,f]$ and $[f,[g,h]] = [[f,g],h] + (-1)^{(|f|+1)(|g|+1)}[g,[f,h]]$. (Proofs of these formulas may be found in [11].)

The Lie subalgebra $\mathcal{L}^1$ of $\mathcal{L}^\bullet$ is isomorphic to the Lie algebra of derivations of $\mathcal{A}$ commuting with ∂, under the map

$$\mathcal{X} = \textstyle\int (f\,\theta_u + g\,\theta_v)\,dt \in \mathcal{L}^1 \mapsto \displaystyle\sum_{n=0}^{\infty}\left(\partial^n f\,\frac{\partial}{\partial(\partial^n u)} + \partial^n g\,\frac{\partial}{\partial(\partial^n v)}\right) \in \mathrm{Der}(\mathcal{A}).$$

For example, the vector fields e and $\mathcal{E}$ of (2.1) correspond to $\int \theta_v\,dt$ and $\int (v\theta_v + 2\theta_u)\,dt$; we will denote these elements of $\mathcal{L}^1$ by e and $\mathcal{E}$ as well.

An element $\mathcal{H}$ of $\mathcal{L}^2$ defines a graded derivation $\delta_{\mathcal{H}}$ of degree 1 on the graded Lie algebra $\mathcal{L}$, by the formula $\delta_{\mathcal{H}} = [\mathcal{H}, -]$. Let $f \in \mathcal{L}^0 = \mathcal{R}^0$ be a functional; the derivation of $\mathcal{A}$ commuting with ∂ which corresponds to the element $\delta_{\mathcal{H}} f$ of $\mathcal{L}^1$ is called the **Hamiltonian vector field** associated to f. The **Poisson bracket** on the space of functionals $\mathcal{L}^0 \cong \mathcal{R}^0$ is defined by the formula

$$\{f,g\}_{\mathcal{H}} = [\delta_{\mathcal{H}} f, g].$$

If $[\mathcal{H},\mathcal{H}] = 0$, $\mathcal{H}$ is called a **Hamiltonian operator**.

Proposition 3.1. *If $\mathcal{H}$ is a Hamiltonian operator, $\delta_{\mathcal{H}}$ is a differential, the bracket $\{f,g\}_{\mathcal{H}}$ is a Lie bracket, and $[\delta_{\mathcal{H}} f, \delta_{\mathcal{H}} g] = \delta_{\mathcal{H}}\{f,g\}_{\mathcal{H}}$.*

Proof. By the graded Jacobi rule,

$$\delta_{\mathcal{H}}(\delta_{\mathcal{H}} f) = [\mathcal{H},[\mathcal{H},f]] = \tfrac{1}{2}[[\mathcal{H},\mathcal{H}],f] = 0,$$

showing that $\delta_{\mathcal{H}}$ is a differential.

We have

$$\{f,g\}_{\mathcal{H}} + \{g,f\}_{\mathcal{H}} = [\delta_{\mathcal{H}} f, g] + [\delta_{\mathcal{H}} g, f] = \delta_{\mathcal{H}}[f,g].$$

Since $[f,g]$ vanishes, we see that $\{f,g\}_{\mathcal{H}}$ is antisymmetric.

By the graded Jacobi rule, we have

$$\begin{aligned}\{\{f,g\}_{\mathcal{H}},h\}_{\mathcal{H}} &= [\delta_{\mathcal{H}}[\delta_{\mathcal{H}} f,g],h] = [[\delta_{\mathcal{H}} f,\delta_{\mathcal{H}} g],h] \\ &= [\delta_{\mathcal{H}} f,[\delta_{\mathcal{H}} g,h]] - [\delta_{\mathcal{H}} g,[\delta_{\mathcal{H}} f,h]] = \{f,\{g,h\}_{\mathcal{H}}\}_{\mathcal{H}} - \{g,\{f,h\}_{\mathcal{H}}\}_{\mathcal{H}}.\end{aligned}$$

Finally, we have $[\delta_{\mathcal{H}} f, \delta_{\mathcal{H}} g] = \delta_{\mathcal{H}}[\delta_{\mathcal{H}} f, g] = \delta_{\mathcal{H}}\{f,g\}_{\mathcal{H}}$. □

A **bihamiltonian structure** is a pair of Hamiltonian operators $(\mathcal{H}, \mathcal{H}_0)$ such that $\mathcal{H} + \lambda\mathcal{H}_0$ is a Hamiltonian operator for all λ, or equivalently, such that $[\mathcal{H}, \mathcal{H}_0] = 0$. We now reformulate Theorem 2.7 in terms of a bihamiltonian structure.

Theorem 3.2. *The operators*

$$\begin{aligned}\mathcal{H} &= \tfrac{1}{2}\textstyle\int \begin{bmatrix}\theta_v & \theta_u\end{bmatrix} \begin{bmatrix}\mathsf{C}_u & v\nabla \\ \nabla v & \varepsilon^{-1}(\mathsf{E} - \mathsf{E}^{-1})\end{bmatrix} \begin{bmatrix}\theta_v \\ \theta_u\end{bmatrix} dt \\ &= \textstyle\int \big(\varepsilon^{-1}\theta_u \, \mathsf{E}\theta_u + v\theta_v \nabla\theta_u + \varepsilon^{-1} q e^u (\mathsf{E}^{-1/2}\theta_v)(\mathsf{E}^{1/2}\theta_v)\big)\, dt, \text{ and} \\ \mathcal{H}_0 &= \tfrac{1}{2}\textstyle\int \begin{bmatrix}\theta_v & \theta_u\end{bmatrix} \begin{bmatrix}0 & \nabla \\ \nabla & 0\end{bmatrix} \begin{bmatrix}\theta_v \\ \theta_u\end{bmatrix} dt = \textstyle\int \theta_v \nabla\theta_u \, dt\end{aligned}$$

give a bihamiltonian structure.

Proof. It is clear that $\delta_v\mathcal{H} = \theta_v(\nabla\theta_u)$. Using the formula

$$\delta^v = \sum_{i=-\infty}^{\infty} \mathsf{E}^{-i/2} \frac{\partial}{\partial(\mathsf{E}^{i/2}\theta_v)},$$

we see that $\delta^v\mathcal{H} = v\nabla\theta_u + \mathsf{C}_u\theta_v$. It follows that

$$\begin{aligned}\delta^v\mathcal{H}\,\delta_v\mathcal{H} &= \mathsf{C}_u\theta_v\,\theta_v(\nabla\theta_u) \\ &= \varepsilon^{-1}q(\mathsf{E}^{1/2}e^u)(\mathsf{E}\theta_v)\,\theta_v(\nabla\theta_u) - \varepsilon^{-1}q(\mathsf{E}^{-1/2}e^u)(\mathsf{E}^{-1}\theta_v)\,\theta_v(\nabla\theta_u).\end{aligned}$$

Likewise, since $\delta_u\mathcal{H} = \varepsilon^{-1}qe^u(\mathsf{E}^{-1/2}\theta_v)(\mathsf{E}^{1/2}\theta_v)$ and $\delta^u\mathcal{H} = \nabla(v\theta_v) + [2]\,\nabla\theta_u$, it follows that

$$\delta^u\mathcal{H}\,\delta_u\mathcal{H} = \varepsilon^{-1}qe^u\,[2](\nabla\theta_u)(\mathsf{E}^{-1/2}\theta_v)(\mathsf{E}^{1/2}\theta_v).$$

We conclude that

$$\begin{aligned}\delta^v\mathcal{H}\,\delta_v\mathcal{H} + \delta^u\mathcal{H}\,\delta_u\mathcal{H} = \varepsilon^{-1}q\big((\mathsf{E} - \mathsf{E}^{-1/2})&(e^u(\mathsf{E}^{1/2}\theta_v)(\mathsf{E}^{-1/2}\theta_v)\theta_u) \\ + (1 - \mathsf{E}^{-1/2})&(e^u(\mathsf{E}^{1/2}\theta_v)(\mathsf{E}^{-1/2}\theta_v)(\mathsf{E}\theta_u)) \\ + (\mathsf{E}^{1/2} - 1)&(e^u(\mathsf{E}^{1/2}\theta_v)(\mathsf{E}^{-1/2}\theta_v)(\mathsf{E}^{-1}\theta_u))\big),\end{aligned}$$

and hence that $[\mathcal{H}, \mathcal{H}] = 0$. The proof that $[\mathcal{H}+\lambda\mathcal{H}_0, \mathcal{H}+\lambda\mathcal{H}_0] = 0$ is a formal consequence of this formula, since $\mathcal{H}+\lambda\mathcal{H}_0$ is obtained from $\mathcal{H}$ by translating v by λ. □

The Hamiltonian operators $\mathcal{H}_0$ and $\mathcal{H}$ have nice commutation relations with the vector fields e and $\mathcal{E}$: it is easily checked that $[e, \mathcal{H}] = \mathcal{H}_0$ and $[e, \mathcal{H}_0] = 0$, and that $[\mathcal{E}, \mathcal{H}] = 0$ and $[\mathcal{E}, \mathcal{H}_0] = -\mathcal{H}_0$.

Denote the Poisson bracket associated to $\mathcal{H}$ by $\{f,g\}_{\mathcal{H}}$, and the Poisson bracket associated to $\mathcal{H}_0$ by $\{f,g\}_0$; then Theorem 2.7 may be reformulated as the identity

$$[\mathcal{H}_0, h_n] = [\mathcal{H}, h_{n-1}], \tag{3.1}$$

or equivalently, as the pair of equations

$$\nabla(\delta_v h_n) = \nabla(v\delta_v h_{n-1}) + \varepsilon^{-1}(\mathsf{E} - \mathsf{E}^{-1})\delta_u h_{n-1} \tag{3.2}$$

$$\nabla(\delta_u h_n) = \mathsf{C}_u \delta_v h_{n-1} + v\nabla\delta_u h_{n-1}. \tag{3.3}$$

Using (3.1), we obtain another proof that the flows associated to the Hamiltonians h_n commute; this proof has the advantage that it does not depend on the Lax equation, and so applies to prove the involutivity of more general hierarchies.

Proposition 3.3. *Let f_n and g_n be sequences of Hamiltonians such that $[\mathcal{H}_0, f_n] = [\mathcal{H}, f_{n-1}]$ and $[\mathcal{H}_0, g_n] = [\mathcal{H}, g_{n-1}]$ for $n > 0$, and $\{f_0, f\} = 0$ for all f. Then $\{f_m, g_n\}_0 = 0$ for all $m, n \geq 0$. In particular, the Hamiltonian vector fields associated to f_m and g_n commute.*

Proof. If $m > 0$, we have

$$\begin{aligned}\{f_m, g_n\}_0 &= [[\mathcal{H}_0, f_m], g_n] = [[\mathcal{H}, f_{m-1}], g_n] = -[[\mathcal{H}, g_n], f_{m-1}] \\ &= -[[\mathcal{H}_0, g_{n+1}], f_{m-1}] = \{f_{m-1}, g_{n+1}\}_0.\end{aligned}$$

Thus, it suffices to prove the proposition for $m = 0$, for which it is clear. □

It is an immediate consequence of Proposition 2.2 that

$$[e, h_k] = kh_{k-1} \text{ and } [\mathcal{E}, h_k] = (k+1)h_k. \tag{3.4}$$

If $a \in \mathbb{Z}$, let $\mathcal{L}^i(a)$ be the generalized eigenspace

$$\mathcal{L}^i(a) = \bigcup_{n>0} \ker\big((\mathrm{ad}(\mathcal{E}) + i - a - 1)^n\big);$$

then $[\mathcal{L}(a), \mathcal{L}(b)] = \mathcal{L}(a+b)$. Since $[\mathcal{E}, \mathcal{H}_0] = -\mathcal{H}_0$, we see that $\mathcal{H}_0 \in \mathcal{L}(0)$, hence the differential δ_0 preserves the graded subspace $\mathcal{L}(a)$. Also, we see that $\mathcal{H} \in \mathcal{L}(1)$.

Theorem 3.4. *The complex $\mathcal{L}(a)$ has vanishing cohomology unless $a = -1$ or 0, while the nonzero cohomology groups of $\mathcal{L}(-1)$ and $\mathcal{L}(0)$ are as follows:*

$$H^0(\mathcal{L}(-1), \delta_0) = \langle \textstyle\int 1\, dt, \int u\, dt\rangle \qquad H^0(\mathcal{L}(0), \delta_0) = \langle \textstyle\int v\, dt\rangle$$

$$H^1(\mathcal{L}(-1), \delta_0) = \langle \textstyle\int \theta_v\, dt\rangle \qquad H^1(\mathcal{L}(0), \delta_0) = \langle \textstyle\int \theta_u\, dt, \int (u\theta_u - v\theta_v)\, dt\rangle$$

$$H^2(\mathcal{L}(0), \delta_0) = \langle \textstyle\int \theta_u \theta_v\, dt\rangle$$

Proof. Write elements of the cone $\mathcal{M}$ of the linear map $\nabla : \Lambda_\infty/\mathbb{Q}_{\varepsilon,q} \to \Lambda_\infty$ as (F,G), where $F \in \Lambda_\infty$ and $G \in \Lambda_\infty/\mathbb{Q}_{\varepsilon,q}$. If D is an linear operator from $\Lambda_\infty/\mathbb{Q}_{\varepsilon,q}$ to Λ_∞, let $\iota D : \mathcal{M} \to \mathcal{M}$ be the operator

$$\iota D(F,G) = (DG, 0).$$

The differential of $\mathcal{M}$ equals $\iota\nabla$.

The differential δ_0 on $\mathcal{L}$ lifts to a differential

$$\delta = \sum_{n=0}^{\infty} \big((\nabla\partial^n\theta_u)\partial_{\partial^n v} + (\nabla\partial^n\theta_v)\partial_{\partial^n u} \big)$$

on $\mathcal{M}$, and the map $\tau : \mathcal{M} \to \mathcal{L}$ which sends (F,G) to $\int F\,dt$ is a chain homotopy equivalence between the complexes $(\mathcal{M}, \delta + \iota\nabla)$ and $(\mathcal{L}, \delta_0)$.

Let S be the chain homotopy

$$S = \sum_{n=0}^{\infty} \big((\mathsf{P}\partial^n v)\partial_{\partial^{n+1}\theta_u} + (\mathsf{P}\partial^n u)\partial_{\partial^{n+1}\theta_v} \big),$$

let $U = \iota(v\partial_{\theta_u} + u\partial_{\theta_v})$, and let P be the semisimple operator

$$P = \sum_{n=0}^{\infty} \big((\partial^n u)\partial_{\partial^n u} + (\partial^n v)\partial_{\partial^n v} \big) + \sum_{n=1}^{\infty} \big((\partial^n\theta_u)\partial_{\partial^n\theta_u} + (\partial^n\theta_v)\partial_{\partial^n\theta_v} \big).$$

We have $[\delta + \iota\nabla, S] = e^{-U} P e^{U} = P + U$. Thus, the cohomology of the complex $(\mathcal{M}, d + \iota\nabla)$ equals the kernel

$$\langle (1,0), (\theta_u, 0), (\theta_v, 0), (\theta_u\theta_v, 0), (u, -\theta_v), (v, -\theta_u), (u\theta_u - v\theta_v, \theta_u\theta_v) \rangle \subset \mathcal{M}$$

of $[\delta + \iota\nabla, S]$. Applying τ, the theorem follows. □

4 The Toda conjecture

Having introduced the Toda lattice hierarchy in the last two sections, we can now formulate the Toda conjecture. Recall the definition of the Gromov-Witten invariants of $\mathbb{CP}^1$. (A good review of the subject is Manin [21].) Let $\overline{\mathcal{M}}_{g,n,d}$ be the moduli stack of stable maps of genus g and degree d, with n marked points, to $\mathbb{CP}^1$. Let $\mathrm{ev}_i : \overline{\mathcal{M}}_{g,n,d} \to \mathbb{CP}^1$, $1 \le i \le n$, be the map

$$\mathrm{ev}_i(f : C \to \mathbb{CP}^1, z_1, \dots, z_n) = f(z_i).$$

defined by evaluating a stable map $f : C \to \mathbb{CP}^1$ at the ith marked point $z_i \in C$. Let

$$[\overline{\mathcal{M}}_{g,n,d}]^{\mathrm{virt}} \in H_{2(2g-2+2d+n)}(\overline{\mathcal{M}}_{g,n,d}, \mathbb{Q})$$

be the virtual fundamental class.

Definition 4.1. *Let $\mathcal{L}_i$ be the line bundle on $\overline{\mathcal{M}}_{g,n,d}$ whose fibre at the stable map $(f : C \to \mathbb{CP}^1, z_1, \ldots, z_n)$ is the line $T^*_{z_i}C$, and let $\psi_i = c_1(\mathcal{L}_i)$ be its first Chern class.*

Let $\gamma_P \in H^0(\mathbb{CP}^1, \mathbb{Z})$ and $\gamma_Q \in H^2(\mathbb{CP}^1, \mathbb{Z})$ be the cohomology classes Poincaré dual to the fundamental class and to a point. Given $k_i \in \mathbb{N}$ and $a_i \in \{P, Q\}$, define

$$\langle \tau_{k_1,a_1} \ldots \tau_{k_n,a_n} \rangle_g = \sum_{d=0}^{\infty} q^d \int_{[\overline{\mathcal{M}}_{g,n,d}]^{\text{virt}}} \text{ev}_1^* \gamma_{a_1} \ldots \text{ev}_n^* \gamma_{a_n} \cup \psi_1^{k_1} \ldots \psi_n^{k_n} \in \mathbb{Q}[q].$$

We write P and Q instead of $\tau_{0,P}$ and $\tau_{0,Q}$.

The **large phase space** M is the formal manifold with coordinates $\{s_k, t_k\}_{k \geq 0}$. Define

$$t_k^a = \begin{cases} s_k, & a = P, \\ t_k, & a = Q. \end{cases}$$

The genus g **Gromov-Witten potential** F_g of $\mathbb{CP}^1$ is the generating function on the large phase space given by the formula

$$F_g = \sum_{n=0}^{\infty} \frac{1}{n!} \sum_{\substack{k_1,\ldots,k_n \\ a_1,\ldots,a_n}} \prod_{i=1}^{n} t_{k_i}^{a_i} \int_{[\overline{\mathcal{M}}_{g,n,d}]^{\text{virt}}} \langle \tau_{k_1,a_1} \ldots \tau_{k_n,a_n} \rangle_g,$$

and $F = \sum_{g=0}^{\infty} \varepsilon^{2g} F_g$ is the total Gromov-Witten potential.

Denote the constant vector fields $\partial/\partial t_k^a$ on the large phase space by $\partial_{k,a}$; in particular, write ∂ and ∂_Q for $\partial_{0,P}$ and $\partial_{0,Q}$. Just as in the theory of the Toda lattice, denote the operator $e^{\varepsilon\partial}$ by E, and the operators $\varepsilon^{-1}(\mathsf{E}^{1/2} - \mathsf{E}^{-1/2})$ and $\mathsf{E}^{1/2} + \mathsf{E}^{-1/2}$ by ∇ and $[2]$. The partial derivatives of F are denoted $\langle\langle \tau_{k_1,a_1} \ldots \tau_{k_n,a_n} \rangle\rangle = \partial_{k_1,a_1} \ldots \partial_{k_n,a_n} F$.

The potential F satisfies the **string equation** $\mathcal{L}_{-1}F + s_0 t_0 = 0$ (Witten [27]) and **Hori's equation** $\mathcal{L}_0 F + s_0^2 = 0$ (Hori [14]), where $\mathcal{L}_{-1}$ and $\mathcal{L}_0$ are the vector fields

$$\mathcal{L}_{-1} = \sum_{k=0}^{\infty} \big(t_{k+1} \partial_{k,Q} + s_{k+1} \partial_{k,P} \big) - \partial \tag{4.1}$$

$$\mathcal{L}_0 = \sum_{k=0}^{\infty} \big((k+1)\, t_k \partial_{k,Q} + k\, s_k \partial_{k,P} + 2\, s_{k+1} \partial_{k,Q} \big) - \partial_{1,P} - 2\, \partial_Q. \tag{4.2}$$

The analogue of Proposition 1.1 holds, with essentially the same proof.

Proposition 4.1. *If f is a function on the large phase space such that ∇f and $\mathcal{L}_{-1} f$ are constant, then f is constant. If in addition, $\lambda \mathcal{L}_0 f = f$ for some constant λ, then $f = 0$.*

There is also an analogue for $\mathbb{CP}^1$ of Theorem 1.2.

Theorem 4.2. *Let $u = \nabla^2 F$ and $v = \nabla\partial_Q F$. The functions $\{\partial^n u, \partial^n v\}_{n\geq 0}$ form a coordinate system on the large phase space. The origin $s_k = t_k = 0$ of the large phase space has coordinates*

$$\partial^n v = \begin{cases} 1, & n = 1, \\ 0, & \textit{otherwise}, \end{cases} \qquad \partial^n u = 0.$$

Proof. The string equation, in conjunction with the genus 0 topological recursion relation

$$\langle\langle\tau_{k,a}\tau_{\ell,b}\tau_{m,c}\rangle\rangle_0 = \langle\langle\tau_{k-1,a}P\rangle\rangle_0\langle\langle Q\tau_{\ell,b}\tau_{m,c}\rangle\rangle_0 + \langle\langle\tau_{k-1,a}Q\rangle\rangle_0\langle\langle P\tau_{\ell,b}\tau_{m,c}\rangle\rangle_0,$$

implies that $\partial_{k,P}(\partial^n u)|_{s_*=t_*=0} = \partial_{k,Q}(\partial^n v)|_{s_*=t_*=0} = \delta_{kn} + O(\varepsilon)$.

At the origin of the large phase space, we have

$$\partial^n v = \sum_{k=1}^{\infty}\sum_{g=0}^{\infty} \frac{\varepsilon^{2g+2k-1}}{2^{2k-2}(2k-1)!}\langle P^{2k-1+n}Q\rangle_g.$$

The moduli space $\overline{\mathcal{M}}_{g,2k+n,d}$ only contributes to this sum if it has virtual dimension 1, that is, if $2(g-1+d+k)+n = 1$. The only solution of this equation is $\overline{\mathcal{M}}_{0,3,0} \cong \mathbb{CP}^1$, which contributes the coefficient 1 to ∂v. The argument for $\partial^n u$ is similar: at the origin in the large phase space,

$$\partial^n u = \sum_{k=1}^{\infty}\sum_{g=0}^{\infty} \frac{\varepsilon^{2g+2k-2}}{2^{2k-1}(2k)!}\langle P^{2k+n}\rangle_g.$$

The only contributions to this sum come from moduli spaces $\overline{\mathcal{M}}_{g,2k+n,d}$ such that $2(g-1+d+k)+n = 0$; this equation has no solutions. □

The vector fields $\mathcal{L}_{-1}$ and $\mathcal{L}_0$ commute with ∂; written in the coordinate system $\{\partial^n u, \partial^n v\}_{n\geq 0}$, they may be identified with $-e$ and $-\mathcal{E}$ (see (2.1)).

Introduce the constraints

$$\alpha^v_{k,Q} = \nabla\langle\langle\tau_{k-1,Q}\rangle\rangle - \tfrac{1}{k!}\delta_v h_k, \qquad \alpha^u_{k,Q} = \partial_Q\langle\langle\tau_{k-1,Q}\rangle\rangle - \tfrac{1}{k!}\delta_u h_k.$$

Observe that $\alpha^v_{1,Q}$ and $\alpha^u_{0,Q}$ vanish. By Proposition 2.2, we have

$$\mathcal{L}_{-1}\alpha^*_{k,Q} = -\alpha^*_{k-1,Q}, \quad \mathcal{L}_0\alpha^v_{k,Q} = -k\alpha^v_{k,Q}, \quad \mathcal{L}_0\alpha^u_{k,Q} = -(k+1)\alpha^u_{k,Q}. \tag{4.3}$$

The Toda conjecture describes the vector fields $\partial_{k,Q}$ in the coordinate system $\{\partial^n u, \partial^n v\}_{n\geq 0}$. It may be formulated as any one of the equivalent conditions in the following theorem.

Theorem 4.3. *Let $\mathcal{D}$ be the differential operator*
$$\mathcal{D} = v\nabla + [2]\,\partial_Q.$$
The following are equivalent:

$$\boxed{\text{for all } k > 0,\ \mathcal{D}\langle\langle\tau_{k-1,Q}\rangle\rangle = (k+1)\,\nabla\langle\langle\tau_{k,Q}\rangle\rangle, \text{ and } \partial_Q^2 F = qe^u} \tag{i}$$

$$\boxed{\text{for all } k > 0,\ k!\,\nabla\langle\langle\tau_{k-1,Q}\rangle\rangle = \delta_v h_k \text{ and } k!\,\partial_Q\langle\langle\tau_{k-1,Q}\rangle\rangle = \delta_u h_k} \tag{ii}$$

$$\boxed{\text{for all } k > 0, \text{ the vector field } k!\,\partial_{k-1,Q} \text{ equals } \delta_k} \tag{iii}$$

Proof. Introduce the constraint
$$y_k = \mathcal{D}\langle\langle\tau_{k,Q}\rangle\rangle - (k+2)\nabla\langle\langle\tau_{k+1,Q}\rangle\rangle.$$
We may reformulate the conditions of the theorem in terms of the constraints y_k, $\alpha^u_{k,Q}$ and $\alpha^v_{k,Q}$ as follows:

$$\boxed{\text{for all } k \geq 0,\ y_k = 0, \text{ and } \alpha^u_{1,Q} = 0} \tag{i}$$

$$\boxed{\text{for all } k > 0,\ \alpha^*_{k,Q} = 0} \tag{ii}$$

$$\boxed{\text{for all } k > 0,\ \nabla\alpha^*_{k,Q} = 0} \tag{iii}$$

The third of these reformulations follows from the formulas $k!\,\nabla\alpha^v_{k,Q} = (k!\,\partial_{k-1,Q} - \delta_k)u$ and $k!\,\nabla\alpha^u_{k,Q} = (k!\,\partial_{k-1,Q} - \delta_k)v$.

Lemma 4.4. *We have $y_{k-1} = v\alpha^v_{k,Q} + [2]\,\alpha^u_{k,Q} - (k+1)\alpha^v_{k+1,Q}$ and*
$$\partial_Q y_{k-1} = \mathsf{C}_u\alpha^v_{k,Q} + v\nabla\alpha^u_{k,Q} - (k+1)\nabla\alpha^u_{k+1,Q}.$$

Proof. The first formula follows from the calculation
$$\begin{aligned}(k+1)\alpha^v_{k+1,Q} &= (k+1)\nabla\langle\langle\tau_{k,Q}\rangle\rangle - \tfrac{1}{k!}\delta_v h_{k+1}\\ &= \mathcal{D}\langle\langle\tau_{k-1,Q}\rangle\rangle - \tfrac{1}{k!}(v\delta_v h_k + [2]\,\delta_u h_k) - y_{k-1}.\end{aligned}$$
To prove the second formula, observe that
$$\begin{aligned}(k+1)\nabla\alpha^u_{k+1,Q} &= (k+1)\nabla\partial_Q\langle\langle\tau_{k,Q}\rangle\rangle - \tfrac{1}{k!}\nabla\delta_u h_{k+1}\\ &= \partial_Q\mathcal{D}\langle\langle\tau_{k-1,Q}\rangle\rangle - \tfrac{1}{k!}\mathsf{C}_u\delta_v h_k - \tfrac{1}{k!}v\nabla\delta_u h_k - \partial_Q y_{k-1}.\end{aligned}$$
Since $\partial_Q^2\langle\langle\tau_{k-1,Q}\rangle\rangle = \partial_{k-1,Q}e^u = e^u\nabla^2\langle\langle\tau_{k-1,Q}\rangle\rangle$ and $\partial_Q v = \nabla e^u$, we see that
$$\begin{aligned}\partial_Q\mathcal{D}\langle\langle\tau_{k-1,Q}\rangle\rangle &= (\partial_Q v)\nabla\langle\langle\tau_{k-1,Q}\rangle\rangle + v\partial_Q\nabla\langle\langle\tau_{k-1,Q}\rangle\rangle + [2]\,\partial_Q^2\langle\langle\tau_{k-1,Q}\rangle\rangle\\ &= (\nabla e^u)\nabla\langle\langle\tau_{k-1,Q}\rangle\rangle + v\partial_Q\nabla\langle\langle\tau_{k-1,Q}\rangle\rangle + [2](e^u\nabla^2\langle\langle\tau_{k-1,Q}\rangle\rangle).\end{aligned}$$

A short calculation shows that $(\nabla e^u)f + [2](e^u \nabla f) = \mathsf{C}_u f$, proving the second formula. □

By (4.3) and Proposition 4.1, we see that if $\nabla\alpha^*_{i,Q} = 0$ for $i \le k$, then $\alpha^*_{k,Q} = 0$; in particular, (iii) implies (ii). By Lemma 4.4, it is clear that (ii) implies (i) and that (i) implies (iii). □

Condition (iii) of Theorem 4.3 has recently been proved by Okounkov and Pandharipande [23] on the submanifold $\{s_k = 0\}_{k>1}$ of the large phase space; as we will see in Section 6, in conjunction with the Virasoro conjecture, this establishes the Toda conjecture.

5 The Toda conjecture and the Virasoro conjecture

The Virasoro conjecture for $\mathbb{CP}^1$ says that the functions z_k, $k \ge -1$, vanish, where z_k is given by the formula

$$z_k = \sum_{m=0}^{\infty} \left(c_k^{m+1} t_m \langle\langle \tau_{m+k,Q} \rangle\rangle + c_k^m \tilde{s}_m \langle\langle \tau_{m+k,P} \rangle\rangle + 2 d_k^m \tilde{s}_m \langle\langle \tau_{m+k-1,Q} \rangle\rangle \right)$$

$$+ \sum_{i+j=k} i!\, j! \left(\varepsilon^2 \langle\langle \tau_{i-1,Q} \tau_{j-1,Q} \rangle\rangle + \langle\langle \tau_{i-1,Q} \rangle\rangle \langle\langle \tau_{j-1,Q} \rangle\rangle \right) + \delta_{k,-1} s_0 t_0 + \delta_{k,0} s_0^2 .$$

Here, $c_k^m = m(m+1)\dots(m+k)$, and $d_k^m = e_k(m,\dots,m+k)$, where $e_k(x_0,\dots,x_k)$ is the kth elementary symmetric function, and $\tilde{s}_m = s_m - \delta_{m,1}$. This conjecture was made by Eguchi, Hori and Yang [5], motivated by their matrix integral representation of the Gromov-Witten potential of $\mathbb{CP}^1$. The string equation and Hori's equation are the special cases with $k = -1$ and $k = 0$ respectively. The formulas

$$\mathcal{L}_{-1} z_k = -(k+1) z_{k-1}, \qquad \mathcal{L}_0 z_k = -k z_k, \tag{5.1}$$

may be proved by direct calculation.

The Virasoro conjecture for $\mathbb{CP}^1$ has been proved by Givental [13]; his proof uses Kontsevich's localization theorem for Gromov-Witten invariants of toric varieties [17], together with results from Dubrovin's theory of semisimple Frobenius manifolds [2].

In this section, we study the constraints

$$x_k = \mathcal{D} \langle\langle \tau_{k,P} \rangle\rangle - (k+1) \nabla \langle\langle \tau_{k+1,P} \rangle\rangle - 2 \nabla \langle\langle \tau_{k,Q} \rangle\rangle .$$

Our main result is that if the Toda and Virasoro conjectures hold, then the constraints x_k vanish.

Theorem 5.1.

$$\mathcal{D}z_k - \nabla z_{k+1} = \sum_{m=0}^{\infty}\big(c_k^{m+1}\, t_m\, y_{m+k} + c_k^m\, \tilde{s}_m\, x_{m+k} + 2\, d_k^m\, \tilde{s}_m\, y_{m+k-1}\big)$$
$$+ \sum_{i+j=k} i!\, j!\big(\varepsilon^2 \partial_{i-1,Q} + [2]\langle\langle\tau_{i-1,Q}\rangle\rangle\big) y_{j-1}$$

Proof. We make use of the following formulas: $[\partial_{i,Q}, \mathcal{D}] = \nabla\langle\langle\tau_{i,Q}Q\rangle\rangle\nabla$,

$$[\mathcal{D}, t_k] = \delta_{k,0}\,[2], \quad [\mathcal{D}, s_k] = \tfrac{1}{2}\delta_{k,0}\big(\varepsilon^2\nabla\partial_Q + v\,[2]\big),$$
$$\nabla(fg) = \tfrac{1}{2}\,[2]\,f\,\nabla g + \tfrac{1}{2}\nabla f\;[2]\,g,$$
$$\mathcal{D}(fg) = \tfrac{1}{2}\,[2]\,f\,\mathcal{D}g + \tfrac{1}{2}\mathcal{D}f\;[2]\,g + \tfrac{1}{2}\varepsilon^2\partial_Q(\nabla f\,\nabla g).$$

It follows that

$$\mathcal{D}\sum_{m=0}^{\infty}(c_k^{m+1}t_m\langle\langle\tau_{m+k,Q}\rangle\rangle + c_k^m\tilde{s}_m\langle\langle\tau_{m+k,P}\rangle\rangle + 2d_k^m\tilde{s}_m\langle\langle\tau_{m+k-1,Q}\rangle\rangle)$$
$$= \sum_{m=0}^{\infty}(c_k^{m+1}t_m\mathcal{D}\langle\langle\tau_{m+k,Q}\rangle\rangle + c_k^m\tilde{s}_m\mathcal{D}\langle\langle\tau_{m+k,P}\rangle\rangle + 2d_k^m\tilde{s}_m\mathcal{D}\langle\langle\tau_{m+k-1,Q}\rangle\rangle)$$
$$+ c_k^1\,[2]\langle\langle\tau_{k,Q}\rangle\rangle + d_k^0(\varepsilon^2\nabla\partial_Q + v\,[2])\langle\langle\tau_{k-1,Q}\rangle\rangle$$
$$= \sum_{m=0}^{\infty}(c_{k+1}^{m+1}t_m\nabla\langle\langle\tau_{m+k+1,Q}\rangle\rangle + c_{k+1}^m\tilde{s}_m\nabla\langle\langle\tau_{m+k+1,P}\rangle\rangle + 2d_{k+1}^m\tilde{s}_m\nabla\langle\langle\tau_{m+k,Q}\rangle\rangle)$$
$$+ \sum_{m=0}^{\infty}(c_k^{m+1}t_m y_{m+k} + c_k^m\tilde{s}_m x_{m+k} + 2d_k^m\tilde{s}_m y_{m+k-1})$$
$$+ c_k^1\,[2]\langle\langle\tau_{k,Q}\rangle\rangle + d_k^0(\varepsilon^2\nabla\partial_Q + v\,[2])\langle\langle\tau_{k-1,Q}\rangle\rangle$$
$$= \nabla\sum_{m=0}^{\infty}(c_{k+1}^{m+1}t_m\langle\langle\tau_{m+k+1,Q}\rangle\rangle + c_{k+1}^m\tilde{s}_m\langle\langle\tau_{m+k+1,P}\rangle\rangle + 2d_{k+1}^m\tilde{s}_m\langle\langle\tau_{m+k,Q}\rangle\rangle)$$
$$+ \sum_{m=0}^{\infty}(c_k^{m+1}t_m y_{m+k} + c_k^m\tilde{s}_m x_{m+k} + 2d_k^m\tilde{s}_m y_{m+k-1})$$
$$+ d_k^0(\varepsilon^2\nabla\partial_Q + v\,[2])\langle\langle\tau_{k-1,Q}\rangle\rangle$$

$$\begin{aligned}
&\mathcal{D} \sum_{i+j=k} i!\,j!(\varepsilon^2\langle\langle\tau_{i-1,Q}\tau_{j-1,Q}\rangle\rangle + \langle\langle\tau_{i-1,Q}\rangle\rangle\langle\langle\tau_{j-1,Q}\rangle\rangle) \\
&\quad = \sum_{i+j=k} i!\,j!(\varepsilon^2\partial_{i-1,Q}\mathcal{D}\langle\langle\tau_{j-1,Q}\rangle\rangle - \varepsilon^2[\partial_{i-1,Q},\mathcal{D}]\langle\langle\tau_{j-1,Q}\rangle\rangle \\
&\qquad + [2]\langle\langle\tau_{i-1,Q}\rangle\rangle\mathcal{D}\langle\langle\tau_{j-1,Q}\rangle\rangle + \tfrac{1}{2}\varepsilon^2\partial_Q(\nabla\langle\langle\tau_{i-1,Q}\rangle\rangle\nabla\langle\langle\tau_{j-1,Q}\rangle\rangle)) \\
&\quad = \sum_{i+j=k} i!\,(j+1)!(\varepsilon^2\nabla\langle\langle\tau_{i-1,Q}\tau_{j,Q}\rangle\rangle + [2]\langle\langle\tau_{i-1,Q}\rangle\rangle\nabla\langle\langle\tau_{j,Q}\rangle\rangle) \\
&\qquad + \sum_{i+j=k} i!\,j!(\varepsilon^2\partial_{i-1,Q} + [2]\langle\langle\tau_{i-1,Q}\rangle\rangle)y_{j-1} \\
&\quad = \nabla \sum_{i+j=k+1} i!\,j!(\varepsilon^2\langle\langle\tau_{i-1,Q}\tau_{j-1,Q}\rangle\rangle + \langle\langle\tau_{i-1,Q}\rangle\rangle\langle\langle\tau_{j-1,Q}\rangle\rangle) \\
&\qquad - k!(\varepsilon^2\nabla\partial_Q + v\,[2])\langle\langle\tau_{k-1,Q}\rangle\rangle + \sum_{i+j=k} i!\,j!(\varepsilon^2\partial_{i-1,Q} + [2]\langle\langle\tau_{i-1,Q}\rangle\rangle)y_{j-1}.
\end{aligned}$$

The theorem follows for $k > 0$ on adding the results of these two calculations; the cases $k = -1$ and $k = 0$ are similar, and we leave them to the reader. □

Corollary 5.2. *If $y_k = 0$ for all $k \geq 0$ and $z_k = 0$ for all $k \geq -1$, then $x_k = 0$ for all $k \geq 0$.*

Proof. If y_k and z_k vanish for all k, then the formula of Theorem 5.1 becomes, for $k \geq -1$,

$$x_{k+1} = \sum_{m=1}^{\infty} \binom{k+m}{m-1} s_m\, x_{k+m}.$$

The result follows by induction on the order of vanishing of the constraints x_k at the origin of the large phase space. □

Assuming the Toda and Virasoro conjectures, we will now show that there are Hamiltonians $g_k \in \mathcal{R}^0$ such that $\partial_{k,P}u = \frac{1}{k!}\nabla\delta_v g_k$ and $\partial_{k,P}v = \frac{1}{k!}\nabla\delta_u g_k$. We may construct the Hamiltonian g_0 explicitly: the equations $\partial u = \nabla\delta_v g_0$ and $\partial v = \nabla\delta_u g_0$ have the solution

$$g_0 = \textstyle\int u \mathsf{P} v\, dt = \displaystyle\sum_{g=0}^{\infty} \frac{\varepsilon^{2g}(2^{1-2g}-1)B_{2g}}{(2g)!} \textstyle\int u\,\partial^{2g} v\, dt. \tag{5.2}$$

Note that $\delta_0 g_0$ lies in the centre of $\mathcal{L}$, since

$$[\delta_0 g_0, \textstyle\int F\, dt] = [\int(\partial u\,\theta_u + \partial v\,\theta_v)\, dt, \int F\, dt] = \int \partial F\, dt = 0. \tag{5.3}$$

We have not been able to find an explicit formula for the Hamiltonians g_k; instead, we construct them using a method due to Gelfand and Dorfman [9].

Theorem 5.3. *There is a unique sequence of Hamiltonians $g_k \in \mathcal{L}^0(k)$ starting with $g_0 = \int u\mathsf{P}v\,dt$ such that $[\mathcal{H}_0, g_k] = [\mathcal{H}, g_{k-1} - \frac{2}{k}h_{k-1}]$.*

Proof. Uniqueness is clear: if $k > 0$, the recursion $[\mathcal{H}_0, g_k] = [\mathcal{H}, g_{k-1} - \frac{2}{k}h_{k-1}]$ determines g_k up to an element of $H^0(\mathcal{L}(k), \delta_0)$, and this cohomology group vanishes by Theorem 3.4.

We will construct g_k by induction.

Lemma 5.4. *The vector field $\mathcal{X}_k = [\mathcal{H}, g_{k-1} - \frac{2}{k}h_{k-1}] \in \mathcal{L}^1(k)$ satisfies $\delta_0\mathcal{X}_k = 0$.*

Proof. We have

$$\delta_0\mathcal{X}_k = [\mathcal{H}_0, [\mathcal{H}, g_{k-1} - \tfrac{2}{k}h_{k-1}]] = -[\mathcal{H}, [\mathcal{H}_0, g_{k-1} - \tfrac{2}{k}h_{k-1}]].$$

If $k > 1$, we have

$$[\mathcal{H}, [\mathcal{H}_0, g_{k-1} - \tfrac{2}{k}h_{k-1}]] = [\mathcal{H}, [\mathcal{H}, g_{k-2} - 2(\tfrac{1}{k} + \tfrac{1}{k-1})h_{k-2}]] = 0.$$

If $k = 1$, we see that $\delta_0\mathcal{X}_1 = -[\mathcal{H}, [\mathcal{H}_0, g_0 - 2h_0]] = [\delta_0 g_0, \mathcal{H}]$, which vanishes by (5.3). $\square$

It follows from this lemma that the cohomology class of $\mathcal{X}_k$ is an element of $H^1(\mathcal{L}(k), \delta_0)$. For $k > 0$, this cohomology group vanishes by Theorem 3.4, hence there is an element g_k of $\mathcal{L}^0(k)$ such that $\delta_0 g_k = \mathcal{X}_k$. $\square$

Corollary 5.5. *Denote the Hamiltonian vector field $\delta_0 g_k$ associated to the Hamiltonian g_k by $\tilde{\delta}_k$:*

$$\tilde{\delta}_k v = \nabla\delta_u g_k, \qquad \tilde{\delta}_k u = \nabla\delta_v g_k.$$

The flows $\tilde{\delta}_k$ commute with each other, and with the flows δ_k.

Proof. Apply Theorem 3.3 to the sequences of Hamiltonians h_k and g_k. $\square$

It is not hard to find an explicit formula for g_1:

$$g_1 = \int \big(\tfrac{1}{2}u\mathsf{P}(v^2 + [2]\,e^u) + \tfrac{1}{2}v([2]\,\mathsf{P} - 2)v - 2qe^u\big)\,dt. \tag{5.4}$$

The equation

$$[\mathcal{H}_0, g_1] = [\mathcal{H}, g_0 - 2h_0] = [\mathcal{H}, g_0] - 2[\mathcal{H}_0, h_1]$$

amounts by (3.2) and (3.3) to the pair of equations

$$\begin{aligned}\nabla(\delta_v g_1) &= \nabla(v\delta_v g_0 + [2]\,\delta_u g_0 - 2\delta_v h_1) = \nabla(v\mathsf{P}u + [2](\mathsf{P}-2)v)\\ \nabla(\delta_u g_1) &= \mathsf{C}_u\delta_v g_1 + v\nabla\delta_u g_1 - 2\nabla\delta_u h_1\\ &= \tfrac{1}{2}q\nabla(e^u\,[2]\,\mathsf{P}u) + \tfrac{1}{2}q\,[2](e^u\partial u) + v\nabla\mathsf{P}v - 2q\nabla e^u\\ &= q\nabla\big(\tfrac{1}{2}e^u\,[2]\,\mathsf{P}u + \tfrac{1}{2}\mathsf{P}(v^2 + [2]\,e^u) - 2e^u\big),\end{aligned}$$

and it is easily seen that these are satisfied.

Proposition 5.6. *For $k > 0$, $[e, g_k] = kg_{k-1}$. For $k \geq 0$,*

$$[\mathcal{E}, g_k] = (k+1)g_k + 2h_k.$$

Proof. We prove these formulas by induction: it is easily checked, using the explicit formula (5.4) for g_1, that $[e, g_1] = g_0$, and, using the explicit formula (5.2) for g_0, that $[\mathcal{E}, g_0] = g_0 + 2h_0$.

We have

$$\begin{aligned}\delta_0[e, g_k] &= [e, \delta_0 g_k] = [e, [\mathcal{H}, g_{k-1} - \tfrac{2}{k}h_{k-1}]]\\ &= [[e,\mathcal{H}], g_{k-1} - \tfrac{2}{k}h_{k-1}] + [\mathcal{H}, [e, g_{k-1} - \tfrac{2}{k}h_{k-1}]]\\ &= \delta_0(g_{k-1} - \tfrac{2}{k}h_{k-1}) + [\mathcal{H}, (k-1)g_{k-2} - \tfrac{2(k-1)}{k}h_{k-2}]\\ &= \delta_0(g_{k-1} - \tfrac{2}{k}h_{k-1}) + \delta_0((k-1)g_{k-1} + \tfrac{2}{k}h_{k-1}) = \delta_0(kg_{k-1}).\end{aligned}$$

Since the cohomology group $H^0(\mathcal{L}(k-1), \delta_0)$ vanishes for $k > 1$, the formula for $[e, g_k]$ follows.

The argument for $\mathcal{E}$ is similar: we have

$$\begin{aligned}\delta_0[\mathcal{E}, g_k] &= [\delta_0\mathcal{E}, g_k] + [\mathcal{E}, \delta_0 g_k] = \delta_0 g_k + [\mathcal{E}, [\mathcal{H}, g_{k-1} - \tfrac{2}{k}h_{k-1}]]\\ &= \delta_0 g_k + [\mathcal{H}, [\mathcal{E}, g_{k-1} - \tfrac{2}{k}h_{k-1}]]\\ &= \delta_0 g_k + [\mathcal{H}, kg_{k-1}] = \delta_0((k+1)g_k + 2h_k).\end{aligned}$$

Since the cohomology group $H^0(\mathcal{L}(k), \delta_0)$ vanishes for $k > 0$, the formula for $[\mathcal{E}, g_k]$ follows. □

Introduce the constraints

$$\alpha^v_{k,P} = \nabla\langle\langle\tau_{k,P}\rangle\rangle - \tfrac{1}{k!}\delta_v g_k, \qquad \alpha^u_{k,P} = \partial_Q\langle\langle\tau_{k,P}\rangle\rangle - \tfrac{1}{k!}\delta_u g_k.$$

By the definitions of u and v, we see that $\alpha^v_{0,P}$ and $\alpha^u_{0,P}$ vanish.

Lemma 5.7. *For $k > 0$, $\mathcal{L}_{-1}\alpha^*_{k,P} = -\alpha^*_{k-1,P}$ and*

$$\mathcal{L}_0\alpha^v_{k,P} = -k\alpha^v_{k,P} - 2\alpha^v_{k,Q}, \qquad \mathcal{L}_0\alpha^u_{k,P} = -(k+1)\alpha^u_{k,P} - 2\alpha^u_{k,Q}.$$

Proof. By the string equation, $\mathcal{L}_{-1}\langle\langle\tau_{k,P}\rangle\rangle = -\langle\langle\tau_{k-1,P}\rangle\rangle$. Since $[e,\delta_v] = [e,\delta_u] = 0$, we see that $\mathcal{L}_{-1}\delta_v g_k = -k\delta_v g_{k-1}$ and $\mathcal{L}_{-1}\delta_u g_k = -k\delta_u g_{k-1}$. This shows that $\mathcal{L}_{-1}\alpha^*_{k,P} = -\alpha^*_{k,P}$.

By Hori's equation $z_0 = 0$, we see that

$$\mathcal{L}_0\nabla\langle\langle\tau_{k,P}\rangle\rangle = -k\nabla\langle\langle\tau_{k,P}\rangle\rangle - 2\nabla\langle\langle\tau_{k-1,Q}\rangle\rangle,$$
$$\mathcal{L}_0\partial_Q\langle\langle\tau_{k,P}\rangle\rangle = -(k+1)\partial_Q\langle\langle\tau_{k,P}\rangle\rangle - 2\partial_Q\langle\langle\tau_{k-1,Q}\rangle\rangle.$$

Since $[\mathcal{E},\delta_v] = -\delta_v$ and $[\mathcal{E},\delta_u] = 0$, we see that $\mathcal{L}_{-1}\delta_v g_k = -k\delta_v g_k - 2\delta_v h_k$ and $\mathcal{L}_0\delta_u g_k = -(k+1)\delta_u g_k - 2\delta_u h_k$. This yields the formulas for $\mathcal{L}_0\alpha^*_{k,P}$. □

Theorem 5.8. *Assume that the Toda conjecture holds. Each of the following conditions are equivalent to the Virasoro conjecture:*

$$\boxed{\text{for all } k > 0,\ \mathcal{D}\langle\langle\tau_{k-1,P}\rangle\rangle = k\nabla\langle\langle\tau_{k,P}\rangle\rangle + 2\langle\langle\tau_{k-1,Q}\rangle\rangle} \tag{i}$$

$$\boxed{\text{for all } k > 0,\ k!\,\nabla\langle\langle\tau_{k,P}\rangle\rangle = \delta_v g_k \text{ and } k!\,\partial_Q\langle\langle\tau_{k,P}\rangle\rangle = \delta_u g_k} \tag{ii}$$

$$\boxed{\text{for all } k \geq 0,\ \text{the vector field } k!\,\partial_{k,P} \text{ equals } \tilde{\delta}_k} \tag{iii}$$

Proof. We may reformulate the conditions of the theorem in terms of the constraints x_k, $\alpha^u_{k,P}$ and $\alpha^v_{k,P}$ as follows:

$$\boxed{\text{for all } k \geq 0,\ x_k = 0} \tag{i}$$

$$\boxed{\text{for all } k \geq 0,\ \alpha^v_{k,P} = \alpha^u_{k,P} = 0} \tag{ii}$$

$$\boxed{\text{for all } k \geq 0,\ \nabla\alpha^v_{k,P} = \nabla\alpha^u_{k,P} = 0} \tag{iii}$$

We have already shown that the Virasoro conjecture implies (i); let us prove the converse. We argue by induction that z_k vanishes, starting with Hori's equation z_0. If $z_{k-1} = 0$, then Theorem 5.1 together with (i) implies that $\nabla z_k = 0$. We conclude by (5.1) and Proposition 4.1 that $z_k = 0$.

The proof of the following lemma is analogous to that of Lemma 4.4.

Lemma 5.9. *We have* $x_{k-1} = v\alpha^v_{k-1,P} + [2]\,\alpha^u_{k-1,P} - k\alpha^v_{k,P} - 2\alpha^v_{k,Q}$ *and*

$$\partial_Q x_{k-1} = \mathsf{C}_u\alpha^v_{k-1,P} + v\nabla\alpha^u_{k-1,P} - k\nabla\alpha^u_{k,P} - 2\nabla\alpha^v_{k,Q}.$$

By Lemma 5.7 and Proposition 4.1, we see that if $\nabla\alpha^*_{i,P} = 0$ for $i \leq k$, then $\alpha^*_{k,P} = 0$; in particular, (iii) implies (ii). Lemma 5.9 shows that (ii) implies (i) and that (i) implies (iii). □

Corollary 5.10. *The Toda and Virasoro conjectures determine the Gromov-Witten potential F of $\mathbb{CP}^1$ up to a constant (that is, an element of $\mathbb{Q}_{\varepsilon,q}$).*

Proof. The Toda and Virasoro conjectures determine the vector fields $\partial_{k,Q}$ and $\partial_{k,P}$ in the coordinate system $\{\partial^n u, \partial^n v\}_{n\geq 0}$ on the large phase space. It follows that the coordinates $\{s_k, t_k\}_{k\geq 0}$ are determined up to constants of integration; but these constants are fixed by Theorem 4.2. By inversion, we see that u is determined as a function of $\{s_k, t_k\}_{k\geq 0}$. Integrating twice, using Lemma 4.1 and the string equation, we see that F is determined up to a constant. □

In order to determine the constant term of F, we may use the divisor equation

$$q\frac{\partial F}{\partial q} + \sum_{k=0}^{\infty} s_{k+1}\frac{\partial F}{\partial t_k} = \partial_Q F.$$

This fixes the constant up to an element of $\mathbb{Q}_\varepsilon$; however, such constant terms correspond to Gromov-Witten invariants of degree 0, and no moduli space $\overline{\mathcal{M}}_{g,n,d}$ has vanishing virtual dimension if $d = 0$. Thus, this constant vanishes.

6 Propagating the Toda conjecture

Consider the submanifold $\mathsf{L} \subset \mathsf{M}$ of the large phase space on which $s_k = 0$, $k > 1$. Okounkov and Pandharipande [23] have proved the Toda conjecture on this submanifold; that is, they prove that $k!\,\partial_{k-1,Q}$ equals δ_k along L, for all $k > 0$. Our results allow us to prove that this, in conjunction with the Virasoro conjecture, implies the full Toda conjecture.

Suppose that the constraints $\nabla\alpha^v_{k,Q}$ and $\nabla\alpha^u_{k,Q}$ vanish to order N along L; the theorem of Okounkov and Pandharipande is the case $N = 1$. We now argue by induction. The proof of Theorem 4.3 shows that the constraints y_k vanish to order N along L. (Here, we use the fact that the vector fields $\mathcal{L}_{-1}$ and $\mathcal{L}_0$ are tangential to L, as may be seen by inspection of the explicit formulas (4.1) and (4.2).)

Applying Theorem 5.1, we see that the constraints x_k vanish to order N along L. The proof of Theorem 5.8 shows that the constraints $\nabla\alpha^v_{k,P}$ and $\nabla\alpha^u_{k,P}$ vanish to order N along L, in other words, that the vector fields $k!\,\partial_{k,P}$ and $\tilde{\delta}_k$ are equal to order N along L.

To prove the induction step, we must show that the vector field $k!\,\partial_{k-1,Q} - \delta_k$ vanishes to order $N+1$ along L, in other words, that $[\ell!\,\partial_{\ell,P}, k!\,\partial_{k-1,Q} - \delta_k]$ vanishes to order N along L for all $\ell > 1$. We have

$$\begin{aligned}[\ell!\,\partial_{\ell,P}, k!\,\partial_{k-1,Q} - \delta_k] &= [\ell!\partial_{\ell,P}, k!\,\partial_{k-1,Q}] - [\tilde{\delta}_\ell, \delta_k] \\ &\quad - [\ell!\,\partial_{\ell,P} - \tilde{\delta}_\ell, k!\,\partial_{k-1,Q}] + [\ell!\,\partial_{\ell,P} - \tilde{\delta}_\ell, k!\,\partial_{k-1,Q} - \delta_k].\end{aligned}$$

Obviously, $[\ell!\partial_{\ell,P}, k!\,\partial_{k-1,Q}]$ vanishes; the commutator $[\tilde{\delta}_\ell, \delta_k]$ vanishes by Corollary 5.5; the vector field $[\ell!\,\partial_{\ell,P} - \tilde{\delta}_\ell, \delta_k]$ vanishes to order N along L, while the vector field $[\ell!\,\partial_{\ell,P} - \tilde{\delta}_\ell, k!\,\partial_{k-1,Q} - \delta_k]$ vanishes to order $2N-1 \geq N$ along L.

Acknowledgments

I thank B. Dubrovin, T. Eguchi, B. Feigin, A. Orlov, R. Pandharipande, T. Shiota, C.-S. Xiong, Y. Zhang and the referee for stimulating my interest in this subject and for their helpful suggestions.

I wish to thank Kyoji Saito and Masa-Hiko Saito, and all of the other organizers and participants in the memorable year 1999–2000 at RIMS, Kyoto University devoted to "Geometry of String Theory."

The research of the author is supported in part by NSF grants DMS-9704320 and DMS-0072508.

References

1. I. Dorfman, "Dirac structures and integrability of nonlinear evolution equations." John Wiley, Chichester, 1993.
2. B. Dubrovin, *Geometry of 2D topological field theories,* in "Integrable systems and quantum groups, Montecalini Terme, 1993," eds. M. Francaviglia and S. Greco, Lect. Notes Math., vol. 1620, Springer-Verlag, Berlin, 1996, pp. 120–348. `<hep-th/9407018>`
3. B. Dubrovin and Y. Zhang, in preparation.
4. T. Eguchi and S.-K. Yang, *The topological* $C\mathrm{P}^1$ *model and the large-N matrix integral.* Modern Phys. Lett. **A 9** (1994), 2893–2902. `<hep-th/9407134>`
5. T. Eguchi, K. Hori and S.-K. Yang, *Topological* σ *models and large-N matrix integral.* Internat. J. Modern Phys. **A 10** (1995), 4203–4224. `<hep-th/9503017>`
6. C. Faber and R. Pandharipande, *Logarithmic series and Hodge integrals in the tautological ring. With an appendix by D. Zagier.* Michigan Math. J. **48** (2000), 215–252. `<math/0002112>`
7. D.B. Fairlie and I.A.B. Strachan, *The algebraic and Hamiltonian structure of the dispersionless Benney and Toda hierarchies*, Inverse Problems **12** (1996), 885–908. `<math/9606022>`
8. I. M. Gelfand and L. A. Dikii, *Asymptotic properties of the resolvent of Sturm-Liouville equations, and the algebra of Korteweg-de Vries equations.* (Russian) Uspehi Mat. Nauk **30** (1975), no. 5, 67–100. (English

translation: Russian Math. Surveys **30** (1975), no. 5, 77–113.)
9. I. M. Gelfand and I. Ja. Dorfman, *Schouten bracket and Hamiltonian operators.* (Russian) Funktsional. Anal. i Prilozhen. **14** (1980), 71–74.
10. E. Getzler, *The Virasoro conjecture for Gromov-Witten invariants,* "Algebraic geometry: Hirzebruch 70 (Warsaw, 1998)," Contemp. Math. **241**, Amer. Math. Soc., Providence, RI, 1999, pp. 147–176. `<math/9812026>`
11. E. Getzler, *A Darboux theorem for Hamiltonian operators in the formal calculus of variations,* To appear, Duke J. Math. `<math/0002164>`
12. E. Getzler and R. Pandharipande, *Virasoro constraints and the Chern classes of the Hodge bundle,* Nucl. Phys. **B 530** (1998), 701–714. `<math/9805114>`
13. A. Givental, *Gromov - Witten invariants and quantization of quadratic hamiltonian,* to appear, Moscow Mathematical Journal. `<math/0108100>`
14. K. Hori, *Constraints for topological strings in $D \geq 1$,* Nucl. Phys. **B439** (1995) 395–420. `<hep-th/9411135>`
15. C. Itzykson and J.-B. Zuber, *Combinatorics of the modular group. II. The Kontsevich integrals,* Internat. J. Modern Phys. **A 7** (1992), 5661–5705. `<hept-th/9201001>`
16. M. Kontsevich, *Intersection theory on moduli spaces of curves and the matrix Airy function*, Commun. Math. Phys. **147** (1992), 1–23.
17. M. Kontsevich, *Enumeration of rational curves via torus actions.* In "The moduli space of curves (Texel Island, 1994)," 335–368, Progr. Math. **129**, Birkhäuser Boston, Boston, MA, 1995.
18. B. A. Kupershmidt, *Discrete Lax equations and differential-difference calculus,* Astérisque **123** (1985).
19. E. Looijenga, *Intersection theory on Deligne-Mumford compactifications (after Witten and Kontsevich).* Séminaire Bourbaki, Vol. 1992/93. Astérisque No. 216 (1993), Exp. No. 768, pp. 187–212.
20. Y. I. Manin, *Algebraic aspects of nonlinear differential equations.* (Russian) In "Current problems in mathematics," Vol. 11, pp. 5–152. Akad. Nauk SSSR Vsesojuz. Inst. Naučn. i Tehn. Informacii, Moscow, 1978.
21. Y. I. Manin, "Frobenius manifolds, quantum cohomology, and moduli spaces." American Mathematical Society Colloquium Publications **47**. American Mathematical Society, Providence, RI, 1999.
22. A. Okounkov and R. Pandharipande, *Gromov-Witten theory, Hurwitz numbers, and matrix models, I.* `<math/0101147>`
23. A. Okounkov and R. Pandharipande, private communication.
24. P. Olver, "Applications of Lie groups to differential equations." Graduate Texts in Mathematics, v. 107. Springer-Verlag, New York, 1993.

25. R. Pandharipande, *The Toda equations and the Gromov-Witten theory of the Riemann sphere.* Lett. Math. Phys. **53** (2000), 59–74. `<math/9912166>`
26. K. Takasaki and T. Takebe, *Quasi-classical limit of Toda hierarchy and W-infinity symmetries.* Lett. Math. Phys. **28** (1993), 165–176.
27. E. Witten, *Two dimensional gravity and intersection theory on moduli space*, Surveys in Differential Geom. **1** (1991), 243–310.

EXAMPLES OF SPECIAL LAGRANGIAN FIBRATIONS.

MARK GROSS
Mathematics Institute, University of Warwick, Coventry, CV4 7AL, USA
E-mail: mgross@maths.warwick.ac.uk

We explore a number of examples of special Lagrangian fibrations on non-compact Calabi-Yau manifolds invariant under torus actions. These include fibrations on crepant resolutions of canonical toric singularities (already found by Goldstein), proper versions of these fibrations, and fibrations on flat deformations of canonical toric singularities. We do this with an eye towards understanding the global structure and discriminant loci of such fibrations. The paper ends with some speculation, both about local mirror symmetry, and the connections of this work with the work of W.D. Ruan and D. Joyce. In the last section, we discuss the philosophy of the Strominger-Yau-Zaslow conjecture in the light of a number of recent ideas of myself and Wilson, Kontsevich and Soibelman, and Joyce.

Introduction

Of late there has been a great deal of interest in special Lagrangian submanifolds and manifolds fibred in special Lagrangian submanifolds, motivated by the Strominger-Yau-Zaslow conjecture [29]. One of the basic approaches to finding examples is to exploit symmetries of the ambient manifold. If X is a (non-compact) n-dimensional Calabi-Yau manifold with Kähler form ω and nowhere vanishing holomorphic n-form Ω, and if there is an action of a Lie group G on X preserving these two forms, one can look for G-invariant special Lagrangian submanifolds of X. For us, G will be a torus T^m. Using this sort of symmetry to search for examples reduces the special Lagrangian equations to simpler ones which can be solved.

This technique has been used independently by M. Haskins and D. Joyce ([12] and [18]) to find new examples of special Lagrangian cones and submanifolds, while it has been used independently by E. Goldstein and myself to construct examples of special Lagrangian fibrations on non-compact Calabi-Yau manifolds with T^{n-1} actions. This paper is an extended version of an informally distributed preprint (essentially just the first two sections of this paper) released at the same time as Goldstein's preprint [6]. Goldstein has developed his examples in some very interesting directions somewhat orthogonal to the ones taken here. While there is some overlap between the examples considered here (especially in §2) and those in [6], our goal will be to develop more global information about these fibrations.

The first set of examples, which we discuss in §2, are special Lagrangian fibrations on crepant resolutions of toric Gorenstein singularities. Such examples were already mentioned in [6]. However, earlier, in [9], I gave topological fibrations on such crepant resolutions, with the belief that these would resemble the actual special Lagrangian fibrations on these manifolds. What is new here is that we show this is indeed the case; with some mild hypotheses on the Kähler metric ω (which we do not require to be Ricci-flat) we find that the topological construction of §3 of [9] coincides with the special Lagrangian fibrations we give here. This gives us a global understanding of the structure of these fibrations, complementing the results in [6]. We also give a new variant of this construction which yields proper fibrations.

The other main set of examples, which we consider in §3, are smoothings (by flat deformations) of isolated toric Gorenstein singularities. The geometry of such smoothings are controlled by combinatorics of the toric data, by results of K. Altmann [1]. These are new examples of special Lagrangian fibrations not discussed elsewhere in the literature. It is great fun to see how special Lagrangian fibrations change if one starts with a crepant resolution of a toric singularity, degenerates by contracting down to the toric singularity, and then smooths (this process is often called an extremal transition). We discuss some examples of this in §3.

§§4 and 5 are more speculative in nature. In §4, we amplify a brief discussion from [9] about the connections between special Lagrangian fibrations on crepant resolutions of toric singularities and the local mirror symmetry of [4]. In doing so, we make a connection with the work of W.-D. Ruan, who came to the description of torus fibrations on Calabi-Yau hypersurfaces in toric varieties via the dual picture to the fibrations on crepant resolutions. We discuss how one might use an S^1 symmetry to construct special Lagrangian fibrations dual to the fibrations on crepant resolutions. It is likely, however, that this construction will come up against a phenomenon revealed in a very recent preprint of Joyce [19], showing that in the S^1-invariant case, we might expect to have codimension one rather than codimension two discriminant loci. This is a serious issue for the SYZ conjecture. In the last section, we propose a weaker version of the SYZ conjecture which will hopefully sidestep these issues.

1 The Basic Construction

Recall from [11]:

Definition 1.1. *Let X be a complex n-dimensional manifold, with a Hermitian metric with Kähler form ω and nowhere vanishing holomorphic n-form Ω.*

Then we say $M \subseteq X$ is special Lagrangian with respect to ω, Ω if $\dim_{\mathbf{R}} M = n$ *and* $\omega|_M = 0$, $\operatorname{Im}\Omega|_M = 0$.

Note we do not assume that either $d\omega = 0$ or the metric is Ricci-flat. However, if the volume form $\omega^n/n!$ is proportional to $\Omega \wedge \bar{\Omega}$, then special Lagrangian submanifolds are volume minimizing as remarked in [7]; see also [5]. Following Joyce, if $d\omega = 0$, we will call the triple (X, ω, Ω) an *almost Calabi-Yau manifold.* Also, we typically want to allow singularities in M; technically this should be done in the language of currents, but we won't worry about such technicalities. Instead, just view M as being a closed set which is a manifold on an open dense subset of M, and the special Lagrangian condition then is required to hold where M is a manifold.

The main tool we will use for constructing examples of special Lagrangian fibrations will be the following result. This theorem first appeared in print in [6], and similar results appear in [18]. For completeness, we also give the proof here.

Theorem 1.2. *Let (X, ω, Ω) be an almost Calabi-Yau manifold, and suppose there is an effective action of $T := T^m$ on X preserving ω and Ω. Let $\mu_0 : X \to \mathbf{t}^* = \mathbf{R}^m$ be the moment map associated to this action.*

1. *Let $X_1, \ldots, X_m$ be a basis for the vector fields generating the action of T. Then $\Omega_{red} = \iota(X_1, \ldots, X_m)\Omega$ descends to an $n-m$-form on non-singular points of $Z_p := \mu_0^{-1}(p)/T$ for $p \in \mu_0(X)$.*

2. *Let ω_{red} be the induced symplectic form on Z_p via symplectic reduction. If $M_{red} \subseteq Z_p$, let M denote the pull-back of M_{red} to $\mu_0^{-1}(p) \subseteq X$. Suppose M is not contained in the set of critical points of μ_0. Then if $M_{red} \subseteq Z_p$ is special Lagrangian with respect to $\omega_{red}, \Omega_{red}$, $M \subseteq X$ is special Lagrangian with respect to ω, Ω.*

3. *Suppose that $g : X \to Y$ is a continuous map to an $n-m$-dimensional real manifold Y, satisfying $g(t \cdot x) = g(x)$ for $t \in T$. Then the map $f = (\mu_0, g) : X \to \mathbf{t}^* \times Y$ has special Lagrangian fibres with respect to ω, Ω if the induced maps $g : Z_p \to Y$ have special Lagrangian fibres with respect to $\omega_{red}, \Omega_{red}$ for p in a dense subset of $\mu_0(X)$.*

Proof. Let $x \in \mu_0^{-1}(p)$, and suppose x is not a critical point for μ_0. Then the tangent space $T_x Z_p$ of Z_p at the point represented by x is identified with

$$\frac{T_x \mu_0^{-1}(p)}{(T_x \mu_0^{-1}(p))^\omega}.$$

Now $(T_x\mu_0^{-1}(p))^\omega$ is the tangent space to the orbit $T \cdot x$, and this is generated by the tangent vectors $X_1, \ldots, X_m$. Thus Ω_{red} vanishes on $(T_x\mu_0^{-1}(p))^\omega$. Since Ω_{red} is invariant under the action of T, Ω_{red} descends to an $n-m$-form on $T_x Z_p$, proving (1).

Now let $M_{red} \subseteq Z_p$ be special Lagrangian with respect to $\omega_{red}, \Omega_{red}$ with M not contained in the set of critical points of μ_0. Let $x \in M$ be a regular point for μ_0. Let $Y_1, \ldots, Y_{n-m}$ be lifts of a basis of tangent vectors to M_{red} at $x \bmod T$. Then $T_x M$ has a basis $X_1, \ldots, X_m, Y_1, \ldots, Y_{n-m}$. By assumption $\omega(X_i, X_j) = 0$, and $\omega(Y_i, Y_j) = \omega_{red}(Y_i, Y_j) = 0$ since M_{red} is Lagrangian. Finally, since $Y_i \in T_x\mu_0^{-1}(p)$, $\omega(X_i, Y_j) = 0$ since $X_i \in (T_x\mu_0^{-1}(p))^\omega$. Thus $\omega|_{T_x M} = 0$.

To show the fibre is special Lagrangian at x, we just observe that

$$\operatorname{Im}\Omega(X_1, \ldots, X_m, Y_1, \ldots, Y_{n-m}) = \operatorname{Im}\Omega_{red}(Y_1, \ldots, Y_{n-m}) = 0.$$

Thus shows M is special Lagrangian. Item (3) now follows immediately from (2). •

As a basic application of this, let $\varphi(x_1, \ldots, x_n)$ be a real-valued function on an open subset of $\mathbf{R}^n$ such that $\varphi(|z_1|^2, \ldots, |z_n|^2)$ is pluri-subharmonic on an open subset U of $\mathbf{C}^n$. Let $\varphi_i = \partial\varphi/\partial x_i$, $\varphi_{ij} = \partial^2\varphi/\partial x_i \partial x_j$. Then the induced Kähler form is

$$\begin{aligned}
\omega &= \frac{i}{2}\partial\bar{\partial}\varphi(|z_1|^2, \ldots, |z_n|^2) \\
&= \frac{i}{2}\partial\left(\sum_i \varphi_i z_i d\bar{z}_i\right) \\
&= \frac{i}{2}\left(\sum_i \varphi_i dz_i \wedge d\bar{z}_i + \sum_{i,j} \varphi_{ij}\bar{z}_j z_i dz_j \wedge d\bar{z}_i\right).
\end{aligned}$$

Corollary 1.3. *The fibres of the map $f : U \to \mathbf{R}^n$ given by*

$$f = (\varphi_1|z_1|^2 - \varphi_2|z_2|^2, \ldots, \varphi_1|z_1|^2 - \varphi_n|z_n|^2, \operatorname{Im}(i^{n+1}\prod_j z_j))$$

are special Lagrangian with respect to ω and $\Omega = dz_1 \wedge \cdots \wedge dz_n$.

Proof. ω is invariant under the natural T^n action on $\mathbf{C}^n$. This T^n action is induced by the vector fields

$$X_j = 2i\left(\bar{z}_j \frac{\partial}{\partial \bar{z}_j} - z_j \frac{\partial}{\partial z_j}\right), \quad j = 1, \ldots, n.$$

Now

$$
\begin{aligned}
\iota(X_j)\omega &= \varphi_j(z_j d\bar{z}_j + \bar{z}_j dz_j) \\
&\quad + \sum_i (\varphi_{ij}|z_j|^2 z_i d\bar{z}_i + \varphi_{ij}|z_j|^2 \bar{z}_i dz_i) \\
&= d(\varphi_j |z_j|^2).
\end{aligned}
$$

Thus X_j is a Hamiltonian vector field with Hamiltonian $\varphi_j|z_j|^2$. In particular, $X_1 - X_2, \ldots, X_1 - X_n$ generate a T^{n-1} action with moment map

$$
\mu_0 = (f_1, \ldots, f_{n-1}) = (\varphi_1|z_1|^2 - \varphi_2|z_2|^2, \ldots, \varphi_1|z_1|^2 - \varphi_n|z_2|^2).
$$

Furthermore, $f_n = \mathrm{Im}(i^{n+1} \prod z_j)$ is constant on the Hamiltonian trajectories of the first $n-1$ functions. We can now apply Theorem 1.2 with $\mu_0 = (f_1, \ldots, f_{n-1})$ and $g = f_n$, as the T^{n-1}-action preserves Ω also. Now

$$
\begin{aligned}
\iota(X_{f_1}, \ldots, X_{f_{n-1}})\Omega &= (-2i)^{n-1}\iota(z_1\frac{\partial}{\partial z_1} - z_2\frac{\partial}{\partial z_2}, \ldots, z_1\frac{\partial}{\partial z_1} - z_n\frac{\partial}{\partial z_n}) dz_1 \wedge \cdots \wedge dz_n \\
&= \pm(-2i)^{n-1} d(z_1 \cdots z_n).
\end{aligned}
$$

It is then clear that g induces a special Lagrangian fibration on the surface $\mu_0^{-1}(p)/T^{n-1}$ for all p. •

2 Resolutions of Toric Singularities

Let $N \cong \mathbf{Z}^n$, and let $M = \mathrm{Hom}(N, \mathbf{Z})$ be the dual lattice. Put $N_{\mathbf{R}} := N \otimes_{\mathbf{Z}} \mathbf{R}$, $T_{\mathbf{C}}(N) = N \otimes_{\mathbf{Z}} \mathbf{C}^*$, $T(N) = N \otimes_{\mathbf{Z}} \mathbf{R}/N$. Then M can be naturally identified with the group of characters $\mathrm{Hom}(T_{\mathbf{C}}(N), \mathbf{C}^*)$, and we will often identify elements of M with such functions.

Let $\sigma \subseteq N_{\mathbf{R}}$ be a strongly convex rational polyhedral cone. We will assume that σ is a Gorenstein canonical cone. This means that if $n_1, \ldots, n_s \in N$ are the set of generators of 1-dimensional faces of σ, then there exists an $m_0 \in M$ such that $\langle m_0, n_i \rangle = 1$ for all i and $\langle m_0, n \rangle \geq 1$ for all $n \in \sigma \cap (N - \{0\})$. Denote by Y_σ the corresponding affine toric variety. Y_σ has Gorenstein canonical singularities. Let P be the convex hull of $n_1, \ldots, n_s$ in the hyperplane $\langle m_0, \cdot \rangle = 1$. From now on we will assume there is a triangulation of P such that the fan Σ obtained as the cone over this triangulation yields a non-singular toric variety Y_Σ. Then $Y_\Sigma \to Y_\sigma$ is a crepant resolution, and $K_{Y_\Sigma} = 0$.

In [9], we constructed a topological fibration on Y_Σ. We recall the construction here. Note that $T(N)$ acts naturally on Y_Σ. If $N_{m_0} = \{n \in N | \langle m_0, n\rangle = 0\}$, then the subtorus $T(N_{m_0})$ of $T(N)$ also acts on Y_Σ. Then one chooses a commutative diagram

$$
\begin{array}{ccc}
Y_\Sigma & \xrightarrow{=} & Y_\Sigma \\
\Big\downarrow {\scriptstyle q_1} & & \Big\downarrow {\scriptstyle \alpha_1 \circ q_1} \\
Y_\Sigma / T(N_{m_0}) & \xrightarrow{\alpha_1} & \mathbf{C} \times \mathbf{R}^{n-1} \\
\Big\downarrow {\scriptstyle q_2} & & \Big\downarrow {\scriptstyle (z,x) \mapsto (|z|, x)} \\
Y_\Sigma / T(N) & \xrightarrow{\alpha_2} & \mathbf{R}_{\geq 0} \times \mathbf{R}^{n-1}
\end{array}
\tag{2.1}
$$

of homeomorphisms α_1, α_2, with q_1, q_2 the quotient maps. If one composes $\alpha_1 \circ q_1$ with the map $(z, x) \mapsto (\operatorname{Im} z, x)$, one obtains a map $f : Y_\Sigma \to \mathbf{R}^n$. This is a topological $T^{n-1} \times \mathbf{R}$ fibration, and its discriminant locus was analyzed in [9].

Choose a basis $e_1, \ldots, e_n$ of N with dual basis $e_1^*, \ldots, e_n^*$ such that $m_0 = e_1^* + \cdots + e_n^*$. The dual basis corresponds to coordinates $z_1, \ldots, z_n$ on $T_{\mathbf{C}}(N) = N \otimes_{\mathbf{Z}} \mathbf{C}^* \subseteq Y_\Sigma$.

Proposition 2.1. *If Ω is a nowhere vanishing holomorphic n-form on Y_Σ, and $T_{\mathbf{C}}(N)$ is identified with the unique dense orbit of $T_{\mathbf{C}}(N)$ acting on Y_Σ, then $\Omega|_{T_{\mathbf{C}}(N)} = C dz_1 \wedge \cdots \wedge dz_n$, where $C \in \mathbf{C}$ is a constant.*

Proof. The proof is standard: see [25]. We give the complete proof here. Clearly $\Omega|_{T_{\mathbf{C}}(N)} = f dz_1 \wedge \cdots \wedge dz_n$ for some holomorphic function f. Furthermore, to guarantee that Ω has no zeroes on $T_{\mathbf{C}}(N)$, f must be a monomial, i.e. a constant times a character. It is also easy to check that the expression $dz_1 \wedge \cdots \wedge dz_n$ is independent of the choice of basis $e_1, \ldots, e_n$ subject to the constraint that $m_0 = e_1^* + \cdots + e_n^*$. So we can take $e_1, \ldots, e_n$ to be chosen to be edges of an n-dimensional cone τ in Σ; since Y_Σ is smooth, this forms a basis, and since each e_i then satisfies $\langle m_0, e_i\rangle = 1$, $m_0 = e_1^* + \cdots + e_n^*$. Then $\tau^\vee$ is generated by $e_1^*, \ldots, e_n^*$, so Y_Σ contains an open affine subset $\operatorname{Spec} \mathbf{C}[\tau^\vee \cap M] = \operatorname{Spec} \mathbf{C}[z_1, \ldots, z_n]$. Then $f dz_1 \wedge \cdots \wedge dz_n$ extends to a non-zero n-form on this open affine subset if and only if f is constant, as desired. •

Theorem 2.2. *Let ω be the Kähler form of a Kähler metric on Y_Σ, invariant under the action of $T(N)$. Let Ω be the nowhere vanishing n-form on Y_Σ which restricts to $dz_1 \wedge \cdots \wedge dz_n$ on $T_{\mathbf{C}}(N)$. Let $\mu : Y_\Sigma \to \mathbf{R}^n$ be the moment map associated to this $T(N)$ action, and let $\mu_0 : Y_\Sigma \to \mathbf{R}^{n-1}$ be the moment map associated to the $T(N_{m_0})$ action. Then the function $g : T_{\mathbf{C}}(N) \to \mathbf{R}$*

given by

$$g(z_1, \ldots, z_n) = \operatorname{Im} i^{n+1} \prod z_i$$

extends to a map $g : Y_\Sigma \to \mathbf{R}$. *Furthermore,* $f = (g, \mu_0) : Y_\Sigma \to \mathbf{R}^n$ *is a special Lagrangian fibration with respect to* ω, Ω. *If* μ *is proper, then* f *coincides topologically with the construction given in [9], §3.*

Proof. We can identify m_0 with the character $\prod z_i$ on $T_{\mathbf{C}}(N)$. A priori m_0 extends to only a rational function on Y_Σ. In fact, it extends to a regular function. To show this, we need to show it extends across every prime divisor of Y_Σ contained in $Y_\Sigma - T_{\mathbf{C}}(N)$. Let n generate a ray in the fan Σ, corresponding to some such divisor D_n. The dual cone to $\tau = \mathbf{R}_{\geq 0} n$ is the half-plane

$$\tau^\vee = \{m \in M | \langle m, n\rangle \geq 0\}.$$

Now $\langle m_0, n\rangle = 1$, so $m_0 \in \tau^\vee, -m_0 \notin \tau^\vee$. The open affine piece of Y_Σ corresponding to the cone τ is $\operatorname{Spec} \mathbf{C}[\tau^\vee \cap M]$. Thus m_0 is a regular function on this open set. Since this open set contains a dense subset of D_n, m_0 extends across D_n, and in fact takes the value zero on D_n, since $-m_0$ (corresponding to the character $\prod z_j^{-1}$) is not in $\tau^\vee$.

Thus $i^{n+1} m_0$ gives a map $Y_\Sigma \to \mathbf{C}$, from which the first claim follows. Using the moment map μ_0 of the $T(N_{m_0})$-action, we obtain a map $(i^{n+1} m_0, \mu_0) : Y_\Sigma \to \mathbf{C} \times \mathbf{R}^{n-1}$. Composing this map with $(z, x) \mapsto (\operatorname{Im} z, x)$, we obtain $f : Y_\Sigma \to \mathbf{R}^n$, an extension of the special Lagrangian fibration $f : T_{\mathbf{C}}(N) \to \mathbf{R}^n$ arising in Corollary 1.3.

To finish, we show f coincides with the construction of [9] if μ is proper.

Because we are assuming μ is proper, the following facts follow from [14] Theorem 4.1: $\mu(Y_\Sigma)$ is convex, and μ has connected fibres. Thus, in particular, μ identifies $\mu(Y_\Sigma)$ with $Y_\Sigma / T(N)$. Furthermore, $\mu(Y_\Sigma)$ is a closed, locally polyhedral convex set, and the extremal points of $\mu(Y_\Sigma)$ are images of fixed points of the $T(N)$-action. Finally the tangent "wedge" to $\mu(Y_\Sigma)$ at such a point $\mu(x)$ is generated by the weights of the $T(N)$-action on $T_x Y_\Sigma$. There is a 1-1 correspondence between fixed points of the $T(N)$-action on Y_Σ and maximal cones τ of the fan Σ. Now since Y_Σ is non-singular, each such cone τ is generated by a basis $v_1, \ldots, v_n$ of N, and the weights of the $T(N)$-representation on $T_x Y_\Sigma$, x the point corresponding to τ, are $v_1^*, \ldots, v_n^*$. Furthermore, since $\langle m_0, v_i\rangle = 1$, it follows that $m_0 = \sum v_i^*$. In particular, m_0 is in the interior of the tangent "wedge" of each such extremal point.

Let $r : \mathbf{R}^n \to \mathbf{R}^{n-1}$ be given by $(x_1, \ldots, x_n) \mapsto (x_1 - x_2, \ldots, x_1 - x_n)$. The composition $r \circ \mu : Y_\Sigma \to \mathbf{R}^{n-1}$ is the moment map μ_0 of the $T(N_{m_0})$ action on Y_Σ.

We will now show that if $L = r^{-1}(c)$, $c \in \mathbf{R}^{n-1}$, is a line in $\mathbf{R}^n = M \otimes_{\mathbf{Z}} \mathbf{R}$ parallel to m_0, then $L \cap \mu(Y_\Sigma)$ is a ray. Indeed, if $L \cap \partial\mu(Y_\Sigma)$ is non-empty, the description above of the tangent wedges of the extremal points shows that $L \cap \partial\mu(Y_\Sigma)$ consists of one point. Thus $L \cap \mu(Y_\Sigma)$ is closed (as μ is proper) and has one boundary point, so it is a ray. If $L \cap \partial\mu(Y_\Sigma)$ is empty, then either $L \cap \mu(Y_\Sigma) = \phi$ or $L \subseteq \mu(Y_\Sigma)$. In either case, choose a line $l \subseteq \mathbf{R}^{n-1}$ such that the plane $r^{-1}(l)$ contains L, $r^{-1}(l) \cap \mu(Y_\Sigma) \neq \phi$, and $r^{-1}(l) \not\subseteq \mu(Y_\Sigma)$. Then $S = r^{-1}(l) \cap \mu(Y_\Sigma)$ is a closed convex set. It can only contain L if S is a half-plane with edge parallel to L, contradicting the description of the tangent wedges to $\mu(Y_\Sigma)$. If $L \cap S = \phi$, then there is a supporting line to S, L', parallel to L. This also contradicts the description of the tangent wedges. Thus $L \cap \mu(Y_\Sigma)$ is a ray.

We now replace r with its restriction to $\mu(Y_\Sigma)$. We have $r : \mu(Y_\Sigma) \to \mathbf{R}^{n-1}$ is surjective, with each fibre being a ray parallel to m_0.

We now wish to define homeomorphisms α_1, α_2 as in the diagram (2.1) so that $f = \mathrm{Im} \circ \alpha_1 \circ q_1$, where $\mathrm{Im} : \mathbf{C} \times \mathbf{R}^{n-1} \to \mathbf{R}^n$ is the map $(z, x) \mapsto (\mathrm{Im}\, z, x)$. This will show that f coincides with the construction of [9], §3.

Note that the value of $|m_0|$ only depends on the $T(N)$-orbit of $(z_1, \ldots, z_n)$, so $|m_0|$ descends to a map $|m_0| : \mu(Y_\Sigma) \to \mathbf{R}_{\geq 0}$. We define α_2 as the product map $(|m_0|, r) : \mu(Y_\Sigma) \to \mathbf{R}_{\geq 0} \times \mathbf{R}^{n-1}$. To show that α_2 is a homeomorphism, it is enough to show that $|m_0| : r^{-1}(c) \to \mathbf{R}_{\geq 0}$ for $c \in \mathbf{R}^{n-1}$ is a homeomorphism. Now $r^{-1}(c)$ is a ray whose endpoint is in $\partial\mu(Y_\Sigma)$, and $\mu^{-1}(\partial\mu(Y_\Sigma)) = Y_\Sigma - T_{\mathbf{C}}(N)$. Thus $|m_0|$ takes the value zero on the endpoint of $r^{-1}(c)$. So all we need to know is that $|m_0|$ is monotonically increasing on $r^{-1}(c)$, and goes to ∞.

To see this, note that $Z_c := \mu_0^{-1}(c)/T(N_0)$ is in fact isomorphic to $\mathbf{C}$, and m_0 descends to a holomorphic function on Z_c. The maximum modulus theorem then tells us that $|m_0|$ is monotonically increasing on $r^{-1}(c)$, and Liouville's theorem tells us $|m_0|$ goes to ∞.

Thus α_2 is an homeomorphism.

Similarly, since $(i^{n+1}m_0, \mu_0)$ is constant on $T(N_{m_0})$-orbits, it descends to give a map $\alpha_1 : Y_\Sigma / T(N_{m_0}) \to \mathbf{C} \times \mathbf{R}^{n-1}$. It is easy to check now that α_1 is a homeomorphism making the diagram (2.1) commute, and that $f = \mathrm{Im} \circ \alpha_1 \circ q_1$, as desired. •

We next comment as to when the hypotheses of Theorem 2.2 can be achieved.

First, we recall the standard construction of Y_Σ as a symplectic quotient: see [3] for details. Here as always Y_Σ is assumed to be non-singular. Let $\Sigma(1)$ denote the set of one-dimensional faces of σ. For each $\sigma \in \Sigma(1)$ there is a toric divisor D_σ corresponding to σ, and $\{D_\sigma|\sigma \in \Sigma(1)\}$ generates PicY_Σ. Define a map

$$\mathbf{Z}^{\Sigma(1)} \xrightarrow{\pi} N$$

taking a standard basis vector e_σ of $\mathbf{Z}^{\Sigma(1)}$, $\sigma \in \Sigma(1)$, to the generator of the corresponding one-dimensional face of Σ. Let $K = \ker \pi$, and assume (as will always be the case in our examples of interest) that π is surjective. For $I \subseteq \Sigma(1)$, define $e_I \subseteq \mathbf{C}^{\Sigma(1)}$ by

$$\{(z_j)_{j\in\Sigma(1)}|z_i = 0 \text{ for } i \notin I\}$$

and let $\bar{I} = \Sigma(1) \setminus I$. Let

$$S = \{I \subseteq \Sigma(1)|\bar{I} \text{ does not span a cone in } \Sigma\}.$$

Define $\mathcal{U}_\Sigma \subseteq \mathbf{C}^{\Sigma(1)}$ by

$$\mathcal{U}_\Sigma = \mathbf{C}^{\Sigma(1)} \setminus \bigcup_{I\in S} e_I.$$

$\mathcal{U}_\Sigma$ inherits the standard Kähler form

$$\omega = \frac{i}{2} \sum_{j\in\Sigma(1)} dz_j \wedge d\bar{z}_j.$$

Let $\mathbf{t}$ denote the Lie algebra of $T(\mathbf{Z}^{\Sigma(1)})$ and $\mathbf{k}$ denote the Lie algebra of $T(K)$. There is a well-known correspondence between elements of $\mathbf{t}^*$ and real divisors: an element $\alpha \in \mathbf{t}^*$ corresponds to the real divisor

$$D_\alpha = \sum_{\sigma\in\Sigma(1)} \alpha(e_\sigma) D_\sigma.$$

On the other hand, such an α defines a piecewise linear function on the support of the fan Σ. Indeed, for each cone τ of Σ, choose $m_\tau \in M$ so that $\langle m_\tau, \pi(e_\sigma)\rangle = -\alpha(e_\sigma)$ for each one-dimensional face σ of τ. Then define $\phi_\alpha : |\Sigma| \to \mathbf{R}$ by $\phi_\alpha(x) = \langle m_\tau, x\rangle$ for $x \in \tau$. Then the $\mathbf{R}$-divisor D_α is ample if and only if ϕ_α is strictly upper convex, i.e.

$$\langle m_\tau, \pi(e_\sigma)\rangle > -\alpha(e_\sigma)$$

whenever σ is not a face of an n-dimensional cone τ. Two divisors are linearly equivalent if the corresponding piecewise linear functions differ by a linear function; thus Pic$Y_\Sigma \otimes \mathbf{R} \cong \mathbf{k}^*$ naturally via the projection $p : \mathbf{t}^* \to \mathbf{k}^*$.

Now the action of $T(\mathbf{Z}^{\Sigma(1)})$ on $\mathcal{U}_\Sigma$ induces the standard moment map $\nu : \mathcal{U}_\Sigma \to \mathbf{t}^*$ and $\nu' = p \circ \nu : \mathcal{U}_\Sigma \to \mathbf{k}^*$ is the moment map for the $T(K)$ action on $\mathcal{U}_\Sigma$.

Let $\alpha \in \mathbf{k}^*$ be the class of an ample $\mathbf{R}$-divisor on Y_Σ. Then by Proposition 3.1.1 of [3], $\nu'^{-1}(\alpha)/T(K)$ is homeomorphic to Y_Σ, and the induced, reduced symplectic form ω_α on Y_Σ has cohomology class equal to D_α. This symplectic form is a Kähler form.

The residual $T(N)$ action on Y_Σ preserves this Kähler form, so this gives an example of a $T(N)$-invariant Kähler form in each Kähler class. Furthermore, this action induces the moment map $\mu : Y_\Sigma \to p^{-1}(\alpha)$, the latter being a translation of $\mathbf{n}^*$, where $\mathbf{n}$ is the Lie algebra of $T(N)$. Here we use the exact sequence

$$0 \longrightarrow \mathbf{n}^* \longrightarrow \mathbf{t}^* \stackrel{p}{\longrightarrow} \mathbf{k}^* \longrightarrow 0.$$

By [3], §3.2, the image of μ is $P_\alpha = p^{-1}(\alpha) \cap \mu(\mathcal{U}_\Sigma)$. It is then not difficult to see that P_α is closed and $\mu : Y_\Sigma \to \mathbf{n}^*$ is proper. Furthermore, knowing the image of μ allows us to determine the discriminant locus of the special Lagrangian fibration precisely.

Example 2.3. Let $N = \mathbf{Z}^3 + \frac{1}{3}(1,1,1)$, and let σ be the cone spanned by $(1,0,0)$, $(0,1,0)$ and $(0,0,1)$. Σ is then obtained by subdividing σ at $\frac{1}{3}(1,1,1)$, giving

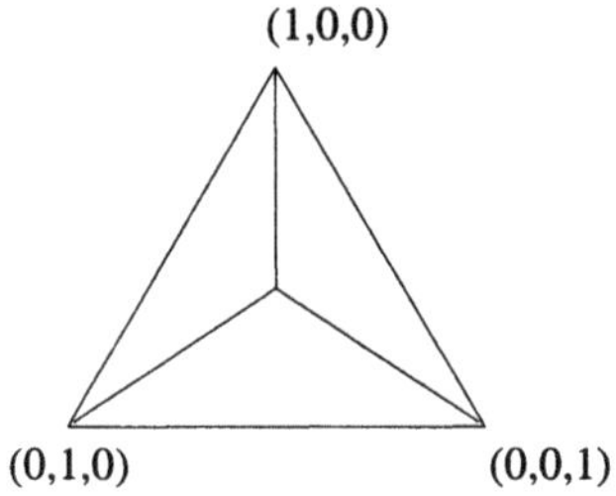

Then

$$\Sigma(1) = \{(1,0,0), (0,1,0), (0,0,1), \frac{1}{3}(1,1,1)\}$$

and the matrix for π is

$$\begin{pmatrix} 1 & 0 & 0 & 1/3 \\ 0 & 1 & 0 & 1/3 \\ 0 & 0 & 1 & 1/3 \end{pmatrix}$$

with kernel generated by $(1,1,1,-3)$. Thus

$$\mathcal{U}_\Sigma = \mathbf{C}^4 \setminus \{z_1 = z_2 = z_3 = 0\}.$$

The moment map ν' is given by

$$\nu'(z_1, z_2, z_3, z_4) = |z_1|^2 + |z_2|^2 + |z_3|^2 - 3|z_4|^2.$$

For $a > 0$, let $(Y_\Sigma, \omega_\Sigma)$ be given by $\nu'^{-1}(a)/T(K)$. Now

$$p^{-1}(a) = \{(r_1, r_2, r_3, (r_1 + r_2 + r_3 - a)/3) | r_1, r_2, r_3 \in \mathbf{R}\},$$

so P_a can be identified with the set

$$\{(r_1, r_2, r_3) \in \mathbf{R}^3 | r_1, r_2, r_3 \geq 0, \quad r_1 + r_2 + r_3 \geq a\}.$$

Finally, the discriminant locus of the induced special Lagrangian fibration $f = (g, \mu_0) : Y_\Sigma \to \mathbf{R}^3$ is the planar graph which is the image of the 1-skeleton of the boundary of P_a under the projection $\mathbf{R}^3 \to \mathbf{R}^2$ given by $(r_1, r_2, r_3) \mapsto (r_1 - r_2, r_1 - r_3)$. This is

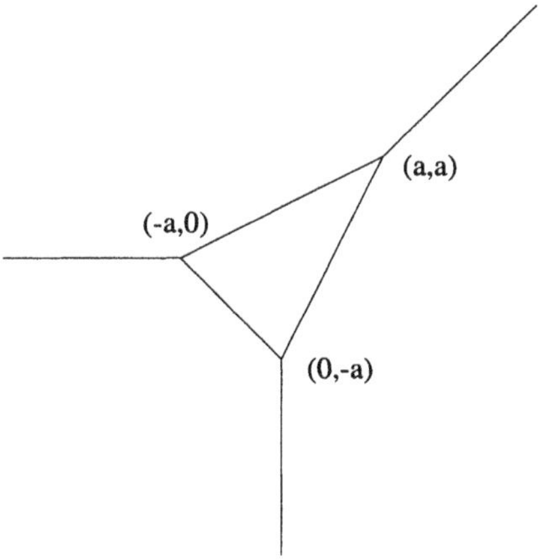

In these examples, the moment map μ is always proper. More generally, however, this need not be the case. Since the above metrics are not, in general, Ricci-flat, we might be interested in a wider range of metrics. Now in any event the action of $T(N)$ is induced by the action of $T_{\mathbf{C}}(N)$, so it follows from [13], Convexity Theorem, §7, that the image of the moment map μ is convex with connected fibres, so μ identifies $\mu(Y_\Sigma)$ with $Y_\Sigma/T(N)$ in general. However, $\mu(Y_\Sigma)$ may not be closed, in which case μ is not proper. Thus we need some additional asymptotic conditions on ω.

For example in [31,32], existence of complete Ricci-flat metrics on some non-compact manifolds was proven. More precise information in certain cases was given in [16,17], where it was proved that there exists ALE or quasiALE Ricci-flat metrics on crepant resolutions of $\mathbf{C}^n/G$, with $G \subseteq SU(n)$. If G is abelian, then $\mathbf{C}^n/G$ is toric. By the uniqueness results of [16,17], these metrics are invariant under the induced $T(N)$ action, and the above theorem applies. Furthermore, the ALE or quasi-ALE conditions guarantee the moment map is proper, being asymptotic to the moment map with respect to the Euclidean metric.

In general, it is not known when Ricci-flat metrics exist on Y_Σ for general Gorenstein cones σ. However, in some cases it is possible to use the results of [31,32] to find further examples. In any event, as long as the moment map with respect to ω is proper, its image will coincide with the image of the moment map induced by ω' given by the symplectic reduction method above when the classes $[\omega], [\omega'] \in H^2(Y_\Sigma, \mathbf{R})$ coincide. Thus we always get a precise description of the discriminant locus.

Since we are interested in special Lagrangian fibrations because of the SYZ conjecture, it is actually more interesting to construct proper special Lagrangian fibrations. In [9], Remark 3.5, we noted we can construct a topological "properification" of the map $f : Y_\Sigma \to \mathbf{R}^3$. Here, we take an alternative route and construct proper special Lagrangian fibrations as a generalization of [7], Example 1.2.

Theorem 2.4. *Let ω be the Kähler form of a Kähler metric on Y_Σ, invariant under the action of $T(N)$. Let Ω be the holomorphic n-form on Y_Σ which restricts to $dz_1 \wedge \cdots \wedge dz_n$ on $T_{\mathbf{C}}(N)$. Let $\mu : Y_\Sigma \to \mathbf{R}^n$ be the moment map of the $T(N)$ action, and assume μ is proper. Let $\mu_0 : Y_\Sigma \to \mathbf{R}^{n-1}$ be the moment map associated to the $T(N_{m_0})$-action. Set*

$$Y'_\Sigma = Y_\Sigma \setminus \{1 + \prod_{i=1}^n z_i = 0\}.$$

Here $\prod_{i=1}^n z_i$ is identified with the character m_0, and as such, defines a regular function on Y_Σ, as we saw in the proof of Theorem 2.2. Let

$$\Omega' = \frac{\Omega}{i^n(1 + \prod_{i=1}^n z_i)}$$

be a nowhere vanishing holomorphic n-form on Y'_Σ. Then $f' = (\log|1 + \prod_{i=1}^n z_i|, \mu_0) : Y'_\Sigma \to \mathbf{R}^n$ is a proper special Lagrangian fibration with respect to ω, Ω', with the same discriminant locus as that of $f : Y_\Sigma \to \mathbf{R}^n$ constructed in Theorem 2.2. Furthermore, the general fibre is an n-torus.

Proof. Let $\mu_0 = (f_1, \ldots, f_{n-1})$ be as in the proof of Corollary 1.3. Then

$$\begin{aligned} \iota(X_{f_1}, \ldots, X_{f_{n-1}})\Omega' &= \pm \frac{(-2i)^{n-1} d(z_1 \cdot \cdots \cdot z_n)}{i^n(1 + \prod_{i=1}^n z_i)} \\ &= \pm(-2)^{n-1} d(i \log(1 + \prod_{i=1}^n z_i)), \end{aligned}$$

so by Theorem 1.2,

$$f' = (\mathrm{Im}(i\log(1+\prod_{i=1}^{n} z_i)), \mu_0) = (\log|1+\prod_{i=1}^{n} z_i|, \mu_0)$$

is a special Lagrangian fibration with respect to ω, Ω'.

Now consider $c \in \mathbf{R}^{n-1}$ and $\mu_0^{-1}(c)/T(N_{m_0})$, with $\mu_0 : Y_\Sigma \to \mathbf{R}^{n-1}$ (rather than its restriction to Y'_Σ). Now in the proof of Theorem 2.2, it was shown that m_0 (as a regular function) descended to $\mu_0^{-1}(c)/T(N_{m_0})$ to give an isomorphism $m_0 : \mu_0^{-1}(c)/T(N_{m_0}) \to \mathbf{C}$. Then m_0 induces an isomorphism $m_0 : (\mu_0^{-1}(c) \cap Y'_\Sigma)/T(N_{m_0}) \to \mathbf{C} \setminus \{-1\}$. Furthermore $\log|1+m_0|$ gives an S^1-fibration $(\mu_0^{-1}(c) \cap Y'_\Sigma)/T(N_{m_0}) \to \mathbf{R}$. The inverse image of the general fibre of this map in $\mu_0^{-1}(c)$ is then $T^{n-1} \times S^1 = T^n$. Thus the general fibre of f' is an n-torus. Now orbits of $T(N_{m_0})$ only drop dimension when $m_0 = 0$. But when $m_0 = 0$, $\log|1+m_0| = 0$, and this makes it clear the discriminant locus coincides with that of f. •

Remark 2.5. (1) While Y'_Σ is certainly not a holomorphic partial compactification of Y_Σ (being contained in Y_Σ), it is possible to prove $Y'_\Sigma \to \mathbf{R}^n$ does coincide with a topological partial compactification of $Y_\Sigma \to \mathbf{R}^n$.

(2) All singular fibres $f'^{-1}(b)$ of the above fibration have the same basic structure: there is a fibration $f'^{-1}(b) \to S_1$ with all but one fibre a T^{n-1}, with the remaining fibre being a torus of dimension between 0 and $n-2$.

(3) Unlike the non-proper case, we should not expect there to exist a complete Kähler metric ω on Y'_Σ satisfying ω^n proportional to $\Omega' \wedge \bar{\Omega}'$.

3 Deformations of toric singularities

The deformation theory of toric Gorenstein singularities is controlled by the combinatorics of the corresponding cones. In particular, if Y_σ is an isolated toric singularity, then [1] gives a beautiful description of the versal deformation space of Y_σ. Irreducible components of this versal deformation space are in one-to-one correspondence with maximal Minkowski decompositions of the polytope P (P as in §2, with σ the cone over P).

Altmann's construction is as follows. Let N, M and σ be as in §2, and assume Y_σ has only an isolated singularity. Now σ is a cone over a polytope P contained in the affine hyperplane $\langle m_0, \cdot\rangle = 1$. By choosing some element n_0 such that $\langle m_0, n_0\rangle = 1$, we can identify P with $P - n_0$ in the hyperplane $L_\mathbf{R} \subseteq N_\mathbf{R}$ given by $L = m_0^\perp$. Let $P = R_0 + \cdots + R_p$ be a Minkowski decomposition of P inside $L_\mathbf{R}$. What this means is that $R_0, \ldots, R_p$ are convex subsets of L such that

$$P = \{r_0 + \cdots + r_p | r_i \in R_i\}.$$

Example 3.1. We focus on the prettiest example, a cone over a del Pezzo surface of degree 6. We can take σ to be generated by

$$n_1, \ldots, n_6 = (0,0,1), (1,0,1), (2,1,1), (2,2,1), (1,2,1), (0,1,1)$$

so that P in $\mathbf{R}^2$ is

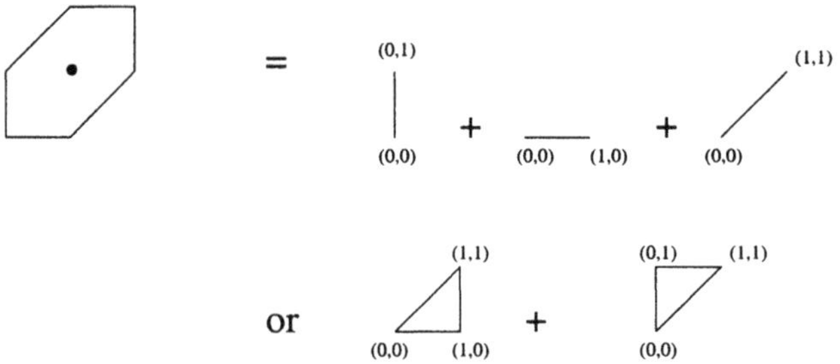

with two different Minkowski decompositions.

Now to each such Minkowski decomposition $P = R_0 + \cdots + R_p$, Altmann constructs a flat deformation of Y_σ as follows. Let $N' = L \oplus \mathbf{Z}^{p+1}$, and let $e_0, \ldots, e_p$ denote the standard basis of $\mathbf{Z}^{p+1}$ and define

$$\tilde{\sigma} = Cone\left(\bigcup_{k=0}^{p} (R_k \times \{e_k\})\right) \subseteq N'_{\mathbf{R}}$$

where $Cone(S)$ denotes the cone generated by the set $S \subseteq N'_{\mathbf{R}}$. If one writes $N = L \oplus \mathbf{Z}n_0$, there is a diagonal embedding $N \hookrightarrow N'$ given by $l + an_0 \mapsto l + a(e_0 + \cdots + e_p)$. Under this embedding $\sigma = \tilde{\sigma} \cap N_{\mathbf{R}}$, and hence we obtain a closed embedding $Y_\sigma \hookrightarrow Y_{\tilde{\sigma}}$. On the other hand, under the projection $N' \to \mathbf{Z}^{p+1}$, $\tilde{\sigma}$ maps to the cone generated by $e_0, \ldots, e_p$, and this induces a morphism $Y_{\tilde{\sigma}} \to \mathbf{C}^{p+1}$. Altmann proves the composed morphism $f : Y_{\tilde{\sigma}} \to \mathbf{C}^{p+1}/\mathbf{C}(1, \ldots, 1)$ is a flat deformation of Y_σ, with $f^{-1}(0) = Y_\sigma$. More explicitly, the surjection $N' \to \mathbf{Z}^{p+1}$ gives an inclusion $\mathbf{Z}^{p+1} \hookrightarrow M'$, with $e_0^*, \ldots, e_p^*$ mapping to elements of M' corresponding to characters $t_0, \ldots, t_p$. These characters extend to regular functions on $Y_{\tilde{\sigma}}$, and f is given by $(t_0 - t_1, \ldots, t_0 - t_p)$.

The main point for us then is that the functions $t_0, \ldots, t_p$ are invariant under the action of $T_{\mathbf{C}}(L) \subseteq T_{\mathbf{C}}(N')$, and thus $T_{\mathbf{C}}(L)$ acts on the fibres of f. This gives the desired T^{n-1}-action on deformations of Y_σ.

What about a $T(L)$-invariant holomorphic n-form on the fibres of f? Well note that $\tilde{\sigma}$ is a Gorenstein canonical cone; if $m_0' = e_0^* + \cdots + e_p^* \in M'$, then all generators of $\tilde{\sigma}$ evaluate to 1 on m_0'. Thus there is a nowhere vanishing holomorphic $n+p$-form Ω on the smooth part of $Y_{\tilde{\sigma}}$ (whose restriction to $T_{\mathbf{C}}(N') \subseteq Y_{\tilde{\sigma}}$ is described by Proposition 2.1). We then have

Proposition 3.2. *If $t_0, \ldots, t_p$ are coordinates on $\mathbf{C}^{p+1}$, and $\partial_{t_0}, \ldots, \partial_{t_p}$ are lifts of the corresponding vector fields to $Y_{\tilde{\sigma}}$, then for $x \in \mathbf{C}^{p+1}/\mathbf{C}(1, \ldots, 1)$,*

$$\Omega_x = (\iota(\partial_{t_1}, \ldots, \partial_{t_p})\Omega)|_{f^{-1}(x)}$$

is a well-defined nowhere vanishing holomorphic n-form on the non-singular part of $f^{-1}(x)$, which we write as $Y^{ns}_{\tilde{\sigma},x}$. In addition, Ω_x is $T(L)$-invariant. Finally, if $z_1, \ldots, z_{n-1}$ are a basis of characters for $T_{\mathbf{C}}(L)$, then $z_1, \ldots, z_{n-1}, t_0, \ldots, t_p$ form a basis of characters for $T_{\mathbf{C}}(N')$ and

$$\Omega|_{T_{\mathbf{C}}(N')} = \frac{dz_1 \wedge \cdots \wedge dz_{n-1}}{\prod z_i} \wedge dt_0 \wedge \cdots \wedge dt_p$$

so up to sign

$$\Omega_x|_{T_{\mathbf{C}}(N) \cap f^{-1}(x)} = \frac{dz_1 \wedge \cdots \wedge dz_{n-1}}{\prod z_i} \wedge dt_0.$$

Proof. That Ω_x is well-defined, independent of the lifts of the ∂_{t_i}'s is standard, and since $\partial_{t_1}, \ldots, \partial_{t_p}$ are linearly independent at a non-singular point of $f^{-1}(x)$, Ω_x is non-zero. Also, Ω is invariant under the action of $T(m_0'^{\perp})$, and $L \subseteq m_0'^{\perp}$, so Ω is invariant under $T(L)$. Since $t_1, \ldots, t_p$ are also invariant under $T(L)$, so is Ω_x. Finally, the explicit form for Ω follows from Proposition 2.1 and the explicit value for m_0'. •

Proposition 3.3. *If ω is a $T(L)$-invariant Kähler form on $Y^{ns}_{\tilde{\sigma},x}$, let $\mu : Y^{ns}_{\tilde{\sigma},x} \rightarrow L^*_{\mathbf{R}}$ be the moment map associated to the $T(L)$-action. Then $f : Y^{ns}_{\tilde{\sigma},x} \rightarrow \mathbf{R} \times L^*_{\mathbf{R}}$ given by $f = (\mathrm{Im}(i^{n+1}t_0), \mu)$ is a special Lagrangian fibration. Furthermore, f is surjective if μ is and the general fibre is diffeomorphic to $\mathbf{R} \times T(L)$. If x is represented by $(x_0, \ldots, x_p) \in \mathbf{C}^{p+1}$, then the discriminant locus is contained in the union of $p+1$ hyperplanes $\{\mathrm{Im}(i^{n+1}(x_0 - x_k)) | k = 0, \ldots, p\} \times L^*_{\mathbf{R}}$.*

Proof. The fact that f is special Lagrangian follows immediately from the form of Ω_x given in Proposition 3.1 and the same type of calculation as performed in Corollary 1.3. Now if $y \in Y^{ns}_{\tilde{\sigma},x}$ then $\mu^{-1}(\mu(y))/T(L)$ is isomorphic to the categorical quotient $Y^{ns}_{\tilde{\sigma},x}//T_{\mathbf{C}}(L)$, which is isomorphic to $\mathbf{C}$ with holomorphic coordinate t_0. Thus the fibre $f^{-1}(f(y))$ is an inverse image of a straight line in $\mathbf{C}$ under the quotient map $\mu^{-1}(\mu(y)) \to \mathbf{C}$. Thus the general fibre is $T(L) \times \mathbf{R}$. Clearly also f is surjective if μ is.

The discriminant locus Δ is the image of the union of $T(L)$ orbits of dimension $< n-1$. Now a subcone τ of dimension k of $\tilde{\sigma}$ corresponds to a codimension k orbit of $T_{\mathbf{C}}(N')$ which is fixed by $T_{\mathbf{C}}(\mathbf{R}\tau \cap N')$. Thus the $T(L)$ orbits on this stratum drop dimension if $\mathbf{R}\tau \cap L_{\mathbf{R}} \neq 0$. The one-dimensional faces of $\tilde{\sigma}$ are generated by $n \times e_k$ where n is a vertex of the polytope R_k. Thus a face τ of $\tilde{\sigma}$ has $\mathbf{R}\tau \cap L_{\mathbf{R}} \neq 0$ if and only if it contains two one-dimensional faces generated by $n_1 \times e_k$ and $n_2 \times e_k$ for some k, for n_1, n_2 two vertices of R_k. Necessarily n_1 and n_2 are the endpoints of an edge of R_k. Thus all minimal faces τ such that $\mathbf{R}\tau \cap L_{\mathbf{R}} \neq 0$ are two-dimensional faces spanned by $n_1 \times e_k$ and $n_2 \times e_k$. Now the function t_k is necessarily zero on the corresponding codimension 2 stratum. If this codimension 2 stratum is called $D_\tau \subseteq Y_{\tilde{\sigma}}$, then on $D_\tau \cap Y_{\tilde{\sigma},x}$, $t_k - t_0 = x_k - x_0$ so $t_0 = x_0 - x_k$. Thus $f(D_\tau \cap Y_{\tilde{\sigma},x})$ is contained in the hyperplane given by $\{\mathrm{Im}(i^{n+1}(x_0 - x_k))\} \times L^*_{\mathbf{R}}$. •

Example 3.4. Continuing with Example 3.1, the discriminant locus depends on the choice of the two decompositions. For general choice of x, the discriminant locus in the first splitting is contained in 3 different planes. There are three choices of two-dimensional τ yielding components of the discriminant locus, and for each τ, $D_\tau \cap Y_{\tilde{\sigma},x}$ (for general x, $Y_{\tilde{\sigma},x}$ is already non-singular) consists just of a $\mathbf{C}^*$. The image of this $\mathbf{C}^*$ under the moment map is a straight line, and the fibres of this map are connected (by [13]). Depending on the properties of μ, this image is either a line segment, a ray, or a line infinite in both directions. So Δ looks like

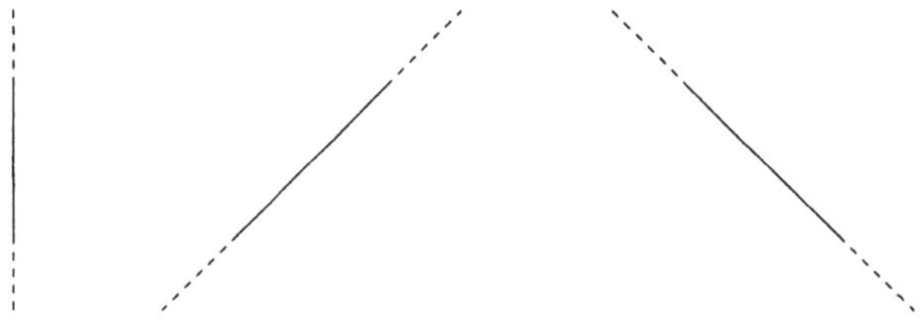

where each line is in a parallel plane. As $x \to 0$, these planes will converge to the same plane, producing, for suitable choice of ω, a discriminant locus for

$x = 0$ of

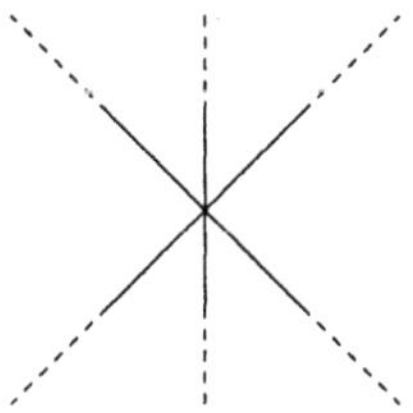

If instead we choose the second smoothing, then similar arguments show that Δ looks like

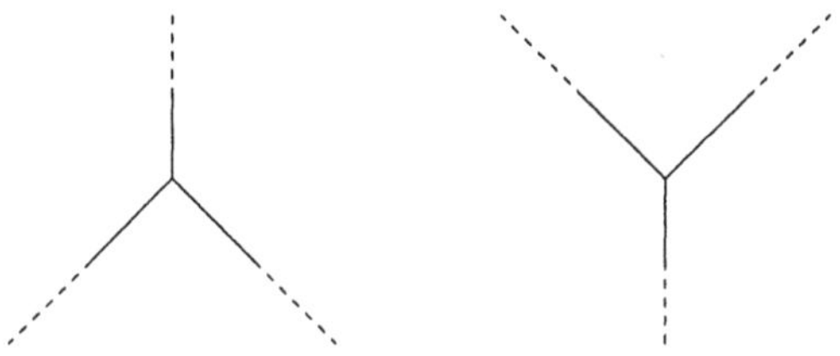

Again, we get the same picture as above as $x \to 0$.

If we take a crepant resolution of $Y_{\tilde{\sigma},0} = Y_\sigma$, then by §2 we obtain a discriminant locus which looks like

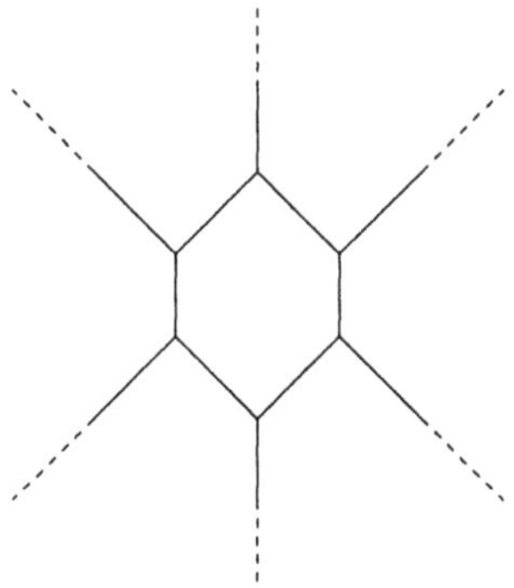

Another, simpler, example, is the ordinary double point, given by

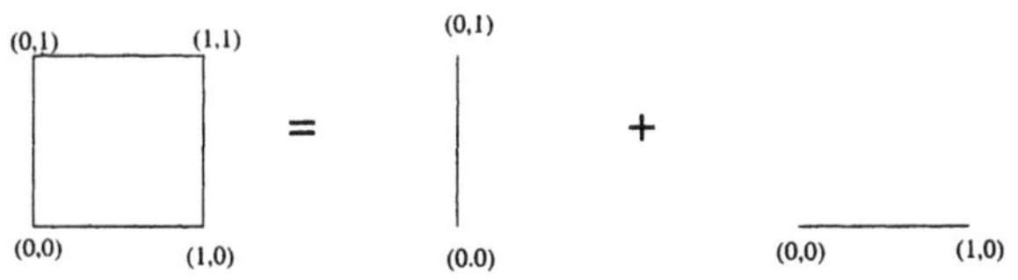

The discriminant locus of the smoothing consists of two lines in different planes. As $x \to 0$ these planes converge, and then there are two different small resolutions of the ordinary double point. This gives a picture

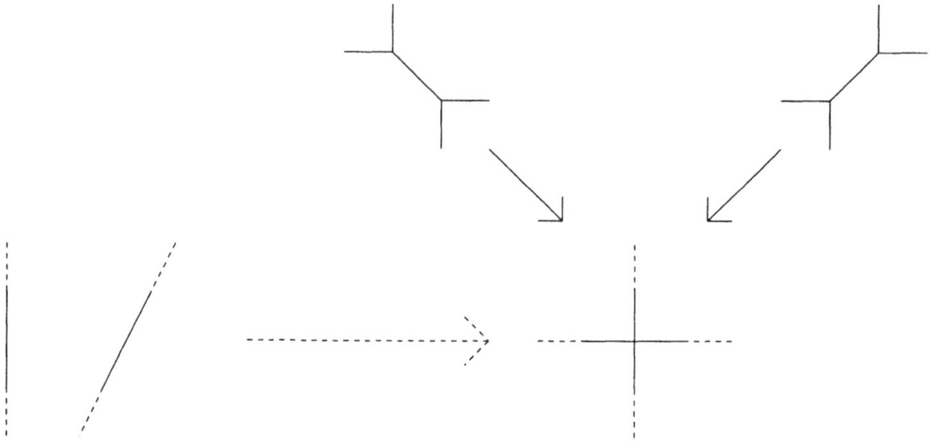

which shows how the discriminant locus changes under smoothing and resolution.

We have not discussed the choice of the metrics on these smoothings, but in some examples one can find Ricci-flat metrics. In the ordinary double point case, Stenzel [29] has constructed an explicit Ricci-flat metric, while in some examples [31,32] apply.

4 Local mirror symmetry and connections with the work of Ruan and Joyce

Let us begin this more informal discussion by asking the question: how do we construct the mirror to Y_Σ, where Y_Σ is as in §2? We shall focus on the three-dimensional case. This should make sense in the context of local mirror symmetry as developed in [4]. If we follow the SYZ philosophy, then we would need to construct duals of the special Lagrangian fibrations $f : Y_\Sigma \to \mathbf{R}^3$ of Theorem 2.2. The difficulty is that the general fibre of f is $T^2 \times \mathbf{R}$, which we can't dualize. However, we have at least constructed, via Theorem 2.4, a topological "properification" $f' : Y'_\Sigma \to \mathbf{R}^3$ of f. (We will ignore the metric properties of Y_Σ and Y'_Σ for the moment). We can then dualize f' topologically. Indeed, from the description of the singular fibres in Remark 2.5, (2), as well as the calculation of monodromy in [7], Example 1.2, we can see that f' has only semi-stable fibres, and in fact satisfies the hypotheses of Corollary 2.2 of [9]. Thus a topological dual $\check{f}: \check{Y}_\Sigma \to \mathbf{R}^3$ of $f': Y'_\Sigma \to \mathbf{R}^3$ exists.

However, it is worthwhile describing this dual explicitly. First, observe that the discriminant locus Δ of f' is a trivalent graph (homeomorphic to the 1-skeleton of $\partial\mu(Y_\Sigma)$). The fibres over the edges of the graph are of type $(2,2)$ in the notation of [8] and [9] (i.e. a product of a circle with a Kodaira type I_1 fibre) and type $(1,2)$ at all vertices[a]. Thus the dual $\check{f}: \check{Y}_\Sigma \to \mathbf{R}^3$ has only type $(2,2)$ and type $(2,1)$ fibres.

Let's describe the monodromy of the fibration f'. Fix a basis e_1, e_2, e_3 of N so that that $m_0 = e_1^* + e_2^* + e_3^*$ and take $f_1 = e_1 - e_2, f_2 = e_1 - e_3$ to be a basis for N_{m_0}. If we take any point $b \in \mathbf{R}^3 \setminus \Delta$, then the fibre $f'^{-1}(b)$ has a $T(N_{m_0})$-action which allows us to identify N_{m_0} with a sublattice of $H_1(f'^{-1}(b), \mathbf{Z})$. We can then choose an element $f_3 \in H_1(f'^{-1}(b), \mathbf{Z})$ such that f_1, f_2, f_3 form a basis for $H_1(f'^{-1}(b), \mathbf{Z})$. It is clear f_1 and f_2 will be monodromy invariant 1-cycles.

An edge l of Δ is the image under f' of a codimension-two $T_{\mathbf{C}}(N)$-orbit, which in turn corresponds to a dimension 2 face τ of Σ with generators n_1, n_2. It then follows from [9], Proposition 3.3 and Example 2.8, that if b is chosen near the edge l, $\gamma: S^1 \to \mathbf{R}^3 \setminus \Delta$ a suitably oriented simple loop about l based at b, and $n_1 - n_2 = a_1 f_1 + a_2 f_2$, then the monodromy transformation $T: H_1(f'^{-1}(b), \mathbf{Z}) \to H_1(f'^{-1}(b), \mathbf{Z})$ about γ is, in the basis f_1, f_2, f_3,

$$T = \begin{pmatrix} 1 & 0 & a_1 \\ 0 & 1 & a_2 \\ 0 & 0 & 1 \end{pmatrix}.$$

The topological dual $\check{f}: \check{Y}_\Sigma \to \mathbf{R}^3$ is constructed as follows. Let $N_{m_0}^* = \mathrm{Hom}_{\mathbf{Z}}(N_{m_0}, \mathbf{Z})$. Set $\bar{X} = T(N_{m_0}^*) \times \mathbf{R}^3$, and let $S \subseteq X$ be a topological surface constructed as follows. S projects to $\Delta \subseteq \mathbf{R}^3$, and for each edge l of Δ, $(T(N_{m_0}^*) \times l) \cap S$ is a cylinder fibering in circles over l, with the circle homotopic to $T((n_1 - n_2)^\perp) \subseteq T(N_{m_0}^*)$. At a vertex, which corresponds to a two-dimensional face of Σ, spanned by n_1, n_2 and n_3, the classes $n_1 - n_2, n_2 - n_3$ and $n_3 - n_1$ add to zero, and thus the three cylinders can be glued above the vertex. This gives a topological manifold S.

Now let $X = \bar{X} \setminus S$, $\pi: Y \to X$ a principal S^1-bundle with Chern class $(0, \pm 1) \in H^2(X, \mathbf{Z}) = H^2(\bar{X}, \mathbf{Z}) \oplus \mathbf{Z}$. Then there is a topological manifold $\bar{Y}$

[a]For generic good T^3-fibrations, which only have singular fibres of type $(2,2)$, $(2,1)$ and $(1,2)$, I am going to second Dave Morrison's suggestion that they be called generic, negative and positive singular fibres respectively, the words negative and positive referring to the sign of the Euler characteristic of the fibre. W.-D. Ruan introduced his own notation of type II and III for these singular fibres.

containing Y and a diagram

$$\begin{array}{ccc} Y & \hookrightarrow & \bar{Y} \\ \downarrow \pi & & \downarrow \bar{\pi} \\ X & \hookrightarrow & \bar{X} \end{array}$$

such that $\bar{\pi}$ is proper and the S^1-action on Y extends to an S^1-action on $\bar{Y}$, with $\bar{\pi}^{-1}(S) \cong S$ (see [9], Proposition 2.5). Taking $\check{Y}_\Sigma = \bar{Y}$ and $\check{f}$ the composition $\bar{Y} \to \bar{X} \to \mathbf{R}^3$, we obtain the topological dual of $f' : Y'_\Sigma \to \mathbf{R}^3$. These are dual in the sense that the monodromy representations are dual, which is the only topological measure of duality.

If we are only interested in the topology, this is the end of the story. But to get further insight into the picture, let us consider the local mirror symmetry picture of [4].

One way to interpret the suggestions of [4] is as follows: the mirror to Y_Σ (not Y'_Σ!) is a curve $C \subseteq (\mathbf{C}^*)^2 = T_\mathbf{C}(N^*_{m_0})$ whose Newton polygon is the translation of the polygon P into the plane $N_{m_0} \otimes \mathbf{R}$. Calculations of certain period integrals should yield the mirror map and predictions for Gromov-Witten invariants on Y_Σ. These period integrals are integrals of $\frac{dz_1 \wedge dz_2}{z_1 z_2}$ on $(\mathbf{C}^*)^2$ over 2-cycles with boundary on C. Such integrals satisfy standard Picard-Fuchs equations.

The basic claim is that the pair $S \subseteq T(N^*_{m_0}) \times \mathbf{R}^2$ contained in $\bar{X}$ (where $\mathbf{R}^2 \subseteq \mathbf{R}^3$ is the plane containing Δ) is the same as the pair $C \subseteq T_\mathbf{C}(N^*_{m_0})$. This will make the connection both between this circle of ideas and local mirror symmetry, as well as the connection with Ruan's work, clear.

To make this connection, we need to be more precise about our choice of the equation for C. First, choose a Kähler class on Y_Σ. (The choice of Kähler class determines $\Delta \subseteq \mathbf{R}^3$.) As in §2, this can be thought of as a strictly convex function $\phi : |\Sigma| \to \mathbf{R}$, which we can restrict to the polygon P. With coordinates z_1, z_2 on $(\mathbf{C}^*)^2$ corresponding to the basis f_1, f_2 of N_{m_0}, we can consider the family of curves C_t given by the equation $h_t = 0$, $t > 0$, where

$$h_t = \sum_{(a,b) \in P \cap N_{m_0}} t^{\phi(a,b)} m_{a,b} z_1^a z_2^b,$$

with $m_{a,b} \in \mathbf{C}$. Let $\nu : T_\mathbf{C}(N^*_{m_0}) \to \mathbf{R}^2$ be the moment map $\nu(z_1, z_2) = (\log|z_1|, \log|z_2|)$.

The following theorem is implicit in Ruan's work ([26,27]) and can also be proved using ideas of Viro [33] and Mikhalkin [24].

Theorem 4.1. *For $|t|$ close to zero, $\nu(C_t)$ is a fattening of the graph Δ, and there is a C^0-isotopy of $T_\mathbf{C}(N^*_{m_0}) = T(N^*_{m_0}) \times \mathbf{R}^2$ identifying C_t and S.*

Ruan uses this to construct torus fibrations on toric hypersurfaces. Thus he was led to his pictures of the discriminant loci of Lagrangian fibrations by looking at the *dual* picture to the fibrations $Y_\Sigma \to \mathbf{R}^3$ developed in [9] and here. These two points of view complement each other nicely.

Example 4.2. Let us continue with Example 2.3. Here $m_0 = (1,1,1)$, and we can take a basis f_1, f_2 of N_{m_0} with $f_1 = (2/3, -1/3, -1/3)$, $f_2 = (-1/3, 2/3, -1/3)$, so that if P is translated to N_{m_0}, we can take it to be the convex hull of f_1, f_2 and $-f_1 - f_2$. The set of integral points of P then corresponds to the monomials $z_1, z_2, 1$ and $z_1^{-1}z_2^{-1}$. We can take

$$h_t = t(z_1 + z_2 + z_1^{-1}z_2^{-1}) + 1,$$

in which case, for t small, the image of C_t under ν looks like

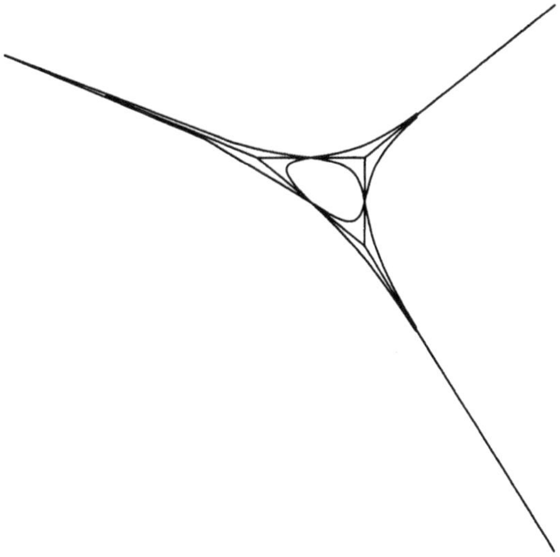

We have in fact only drawn the boundary of $\nu(C_t)$, and superimposed the codimension two discriminant locus Δ of $f : Y_\Sigma \to \mathbf{R}^3$ with symplectic form given by the monomial-divisor mirror map. This discriminant locus Δ is lying in the interior of $\nu(C_t)$. (The actual shape of Δ disagrees with the one drawn in Example 2.3, because we have used a different basis for N_{m_0}.)

Now, so far we have produced a topological fibration $\check{f} : \check{Y}_\Sigma \to \mathbf{R}^3$. How might we construct a special Lagrangian fibration? To do this we must first realise $\check{Y}_\Sigma$ as an (almost) Calabi-Yau manifold. We should expect this structure to be invariant under an S^1 action with fixed locus isomorphic to

C_t. Furthermore, we might expect this S^1 action to extend to a $\mathbf{C}^*$ action on $\check{Y}_\Sigma$. One way to accomplish this is as follows. Let h be a regular function on $(\mathbf{C}^*)^2$, and let

$$Y_h = \{(x, y, z_1, z_2) \in \mathbf{C}^2 \times (\mathbf{C}^*)^2 | xy = h(z_1, z_2)\}.$$

Y_h has a $\mathbf{C}^*$-action given by $(x, y, z_1, z_2) \mapsto (\lambda x, \lambda^{-1} y, z_1, z_2)$ for $\lambda \in \mathbf{C}^*$, and the fixed locus is the curve $x = y = h = 0$. We also need to choose a Kähler and holomorphic 3-form on Y_h. The holomorphic form will be

$$\Omega = i\frac{dx \wedge dz_1 \wedge dz_2}{xz_1z_2} = -i\frac{dy \wedge dz_1 \wedge dz_2}{yz_1z_2}$$

on Y_h. We have more choice for ω, but for convenience we will take

$$\omega = \frac{i}{2}\left(dx \wedge d\bar{x} + dy \wedge d\bar{y} + \frac{dz_1 \wedge d\bar{z}_1}{|z_1|^2} + \frac{dz_2 \wedge d\bar{z}_2}{|z_2|^2}\right).$$

(The form of the part of ω which is a pull-back from $(\mathbf{C}^*)^2$ is crucial.)

Now try to construct a special Lagrangian fibration on Y_h using Theorem 1.2. The moment map of the S^1 action is $\mu : Y_h \to \mathbf{R}$ given by $\mu(x, y, z_1, z_2) = |x|^2 - |y|^2$. We need, for each $c \in \mathbf{R}$, to find a special Lagrangian fibration on $\mu^{-1}(c)/S^1$. Now $\mu^{-1}(c)/S^1$ is canonically isomorphic to $(\mathbf{C}^*)^2$ as a complex manifold, and

$$\begin{aligned}\Omega_{red} &= \iota(2i(y\partial_y - x\partial_x))\Omega \\ &= 2\frac{dz_1 \wedge dz_2}{z_1z_2},\end{aligned}$$

while ω_{red} can be calculated with some effort to be

$$\omega_{red} = \frac{i}{2}\left(\frac{dz_1 \wedge d\bar{z}_1}{|z_1|^2} + \frac{dz_2 \wedge d\bar{z}_2}{|z_2|^2} + \frac{1}{\sqrt{c^2 + 4|h|^2}} dh \wedge d\bar{h}\right).$$

We need to find a special Lagrangian fibration on $((\mathbf{C}^*)^2, \omega_{red}, \Omega_{red})$ depending on the value of c. Now if we took the limit $c \to \infty$, we get to the case where

$$\omega_{red} = \frac{i}{2}\left(\frac{dz_1 \wedge d\bar{z}_1}{|z_1|^2} + \frac{dz_2 \wedge d\bar{z}_2}{|z_2|^2}\right),$$

and the moment map $\nu : (\mathbf{C}^*)^2 \to \mathbf{R}^2$ is then a special Lagrangian fibration. Thus the special Lagrangian fibrations for finite c, if they exist, should be viewed as a deformation of ν. For finite $c \neq 0$ one might hope to prove the existence of such a deformation using pseudo-holomorphic curve techniques and Gromov compactness. However, at $c = 0$ the almost Calabi-Yau structure becomes singular, and it is difficult to predict the structure of a special Lagrangian fibration on Y_h, if it should exist. The conjectural picture however is that there exists a special Lagrangian fibration $f : Y_h \to \mathbf{R}^3$ given by $(|x|^2 - |y|^2, g)$, where $g : Y_h \to \mathbf{R}^2$ induces a special Lagrangian fibration on each reduced space. Furthermore the discriminant locus of f will be contained in the hyperplane in $\mathbf{R}^3$ where the first coordinate is zero, and would be the image of $x = y = h = 0$ under the map f, i.e. the image of $h = 0$ in $(\mathbf{C}^*)^2$ under some deformation of the moment map ν.

Of course, if we are interested in a dual to $f' : Y'_\Sigma \to \mathbf{R}^3$ which is special Lagrangian, we should take $h = h_t$, identifying $\check{Y}_\Sigma$ with Y_{h_t} via Thereom 4.1. Getting more speculative, we recall that dualizing should exchange information at a deeper level, i.e. interchange the data of the symplectic structure on Y_Σ with the complex structure on $\check{Y}_\Sigma$. The precise correspondence is understood at the level of the mirror map, as calculated in [4]. I don't want to go into details here, but the main point is that one wants to compute periods of the holomorphic 3-form Ω on $\check{Y}_\Sigma$. Taking $\check{Y}_\Sigma = Y_{h_t}$, note that $H_3(\check{Y}_\Sigma, \mathbf{Z}) = H_3(\mu^{-1}(0), \mathbf{Z})$ (as gradient flow gives a retraction of $\check{Y}_\Sigma$ onto $\mu^{-1}(0)$), and in turn $H_3(\mu^{-1}(0), \mathbf{Z}) \cong H_2((\mathbf{C}^*)^2, C_t, \mathbf{Z})$. In other words a 3-cycle is the inverse image under the quotient map of a 2-chain with boundary in C_t. Integrating over such a 3-cycle reduces to integrating $\frac{dz_1 \wedge dz_2}{z_1 z_2}$ over such a 2-chain. These period integrals satisfy the relevant Picard-Fuchs equations and define the right mirror map as explored in [4]. This perhaps explains why $C_t \subseteq (\mathbf{C}^*)^2$ should be considered as the mirror of Y_Σ.

As one further intuitive observation in this direction, let us try to explain why it makes sense to consider $C_t \subseteq (\mathbf{C}^*)^2$ to be the mirror of Y_Σ rather than Y'_Σ. T-duality, to first approximation, should exchange long and short distances in the fibres. Thus the fibres of $f : Y_\Sigma \to \mathbf{R}^3$ should be viewed as a limit of tori with greater and greater radius in one direction, so in the limit the fibre is $\mathbf{R} \times T^2$ rather than T^3. The T-dual fibres should have this radius approaching zero. Intuitively, it then appears natural to divide Y_{h_t} by the S^1-action, as the S^1's should correspond to the "small" direction.

This discussion should not be taken too seriously. It is clear that there is much to understand, but I believe this circle of ideas and examples will prove to be an excellent laboratory for exploring the more intricate questions surrounding the SYZ conjecture.

Perhaps the most pressing question is the following. It was originally my and many others' hope that special Lagrangian fibrations would be reasonably differentiable, and differentiability implies certain conditions on the discriminant locus. For example, the discriminant locus is Hausdorff codimension 2 if the map is C^∞ (see [9], §1). However, Joyce has now given in [19] examples of special Lagrangian fibrations which are only piecewise differentiable, and whose discriminant locus is codimension one. In fact, his basic example is of a very similar flavour to the S^1-invariant setup above. From these examples, I believe the likelihood is that if $f : Y_h \to \mathbf{R}^3$ exists, it is only piecewise smooth, and the discriminant locus is amoeba-like rather than a graph. Thus if there is a special Lagrangian fibration $f : Y_{h_t} \to \mathbf{R}^3$, it is only a perturbation of the topological fibration $\check{f} : \check{Y}_\Sigma \to \mathbf{R}^3$.

5 The Future of the SYZ Conjecture

It is clear that Joyce's picture forces us to reconsider the full strength version of the SYZ conjecture, as opposed to the topological ones considered in [9] and [26]. The notion of dualizing topological torus fibrations as developed in [9] will not be the right one. We would expect that if a mirror pair $X, \check{X}$ possess special Lagrangian fibrations $f : X \to B$, $\check{f} : \check{X} \to B$, they will have different amoeba-like discriminant loci Δ and $\check{\Delta}$. They will presumably be dual only in a relatively crude topological sense in that the monodromy representations $\rho : \pi_1(B\backslash\Delta) \to SL_3(\mathbf{Z})$ and $\check{\rho} : \pi_1(B\backslash\check{\Delta}) \to SL_3(\mathbf{Z})$ are dual representations.

It is my current belief that the SYZ conjecture will make most sense in a limiting picture. First let us recall from [15] certain structures which appear naturally on moduli spaces of special Lagrangian submanifolds. Let (X, ω, Ω) be a Calabi-Yau manifold, B a moduli space of deformations of some special Lagrangian submanifold on X, along with a universal family

$$\begin{array}{ccc} \mathcal{U} & \hookrightarrow & X \times B \\ \downarrow{\scriptstyle f} & & \\ B & & \end{array}$$

Let $p : \mathcal{U} \to X$ be the projection. All fibres of f are assumed to be smooth submanifolds of X. Then from [23], we know B is smooth, with a canonical identification of $\mathcal{T}_{B,b}$ with $\mathcal{H}^1(f^{-1}(b), \mathbf{R})$, the space of $\mathbf{R}$-valued harmonic one-forms, on $f^{-1}(b)$. This is via the map $v \in \mathcal{T}_{B,b} \mapsto \iota(v)p^*\omega$, where v is pulled back to a vector field normal to $f^{-1}(b)$ in $\mathcal{U}$.

There are two important structures on B:

1. an integral affine structure. If $U \subseteq B$ is a contractible open set, with coordinates $t_1, \ldots, t_n$, let $\gamma_1, \ldots, \gamma_n \in H_1(f^{-1}(b), \mathbf{Z})$ be a basis for first

homology varying continuously with b. Then the 1-forms α_i given by

$$\partial/\partial t_j \mapsto \int_{\gamma_i} \iota(\partial/\partial t_j)p^*\omega$$

on B are closed and linearly independent ([15], Proposition 1). Thus there exists a coordinate system $y_1, \ldots, y_n$ on U with $\alpha_i = dy_i$, and these coordinates are well-defined up to integral affine transformations (elements of $\mathbf{R}^n \rtimes GL_n(\mathbf{Z})$). This defines an integral affine structure on B.

2. There is a metric on B (the McLean metric) given by

$$g(\partial/\partial t_i, \partial/\partial t_j) = -\int_{f^{-1}(b)} \iota(\partial/\partial t_i)p^*\omega \wedge \iota(\partial/\partial t_j)p^* \operatorname{Im}\Omega.$$

(One might want to normalize this metric in various ways).

Hitchin showed there is a compatability between the metric and affine structure: locally there exists a function K such that $g(\partial/\partial y_i, \partial/\partial y_j) = \partial^2 K/\partial y_i \partial y_j$. Kontsevich and Soibelman [20] call this structure on B_0 of affine structure plus metric of this form an *affine Kähler* (AK) manifold. Such a manifold was called *Hessian* in earlier work of H. Shima: see [28] and references therein. If in addition the function K satisfies the real Monge-Ampère equation $\det \partial^2 K/\partial y_i \partial y_j = constant$, Kontsevich and Soibelman call such an Hessian manifold a *Monge-Ampère manifold.*

Next recall the definition of Gromov-Hausdorff convergence.

Definition 5.1. *Let (X, d_X), (Y, d_Y) be two compact metric spaces. Suppose there exists maps $f : X \to Y$ and $g : Y \to X$ (not necessarily continuous) such that for all $x_1, x_2 \in X$,*

$$|d_X(x_1, x_2) - d_Y(f(x_1), f(x_2))| < \epsilon$$

and for all $x \in X$,

$$d_X(x, g \circ f(x)) < \epsilon,$$

and the two symmetric properties for Y hold. Then we say the Gromov–Hausdorff distance between X and Y is at most ϵ. The Gromov–Hausdorff distance $d_{GH}(X, Y)$ is the infinum of all such ϵ.

There are two distinct situations we might want to apply this notion. The first was discussed independently by myself and Wilson [10] and Kontsevich and Soibelman in [20]. The second situation follows naturally from these ideas. First, let $\mathcal{X} \to \Delta$ be a flat family of degenerating Calabi-Yau n-folds, with $0 \in \Delta$ a large complex structure limit point (or maximally unipotent boundary point). Let $t_i \in \Delta$ be a sequence of points converging to $0 \in \Delta$, and let g_i on $\mathcal{X}_{t_i}$ be a Ricci-flat metric normalized so that $Diam(\mathcal{X}_{t_i}, g_i)$ remains constant. Then general results about Gromov-Hausdorff convergence tell us that a subsequence of $(\mathcal{X}_{t_i}, g_i)$ converges to a metric space (X_∞, g_∞).

Second, consider another sequence of metric spaces, whose existence (or rather non-emptiness) is currently conjectural. Suppose that for t_i sufficiently close to 0, there is a special Lagrangian T^n whose homology class is invariant under monodromy near 0. (This is a property we expect to find of fibres of a special Lagrangian fibration associated to a large complex structure limit point). Let $B_{0,i}$ be the moduli space of deformations of this torus, every point of $B_{0,i}$ corresponding to a smooth torus in $\mathcal{X}_{t_i}$. The manifold $B_{0,i}$ comes equipped with the McLean metric. We should then compactify $B_{0,i} \subseteq B_i$ in some manner: probably taking the closure of $B_{0,i}$ in the space of special Lagrangian currents on $\mathcal{X}_{t_i}$ is the right thing to do. This should give a series of metric spaces (B_i, d_i), which again, if the McLean metric is normalized properly to keep the diameter constant, may have a convergent subsequence, converging to a compact metric space (B_∞, d_∞).

The following is a slight souping up of the conjectures in [10] and [20].

Conjecture 5.2. *If $(\mathcal{X}_{t_i}, g_i)$ converges to (X_∞, g_∞) and (B_i, d_i) is non-empty for large i and converges to (B_∞, d_∞), then B_∞ and X_∞ are isometric up to scaling. Furthermore, there is a subspace $B_0 \subseteq B_\infty$ with $\Delta = B_\infty \setminus B_0$ of Hausdorff codimension 2 in B_∞ such that B_0 is a Monge-Ampère manifold, with the metric inducing d_∞ on B_0.*

This is a considerably weaker conjecture than the original full-strength SYZ proposal on the existence of special Lagrangian fibrations. But following the philosophy of Kontsevich and Soibelman, this should be sufficient for most purposes.

Remarks 5.2. (1) This conjecture doesn't assume the existence of special Lagrangian fibrations on $\mathcal{X}_{t_i}$ for any i. It would of course be nice if this is the case, but taking Joyce's philosophy seriously means we only see the codimension 2 structure in the limit. We expect that as $i \to \infty$, the area of the critical locus of a special Lagrangian fibration on $\mathcal{X}_{t_i}$ goes to zero, so its image hopefully deforms to something of codimension two.

Even once one finds a single special Lagrangian torus, it could fail to give a fibration either because deformations may not be disjoint from each other, or the deformations simply may not fill out the entire manifold, so that B_i has a boundary. The expectation might be that these sorts of things are more likely to happen near the discriminant locus in B_∞.

(2) We do not expect $B_{0,i}$ to be a Monge-Ampère manifold, but only an affine Kähler manifold. This is because of the existence of examples of moduli of special Lagrangian tori where this is not the case: see the work of Matessi in [22].

(3) Stated in the proper way, [10] proves this conjecture for K3 surfaces.

The philosophy of Kontsevich and Soibelman, which I believe is the right one, is that it may be enough to work purely with the limiting data. Whereas the original form of the SYZ conjecture proposed dualizing torus fibrations, we instead dualize the limiting data. Given a Hessian manifold, one obtains a new affine structure with local affine coordinates $\check{y}_i = \partial K/\partial y_i$, where K is the potential of the metric. The metric remains the same, but the new potential $\check{K}$ is the Legendre transform of K. This was first suggested in the context of mirror symmetry by Hitchin in [15], and this idea was used effectively in [20] and [21][b].

We are left with two fundamental questions:

The Limit Question. *How can one calculate the limit data, or guess it conjecturally either for general degenerations or for standard cases such as hypersurfaces in toric varieties?*

The Reconstruction Question. *Given a set of limiting data, how do we reconstruct a family of Calabi-Yau manifolds converging to this limit?*

Kontsevich and Soibelman discuss these two questions, suggesting some approaches involving rigid analytic geometry and Berkovich spaces. From my point of view, these questions can be developed at a topological level (where one only preserves the limiting information of monodromy about Δ), a symplectic level (pay attention only to the affine structure) and the full metric level. For the limit question, [9] gives a conjectural limit for the quintic and its mirror on the topological level, while [26] gives it for general toric hypersurfaces. The affine structure can be guessed at from the ideas in [26], and in future work I will give a purely combinatorial description of a conjectural affine structure in the limit for hypersurfaces in toric varieties.

For the reconstruction question, the results of [9] allow a reconstruction of the underlying topological manifold from the limiting data in sufficiently generic cases, while current work in progress of my own explores the symplectic reconstruction problem.

[b] Intriguingly, this duality was mentioned in [28], which gave a reference to a 1985 work in statistics, [2], which makes serious use of this duality between Hessian manifolds.

However, solving these general questions at the metric (or complex structure) level will require some substantial new ideas.

Acknowledgments

I would like to thank N. Hitchin, D. Joyce, D. Morrison, and P.M.H. Wilson for useful conversations. This work has been supported by grants from the NSF and EPSRC

References

1. Altmann, K., "The Versal Deformation of an Isolated Toric Gorenstein Singularity," Inv. Math. **128**, 443 (1997).
2. Amari, S., *Differential-Geometric Methods in Statistics, Lecture Notes in Statistics,* **28**, Springer-Verlag, 1985.
3. Audin, M., *The Topology of Torus Actions on Symplectic Manifolds, Prgress in Mathematics, 93*, Birkhäuser Verlag, Basel, 1991.
4. Chiang, T.-M., Klemm, A., Yau, S.-T. and Zaslow, E., "Local Mirror Symmetry: Calculations and Interpretations," preprint, hep-th/9903053.
5. Goldstein, E., "Calibrated Fibrations," preprint, math/9911093.
6. Goldstein, E., "Calibrated Fibrations on Complete Manifolds via Torus Action," preprint, math/0002097.
7. Gross, M., "Special Lagrangian Fibrations I: Topology," in *Integrable Systems and Algebraic Geometry*, eds. M.-H. Saito, Y. Shimizu and K. Ueno, World Scientific, 1998, 156.
8. Gross, M., "Special Lagrangian Fibrations II: Geometry," *Surveys in Differential Geometry,* Somerville: MA, International Press, 1999, 341.
9. Gross, M., "Topological Mirror Symmetry," preprint, math.AG/9909015 (1999), to appear in *Inv. Math.*
10. Gross, M., and Wilson, P.M.H., "Large Complex Structure Limits of K3 Surfaces," preprint, math.DG/0008018.
11. Harvey, R., and Lawson, H.B. Jr., "Calibrated Geometries," Acta Math. **148**, 47 (1982).
12. Haskins, M., "Special Lagrangian Cones," preprint, math.DG/0005164.
13. Heinzner, P., and Huckleberry, A., "Kählerian Potentials and Convexity Properties of the Moment Map," Inv. Math. **126**, 65 (1996).
14. Hilgert, J., Neeb, K.-H., and Plank, W., "Symplectic Convexity Theorems and Coadjoint Orbits," Comp. Math. **94**, 129 (1994).
15. Hitchin, N., "The Moduli Space of Special Lagrangian Submanifolds," Ann. Scuola Norm. Sup. Pisa Cl. Sci. (4) **25**, 503 (1997).

16. Joyce, D., "Asymptotically Locally Euclidean Metrics with Holonomy $SU(m)$," preprint, math.AG/9905041.
17. Joyce, D., "Quasi-ALE Metrics with Holonomy $SU(m)$ and $Sp(m)$," preprint, math.AG/9905043.
18. Joyce, D., "Special Lagrangian m-folds in $\mathbf{C}^m$ with Symmetries," preprint, math.DG/0008021.
19. Joyce, D., "Singularities of Special Lagrangian Fibrations and the SYZ Conjecture," preprint, math.DG/0011179.
20. Kontsevich, M., and Soibelman, Y., "Homological Mirror Symmetry and Torus Fibrations," preprint, math.SG/0011041.
21. Leung, N.C., "Mirror Symmetry Without Corrections," preprint, math.DG/0009235.
22. Matessi, D., "Some Families of Special Lagrangian Tori," preprint, math.DG/0011061.
23. McLean, R.C., " Deformations of Calibrated Submanifolds," Comm. Anal. Geom. **6**, 705 (1998).
24. Mikhalkin, G., "Real Algebraic Curves, the Moment Map, and Amoebas," Ann. of Math. **151**, 309 (2000).
25. Reid, M., "Decomposition of Toric Morphisms," *Arithemetic and Geometry, Vol. II*, 395, *Progr. Math. 36*, Birkhäuser Boston, 1983.
26. Ruan, W.-D., "Lagrangian Torus Fibration and Mirror Symmetry of Calabi-Yau Hypersurface in Toric Variety," preprint, math.DG/0007028.
27. Ruan, W.-D., "Newton Polygon and String Diagram," preprint, math.DG/0011012.
28. Shima, H., and Yagi, K., "Geometry of Hessian Manifolds," Differential Geom. Appl. **7**, 277 (1997).
29. Stenzel, M., "Ricci-flat Metrics on the Complexification of a Compact Rank One Symmetric Space," Manuscripta Math. **80**, 151 (1993).
30. Strominger, A., Yau, S.-T., and Zaslow, E., "Mirror Symmetry is T-Duality," t Nucl. Phys. B **479**, 243 (1996).
31. Tian, G., and Yau, S.-T., "Complete Kähler manifolds with zero Ricci curvature, I." J. Amer. Math. Soc. **3**, 579 (1990).
32. Tian, G., and Yau, S.-T., "Complete Kähler manifolds with zero Ricci curvature, II." Invent. math. **106**, 27 (1991).
33. Viro, O., "Gluing of Plane Real Algebraic Curves and Constructions of Curves of Degrees 6 and 7," in *Topology (Leningrad, 1982)*, 187, *Lecture Notes in Math., 1060*, Springer-Verlag, 1984.

LINEAR MODELS OF SUPERSYMMETRIC D-BRANES

KENTARO HORI
Jefferson Physical Laboratory, Harvard University
Cambridge, MA 02138, U.S.A.

We construct a class of supersymmetric boundary interactions in $\mathcal{N}=2$ field theories on the half-space, which depend on parameters that are not at all renormalized or not renormalized in perturbation theory beyond one-loop. This can be used to study D-branes wrapped on a certain class of Lagrangian submanifolds as well as holomorphic cycles. The construction of holomorphic D-branes is in close relationship with the background independent open string field theory approach to brane/anti-brane systems. As an application, mirror pairs of Lagrangian and holomorphic D-branes are identified. The mirror pairs are studied by twisting to open topological field theories.

1 Introduction

Let us consider the 1+1 dimensional $U(1)$ gauge theory of charge 1 complex scalar fields $\phi_1, \ldots, \phi_N$ with the following action

$$S = \frac{1}{2\pi}\int \mathrm{d}^2x \left\{ -\sum_{i=1}^{N} |D_\mu \phi_i|^2 + D\left(\sum_{i=1}^{N} |\phi_i|^2 - r\right) + \theta v_{01} \right\}. \tag{1.1}$$

Here D is an auxiliary field and $v_{01} = \partial_0 v_1 - \partial_1 v_0$ is the fieldstrength of a $U(1)$ gauge potential v_μ. Eliminating the auxiliary field D we obtain the constraint

$$\sum_{i=1}^{N} |\phi_i|^2 = r. \tag{1.2}$$

One can also solve for v_μ and, after modding out by the $U(1)$ gauge group action, we obtain the non-linear sigma model whose target space is the complex projective space $\mathbf{CP}^{N-1}$. The Theta term θv_{01} becomes the B-field term which is non-zero for a topologically non-trivial field configuration. The above gauge theory is called the linear sigma model for $\mathbf{CP}^{N-1}$.

Linear sigma models have played important roles in understanding several dynamical aspects of quantum field theories. In recent years, $\mathcal{N} = 2$ supersymmetric linear sigma models in $1 + 1$ dimensions have been used effectively to understand some of the key aspects of supersymmetric non-linear sigma models and related models. An advantage in the construction [1] is that the parameters that are not renormalized or renormalized only at the one-loop level are explicitly identified and simply realized. For instance, in the supersymmetric generalization of the model (1.1), the complex combination

$$t = r - i\theta \tag{1.3}$$

appears in a twisted superpotential and it is manifest that it is renormalized only at the one-loop level. Sigma model on a hypersurface of $\mathbb{C}\mathrm{P}^{N-1}$ can also be realized as a gauge theory and the complex structure parameters enter into the superpotential term, which is not renormalized and is decoupled from the Kahler class parameters. One can make a precise statement on the theory only after such an identification of parameters is made. The proof of mirror symmetry in [2] makes use of this advantage of the construction.

It is natural to ask whether there is a similar construction for theories formulated on a worldsheet with boundaries. The boundary conditions and interactions are the new ingredient. Such theories are the relevant models in defining and analyzing open strings or D-branes in string theory. Recently, great progress has been made in understanding several aspects of supersymmetric D-branes [3] in Calabi-Yau and non-geometric compactification. In particular, a lot has been understood and clarified from the worldsheet points of view [4–19]. (Earlier relevant works are in [20, 21]. See also [22–24] for some studies from other approaches.) However, we still lack a worldsheet formulation with boundary interactions depending on parameters that are not renormalized or simply renormalized. We expect to learn a lot more by having such a formulation.[1] The aim of this paper is to construct such a formulation.

As a basic example, let us consider a D1-brane located at the circle $|z|^2 = c$ in the complex plane $\mathbf{C}$. The worldsheet theory is described in terms of a

[1] A similar point was also made in [26].

complex scalar field ϕ with the action $S = -\frac{1}{2\pi}\int d^2x|\partial_\mu\phi|^2$. We impose the following condition at the worldsheet boundary

$$\begin{aligned} &(\mathrm{D}) \quad |\phi|^2 = c, \\ &(\mathrm{N}) \quad \partial_n \arg(\phi) = 0, \end{aligned} \tag{1.4}$$

where ∂_n is the normal derivative. One can also consider adding the boundary term in the action

$$S_a = \frac{a}{2\pi}\int_{\partial\Sigma} d\arg(\phi), \tag{1.5}$$

which is non-zero for a topologically non-trivial configuration. The boundary condition breaks the scale invariance of the bulk theory and the parameter c is renormalized at the one loop level as $c(\mu) = c(\mu') + \log(\mu/\mu')$ [27] (as reviewed in Secion 3.3). In the supersymmetric generalization of this model, the axial $U(1)$ R-symmetry is anomalous or the axial rotation shifts the parameter a. This suggests that the parameters c and a are superpartners of each other and it is natural to consider the complex combination

$$s = c - ia. \tag{1.6}$$

We will construct a boundary interaction that induces (1.5) and also yields the boundary condition (1.4) in such a way that the parameter s enters in some kind of superpotential term on the boundary. The basic idea is to introduce auxiliary degrees of freedom, just like D and v_μ in the $\mathbf{CP}^{N-1}$ model, but living only on the worldsheet boundary. When applied to more general bulk theories, this will lead to the construction of supersymmetric boundary theories for D-branes wrapped on a class of Lagrangian submanifolds of the target space. The parameters will be protected from (higher) loop corrections although they can receive non-perturbative corrections.

Another type of supersymmetric boundary theories correspond to D-branes wrapped on holomorphic cycles, or those supporting holomorphic vector bundles. We will construct simple models that realize a certain class of such D-branes, where the parameters characterizing the D-branes enter into a boundary superpotential term. The construction is closely related to the worldsheet approach to the problem of the tachyon condensation of unstable D-brane systems [28–30]. (The approach of the latter references is in the context of the background independent open string field theory [31–33].) In this construction, we will also have boundary degrees of freedom as the essential ingredients. Our construction gives rise to a simple and explicit realization of a certain class of complexes of sheaves that has been argued to be the basic elements in supersymmetric D-branes [17].

The rest of the paper is organized as follows. In Section 2, we introduce superspace formalism on the worldsheet with boundary. In Section 3, the supersymmetric version of the D1-brane discussed above is studied and it is shown that (1.6) is the chiral parameter. We then construct a "linear model" for this basic example. In Section 4, we apply the construction to certain supersymmetric gauge theories. This leads to the worldsheet definition of a certain class of A-type Lagrangian D-branes in toric manifolds. We also find the mirror description of such D-branes. In Section 5 we construct the boundary interactions corresponding to B-type holomorphic D-branes. We start with space-filling brane/anti-brane system and find the condition for $\mathcal{N} = 2$ worldsheet supersymmetry. In Section 6, we study B-type D-branes in LG models. We especially study in detail the properties of D0-branes which are the mirrors of toroidal Lagrangian D-branes.

2 $\mathcal{N} = 2$ Boundary Superspace

In this paper, we consider $1+1$ dimensional field theories with $(2,2)$ supersymmetry in the bulk, a half of which is preserved by the boundary conditions or boundary interactions. In order to make the supersymmetry structure manifest and to identify parameters that are not renormalized or do not receive perturbative renormalization beyond one loop, it is convenient to introduce the superspace formalism on the worldsheet with boundaries.

2.1 (2, 2) Superspace

To fix the notation, we briefly describe here the superspace formalism for (2, 2) supersymmetry in the bulk.

The (2, 2) superspace has four fermionic coordinates θ^+, θ^-, $\overline{\theta}^+$, $\overline{\theta}^-$, in addition to the bosonic coordinates x^0, x^1. The supersymmetry transformations are generated by the following differential operators on the superspace,

$$\mathcal{Q}_\pm = \frac{\partial}{\partial\theta^\pm} + i\overline{\theta}^\pm\,\partial_\pm, \tag{2.1}$$

$$\overline{\mathcal{Q}}_\pm = -\frac{\partial}{\partial\overline{\theta}^\pm} - i\theta^\pm\,\partial_\pm, \tag{2.2}$$

where $\partial_\pm$ are differentiations by $x^\pm = x^0 \pm x^1$;

$$\partial_\pm = \frac{\partial}{\partial x^\pm} = \frac{1}{2}\left(\frac{\partial}{\partial x^0} \pm \frac{\partial}{\partial x^1}\right). \tag{2.3}$$

These differential operators obey the anti-commutation relations $\{\mathcal{Q}_\pm, \overline{\mathcal{Q}}_\pm\} = -2i\partial_\pm$. We introduce another set of differential operators

$$D_\pm = \frac{\partial}{\partial\theta^\pm} - i\overline{\theta}^\pm\,\partial_\pm, \tag{2.4}$$

$$\overline{D}_\pm = -\frac{\partial}{\partial\overline{\theta}^\pm} + i\theta^\pm\,\partial_\pm, \tag{2.5}$$

which anti-commute with $\mathcal{Q}_\pm$ and $\overline{\mathcal{Q}}_\pm$. These obey the similar anti-commutation relations $\{D_\pm, \overline{D}_\pm\} = 2i\partial_\pm$. *Vector R-rotation* and *axial R-rotation* of a superfield are defined by

$$\mathrm{e}^{i\alpha F_V} : \mathcal{F}(x^\mu, \theta^\pm, \overline{\theta}^\pm) \mapsto \mathrm{e}^{i\alpha q_V}\,\mathcal{F}(x^\mu, \mathrm{e}^{-i\alpha}\theta^\pm, \mathrm{e}^{i\alpha}\overline{\theta}^\pm) \tag{2.6}$$

$$\mathrm{e}^{i\beta F_A} : \mathcal{F}(x^\mu, \theta^\pm, \overline{\theta}^\pm) \mapsto \mathrm{e}^{i\beta q_A}\,\mathcal{F}(x^\mu, \mathrm{e}^{\mp i\beta}\theta^\pm, \mathrm{e}^{\pm i\beta}\overline{\theta}^\pm), \tag{2.7}$$

where q_V and q_A are numbers called vector R-charge and axial R-charge of $\mathcal{F}$. A *chiral superfield* Φ is a superfield which satisfies the equations, $\overline{D}_\pm\Phi = 0$. It has the follwing expansion ($y^\pm := x^\pm - i\theta^\pm\overline{\theta}^\pm$)

$$\Phi(x^\mu, \theta^\pm, \overline{\theta}^\pm) = \phi(y) + \theta^\alpha\psi_\alpha(y) + \theta^+\theta^- F(y). \tag{2.8}$$

A ***twisted chiral superfield*** $\widetilde{\Phi}$ is a superfield which satisfies $\overline{D}_+\widetilde{\Phi} = D_-\widetilde{\Phi} = 0$. It has the following expansion ($\widetilde{y}^{\pm} := x^{\pm} \mp i\theta^{\pm}\overline{\theta}^{\pm}$)

$$\widetilde{\Phi}(x^\mu, \theta^\pm, \overline{\theta}^\pm) = \widetilde{\phi}(\widetilde{y}) + \theta^+\overline{\chi}_+(\widetilde{y}) + \overline{\theta}^-\chi_-(\widetilde{y}) + \theta^+\overline{\theta}^- E(\widetilde{y}). \qquad (2.9)$$

There are three kind of action functionals of superfields which are invariant under the supersymmetry transformation $\delta = \epsilon_+ Q_- - \epsilon_- Q_+ - \overline{\epsilon}_+\overline{Q}_- + \overline{\epsilon}_-\overline{Q}_+$. One is the D-term integral

$$\int d^2x\, d^4\theta\, K(\mathcal{F}_i) = \int d^2x\, d\theta^+ d\theta^- d\overline{\theta}^- d\overline{\theta}^+\, K(\mathcal{F}_i), \qquad (2.10)$$

where $K(-)$ is an arbitrary differentiable function of arbitrary superfields $\mathcal{F}_i$. The next is the F-term integral

$$\int d^2x d^2\theta\, W(\Phi_i) = \int d^2x\, d\theta^- d\theta^+\, W(\Phi_i)\Big|_{\overline{\theta}^\pm = 0}, \qquad (2.11)$$

where $W(\Phi_i)$ is a holomorphic function of chiral superfields Φ_i which is called a superpotential. The third is the twisted F-term integral

$$\int d^2x\, d^2\widetilde{\theta}\, \widetilde{W}(\widetilde{\Phi}_i) = \int d^2x\, d\overline{\theta}^- d\theta^+\, \widetilde{W}(\widetilde{\Phi}_i)\Big|_{\overline{\theta}^+ = \theta^- = 0}, \qquad (2.12)$$

where $\widetilde{W}(\widetilde{\Phi}_i)$ is a holomorphic function of twisted chiral superfields $\widetilde{\Phi}_i$ which is called a twisted superpotential.

Chiral superfields cannot enter into twisted F-term and twisted chiral superfields cannot enter into F-term. Using the argument of Seiberg [35], one can also show that the parameters that enters into the superpotential (resp. twisted superpotential) at a high energy scale cannot enter into the twisted superpotential (resp. superpotential) in the effective action at a lower energy.

2.2 Superspace Boundaries

Let us put a (2, 2) supersymmetric field theory on the "left half plane"

$$\Sigma = \mathbf{R} \times (-\infty, 0], \qquad (2.13)$$

where $\mathbf{R}$ and $(-\infty, 0]$ are parametrized by the time and spatial coordinates $-\infty < x^0 < +\infty$, $-\infty < x^1 \le 0$ respectively. The left half plane Σ has its boundary $\partial\Sigma$ at

$$x^1 = 0. \tag{2.14}$$

We suppose that there is an analogous "boundary" in the fermionic coordinates $\theta^\pm, \overline{\theta}^\pm$ as well. We now determine what kind of boundary is possible.[1]

To define a consistent theory, one must impose some boundary condition on the fields. The boundary condition usually relates the left moving modes and the right moving modes. In particular, the left moving and the right moving fermions are related to each other. This suggests that the boundary relates the fermionic coordinates $\theta^+, \overline{\theta}^+$ and the other coordinates $\theta^-, \overline{\theta}^-$. There are essentially two ways to relate them;

$$\text{(A)} \qquad \theta^+ + e^{i\alpha}\overline{\theta}^- = 0, \quad \overline{\theta}^+ + e^{-i\alpha}\theta^- = 0, \tag{2.15}$$

$$\text{(B)} \qquad \theta^+ - e^{i\beta}\theta^- = 0, \quad \overline{\theta}^+ - e^{-i\beta}\overline{\theta}^- = 0. \tag{2.16}$$

In the above expression, $e^{i\alpha}$ and $e^{i\beta}$ are fixed phases. In most of the following discussion, we shall set these phases trivial, $e^{i\alpha} = 1$, $e^{i\beta} = 1$. We call (2.14) and (2.15) *A-boundary* while (2.14) and (2.16) will be called *B-boundary*. As we will see below, these superspace boundaries corresponds to theories in which the following combinations of the supercharges are conserved;

$$\text{(A)} \qquad \overline{Q} = \overline{Q}_+ + e^{i\alpha}Q_-, \quad Q = Q_+ + e^{-i\alpha}\overline{Q}_-, \tag{2.17}$$

$$\text{(B)} \qquad \overline{Q} = \overline{Q}_+ + e^{i\beta}\overline{Q}_-, \quad Q = Q_+ + e^{-i\beta}Q_-. \tag{2.18}$$

We shall call the former A-type supersymmetry and the latter B-type supersymmetry.

---

[1] A part of the construction in this section has been made by E. Martinec [36]. In particular, (2.15) and (2.16) was stated in [36]. We thank him for letting us know about it. Boundary superspace was also considered in [37] for $\mathcal{N} = 1$ supersymmetry.

2.2.1 A-boundary

We develop here the superspace formalism for A-boundary. We set $e^{i\alpha} = 1$ but the generalization is straightforward. We first introduce the fermionic coordinates at the A-boundary as

$$\theta := \theta^+ = -\overline{\theta}^-, \quad \overline{\theta} := \overline{\theta}^+ = -\theta^-. \tag{2.19}$$

The boundary (2.14)-(2.15) is preserved by ∂_0 and the following differential operators

$$\overline{D} = \overline{D}_+ + D_- = -\frac{\partial}{\partial\overline{\theta}} + i\theta\partial_0, \tag{2.20}$$

$$D = D_+ + \overline{D}_- = \frac{\partial}{\partial\theta} - i\overline{\theta}\partial_0, \tag{2.21}$$

$$\overline{\mathcal{Q}} = \overline{\mathcal{Q}}_+ + \mathcal{Q}_- = -\frac{\partial}{\partial\overline{\theta}} - i\theta\partial_0, \tag{2.22}$$

$$\mathcal{Q} = \mathcal{Q}_+ + \overline{\mathcal{Q}}_- = \frac{\partial}{\partial\theta} + i\overline{\theta}\partial_0, \tag{2.23}$$

This indeed shows that the supersymmetry preserved by the full theory should be (2.17). We also note that the boundary (2.15) is invariant under the axial R-rotation whereas the vector R-rotation rotates the phase $e^{i\alpha}$ defining the boundary (2.15).

The differential operators obey the anti-commutation relations $\{D, \overline{D}\} = 2i\partial_0$, $D^2 = \overline{D}^2 = 0$, and $\{\mathcal{Q}, \overline{\mathcal{Q}}\} = -2i\partial_0$, $\mathcal{Q}^2 = \overline{\mathcal{Q}}^2 = 0$. A boundary superfield is a function of the boundary coordinates x^0, θ and $\overline{\theta}$, which transforms under A-type supersymmetry (with parameter $\epsilon_+ = \overline{\epsilon}_- = \epsilon$, $\overline{\epsilon}_+ = \epsilon_- = \overline{\epsilon}$) by $\delta = \epsilon\overline{\mathcal{Q}} - \overline{\epsilon}\mathcal{Q}$. The boundary R-rotation transforms the boundary superfield $\mathcal{F}$ as

$$\mathcal{F}(x^0, \theta, \overline{\theta}) \mapsto e^{iq\gamma}\mathcal{F}(x^0, e^{-i\gamma}\theta, e^{i\gamma}\overline{\theta}), \tag{2.24}$$

where q is called the boundary R-charge of $\mathcal{F}$. A boundary superfield Φ is called a *boundary chiral superfield* if it obeys

$$\overline{D}\Phi = 0. \tag{2.25}$$

A boundary chiral superfield has the following Theta expansion

$$\Phi = \phi(x^0) + \theta\psi(x^0) - i\theta\overline{\theta}\partial_0\phi(x^0). \tag{2.26}$$

We often call a fermionic boundary chiral superfield a ***boundary Fermi superfield***. Given a function $J(\mathcal{F}_i)$ of boundary superfields $\mathcal{F}_i$, the following integral is invariant under the supersymmetry variation δ

$$\int \mathrm{d}x^0 \mathrm{d}\theta \mathrm{d}\overline{\theta}\, J(\mathcal{F}_i). \tag{2.27}$$

Also, given a boundary Fermi superfield Ψ and a holomorphic function $\mathcal{V}(\Phi_i)$ of boundary chiral superfields Φ_i, the integral

$$\int \mathrm{d}x^0 \mathrm{d}\theta \; \Psi \mathcal{V}(\Phi) \Big|_{\overline{\theta}=0}, \tag{2.28}$$

is invariant under the supersymmetry δ. We call $\mathcal{V}(\Phi_i)$ a ***boundary superpotential***. The above term is invariant under the boundary R-rotation if $\Phi\mathcal{V}(\Phi_i)$ has boundary R-charge 1. We shall sometimes refer to (2.27) and (2.28) as boundary D-term and boundary F-term respectively.

A bulk superfield restricted to the boundary (2.14)-(2.15) is a boundary superfield. The boundary R-rotation comes from the axial R-rotation in the bulk. It is easy to see that a bulk twisted chiral superfield restricted on the A-boundary is a boundary chiral superfield. The boundary superpotential must be a holomorphic function of the boundary chiral superfields. This strongly constrains a possible form of quantum corrections, as in [35], as we will see explicitly in several examples.

2.2.2 B-boundary

We briefly repeat the same thing for B-boundary (we again set $\mathrm{e}^{i\beta} = 1$). The fermionic coordinates at the B-boundary are

$$\theta := \theta^+ = \theta^-, \quad \overline{\theta} := \overline{\theta}^+ = \overline{\theta}^-. \tag{2.29}$$

The following differential operators preserve the B-boundary (2.14)-(2.16);

$$\overline{D} = \overline{D}_+ + \overline{D}_- = -\frac{\partial}{\partial \overline{\theta}} + i\theta\partial_0, \tag{2.30}$$

$$D = D_+ + D_- = \frac{\partial}{\partial \theta} - i\overline{\theta}\partial_0, \tag{2.31}$$

$$\overline{Q} = \overline{Q}_+ + \overline{Q}_- = -\frac{\partial}{\partial\overline{\theta}} - i\theta\partial_0, \tag{2.32}$$

$$Q = Q_+ + Q_- = \frac{\partial}{\partial\theta} + i\overline{\theta}\partial_0, \tag{2.33}$$

This shows that the supersymmetry preserved by the full theory should be (2.18). We also note that the boundary (2.16) is invariant under the vector R-rotation whereas the axial R-rotation rotates the phase $e^{i\beta}$ defining the boundary (2.16).

One can develop the boundary superfield formalism identically as in the case of A-boundary. B-type supersymmetry transformation (with parameter $\epsilon_+ = -\epsilon_- = \epsilon, \overline{\epsilon} = \overline{\epsilon}_+ = -\overline{\epsilon}_-$) of the superfields is given by $\delta = \epsilon Q - \overline{\epsilon}\overline{Q}$. A bulk superfield restricted to the boundary (2.14)-(2.16) is a boundary superfield. The boundary R-rotation comes from the vector R-rotation in the bulk. A bulk chiral superfield restricted on the B-boundary is a boundary chiral superfield.

2.3 (1, 1) Superspace and its Boundary

It is useful also to introduce the $(1,1)$ superspace and its boundary. The $(1,1)$ superspace can be defined as a subspace of the $(2,2)$ superspace:

$$\theta^{\pm} = i\theta_1^{\pm}, \quad \theta_1^{\pm} \text{ real}. \tag{2.34}$$

(Again, there is a freedom to change the phase i to $i\,e^{i\nu_\pm}$ but we set $\nu_\pm = 0$ for simplicity.) This subspace can also be defined by the equations $\theta^{\pm} + \overline{\theta}^{\pm} = 0$ which are preserved by the differential operators

$$Q_{\pm}^1 := Q_{\pm} + \overline{Q}_{\pm} = -i\frac{\partial}{\partial\theta_1^{\pm}} + 2\theta_1^{\pm}\partial_{\pm}, \tag{2.35}$$

$$D_{\pm}^1 := D_{\pm} + \overline{D}_{\pm} = -i\frac{\partial}{\partial\theta_1^{\pm}} - 2\theta_1^{\pm}\partial_{\pm}, \tag{2.36}$$

These obey the anti-commutation relations such as $\{Q_{\pm}^1, Q_{\pm}^1\} = -4i\partial_{\pm}$, $\{D_{\pm}^1, D_{\pm}^1\} = 4i\partial_{\pm}$, and $\{Q_{\alpha}^1, D_{\beta}^1\} = 0$.

The boundary of $(1,1)$ superspace can be defined as the subspace with $x^0 = 0, \pi$ and $\theta_1^+ = \pm\theta_1^-$. We take here the plus sign, $\theta_1^+ = \theta_1^-$, so that both A-boundar and B-boundary (with trivial phases) of the (2, 2) superspace includes this $\mathcal{N} = 1$ boundary as the subspace

$$\theta + \overline{\theta} = 0. \tag{2.37}$$

This subspace is preserved by the differential operators

$$\mathcal{Q}^1 := \mathcal{Q}^1_+ + \mathcal{Q}^1_- = \mathcal{Q} + \overline{\mathcal{Q}} = -i\frac{\partial}{\partial\theta_1} + 2\theta_1\partial_0, \tag{2.38}$$

$$D^1 := D^1_+ + D^1_- = D + \overline{D} = -i\frac{\partial}{\partial\theta_1} - 2\theta_1\partial_0, \tag{2.39}$$

where $\theta_1 := \operatorname{Im}\theta$ is the fermionic coordinate of the $\mathcal{N} = 1$ boundary.

It is straighforward to show that for an $\mathcal{N} = 2$ boundary superfield $\mathcal{F}$ and for a boundary Fermi superfield Ψ, we have the identities

$$\int \mathrm{d}\theta \mathrm{d}\overline{\theta}\, \mathcal{F} = -\frac{i}{2}\int \mathrm{d}\theta_1 \left[(D - \overline{D})\mathcal{F}\right]_1, \tag{2.40}$$

$$\int \mathrm{d}\theta\ \Psi\Big|_{\overline{\theta}=0} = -i\int \mathrm{d}\theta_1 [\Psi]_1, \tag{2.41}$$

where $[-]_1$ stands for the restriction on the $\mathcal{N} = 1$ subspace (2.37).

3 A Linear Model: An Example

In this section, we consider the supersymmetric version of the D1-brane in the complex plane $\mathbb{C}$ which was introduced for the bosonic case in Section 1. We first take the standard approach to the worlsheet theory based on supersymmetric boundary condition. There we identify the chiral parameter of the theory. We then move on to construct "a linear model" of such a D-brane. The basic idea is to introduce boundary degrees of freedom and boundary interactions that impose the boundary condition and also induce the Wilson line term. We construct the action so that the chiral parameter enters into a boundary F-term.

The supersymmetric worldsheet theory includes a Dirac fermion fields $\psi_\pm$, $\overline{\psi}_\pm$ in addition to the complex scalar field ϕ. The action of the system is given by

$$S=\frac{1}{2\pi}\int \mathrm{d}^2x\left(|\partial_0\phi|^2-|\partial_1\phi|^2+\frac{i}{2}\overline{\psi}_-(\overleftrightarrow{\partial_0}+\overleftrightarrow{\partial_1})\psi_-+\frac{i}{2}\overline{\psi}_+(\overleftrightarrow{\partial_0}-\overleftrightarrow{\partial_1})\psi_+\right),\tag{3.1}$$

where $\overline{\psi}\overleftrightarrow{\partial_\mu}\psi:=\overline{\psi}\partial_\mu\psi-(\partial_\mu\overline{\psi})\psi$. If the worldsheet has no boundary, the action is invariant under the $(2,2)$ supersymmetry transformations. This is manifest if we express the action in the $(2,2)$ superspace. Let Φ be the chiral superfield which has an expansion

$$\Phi=\phi(y)+\theta^\alpha\psi_\alpha(y)+\theta^+\theta^-F(y).\tag{3.2}$$

The action (3.1) is obtained from

$$S=\frac{1}{2\pi}\int \mathrm{d}^2x\,\mathrm{d}^4\theta\,\overline{\Phi}\Phi,\tag{3.3}$$

after an appropriate partial integration and elimination of the auxiliary field F by its equation of motion.

3.1 The Boundary Condition

We now consider the D-brane located at the circle $|\phi|^2=c$ in this supersymmetric theory. We thus formulate the theory on the left half plane $\Sigma=\mathbf{R}\times(-\infty,0]$, and we will find a boundary condition at $\partial\Sigma$ so that the theory is invariant under A-type supersymmetry. In the bosonic theory, the D-brane was represented by the boundary condition (1.4) at $x^1=0$. We claim that the boundary condition in the supersymmetric theory is

$$\overline{\Phi}\Phi=c\quad\text{at A-boundary.}\tag{3.4}$$

By "at A-boundary", we mean *at A-boundary of the* $(2,2)$ *superspace*: $x^1=0$, $\theta^+=-\overline{\theta}^-=\theta$ and $\overline{\theta}=\overline{\theta}^+=-\theta^-$ (we set $e^{i\alpha}=1$ in (2.15)). In this way of writing, the condition itself is manifestly invariant under A-type supersymmetry.

In terms of the component fields, the condition (3.4) is expressed as

$$\left.\begin{aligned}&|\phi|^2 = c,\\&i\overline{\phi}\overleftrightarrow{\partial_1}\phi + \overline{\psi}_+\psi_+ - \overline{\psi}_-\psi_- + \overline{F}\phi + \overline{\phi}F = 0,\\&\overline{\phi}\psi_- + \overline{\psi}_+\phi = 0,\\&\overline{\psi}_-\phi + \overline{\phi}\psi_+ = 0,\end{aligned}\right. \quad \text{at } x^1 = 0. \tag{3.5}$$

We note that the first equation is identical to the Dirichlet boundary condition for $|\phi|$ in (1.4) while the second equation generalizes the Neumann boundary condition for $\arg(\phi)$ in (1.4). It is straightforward to show that the action (3.1) plus the auxiliary term $\frac{1}{2\pi}\int d^2x|F|^2$ is invariant under A-type supersymmetry $\delta = \epsilon\overline{Q} - \overline{\epsilon}Q$ which acts on the component fields as

$$\begin{aligned}&\delta\phi = \epsilon\psi_- - \overline{\epsilon}\psi_+, && \delta\overline{\phi} = \epsilon\overline{\psi}_+ - \overline{\epsilon}\overline{\psi}_-,\\&\delta\psi_+ = \epsilon(2i\partial_+\phi + F), && \delta\overline{\psi}_+ = \overline{\epsilon}(-2i\partial_+\overline{\phi} + \overline{F}),\\&\delta\psi_- = \overline{\epsilon}(-2i\partial_-\phi + F), && \delta\overline{\psi}_- = \epsilon(2i\partial_-\overline{\phi} + \overline{F}),\\&\delta F = -2i\epsilon\partial_+\psi_- - 2i\overline{\epsilon}\partial_-\psi_+. && \delta\overline{F} = -2i\overline{\epsilon}\partial_+\overline{\psi}_- - 2i\epsilon\partial_-\overline{\psi}_+.\end{aligned} \tag{3.6}$$

Also, one can show that the equation of motion remains the same as the standard one

$$(\partial_0^2 - \partial_1^2)\phi = 0, \quad (\partial_0 \pm \partial_1)\psi_\mp = 0, \quad F = 0, \tag{3.7}$$

under the boundary condition (3.5).[1]

3.2 The Boundary Term

Due to the boundary condition $|\phi|^2 = c$ with non-zero c,

$$\varphi := \arg(\phi) \quad \text{at } x^1 = 0 \tag{3.8}$$

[1]We notice a slight discrepancy of (3.5) from the condition given in [13] which would require $\overline{\phi}\partial_1\phi - \partial_1\overline{\phi}\phi = 0$. This is because the requirements (3.2) and (3.3) in [13] was too strong. The most general condition for the variation is an average of (3.2) and (3.3) but not the separate ones.

is well-defined up to 2π shifts. Then, it is possible to add to the action (3.1) the following boundary term

$$S_a = \int_{\partial\Sigma} \frac{a}{2\pi} \partial_0 \varphi \, \mathrm{d}x^0. \tag{3.9}$$

Since it is a total derivative in the boundary coordinate, it is a topological term. In particular it cannot break the supersymmetry of the system. Thus the system with the action $S+S_a$ is still invariant under A-type supersymmetry. The equation of motion also remains the same as (3.7).

This boundary term represents the interaction of the open string end points and the $U(1)$ gauge field on the D-brane which has holonomy e^{ia} along the worldvolume S^1.

3.3 Renormalization and R-Anomaly

Renormalization of c

As mentioned in the introduction, the boundary condition breaks the scale invariance of the bulk theory and the constant c runs as the scale is varied. The renormalization group flow for the D-brane location was found in [27] to be the mean curvature flow: Let us consider a D-brane whose worldvolume is embedded in the space-time by the map $f^I(\zeta^\alpha)$ (where I and α are the space-time and the worldvolume indices). The one-loop beta functional for the embedding function $f^I(\zeta^\alpha)$ is given by

$$\beta^I = \mu \frac{\mathrm{d}}{\mathrm{d}\mu} f^I = -h^{\alpha\beta} K^I_{\alpha\beta}, \tag{3.10}$$

where $h_{\alpha\beta}$ is the induced metric and $K^I_{\alpha\beta}$ is the extrinsic curvature that appears in the normal coordinate expansion $f^I = \partial_\alpha f^I \zeta^\alpha + \frac{1}{2} K^I_{\alpha\beta} \zeta^\alpha \zeta^\beta + \cdots$. In the present case, the embedding function for our circle $|\phi|^2 = c$ is given by $x = \sqrt{2c}\cos\theta$, $y = \sqrt{2c}\sin\theta$ where x and y are the normal coordinates on the complex ϕ-plane, $\phi = (x+iy)/\sqrt{2}$, and θ is the angular coordinates of the circle. This shows that $h^{\theta\theta} = 1/2c$ and, say at $\theta = 0$, $K^x_{\theta\theta} = -\sqrt{2c}$, $K^y_{\theta\theta} = 0$.

Thus, we have $\mu\frac{\mathrm{d}}{\mathrm{d}\mu}\sqrt{2c} = 1/\sqrt{2c}$. In other words, the parameter c at the cut-off scale Λ_{UV} and the one at a lower energy scale μ are related by

$$c(\Lambda_{\mathrm{UV}}) = c(\mu) + \log(\Lambda_{\mathrm{UV}}/\mu), \tag{3.11}$$

at the one-loop level. This is the story for the bosonic model, but the fermions does not affect the running of c at the one-loop level, as in the case of the RG flow of the metric in the bulk non-linear sigma models [34].

Axial Anomaly

There is a related quntum effect; the anomaly of the boundary R-symmetry. The bulk theory is invariant under both vector and axial R-rotations. As mentioned in the previous section, A-boundary is broken by the vector R-rotation but is preserved by the axial R-rotation. The axial rotation (with the trivial R-charge for Φ) acts trivially on the bosonic fields ϕ and F but non-trivially on the fermions as

$$\psi_\pm \to \mathrm{e}^{\mp i\gamma}\psi_\pm, \quad \overline{\psi}_\pm \to \mathrm{e}^{\pm i\gamma}\overline{\psi}_\pm, \tag{3.12}$$

and it indeed preserves the boundary condition (3.5). Thus, the boundary R-rotation that comes from the axial R-rotation is a symmetry of the classical theory. However, in the quantum theory it is broken by an anomaly. This can be seen by counting, as follwos, the number of fermion zero modes in a topologically non-trivial backgroun.

We note from (3.12) that ψ_- and $\overline{\psi}_+$ has R-charge 1 while ψ_+ and $\overline{\psi}_-$ has R-charge -1. Thus, we are interested in the index which is the difference of the number of $(\psi_-, \overline{\psi}_+)$-zero modes and that of $(\psi_+, \overline{\psi}_-)$-zero modes.

Let Σ be the Euclidean left half plane $\mathrm{Re}(z) \leq 0$ with the canonically flat metric $\mathrm{d}s^2 = |\mathrm{d}z|^2$. We would like to count the above index in a background ϕ in which the worldsheet Σ is mapped to the complex plane so that the image of the boundary $\partial\Sigma$ (the imaginary axis $\mathrm{Re}(z) = 0$) winds k-times around the circle $|\phi|^2 = c$ where the D-brane is located. The left half-plane is mapped by the conformal map $w = (1 + z)/(1 - z)$ to the unit disk $|w| \leq 1$ where the boundary $\text{Re}(z) = 0$ is mapped to the disk boundary $|w| = 1$ with the infinity mapped to $w = -1$. Now, one can choose the configuration to be

$$\phi(z, \bar{z}) = \sqrt{c} w^k. \tag{3.13}$$

We note that the positive and the negative chirality spinors are identified along the boundary $\partial\Sigma$ as $(\mathrm{d}z)^{\frac{1}{2}} = (\mathrm{d}\bar{z})^{\frac{1}{2}}$. [1] In terms of the coordinate w this is translated as $(\mathrm{d}w)^{\frac{1}{2}}/(w+1) = (\mathrm{d}\overline{w})^{\frac{1}{2}}/(\overline{w}+1)$ along $|w| = 1$. The boundary conditions on the fermions

$$\overline{\phi}\psi_- + \overline{\psi}_+\phi = 0, \quad \overline{\psi}_-\phi + \overline{\phi}\psi_+ = 0, \tag{3.14}$$

should be understood under such an identification. The zero modes obey the Cauchy-Riemann equations — ψ_- and $\overline{\psi}_-$ are holomorphic in z or w and ψ_+ and $\overline{\psi}_+$ are anti-holomorphic — and they must be regular in the disc $|w| \leq 1$. Thus, they can be expanded as

$$\psi_- = \sum_{n=0}^{\infty} c_n w^n \, (\mathrm{d}w)^{\frac{1}{2}}, \quad \overline{\psi}_- = \sum_{n=0}^{\infty} b_n w^n \, (\mathrm{d}w)^{\frac{1}{2}}, \tag{3.15}$$

$$\psi_+ = \sum_{n=0}^{\infty} \bar{c}_n \overline{w}^n \, (\mathrm{d}\overline{w})^{\frac{1}{2}}, \quad \overline{\psi}_+ = \sum_{n=0}^{\infty} \bar{b}_n \overline{w}^n \, (\mathrm{d}\overline{w})^{\frac{1}{2}}. \tag{3.16}$$

The boundary condition (3.14) then requires

$$(1 + \mathrm{e}^{i\sigma}) \sum_{n=0}^{\infty} c_n \, \mathrm{e}^{i(n-k)\sigma} + (1 + \mathrm{e}^{-i\sigma}) \sum_{n=0}^{\infty} \bar{b}_n \, \mathrm{e}^{-i(n-k)\sigma} = 0, \tag{3.17}$$

$$(1 + \mathrm{e}^{i\sigma}) \sum_{n=0}^{\infty} b_n \, \mathrm{e}^{i(n+k)\sigma} + (1 + \mathrm{e}^{-i\sigma}) \sum_{n=0}^{\infty} \bar{c}_n \, \mathrm{e}^{-i(n+k)\sigma} = 0, \tag{3.18}$$

where σ is the angular part of the polar coordinates $w = |w| \, \mathrm{e}^{i\sigma}$ so that $\phi = \sqrt{c} \, \mathrm{e}^{ik\sigma}$ along the boundary. It is easy to see that there is only a trivial solution $b_n = \bar{c}_n = 0$ for the second equation but there are $2k$ non-singular solutions to

[1] The reason we do not start from "$\Sigma =$ the flat disk" is that the boundary would have an extrinsic curvature and the Wick rotation of the boundary condition on the fermions would not be straightforward, the point emphasized in [13]. We could of course have started from the disk with the hemi-sphereical metric. It is easy to see that the result is the same as the one given below.

the first one; $c_{i-1} + \bar{b}_{2k-i} = 0$ $(i = 1, 2, \ldots, 2k)$. Thus the index we wanted to know is $2k$ in the background (3.13).

This means that the path-integral measure changes as

$$\mathcal{D}\psi_\pm \mathcal{D}\overline{\psi}_\pm \longrightarrow \mathrm{e}^{-2ki\gamma} \mathcal{D}\psi_\pm \mathcal{D}\overline{\psi}_\pm \tag{3.19}$$

under the R-rotation (3.12). Namely, the classical R-symmetry is anomalously broken. We note here that the boundary term S_a yields the following path-integral weight in this background;

$$\exp\left(i \int_{\partial\Sigma} \frac{a}{2\pi} \mathrm{d}\varphi \right) = \exp(ika). \tag{3.20}$$

Thus, the effect of the R-rotation is the shift of a as

$$a \longrightarrow a - 2\gamma. \tag{3.21}$$

This also shows that the parameter a is not actually a physical parameter of the present theory but can be absorbed by a field redefinition.

The Chiral Parameter

The one-loop running (3.11) of the parameter c and the R-anomaly (3.21) suggests that the parameters c and a are superpartners of each other and can be combined into a complex parameter. The precise combination can be found by identifying the instantons and computing the action. A configuration such that ϕ is holomorphic and $F = 0$ preserves half of the supersymmetry. (See the ϵ-variation in the Euclidean version of (3.6).) The configuration (3.13) is an example of such an instanton. The Euclidean action in such a background is

$$\begin{aligned} S_E &= \frac{1}{2\pi} \int_\Sigma \left(2|\partial_z \phi|^2 + 2|\partial_{\bar{z}} \phi|^2 \right) \mathrm{d}^2 z - i \frac{a}{2\pi} \int_{\partial\Sigma} \mathrm{d}\varphi \\ &= \frac{c}{2\pi} \int_{|w| \leq 1} 2|\partial_w w^k|^2 \mathrm{d}^2 w - i \frac{a}{2\pi} 2\pi k = (c - ia)k. \end{aligned} \tag{3.22}$$

This shows that the right complex combination is

$$s = c - ia. \tag{3.23}$$

3.4 A Linear Model

We now construct a linear model of this A-type D-brane. The general structure of the worldsheet action we would like to have is as follows. It is a sum of two parts,

$$S_{tot} = S(\Phi) + S_{bd}(U, \Phi) \tag{3.24}$$

where $S(\Phi)$ involves only the bulk field Φ and includes the bulk action (3.1) while $S_{bd}(U, \Phi)$ is the boundary interaction of Φ and a boundary superfield U that imposes the boundary condition (3.4) at low enough energies. The boundary condition on Φ is not something we impose by hand at the beginning, but is regarded as derived through the interaction with the boundary fields. We in particular require S_{tot} to be supersymmetric without using boundary condition on Φ nor its equation of motion. Also, we would like $S_{bd}(U, \Phi)$ to be manifestly supersymmetric so that the parameter s in (3.23) appears in the boundary superpotential. This requires $S(\Phi)$ to be supersymmetric by itself.

We start with finding the boundary term. Let us introduce a real bosonic boundary superfield

$$U = u + \theta\overline{\chi} - \overline{\theta}\chi + \theta\overline{\theta}E. \tag{3.25}$$

The Lagrangian

$$L_{bd}^{(1)} = \int \mathrm{d}\theta \mathrm{d}\overline{\theta}\, (\overline{\Phi}\Phi - c)U, \tag{3.26}$$

imposes the boundary condition $\overline{\Phi}\Phi = c$ via the equation of motion for U. This however does not lead to the a-term (3.9) nor does it make clear that $s = c - ia$ is a chiral parameter. We note that one can replace the term $\int \mathrm{d}\theta\mathrm{d}\overline{\theta}\,(-c)U$ by $\frac{1}{2}\int \mathrm{d}\theta c\overline{D}U$ + c.c.. This motivates us to define the "fieldstrength" Υ of U as

$$\Upsilon := \overline{D}U = \chi + \theta(E + i\partial_0 u) - i\theta\overline{\theta}\partial_0\chi, \tag{3.27}$$

which is a boundary Fermi superfield, $\overline{D}\Upsilon = 0$. Now we try the following Lagrangian

$$L_{bd}^{(2)} = \int \mathrm{d}\theta\mathrm{d}\overline{\theta}\, \overline{\Phi}\Phi U + \mathrm{Re}\int \mathrm{d}\theta\, s\Upsilon. \tag{3.28}$$

in which s is manifestly a chiral parameter. In terms of the component fields the boundary superpotential term can be written as

$$\mathrm{Re}\int \mathrm{d}\theta\, s\Upsilon = c\,E\, +\, a\,\partial_0 u. \tag{3.29}$$

We indeed find an a-term. However, we should note that u has no relationship with $\varphi = \arg(\phi)$ at this stage. Furthermore, if u is a single valued field (an ordinary field with values in $\mathbf{R}$), the term $a\partial_0 u$ can be set equal to zero since it is a total derivative. We may declare that u has the same periodicity as φ or that $u - \varphi$ is a single valued field. However, that would make the term $\int \mathrm{d}\theta\mathrm{d}\overline{\theta}\,\overline{\Phi}\Phi U$ ill-defined; it changes its value as $u \to u + 2\pi$.

Putting aside this problem, let us try to determine the term $S(\Phi)$. We would like it to be an action with the bulk part (3.1) that is invariant under A-type supersymmetry without using a boundary condition. It is straighforward to see that the following meets such a requirement;

$$\begin{aligned} S_A(\Phi) &= \frac{1}{2\pi}\int_{\Sigma}\Big(|\partial_0\phi|^2 - |\partial_1\phi|^2 + \frac{i}{2}\overline{\psi}_-(\overleftrightarrow{\partial_0} + \overleftrightarrow{\partial_1})\psi_- + \frac{i}{2}\overline{\psi}_+(\overleftrightarrow{\partial_0} - \overleftrightarrow{\partial_1})\psi_+ + |F|^2\Big)\mathrm{d}^2x \\ &\quad + \frac{1}{4\pi}\int_{\partial\Sigma}\Big(\partial_1|\phi|^2 + i(\overline{F}\phi - \overline{\phi}F)\Big)\mathrm{d}x^0. \end{aligned} \tag{3.30}$$

However, the second term (boundary term) is non-vanishing even if we use the boundary condition $\overline{\Phi}\Phi = c$. This is something we do not want.

Thus, the candidate action $S_A(\Phi) + \frac{1}{2\pi}\int_{\partial\Sigma} L_{bd}^{(2)}\mathrm{d}x^0$ is invariant under A-type supersymmetry but has two problems; $S_A(\Phi)$ contains unwanted boundary interactions and the term $\int_{\partial\Sigma} L_{bd}^{(2)}\mathrm{d}x^0$ either is ill-defined or lacks the a-term. Fortunately, both of these problems can be cancelled by addition of the following boundary term in the Lagrangian

$$\Delta L_{bd} = -\int \mathrm{d}\theta\mathrm{d}\overline{\theta}\,\overline{\Phi}\Phi\mathrm{Im}\log\Phi = \frac{i}{2}\int \mathrm{d}\theta\mathrm{d}\overline{\theta}\,\overline{\Phi}\Phi(\log\Phi - \log\overline{\Phi}). \tag{3.31}$$

The unwanted boundary interaction in $S_A(\Phi)$ is precisely cancelled by this term. Also, $L_{bd}^{(2)} + \Delta L_{bd}$ contains the term $\int \mathrm{d}\theta\mathrm{d}\overline{\theta}\,(u - \varphi)\overline{\Phi}\Phi$, which requires $u - \varphi$ to be a single valued field. Then, the a-term for u becomes the a-term for φ;

$$\frac{a}{2\pi}\int_{\partial\Sigma}\partial_0 u\mathrm{d}x^0=\frac{a}{2\pi}\int_{\partial\Sigma}\Big\{\partial_0\varphi+\partial_0(u-\varphi)\Big\}\mathrm{d}x^0=\frac{a}{2\pi}\int_{\partial\Sigma}\partial_0\varphi\mathrm{d}x^0. \qquad (3.32)$$

It may be appropriate to explain on the single-valuedness of the boundary interaction $L^{(2)}_{bd}+\Delta L_{bd}$, in particular the term $\overline{\Phi}\Phi(U-\operatorname{Im}\log\Phi)$. We can express the field Φ as $\Phi=\mathrm{e}^{\Psi}$ and consider Φ as a gauge invariant field where the gauge symmetry is $\mathbb{Z}$ which acts on Ψ as $\Psi\to\Psi+2\pi in$. We then consider U as the gauge field on which $\mathbb{Z}$ acts by $U\to U+2\pi n$. Then, $u-\varphi$ is gauge invariant and $\partial_0 u\mathrm{d}x^0$ can really be considered as the fieldstrength of this gauge symmetry.

To summarize, the total action we were looking for is

$$S_{tot} = S_A(\Phi)+\frac{1}{2\pi}\int_{\partial\Sigma}\mathrm{d}x^0\left[\int\mathrm{d}\theta\mathrm{d}\overline{\theta}\,\overline{\Phi}\Phi(U-\operatorname{Im}\log\Phi)+\operatorname{Re}\int\mathrm{d}\theta\, s\Upsilon\right], \qquad (3.33)$$

where $S_A(\Phi)$ is given by (3.30). The boundary term appears highly non-linear in Φ but the essential non-linearity resides only in $\overline{\Phi}\Phi\varphi$ which is absorbed by a redefinition of u that simply yields the $a\partial_0\varphi$ term (3.9) when $|\phi|\neq 0$.

We note that the above boundary interaction reproduces the correct one-loop running (3.11) of the parameter c. Up to the topological term, the bosonic part of the action (which is the relevant part in this discussion) is given by

$$S=\frac{1}{2\pi}\int_{\Sigma}\mathrm{d}^2x\,(\,|\partial_0\phi|^2-|\partial_1\phi|^2\,)+\frac{1}{2\pi}\int_{\partial\Sigma}\mathrm{d}x^0\left\{E(|\phi|^2-c)+iu'\,\overline{\phi}\overleftrightarrow{\partial_1}\phi\right\} \qquad (3.34)$$

where $u'=u-\varphi$ is the single valued field. In the effective action at the energy scale $\mu\ll\Lambda_{\rm UV}$, $|\phi|^2$ in $E(|\phi|^2-c)$ is shifted by $\langle|\phi|^2\rangle_\mu$, the one-loop momentum integral in the range $\mu\leq|k|\leq\Lambda_{\rm UV}$. In the present case one real component of ϕ obeys the Neumann boundary condition and the other obeys Dirichlet. Thus, the one point function has the same divergence as the ordinary one point function in the bulk. Thus, the shift is by

$$\int\limits_{\mu\leq|k|\leq\Lambda_{\rm UV}} \frac{d^2k}{(2\pi)^2}\frac{2\pi}{k^2} = \log(\Lambda_{\rm UV}/\mu). \tag{3.35}$$

This divergence (3.35) is absorbed exactly by giving the scale dependence of c as in (3.11).

We stress that the most important aspect of this formulation is that the parameter $s = c - ia$ enters into the boundary F-term. Any correction to the boundary F-term has to be holomorphic in s, periodic under $2\pi i$ shifts of s, and of boundary R-charge 1 if we assign R-charge 2 to e^{-s}. This excludes any perturbative renormalization except at one loop. Also, if we require the correction to be small at large s, we see that no correction is possible at all. In more general examples we will consider in the next section, we will always see this perturbative non-renormalization theorem (except at one-loop). However, non-perturbative non-renormalization still holds in some cases but fails in some other cases.

4 A-Type D-Branes in Linear Sigma Model

In this section, we apply the construction of the linear model to gauge theories. This enables us to define A-type D-branes in toric sigma models so that the D-brane location and the Wilson lines, combined into complex chiral parameters, enter into a boundary F-term. We will also find the dual description of the D-brane in the mirror Landau-Ginzburg model.

4.1 Supersymmetric Gauge theory

Here we record the basic facts on supersymmetric gauge theory in the bulk. We consider the simplest example: $U(1)$ gauge theory with a single chiral matter field of charge 1. The gauge transformation of the vector superfield V and the chiral matter field Φ is given by

$$V \longrightarrow V - iA + i\overline{A}, \quad \Phi \longrightarrow e^{iA}\Phi, \tag{4.1}$$

where A is a chiral superfield. We usually partially fix the gauge so that the vector superfield takes the form

$$\begin{aligned} V &= \theta^-\overline{\theta}^-(v_0 - v_1) + \theta^+\overline{\theta}^+(v_0 + v_1) - \theta^-\overline{\theta}^+\sigma - \theta^+\overline{\theta}^-\overline{\sigma} \\ &\quad + i\theta^-\theta^+(\overline{\theta}^-\overline{\lambda}_- + \overline{\theta}^+\overline{\lambda}_+) + i\overline{\theta}^+\overline{\theta}^-(\theta^-\lambda_- + \theta^+\lambda_+) + \theta^-\theta^+\overline{\theta}^+\overline{\theta}^- D \end{aligned} \quad (4.2)$$

v_0 and v_1 define a one-form field, σ is a complex scalar field, $\lambda_\pm$ and $\overline{\lambda}_\pm$ define a Dirac fermion, and D is a real scalar field. This is called the Wess-Zumino gauge and the residual gauge symmetry is the one with $A = \alpha(x)$ which rotates the phase of Φ and transforms v_μ as

$$v_\mu(x) \to v_\mu(x) - \partial_\mu \alpha(x). \quad (4.3)$$

The supersymmetry variation $\delta = \epsilon_+ Q_- - \epsilon_- Q_+ - \overline{\epsilon}_+ \overline{Q}_- + \overline{\epsilon}_- \overline{Q}_+$ does not in general preserve the Wess-Zumino gauge. In order to find the supersymmetry transformation of the component fields σ, $\lambda_\pm$, v_μ and D, we need to amend it with a gauge transformation that brings δV back into the Wess-Zumino gauge. It turns out that the required gauge transformation is the one with

$$A = i\theta^+\left(\overline{\epsilon}_+\overline{\sigma} + \overline{\epsilon}_-(v_0 + v_1)\right) - i\theta^-\left(\overline{\epsilon}_-\sigma + \overline{\epsilon}_+(v_0 - v_1)\right) + \theta^+\theta^-\left(\overline{\epsilon}_-\overline{\lambda}_+ - \overline{\epsilon}_+\overline{\lambda}_-\right) + \cdots, \quad (4.4)$$

where $+\cdots$ are the derivative terms to make A chiral. In this way, we find that the supersymmetry transformation of the component fields of V is

$$\begin{aligned} &\delta(v_0 \pm v_1) = i\overline{\epsilon}_\pm\lambda_\pm + i\epsilon_\pm\overline{\lambda}_\pm, \\ &\delta\sigma = -i\overline{\epsilon}_+\lambda_- - i\epsilon_-\overline{\lambda}_+, \\ &\delta D = -\overline{\epsilon}_+\partial_-\lambda_+ - \overline{\epsilon}_-\partial_+\lambda_- + \epsilon_+\partial_-\overline{\lambda}_+ + \epsilon_-\partial_+\overline{\lambda}_-, \\ &\delta\lambda_+ = i\epsilon_+(D + iv_{01}) + 2\epsilon_-\partial_+\overline{\sigma}, \\ &\delta\lambda_- = i\epsilon_-(D - iv_{01}) + 2\epsilon_+\partial_-\sigma, \end{aligned}$$

while that of Φ is

$$\begin{aligned} &\delta\phi = \epsilon_+\psi_- - \epsilon_-\psi_+, \\ &\delta\psi_+ = i\overline{\epsilon}_-(D_0 + D_1)\phi + \epsilon_+ F - \overline{\epsilon}_+\overline{\sigma}\phi, \\ &\delta\psi_- = -i\overline{\epsilon}_+(D_0 - D_1)\phi + \epsilon_- F + \overline{\epsilon}_-\sigma\phi, \\ &\delta F = -i\overline{\epsilon}_+(D_0 - D_1)\psi_+ - i\overline{\epsilon}_-(D_0 + D_1)\psi_- \\ &\qquad + \overline{\epsilon}_+\overline{\sigma}\psi_- + \overline{\epsilon}_-\sigma\psi_+ + i(\overline{\epsilon}_-\overline{\lambda}_+ - \overline{\epsilon}_+\overline{\lambda}_-)\phi. \end{aligned}$$

The superfield

$$\Sigma := \overline{D}_+ D_- V \tag{4.5}$$

is invariant under the gauge transformation $V \to V + i(\overline{A} - A)$. It is a twisted chiral superfield which is expressed in the Wess-Zumino gauge as

$$\Sigma = \sigma(\widetilde{y}) + i\theta^+ \overline{\lambda}_+(\widetilde{y}) - i\overline{\theta}^- \lambda_-(\widetilde{y}) + \theta^+\overline{\theta}^-(D - iv_{01})(\widetilde{y}). \tag{4.6}$$

where v_{01} is the field-strength $v_{01} := \partial_0 v_1 - \partial_1 v_0$. The superfield Σ is called the ***super-field-strength*** of V.

We consider the following gauge invariant action

$$S = \frac{1}{2\pi}\int_\Sigma \left[\int \mathrm{d}^4\theta \left(\overline{\Phi}\,\mathrm{e}^V \Phi - \frac{1}{2e^2}\overline{\Sigma}\Sigma \right) + \mathrm{Re}\int d^2\widetilde{\theta}\,(-t\Sigma) \right] \mathrm{d}^2x, \tag{4.7}$$

where

$$t = r - i\theta \tag{4.8}$$

is a dimensionless twisted chiral parameter; r is the Fayet-Illiopoulos parameter and θ is the Theta angle. This Lagrangian is invariant under $(2,2)$ supersymmetry when the worldsheet Σ has no boundary. In terms of the component fields this is written as

$$\begin{aligned} S \;=\; \frac{1}{2\pi}\int_\Sigma \Bigg[& -D^\mu\overline{\phi}D_\mu\phi + \frac{i}{2}\overline{\psi}_-(\overleftrightarrow{D_0} + \overleftrightarrow{D_1})\psi_- + \frac{i}{2}\overline{\psi}_+(\overleftrightarrow{D_0} - \overleftrightarrow{D_1})\psi_+ + D|\phi|^2 + |F|^2 \\ & - |\sigma|^2|\phi|^2 - \overline{\psi}_-\sigma\psi_+ - \overline{\psi}_+\overline{\sigma}\psi_- - i\overline{\phi}\lambda_-\psi_+ + i\overline{\phi}\lambda_+\psi_- + i\overline{\psi}_+\overline{\lambda}_-\phi - i\overline{\psi}_-\overline{\lambda}_+\phi \\ & + \frac{1}{2e^2}\left(-\partial^\mu\overline{\sigma}\partial_\mu\sigma + \frac{i}{2}\overline{\lambda}_-(\overleftrightarrow{\partial_0} + \overleftrightarrow{\partial_1})\lambda_- + \frac{i}{2}\overline{\lambda}_+(\overleftrightarrow{\partial_0} - \overleftrightarrow{\partial_1})\lambda_+ + v_{01}^2 + D^2 \right) \\ & - rD + \theta v_{01} \Bigg] \mathrm{d}^2x, \end{aligned} \tag{4.9}$$

where an appropriate partial integration is made.

This theory is super-renormalizable with respect to the dimensionful gauge coupling constant e. The FI parameter r is renormalized in such a way as

$$r(\Lambda_{\mathrm{UV}}) = r(\mu) + \log(\Lambda_{\mathrm{UV}}/\mu). \tag{4.10}$$

The vector R-symmetry is unbroken but the axial R-symmetry is anomalous; The axial R-rotation shifts the Theta angle as $\theta \to \theta - 2\alpha$.

4.2 The Boundary Interaction

Let us now choose the worldsheet Σ to be the half-plane $\mathbb{R}\times(-\infty,0]$. The (2,2) supersymmetry variantion of the action (4.9) is given by a boundary term, as analyzed in [13]. For A-type supersymmetry one can show that the combination

$$S_A = S + \frac{1}{2\pi}\int\limits_{\partial\Sigma} \mathrm{d}x^0 \left[\frac{1}{2}\partial_1|\phi|^2 + \frac{i}{2}(\overline{F}\phi - \overline{\phi}F) + \frac{i}{4e^2}(\lambda_-\lambda_+ - \overline{\lambda}_+\overline{\lambda}_-) + \theta v_0\right] \tag{4.11}$$

has the following simple transformation property;

$$\delta S_A = \frac{r}{4\pi}\int\limits_{\partial\Sigma} \mathrm{d}x^0 \left\{\epsilon(\overline{\lambda}_+ + \lambda_-) - \overline{\epsilon}(\overline{\lambda}_- + \lambda_+)\right\}. \tag{4.12}$$

We would like to construct a linear model for D-branes in this gauge theory. As in the case without the gauge field, we introduce the superfield U on the A-boundary with the boundary interaction as in (3.33):

$$S_{boundary} = \frac{1}{2\pi}\int\limits_{\partial\Sigma} \mathrm{d}x^0 \left[\int \mathrm{d}\theta \mathrm{d}\overline{\theta}\, \overline{\Phi}\, \mathrm{e}^V \Phi(U - \mathrm{Im}\log\Phi) + \mathrm{Re}\int \mathrm{d}\theta\, s\Upsilon\right], \tag{4.13}$$

where $\Upsilon = \overline{D}U$ is the fieldstrength. For gauge invariance of the first term, we would like the field U to transform in the same way as $\mathrm{Im}\log\Phi$. Thus, it transforms as

$$U \longrightarrow U + \frac{1}{2}(A + \overline{A}), \tag{4.14}$$

under the gauge transformation (4.1). In particular, the supersymmetry transformation of the component field is modified in the Wess-Zumino gauge as

$$\begin{aligned}
\delta u &= \epsilon\chi - \overline{\epsilon\chi},\\
\delta\chi &= -\overline{\epsilon}(E + i(\partial_0 u + v_0)) - i\epsilon\sigma,\\
\delta\overline{\chi} &= -\epsilon(E - i(\partial_0 u + v_0)) + i\overline{\epsilon\sigma},\\
\delta E &= i\epsilon\partial_0\chi + i\overline{\epsilon}\partial_0\overline{\chi} - \frac{1}{2}\epsilon(\lambda_- + \overline{\lambda}_+) + \frac{1}{2}\overline{\epsilon}(\overline{\lambda}_- + \lambda_+).
\end{aligned} \tag{4.15}$$

Under this modified supersymmetry transformation, the boundary superpotential term

$$\frac{1}{2\pi}\int_{\partial\Sigma} dx^0 \,\mathrm{Re}\int d\theta\, s\Upsilon = \frac{1}{2\pi}\int_{\partial\Sigma} dx^0\,(cE + a\partial_0 u) \tag{4.16}$$

is not invariant but varies as

$$\delta\left[\frac{1}{2\pi}\int_{\partial\Sigma} dx^0\,\mathrm{Re}\int d\theta\, s\Upsilon\right] = -\frac{c}{4\pi}\int_{\partial\Sigma} dx^0\left\{\epsilon(\overline{\lambda}_+ + \lambda_-) - \overline{\epsilon}(\overline{\lambda}_- + \lambda_+)\right\}. \tag{4.17}$$

We note that this is proportional to δS_A given in (4.12). Thus, $S_A + S_{boundary}$ is invariant under A-type supersymmetry if and only if

$$c = r. \tag{4.18}$$

The non-trivial variation (4.17) comes from the fact that the Fermi-superfield Υ is not invariant under the gauge transformation (4.14). This reminds us of another constraint; The action must be gauge invariant, or it must be invariant under the residual gauge transformation $A = \alpha(x)$ in the Wess-Zumino gauge we are in. The gauge transformation shifts the component u as $u \to u + \alpha$ while χ and E are invariant. Thus the term (4.16) is shifted by $\frac{a}{2\pi}\int dx^0\partial_0\alpha$. We also notice that the action S_A is not gauge invariant but is shifted by $-\frac{\theta}{2\pi}\int dx^0\partial_0\alpha$ since v_μ is transformed as $v_\mu \to v_\mu - \partial_\mu\alpha$. Thus, the action $S_A + S_{boundary}$ is gauge invariant if and only if

$$a = \theta \pmod{2\pi\mathbb{Z}}. \tag{4.19}$$

Thus, the action

$$S_{tot} = S_A + S_{boundary} \tag{4.20}$$

is supersymmetric and gauge invariant if and only if

$$s = t \pmod{2\pi i\mathbb{Z}}. \tag{4.21}$$

We note that the real part of the condition, $r = c$, makes the derived boundary condition $|\phi|^2 = c$ to be compatible with the D-term constraint $|\phi|^2 = r$. We also note again that no boundary condition on the bulk fields are required for the supersymmetry.

4.3 Construction in Linear Sigma Model

Let us consider a $U(1)$ gauge theory with N chiral multiplets Φ_i of charge Q_i where the action is given by

$$S = \frac{1}{2\pi} \int_{\Sigma} \left[\int \mathrm{d}^4\theta \left(\sum_{i=1}^{N} \overline{\Phi}_i \, \mathrm{e}^{Q_i V} \Phi_i - \frac{1}{2e^2} \overline{\Sigma}\Sigma \right) + \mathrm{Re} \int d^2\widetilde{\theta} \, (-t\Sigma) \right] \mathrm{d}^2 x, \tag{4.22}$$

The theory reduces at low enough energies to non-linear sigma model on a certain toric manifold X. For instance, $X = \mathbb{C}\mathrm{P}^{N-1}$ if $Q_i = 1$ for all i; X is the total space of the line bundle $\mathcal{O}(-d)$ over $\mathbb{C}\mathrm{P}^{N-2}$ if $Q_1 = \cdots = Q_{N-1} = 1$ and $Q_N = -d$; X is the total space of $\mathcal{O}(-1) \oplus \mathcal{O}(-1)$ over $\mathbb{C}\mathrm{P}^1$ if $N = 4$ and $Q_1 = Q_2 = 1$, $Q_3 = Q_4 = -1$. The character of the theory depends on whether the sum of charges

$$b_1 = \sum_{i=1}^{N} Q_i \tag{4.23}$$

is zero or not. If $b_1 \neq 0$, the scale invariance is broken at the one-loop level and the dimensionless FI parameter r is replaced by a dynamically generated scale parameter Λ by $r(\mu) = b_1 \log(\mu/\Lambda)$. Also, the axial $U(1)$ R-symmetry is anomalously broken; the axial R-rotation shifts the Theta angle as $\theta \to \theta - 2b_1\beta$. The sign of r at the cut-off scale is determined by b_1 and hence the target space is uniquely determined by the charges Q_i. On the other hand, if $b_1 = 0$, the scale invariance is preserved at the one-loop level. In particular the FI-Theta parameter $t = r - i\theta$ is the dimensionless parameter of the theory. The toric manifold X in this case is a (non-compact) Calabi-Yau manifold. We can choose the sign of r as we wish; both positive and negative r are possible and the sigma model target space X differs in general.

Let us now formulate the theory on the left half plane $\Sigma = \mathbb{R} \times (-\infty, 0]$. To fix the expression of the bulk action S in terms of the component field, we take the obvious generalization of the standard one (4.9). We also define S_A in the same way as in (4.11). Then we still have the simple supersymmetry variation (4.12). Now, we introduce N boundary real superfields U_i (with fieldstrength

Υ_i) and the boundary interaction term

$$S_{boundary} = \frac{1}{2\pi}\sum_{i=1}^{N}\int_{\partial\Sigma} \mathrm{d}x^0 \left[\int \mathrm{d}\theta\mathrm{d}\bar{\theta}\,\overline{\Phi}_i\,\mathrm{e}^{Q_i V}\Phi_i(U_i - \mathrm{Im}\log\Phi_i) + \mathrm{Re}\int \mathrm{d}\theta\, s_i\Upsilon_i \right], \tag{4.24}$$

One can show as before that the total action $S_{tot} = S_A + S_{boundary}$ is gauge invariant and supersymmetric if and only if

$$\sum_{i=1}^{N} Q_i s_i = t \quad (\mathrm{mod}\ 2\pi i\mathbf{Z}). \tag{4.25}$$

We note that the one-loop renormalization group flows of s_i's and t are compatible with each other: The former runs as $s_i = \log(\mu/\Lambda)$+constant while the latter runs as $t = b_1\log(\mu/\Lambda)$ only if $b_1 = \sum_{i=1}^{N} Q_i$ is non-zero. In any case there are one scale parameter and $N-1$ dimensionless complex parameters.

It is straightforward to generalize the construction to the case of higher rank gauge group $U(1)^k = \prod_{a=1}^{k} U(1)_a$ with N matters with charge Q_{ia}. The condition for supersymmetry and gauge invariance is $\sum_{i=1}^{N} Q_{ia}s_i = t_a$ where t_a is the FI-Theta parameter for $U(1)_a$.

Geometric Interpretation

Let us find out what the above boundary interaction corresponds to in the non-linear sigma model limit $\mu \ll e$. Integrating out the boundary superfields U_i yields the boundary condition

$$\overline{\Phi}_i\,\mathrm{e}^{Q_i V}\Phi_i = c_i \quad \text{at A-boundary}. \tag{4.26}$$

We note that at high enough enrgies $\mu \gg \Lambda$, c_i are all positive and one can solve the constraint (4.26). This boundary condition corresponds to a D-brane wrapped on the $(N-1)$-dimensional torus T in X located at

$$|\phi_i|^2 = c_i. \tag{4.27}$$

It is easy to see that T is a Lagrangian submanifold of X. The parameter a_i parametrizes the holonomy of the flat $U(1)$ gauge field on the D-brane since the boundary term $S_{boundary}$ contains the term

$$\frac{1}{2\pi}\sum_{i=1}^{N}\int_{\partial\Sigma} a_i \mathrm{d}\varphi_i, \tag{4.28}$$

that is obtained through the process (3.32). It may appear that the holonomy in the unphysical gauge orbit direction is $\sum_{i=1}^{N} Q_i a_i = \theta$ which is non-vanishing for a non-zero worldsheet Theta angle. However, it is not the case; there is also a boundary term $\frac{\theta}{2\pi}\int_{\partial\Sigma} \mathrm{d}x^0 v_0$ (see (4.11)). At low enough energies, the worldsheet gauge field v_μ is frozen at

$$v_\mu = \frac{i}{2}\frac{\sum_{i=1}^{N} Q_i \overline{\phi}_i \overleftrightarrow{\partial}_\mu \phi_i}{\sum_{i=1}^{N} Q_i^2 |\phi_i|^2} = -\frac{\sum_{i=1}^{N} Q_i c_i \partial_\mu \varphi_i}{\sum_{i=1}^{N} Q_i^2 c_i}, \tag{4.29}$$

where in the last step we have used the constraint (4.27). Thus, the total holonomy term is

$$\frac{1}{2\pi}\int_{\partial\Sigma}\left[\sum_{i=1}^{N} a_i \mathrm{d}\varphi_i - \theta\left(\sum_{i=1}^{N} Q_i c_i \mathrm{d}\varphi_i \Big/ \sum_{i=1}^{N} Q_i^2 c_i\right)\right]. \tag{4.30}$$

It is easy to see that the holonomy in the gauge orbit direction indeed vanishes if the condition $\theta = \sum_{i=1}^{N} Q_i a_i$ in (4.25) is satisfied.

We recall that the parameters c_i are running coupling constants, irrespective of whether $b_1 = \sum_{i=1}^{N} Q_i$ is zero or not. c_i becomes smaller as the energy is reduced. If $b_1 > 0$, the manifold X itself also becomes smaller at lower energies. Thus, in this case, there is a chance that the D-brane stays in the theory at low energies where the sigma model description breaks down. We will indeed find in the mirror description (which is the better description at low energies) that the D-brane stays in the theory for special values of s_i. If $b_1 = 0$, the size of the manifold X does not change. It is expected that, as in the basic example of Section 3, the D-brane disappears from the theory at extreme low energies.

As usual, it is easy to exclude perturbative renormalization beyond one-loop level. We claim that there is no non-perturbative renormalization either in the case where $b_1 > 0$.. This follows from the requirement that the correction is small at small Σ, at large s_i for any i and at large t.

4.4 Promoting s_i to Chiral Superfields

There is actually an interesting generalization of the above construction. It is to promote the parameters s_i to boundary chiral superfields. The gauge symmetry and the supersymmetry is not spoiled even if we do so, provided the condition (4.25) is obeyed. Thus, we make the replacement

$$s_i \longrightarrow S_i(Z_1, \ldots, Z_\ell), \tag{4.31}$$

where Z_α are boundary chiral superfields and $S_i(Z_\alpha)$ are holomorphic functions obeying $\sum_{i=1}^N Q_i S_i(Z_\alpha) = t$. A simple class of such functions are linear ones. This is motivated by the recent work [19]. Let m_i^α be such that $\sum_{i=1}^N Q_i m_i^\alpha = 0$. Then one can take

$$S_i = \sum_{\alpha=1}^{\ell} m_i^\alpha Z_\alpha + s_i, \tag{4.32}$$

where s_i are parameters obeying $\sum_{i=1}^N Q_i s_i = t$.

When the charges Q_i satisfy the condition $b_1 = \sum_{i=1}^N Q_i = 0$, the bulk theory is scale invariant at the one-loop level and is expected to flow to a non-trivial SCFT in the infra-red limit. In such a case, it is natural to ask under what condition the boundary interaction does not break the scale invariance. We recall that $c_i = \mathrm{Re}(s_i)$ is a running coupling constant:

$$c_i(\Lambda_{\mathrm{UV}}) = c_i(\mu) + \log(\Lambda_{\mathrm{UV}}/\mu). \tag{4.33}$$

However, this running can be absorbed by the shift of the fields Z_α by a certain condition on m_i^α's. The condition is that there are numbers δ_α such that

$$\sum_{\alpha=1}^{\ell} m_i^\alpha \delta_\alpha = 1. \tag{4.34}$$

In such a case, the boundary interaction is scale invariant at the one-loop level.

Geometric Interpretation and a Constraint on the Parameters

Let us see what this boundary interaction corresponds to in the non-linear sigma model. The Z_α-equation of motion yields the constraint $\sum_{i=1}^N m_i^\alpha \overline{D} U_i =$

$\sum_{i=1}^N m_i^\alpha DU_i = 0$ which means

$$\sum_{i=1}^N m_i^\alpha U_i = \text{constant} \quad \text{at A-boundary.} \tag{4.35}$$

Here "constant at A-boundary" means that it does not depend on $x^0, \theta, \overline{\theta}$. Namely, the coefficient of $\theta, \overline{\theta}$ and $\theta\overline{\theta}$ vanishes and the leading term is independent of x^0. On the other hand, the equation of motion for Φ_i identifies $\mathrm{Im}\log\Phi_i$ with U_i. (In the derivation of this statement the bulk and the boundary terms in the action S_A play an important role.) Thus, we obtain the following constarint for Φ_i's:

$$\sum_{i=1}^N m_i^\alpha \mathrm{Im}\log\Phi_i = \text{constant} \quad \text{at A-boundary.} \tag{4.36}$$

The equation of motion for the remaining U_i yields the condition

$$\overline{\Phi}_i e^{Q_i V}\Phi_i = \mathrm{Re}\left(\sum_{\alpha=1}^{\ell} m_i^\alpha Z_\alpha + s_i\right) \quad \text{at A-boundary.} \tag{4.37}$$

The boundary conditions (4.36)-(4.37) are that for a D-brane located at

$$|\phi_i|^2 = \sum_{\alpha=1}^{\ell} m_i^\alpha \zeta_\alpha + c_i, \qquad i = 1, \ldots, N; \tag{4.38}$$

$$\sum_{i=1}^N m_i^\alpha \varphi_i = \text{constant}, \qquad \alpha = 1, \ldots, \ell, \tag{4.39}$$

where ζ_α are real coordinates that can vary. It is a $(N-1)$ dimensional subspace L which is a fibration over the ℓ-dimensional subspace (4.38) in the $|\phi_i|^2$-plane with the $(N-\ell-1)$ dimensional torus (4.39) as its fibre. It is a Lagrangian submanifold of X. We note that ℓ of the c_i parameters can be absorbed by the redefinition of ζ_α. Also, by the constraints (4.39), the a_i's related by the shifts by m_i^α are physically equivalent. There are only $(N-\ell-1)$ physical parameters.

When the charges Q_{ia} satisfy $b_{1,a} = \sum_{i=1}^N Q_{ia} = 0$, the target space is a non-compact Calabi-Yau manifold and the sigma model is scale invariant at the one-loop level (cooresponding, of course, to the scale invariance of the gauge theory). We have seen that the condition for the boundary interaction to prteserve this one-loop scale invariance is given by (4.34). This actually corresponds to the condition that L is a special Lagrangian submanifold.[1] To see this we note that the holomorphic volume form of X is proportional to $\exp(i\sum_{i=1}^{N}\varphi_i)$. Under the condition (4.34), the phase is a constant

$$\sum_{i=1}^{N}\varphi_i=\sum_{i=1}^{N}\sum_{\alpha=1}^{\ell}\delta_\alpha m_i^\alpha\varphi_i=\sum_{\alpha=1}^{\ell}\delta_\alpha\mathrm{const}_\alpha. \tag{4.40}$$

One important thing to notice here is that $\varphi_i = \operatorname{Im}\log\phi_i$ is well-defined only if $|\phi_i|^2 \neq 0$ but the equation (4.38) allows some of $|\phi_i|^2$ to vanish. Generically, the subspace is singular or has a boundary at such a point. It is only in a special case where (4.38)-(4.39) defines a smooth submanifold. For example, let us consider our basic example of a single chiral superfield Φ (where there is no gauge symmetry) and promote S to a boundary chiral superfield. This will yields the D-brane at $\varphi = \operatorname{Im}\log\phi = $ constant. This indeed has an end point at $\phi = 0$ and is not smooth. Similarly, in many cases L is singular for any values of c_i. However, there are some cases where L is smooth for special values of c_i. In such a case, the condition that L is smooth can be considered as a constraint on the parameters c_i. We exhibit this in the examples below.

Special Lagrangian Families in $\mathbb{C}^N$

In this example, we do not consider a gauge theory but a free theory of N chiral superfields Φ_i. One can straightforwardly apply the above construction of boundary interaction to this case. (Simply ignore V and t and omit the constraints such as $\sum_{i=1}^{N}Q_iS_i = t$.) In particular, (4.34) is still the condition of one-loop scale invariance. We will focus on such a case with $\ell = 1$. Namely, the case where m_i are all equal, say to 1. The equation defining the subspace

[1] The numbers m_i^α are related to "charges" q_i^A ($A = 1, \ldots, N-\ell$) in [19] by the relation $\sum_{i=1}^{N} m_i^\alpha q_i^A = 0$. Then, the scale invariance condition (4.34) is equivalent to the condition $\sum_{i=1}^{N} q_i^A = 0$ in [19] for L to be special Lagrangian.

L is

$$|\phi_i|^2 = \zeta + c_i \quad (i = 1, \ldots, N),$$
$$\varphi_1 + \cdots + \varphi_N = 0. \tag{4.41}$$

This is smooth if and only if one can find a pair (i_1, i_2) such that $c_{i_1} = c_{i_2} < c_j$ for $j \neq i_1, i_2$. Otherwise, L has an end at the locus where only one ϕ_i vanishes. If the condition holds, say for $(i_1, i_2) = (1, 2)$, ϕ_1 and ϕ_2 vanishes at the same time, and L is smooth since the defining equation can be written as

$$\overline{\phi}_1 = \phi_2 \times \exp\left(i\textstyle\sum_{j=3}^N \varphi_j\right),$$
$$|\phi_j|^2 = |\phi_1|^2 + c_j - c_1, \quad (j \neq 1, 2).$$

In this case, the circle of φ_1 and φ_2 are topologically trivial in L and thus the holonomy must be trivial; $a_1 = a_2 = 0$. This is relaxed to $a_1 = a_2$ by using the freedom to shift a_i uniformly (coming from the second equation of (4.41)). In general, we must have $a_i = a_j$ in the branch where $c_i = c_j$. The space of (c_i)'s satisfying the constraint is a union of walls in the $(N-1)$-dimensional space (-1 is from the redefinition of ζ). We depict in Fig. 1 the case of $N = 3$. The origin is deleted since L is singular there.

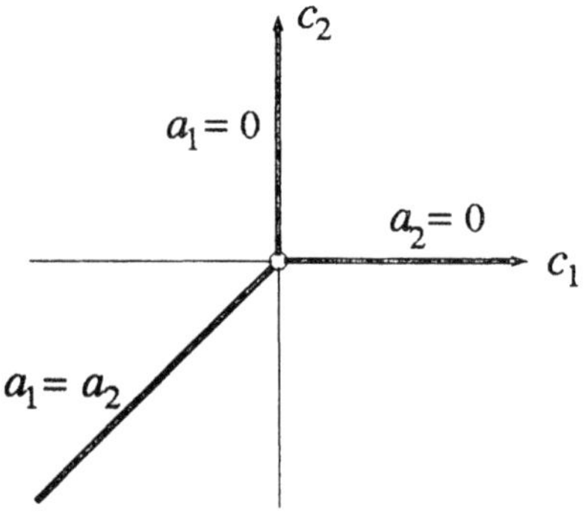

Figure 1: The constraint: The three bold open lines are the locus where L is smooth. We have set $a_3 = c_3 = 0$ to eliminate the shift ambiguity.

$\mathcal{O}(-1) \oplus \mathcal{O}(-1)$ over $\mathbf{CP}^1$

We next consider the $U(1)$ gauge theory with four matters of charge $Q_i = 1, 1, -1, -1$. We stay in the region where r is large positive where the bulk theory reduces to non-compact Calabi-Yau sigma model. We consider the case $\ell = 1$ where $m_i = 1, 1, 1, 1$. This is again the case where (4.34) is met and the boundary interaction is scale invariant at the one-loop level. For the gauge invariance and supersymmetry c_i and a_i are constrained by $c_1 + c_2 - c_3 - c_4 = r$, $a_1 + a_2 - a_3 - a_4 = \theta$. The overall shifts of c_i and a_i are unphysical. Thus space of physical c_i is two-dimensional and so is that of a_i. The equation defining the subspace L is the same as the $N = 4$ case of (4.41). L is smooth if and only if one can find a pair (i, j) such that $c_i = c_j < c_k$ for $k \neq i, j$. In the $c_i = c_j$ branch, $a_i = a_j$ must be satisfied. Thus, the parameter space consists of five disconnected pieces, each being an open cylinder (four of them semi-infinite). There are two points where three of the branches becomes close as in Fig. 1.

In the two examples, we have seen that the smoothness of the submanifold L put a constraint of the parameters c_i and a_i. However, we notice that the geometry itself is derived from the action involving the boundary D-terms. We know that boundary D-terms are not protected even from loop corrections. Accordingly, we expect that the constraint on c_i and a_i is subject to quantum corrections as well. We will see that it is indeed the case.

4.5 The Mirror Description

In [2], a dual description of the linear sigma model was found. Dualizing on the phase of each chiral superfield Φ_i we obtain a twisted chiral superfield Y_i whose real part is related to the gauge invariant composite of Φ_i via

$$\overline{\Phi}_i \, \mathrm{e}^{Q_i V} \Phi_i = \mathrm{Re}\, Y_i. \tag{4.42}$$

The dual theory has the (twisted) superpotential

$$\widetilde{W} = \Sigma \left(\sum_{i=1}^{N} Q_i Y_i - t \right) + \sum_{i=1}^{N} \mathrm{e}^{-Y_i}, \tag{4.43}$$

where Σ-linear term appears at the dualization process and the exponential terms are generated by the effect of the instantons which are vortices in the gauge theory. In the sigma model limit where $e \to \infty$, it is appropriate to integrate out the gauge multiplet and that induces the constraint

$$\sum_{i=1}^{N} Q_i Y_i = t. \tag{4.44}$$

The theory becomes the Landau-Ginzburg model on this $N-1$ dimensional algebraic torus $(\mathbb{C}^\times)^{N-1}$ with the superpotential

$$\widetilde{W} = \sum_{i=1}^{N} \mathrm{e}^{-Y_i}. \tag{4.45}$$

We would now like to ask how the A-type D-brane constructed above is described in the dual theory.

We first give a rough argument which in the end turns out to be the correct one. Let us look at the boundary interaction term (4.24). Here we replace s_i by S_i which can either be a parameter or a boundary chiral superfield. If we use the relation (4.42), the U_i term in (4.24) can be made into a boundary F-term and the total boundary F-term is expressed as

$$\frac{1}{2\pi} \sum_{i=1}^{N} \int_{\partial\Sigma} \mathrm{d}x^0 \,\mathrm{Re} \int \mathrm{d}\theta \,(S_i - Y_i)\Upsilon_i. \tag{4.46}$$

Thus, Υ_i integration yields the constraint

$$Y_i = S_i. \tag{4.47}$$

This argument was not precise for two reasons: First, the relation (4.42) obtained in the bulk theory is used without paying attention to the presence of the boundary. Second, it ignores the other boundary interactions — the boundary term in S_A (see eqn. (4.11)) and the term involving $\mathrm{Im}\log\Phi_i$ in (4.24). We now show that the result (4.46) or (4.47) remains correct (with a different interpretation of Υ_i) even if we take these points into account.

We first note that during the dualization procedure we take $|\phi_i|^2$ to be non-zero and $\varphi_i = \mathrm{Im}\log\phi_i$ is well-defined. This allows us to shift the U_i field as $U_i = \varphi_i + U_i'$ so that U_i' is a single valued superfield. In terms of the shifted variables the boundary term (4.24) is expressed as

$$S_{boundary} = \frac{1}{2\pi}\sum_{i=1}^{N}\int_{\partial\Sigma} \mathrm{d}x^0 \left[\int \mathrm{d}\theta \mathrm{d}\bar{\theta}\, \overline{\Phi}_i \mathrm{e}^{Q_i V} \Phi_i U_i' + \mathrm{Re}\int \mathrm{d}\theta\, S_i \Upsilon_i' + a_i \partial_0 \varphi_i\right]. \tag{4.48}$$

The terms in the action relevant for the dualization are those involving φ_i's:

$$\begin{aligned} &-\frac{1}{2\pi}\sum_{i=1}^{N}\int_{\Sigma} |\phi_i|^2(\partial_\mu\varphi_i + Q_i v_\mu)^2 \mathrm{d}^2x \\ &\qquad + \frac{1}{2\pi}\int_{\partial\Sigma}\left[\sum_{i=1}^{N}\left(-2u_i'|\phi_i|^2(\partial_1\varphi_i + Q_i v_1) + a_i\partial_0\varphi_i\right) + \theta v_0\right]\mathrm{d}x^0\,, \end{aligned} \tag{4.49}$$

where u_i' is the lowest component of U_i'. Here we have included the boundary Theta term $\frac{\theta}{2\pi}\int_{\partial\Sigma} v_0 \mathrm{d}x^0$ (a term in S_A) in order to keep the gauge invariance: Note that (4.49) itself is invariant under the gauge transformation $\varphi_i \to \varphi_i + Q_i\alpha$, $v_\mu \to v_\mu - \partial_\mu\alpha$ provided the condition $\sum_{i=1}^N Q_i a_i = \theta$ from (4.25) holds. Also we have ignored the terms involving fermions. (This is merely for simplicity and there is no obstacle to include them into the discussion below.) Now we consider a system of N one form fields $(B_i)_\mu$ and $N+1$ periodic scalar fields ϑ_i, $\widetilde{u}$ with the following action

$$S' = \frac{1}{2\pi}\sum_{i=1}^{N}\left[\int_{\Sigma}\left(-\frac{1}{2}|B_i|^2\mathrm{d}^2x + B_i\wedge \mathrm{d}\vartheta_i + Q_i\vartheta_i F_v\right) + \int_{\partial\Sigma}(a_i - \vartheta_i)\partial_0\widetilde{u}\mathrm{d}x^0\right], \tag{4.50}$$

where F_v is the curvature of v, $F_v = \mathrm{d}v$. We impose the boundary condition

$$(B_i)_1 = 0. \tag{4.51}$$

If we integrate out the B_i-fields, we obtain the action for twisted chiral superfields $Y_i = |\phi_i|^2 - i\vartheta_i + \cdots$ and Σ with the bulk superpotential $\widetilde{W}_{dual} = \Sigma(\sum_{i=1}^N Q_i Y_i - t)$ and the boundary interaction (4.46) in which Υ_i is the "field-strength" for U_i with the lowest component $\widetilde{u}_i$. Thus, the rest is to show that the integration in the opposite order, ϑ_i first, yields the action (4.49). The variation with respect to ϑ_i gives the constraints

$$\begin{aligned} \mathrm{d}B_i &= Q_i F_v \quad \text{on } \Sigma \\ (B_i)_0 &= \partial_0 \widetilde{u} \quad \text{along } \partial\Sigma. \end{aligned}$$

The first constraint is solved by $B_i = \mathrm{d}\varphi_i + Q_i v$ where φ_i is a periodic scalar field of period 2π. By the second equation plus the boundary condition (4.51) this leads to the following relations on the boundary

$$\partial_0 \varphi_i + Q_i v_0 = \partial_0 \widetilde{u}, \tag{4.52}$$

$$\partial_1 \varphi_i + Q_i v_1 = 0. \tag{4.53}$$

Plugging the first relation back into (4.50) and using the relation $\sum_{i=1}^{N} Q_i a_i = \theta$ we obtain the action (4.49) without the u_i'-dependent terms. The condition (4.53) is actually equivalent to having those u_i'-dependent terms; integrating out u_i' simply imposes (4.53).

Dualization is not the end of the story in finding the mirror description [2]. As mensioned above, the bulk superpotential $\sum_{i=1}^{N} \mathrm{e}^{-Y_i}$ is generated by the instanton effect. Like in that case, one may wonder if the boundary F-term is generated as well. We now show that nothing can be generated. As in [2], we extend the gauge symmetry to $U(1)^N$ where each chiral superfield Φ_i has charge 1 under the i-th $U(1)$ and neutral under the others. We have N FI-Theta parameters t_i which is promoted to a twisted chiral superfield T_i, and $Q_i S_i$ must be the A-boundary value of T_i for gauge invariance and supersymmetry. For an appropriate choice of the D-terms we have N decoupled systems while for another limit of the D-term couplings we recover the original system. Since the deformation of the D-term does not affect the F-terms it is enough to show that (4.46) is not corrected for each i. We first note that Υ_i have boundary R-charge 1 while e^{-Y_i} and e^{-S_i} both have R-charge 2. The boundary F-term must be holomorphic in these superfields, and it must approach the classical expression (4.46) at large Y_i and large S_i. These requirement is satisfied only by (4.46) itself.

4.6 Quantum Deformation of the Constraint

The mirror of our D-brane is thus given by (4.47). This is true not only when S_i are parameters but also when they are boundary chiral superfields. If S_i are parameters, (4.47) means that the boundary value of Y_i is fixed at s_i and we see that the mirror of our D-brane is a D0-brane at a point in $(\mathbf{C}^{\times})^{N-1}$. If S_i are functions of ℓ boundary chiral superfields as in (4.31), the mirror is a D(2ℓ)-brane wrapped on a holomorphic cycle Z defined by

$$Y_i = S_i(Z_1, \dots, Z_\ell). \tag{4.54}$$

These are B-type D-branes in the LG model (in the flipped convention where Y_i are *chiral*). B-type D-branes in LG model were brielfly studied in [15, 13] and will be studied in some more detail in Section 6.

One important constraint of worldsheet supersymmetry is that the bulk superpotential must be a constant on the D-brane. This itself gives no condition in the case where S_i are parameters since the mirror D-brane is at a point. However, the constraint gives a strong condition when S_i are functions of the chiral superfield Z_α so that the mirror D-brane is wrapped on a cycle X defined by (4.54). It constrains the functional form to be

$$\widetilde{W} = \sum_{i=1}^{N} \mathrm{e}^{-S_i(Z)} = \text{constant}. \tag{4.55}$$

This is a very strong constraint and is not satisfied by a generic function $S_i(Z)$.

Let us see whether this constraint is satisfied in the case of linear functions $S_i = \sum_{\alpha=1}^{\ell} m_i^\alpha Z_\alpha + s_i$. (This discussion is motivated by [19] where the same constraint is obtained from a geometric consideration.) In this case, the cycle Z is the mirror of our non-compact Lagrangian subspace L. The condition reads as

$$\sum_{i=1}^{N} \mathrm{e}^{-m_i^\alpha Z_\alpha - s_i} = \text{constant}. \tag{4.56}$$

This requires the following condition. Let us separate the set of i's, $I = \{i\}$, into groups $I = \cup_a I_a$ where i's in each group I_a have the same m_i^α. Then, the condition is that $\sum_{i\in I_a} e^{-s_i} = 0$ for each group I_a. In the case where $\ell = 1$ and $m_i = 1$ for all i, this reads

$$\sum_{i=1}^{N} e^{-s_i} = 0. \tag{4.57}$$

This replaces the classical constraint that L is smooth. In fact, in the asymptotic regions where two of s_i are mush smaller than others, this reduces to the classical constraint found in the geometric analysis (up to a shift in the a-angle). Thus, (4.57) can be considered as the quantum deformation of the classical constraint. To see this let us consider the $N = 4$ case and send r to infinity where X approaches $\mathbb{C}^3$. We will foucus on the region where $s_4 \sim t \to \infty$. Then, the constraint becomes

$$e^{-s_1} + e^{-s_2} + e^{-s_3} = 0. \tag{4.58}$$

It is easy to see that this reduces to the constraint depicted in Fig. 1 as long as one of $c_i = \mathrm{Re}(s_i)$ is large compared to the other two. One important point is that the region where the branches meet was excluded in the classical description but the distinct branches are smoothly connected in the quantum description.

There is actually a subtlety associated with the a-angles. Let us focus on the region $c_1, c_2 \ll c_3$. In the geometric discussion we have found that $c_1 = c_2$ and $a_1 = a_2$. On the other habd, the condition (4.58) is approximately $e^{-c_1+ia_1} + e^{-c_2+ia_2} = 0$ in this asymptotic region. This means that $c_1 = c_2$ but $a_1 = a_2 + \pi$. This should be understood in the original linear sigma model. Including this point, many things has to be clarified. For instance, it is important to understand whether there is a singularity in the parameter space. This requires a more careful dynamical study of this gauge system on the half-plane (or the strip). Also, as computed in [19] in the geometric model related to this LG model in the weak sense of [2], the space-time superpotential as a function of s_i's is generated. This means that a non-perturbative correction to the beta function for s_i is non-zero. How it is computed directly in the LG model should also be clarified. We hope to discuss those points elsewhere.

The Mirror of the Compact Torus T

Let us consider the case where S_i are parameters s_i and the original A-type D-brane is wrapped on a compact torus T. In this case, the constraint of Y_i is

$$Y_i = s_i. \tag{4.59}$$

Namely, the mirror D-brane is the D0-brane at a point (s_i) in $(\mathbb{C}^\times)^{N-1}$. Then, the condition that the superpotential is a constant on the brane is satisfied for any value of s_i. However, as we will discuss in Section 6, there is a further constraint that (s_i) must be at the critical point of $\widetilde{W}$. In the case where X is a compact toric manifold of positive first Chern class, one can find such a critical point as many as $\chi(X)$, the Euler number of X. Thus, we find $\chi(X)$ D-branes. For example, for $X = \mathbb{CP}^1 = S^2$, the mirror theory is the $\mathcal{N} = 2$ sine-Gordon model with the superpotential $\widetilde{W} = \mathrm{e}^{-Y} + q\,\mathrm{e}^{Y}$ where $q = \mathrm{e}^{-t}$. The critical points are $\mathrm{e}^{-Y} = \pm\sqrt{q}$. Thus, we have two kinds of D-branes with

$$\mathrm{e}^{-s_1} = \mathrm{e}^{-s_2} = \pm\sqrt{q}. \tag{4.60}$$

This means that $c_1 = c_2 = r/2$ and $a_1 = a_2 = 0$ or π. This corresponds to the D-branes at the equator of S^2 with the holonomy ± 1.

In the case where X is a non-compact Calabi-Yau manifold, $\widetilde{W}$ is of Liouville type and there is no critical point at finite Y_i. Therefore we cannot find a supersymmetric D0-brane.

5 B-type D-Branes and Tachyon Condensation

We turn to D-branes which preserve B-type supersymmetry. We set the phase trivial $\mathrm{e}^{i\beta} = 1$ unless otherwise stated. One of the basic examples is the space filling D-brane which is described in the non-linear sigma model (with the trivial B-field) by the full Neumann boundary condition for the bosonic fields, or more completely by

$$D_+\Phi^i = D_-\Phi^i \quad \text{at B-boundary}, \tag{5.1}$$

where Φ^i are the chiral superfields representing the complex coordinates of the target space. We would like to study more non-trivial examples in what follows.

5.1 The System of a D-Brane and an Anti-D-Brane

The first example we consider is the D0-brane in the complex plane. As before we realize the supersymmetric sigma model on the complex plane by the theory of a single chiral superfield $\Phi = \phi + \theta^\alpha \psi_\alpha + \theta^+\theta^- F + \cdots$. We use the following action

$$\begin{aligned} S_B &= \frac{1}{2\pi}\int_\Sigma \mathrm{d}^2x \left(|\partial_0\phi|^2 - |\partial_1\phi|^2 + \frac{i}{2}\overline{\psi}_-(\overleftrightarrow{\partial_0} + \overleftrightarrow{\partial_1})\psi_- + \frac{i}{2}\overline{\psi}_+(\overleftrightarrow{\partial_0} - \overleftrightarrow{\partial_1})\psi_+ + |F|^2 \right) \\ &\quad + \frac{i}{4\pi}\int_{\partial\Sigma} \mathrm{d}x^0 \left(\overline{\psi}_-\psi_+ - \overline{\psi}_+\psi_-\right), \end{aligned} \tag{5.2}$$

which is invariant under B-type supersymmetry without using equation of motion nor any boundary condition. In the standard approach, the D0-brane at $\phi = \phi_0$ is described by the supersymmetric Dirichlet boundary condition for the fields which is conveniently summarized as

$$\Phi = \phi_0 \quad \text{at B-boundary.} \tag{5.3}$$

In the "linear model approach" the same D-brane can be represented by the theory involving a boundary Fermi superfield Γ and the boundary interaction

$$S_{boundary} = \frac{1}{2\pi}\int_{\partial\Sigma} \mathrm{d}x^0 \,\mathrm{Re}\int \mathrm{d}\theta\, \Gamma(\Phi - \phi_0), \tag{5.4}$$

where of course the integration is along the B-type boundary $\theta^+ = \theta^-$. In this way of writing, it is manifest that ϕ_0 is the boundary chiral parameter. In what follows, we show that this latter formulation appears very naturally as the infra-red limit of the system of D2 and anti-D2 branes with a specific tachyon configuration. Basically, $\Phi - \phi_0$ that appears in (5.4) is the tachyon configuration.

5.1.1 The $\mathcal{N}=1$ Boundary Interaction

We consider the system of a D-brane and an anti-D-brane filling the target space (the complex plane in the above example). As is well-known e.g. [38], the low lying spectrum of this system consists of four parts — gauge fields A^1 and A^2 from p-p and $\bar{p}$-$\bar{p}$ strings[1] and tachyon fields T and $\overline{T}$ from p-$\bar{p}$ and $\bar{p}$-p strings — which constitute the Chan Paton matrix

$$\begin{pmatrix} A^1 & T \\ \overline{T} & A^2 \end{pmatrix}. \tag{5.5}$$

Thus, the worldsheet path-integral receives a factor

$$\mathrm{Tr}\, P \exp\left[-\oint_C \left\{ iA^1_\tau \begin{pmatrix} 1 & 0 \\ 0 & 0 \end{pmatrix} + iA^2_\tau \begin{pmatrix} 0 & 0 \\ 0 & 1 \end{pmatrix} + T_{(0)} \begin{pmatrix} 0 & 1 \\ 0 & 0 \end{pmatrix} + \overline{T}_{(0)} \begin{pmatrix} 0 & 0 \\ 1 & 0 \end{pmatrix} \right\} \mathrm{d}\tau \right] \tag{5.6}$$

from each boundary component C with coordinate τ. Here $A^i_\tau \mathrm{d}\tau$ is the pull-back of the gauge field A^i to the worldline C, and $T_{(0)}$ is the tachyon vertex operator in the zero picture, $T_{(0)} = \psi^I D_I T$ in which $D_I T = (\partial_I + iA^1_I - iA^2_I)T$ and $\psi^I = \psi^I_+ + \psi^I_-$. There is a convenient representation of the Chan-Paton factor using the complex Clifford algebra [39]. The algebra is generated by η and $\overline{\eta}$ obeying

$$\{\eta, \overline{\eta}\} = 1, \quad \eta^2 = \overline{\eta}^2 = 0, \tag{5.7}$$

and has the spinor representation spanned by vectors $|0\rangle$, $\overline{\eta}|0\rangle$ (where $|0\rangle$ is annihilated by η) on which Chan-Paton matrices are realized as

$$\begin{pmatrix} 1 & 0 \\ 0 & 0 \end{pmatrix} = \eta\overline{\eta}, \quad \begin{pmatrix} 0 & 0 \\ 0 & 1 \end{pmatrix} = \overline{\eta}\eta, \quad \begin{pmatrix} 0 & 1 \\ 0 & 0 \end{pmatrix} = \eta, \quad \begin{pmatrix} 0 & 0 \\ 1 & 0 \end{pmatrix} = \overline{\eta}. \tag{5.8}$$

Then, the Chan-Paton factor (5.6) can be considered as the partition function of the quantum mechanics represented on the ($|0\rangle$, $\overline{\eta}|0\rangle$) space with the Hamiltonian

$$H = iA^1_\tau \eta\overline{\eta} + iA^2_\tau \overline{\eta}\eta + T_{(0)}\eta + \overline{\eta}\,\overline{T}_{(0)}. \tag{5.9}$$

[1]We refer to Dp and anti-Dp branes as p and $\bar{p}$ respectively. "A p-p string" for example stands for an open string stretched between Dp and Dp.

Therefore, it has the path-integral representation

$$\int \mathcal{D}\eta\mathcal{D}\overline{\eta}\ \exp\left(-\oint_C \left[\overline{\eta}\mathcal{D}_\tau\eta + T_{(0)}\eta + \overline{\eta}\,\overline{T}_{(0)}\right] \mathrm{d}\tau\right), \tag{5.10}$$

where $\mathcal{D}_\tau = \mathrm{d}/\mathrm{d}\tau - iA^1_\tau + iA^2_\tau$. Note that the gauge transformation $A^1 \to A^1 - \mathrm{d}(\arg g_1)$, $A^2 \to A^2 - \mathrm{d}(\arg g_2)$, $T \to g_1 T g_2^{-1}$, and $\overline{T} \to g_2\overline{T}g_1^{-1}$ is compasated by the transformation

$$\eta \to g_2\eta g_1^{-1}, \quad \overline{\eta} \to g_1\overline{\eta}g_2^{-1}. \tag{5.11}$$

In other words, this can be considered as the gauge transformation property of the boundary fields η and $\overline{\eta}$.

The supersymmetric completion of this system can be described by introducing the $\mathcal{N} = 1$ boundary superfields including the tachyon T and the boundary fermion $\eta, \overline{\eta}$. (We come back to the Minkowski signature.) The former is a bosonic complex boundary superfield $\mathbf{T}$ which has an expansion

$$\mathbf{T} = T + i\theta_1\psi^I D_I T. \tag{5.12}$$

The latter is a fermionic superfield

$$\mathbf{\Gamma} = \eta + i\theta_1 G. \tag{5.13}$$

The boundary interaction that completes the one that appears in (5.10) is given by

$$S_{boundary} = \frac{1}{2\pi}\int_{\partial\Sigma} \mathrm{d}x^0\mathrm{d}\theta_1 \left(\overline{\mathbf{\Gamma}}\mathcal{D}^1\mathbf{\Gamma} + \mathbf{\Gamma}\mathbf{T} + \overline{\mathbf{T}\mathbf{\Gamma}}\right). \tag{5.14}$$

In the above expression, $\mathcal{D}\mathbf{\Gamma}$ is defined by $\mathcal{D}^1\mathbf{\Gamma} = \left(-i\frac{\partial}{\partial\theta_1} - 2\theta_1(\partial_0 - iA_0)\right)\mathbf{\Gamma}$ with

$$A_0 := A_I\partial_0\phi^I - \frac{i}{4}F_{IJ}\psi^I\psi^J, \tag{5.15}$$

where $A_I = A^1_I - A^2_I$ is the gauge field of the relative gauge group $U(1)^{rel}$ and F_{IJ} is its curvature $\partial_I A_J - \partial_J A_I$. This boundary interaction is invariant under the $U(1)^{rel}$ gauge symmetry

$$\begin{aligned} A_I &\to A_I - \partial_I \alpha(\phi), \\ \mathbf{\Gamma} &\to e^{-i\alpha(\phi)}\mathbf{\Gamma}, \\ \mathbf{T} &\to e^{i\alpha(\phi)}\mathbf{T}. \end{aligned} \tag{5.16}$$

It is also invariant under the gauge-modified supersymmetry transformation

$$\mathcal{Q}^1 = -i\frac{\partial}{\partial\theta_1} + 2\theta_1(\partial_0 + iqA_0) - iqA_I\psi^I, \tag{5.17}$$

where q is the $U(1)^{rel}$ charge of the field on which $\mathcal{Q}^1$ is acting. For example, $\mathbf{\Gamma}$ has $q = -1$, $\mathbf{T}$ has $q = 1$, while $\mathbf{\Phi}^I = \phi^I + i\theta_1\psi^I$ (the restriction of the bulk superfield on the boundary) has $q = 0$. One can check that the superfield $\mathbb{T}$, which is a function of the fields $\mathbf{\Phi}^I$, has the right transformation property.

Remark 1. In the above argument we have assumed that the path-integral (5.10) leads to the Hamiltonian (5.9). However, there is a standard operator ordering ambiguity that is fixed by an explicit regularization scheme; the first two terms in (5.9) could be replaced by $i(A^1_\tau - A^2_\tau)\eta\overline{\eta}$ or $i(A^2_\tau - A^1_\tau)\overline{\eta}\eta$, or a combination of them. This ambiguity would be annoying in the following discussion where we start with the Lagrangian. In what follows, instead of the above choice, we take the ordering where the action (5.14) corresponds to $A^2 = 0$ and $A^1 = A$. In this ordering, if we would like to have a non-zero A^2, we need to add the term

$$\Delta S_{boundary} = -\int\limits_{\partial\Sigma} dx^0 \left(A^2_I \partial_0 \phi^I - \frac{i}{4} F^2_{IJ}\psi^I\psi^J \right) \tag{5.18}$$

to (5.14) in which $A = A^1 - A^2$.

Remark 2. From the above result, one can also obtain the boundary interaction for the non-BPS D-branes [40, 41] in Type II string theory. In fact the latter is defined as the $(-1)^{F_L}$ orbifold of the brane-anti-brane system, where the orbifolding yields the reality constraint on $\mathbb{\Gamma}$ and $\mathbb{T}$ and also projects out the relative gauge field A_I. The resulting boundary interaction is nothing but the one used in [28–30]. The real boundary fermion η was originally introduced in [42] to reproduce the interaction rule of [43] for Type I D0-brane [44].

5.1.2 The Condition of $\mathcal{N}=2$ Supersymmetry

We would now like to find the $\mathcal{N}=2$ extension of the above result. Thus, we consider a supersymmetric sigma model on a Kahler manifold X formulated on the half space $\Sigma = \mathbf{R}\times(-\infty,0]$. On the B-boundary, the chiral superfields Φ^i representing the complex coordinates of X become boundary chiral superfields that are expanded as

$$\Phi^i = \phi^i + \theta\psi^i - i\theta\overline{\theta}\partial_0\phi^i, \tag{5.19}$$

where $\psi^i = \psi^i_+ + \psi^i_-$. The anti-chiral superfields $\overline{\Phi}^{\bar{i}}$ become boundary anti-chiral superfields.

Boundary Gauge Symmetry

We start with introducing boundary chiral gauge symmetry. Let $\Xi = \xi + \theta J - i\theta\overline{\theta}\partial_0\xi$ be a boundary Fermi superfield. We would like to construct a supersymmetric boundary Lagrangian that is invariant under the gauge transformation

$$\Xi \to \mathrm{e}^{iqA}\Xi, \tag{5.20}$$

where A is a boundary chiral superfield and q is the charge of Ξ. As in the bulk, it is appropriate to introduce a real boundary superfield V_b which transforms as

$$V_b \to V_b - iA + i\overline{A}. \tag{5.21}$$

Then a gauge invariant and supersymmetric Lagrangian is given by

$$L = \frac{1}{2}\int \mathrm{d}\theta\mathrm{d}\overline{\theta}\;\overline{\Xi}\mathrm{e}^{qV_b}\Xi. \tag{5.22}$$

One can choose a "Wess-Zumino gauge" where V_b has only the highest component

$$V_b = 2\theta\overline{\theta}A_0. \tag{5.23}$$

The residual gauge symmetry in the Wess-Zumino gauge is the one with $A = \alpha - i\theta\overline{\theta}\partial_0\alpha$ with real valued α which acts on the component fields as

$$\xi \to \mathrm{e}^{iq\alpha}\xi, \quad J \to \mathrm{e}^{iq\alpha}J,$$
$$A_0 \to A_0 - \partial_0\alpha.$$

In this gauge, the Lagrangian (5.22) is expressed as

$$L = \frac{i}{2}\bar{\xi}\mathcal{D}_0\xi - \frac{i}{2}\mathcal{D}_0\bar{\xi}\xi + \frac{1}{2}|J|^2, \tag{5.24}$$

where $\mathcal{D}_0\xi = (\partial_0 + iqA_0)\xi$. The ordinary supersymmetry transformation $\delta = \epsilon Q - \bar{\epsilon}\overline{Q}$ does not preserve the Wess-Zumino gauge. To find the supersymmetry transformation of the component fields ξ, J, A_0, we must modify it with a gauge transformation. It turns out that the required gauge transformation is the one with $iA = 2\theta\bar{\epsilon}A_0$ and we find

$$\delta_{\rm tot}\xi = \epsilon J + iq\alpha_*\xi, \tag{5.25}$$

$$\delta_{\rm tot}J = -2i\bar{\epsilon}\mathcal{D}_0\xi + iq\alpha_* J, \tag{5.26}$$

$$\delta_{\rm tot}A_0 = -\partial_0\alpha_*, \tag{5.27}$$

where α_* comes from an ambiguity in the choice of iA.

Gauge Field and Tachyon on the Brane

We would like to embed the above construction of gauge invariant interaction to the system of a D-brane and an anti-D-brane. We extend the $\mathcal{N} = 1$ superfield $\mathbf{\Gamma}$ to a boundary Fermi superfield Γ by replacing $i\theta_1$ by θ and adding the top component as

$$\Gamma = \eta + \theta G - i\theta\bar{\theta}\partial_0\eta. \tag{5.28}$$

We assign gauge charge $q = -1$ to Γ. Here, unlike in the above construction, the boundary gauge symmetry is linked to the gauge symmetry on the branes. That is, the field A_0 in (5.23) is a function of ϕ^I and ψ^I defined in (5.15) and the gauge transformation parameter α is also considered as a function of ϕ^I. There is also a tachyon field (5.12) that is a function of ϕ^I and ψ^I. The supersymmetry transformation of these fields are dictated by that of Φ^i and $\overline{\Phi}^{\bar{i}}$, which are in components given by

$$\delta\phi^i = \epsilon\psi^i, \quad \delta\psi^i = -2i\bar{\epsilon}\partial_0\phi^i, \tag{5.29}$$

$$\delta\overline{\phi}^{\bar{i}} = -\bar{\epsilon}\overline{\psi}^{\bar{i}}, \quad \delta\overline{\psi}^{\bar{i}} = 2i\epsilon\partial_0\overline{\phi}^{\bar{i}}. \tag{5.30}$$

It is easy to see that the transformation of A_0 takes the form (5.27) if and only if

$$F_{ij} = F_{\bar{\imath}\bar{\jmath}} = 0. \tag{5.31}$$

Namely if and only if the operator $\overline{\partial}_A = \mathrm{d}\bar{z}^{\bar{\jmath}} D_{\bar{\jmath}}$ is nilpotent and defines a holomorphic structure on the associated complex line bundle. The gauge transformation parameter α_* that appears in (5.27) is given by $\alpha_* = -\epsilon\psi^i A_i + \bar{\epsilon}\overline{\psi}^{\bar{\jmath}} A_{\bar{\jmath}}$. The tachyon field $\mathbf{T}$ has an extension to a boundary chiral superfield if and only if T is holomorphic,

$$D_{\bar{\jmath}} T = 0. \tag{5.32}$$

The chiral extension is then denoted by

$$\mathcal{T} = T + \theta\psi^i D_i T - i\theta\bar{\theta}\partial_0 T. \tag{5.33}$$

One can check that it has the right gauge and supersymmetry transformation property. A manifestly $\mathcal{N} = 2$ invariant boundary interaction that reduces to (5.14) is now given by

$$S_{boundary} = \frac{1}{2\pi}\int_{\partial\Sigma} \mathrm{d}x^0 \left[\frac{1}{2}\int \mathrm{d}\theta\mathrm{d}\bar{\theta}\, \overline{\Gamma}\, \mathrm{e}^{-V_b}\Gamma + \mathrm{Re}\int \mathrm{d}\theta\, i\Gamma\,\mathcal{T} \right]. \tag{5.34}$$

In this formulation, Γ and $\mathcal{T}$ have mass dimensions 0 and 1/2 respectively.

Like A_0, the supersymmetry transformation of $A_0^2 = A_I^2 \partial_0\phi^I - \frac{i}{4}F_{IJ}^2\psi^I\psi^J$ is a pure gauge if and only if $F_{ij}^2 = F_{\bar{\imath}\bar{\jmath}}^2 = 0$. Namely, the latter is the condition of $\mathcal{N} = 2$ supersymmetry of the anti-brane Wilson line term (5.18).

To summarize, *the system of a D-brane and an anti-D-brane has (B-type) $\mathcal{N} = 2$ worldsheet supersymmetry if and only if each of the the gauge fields defines a holomorphisc structure for the associated line bundle, and the tachyon field is a holomorphic section of the relative line bundle.*

5.1.3 The D0-Branes

Let us come back to the sigma model on the complex plane $X = \mathbb{C}$. We consider the following configuration:

$$A_\phi = A_{\overline{\phi}} = 0, \tag{5.35}$$

$$T = \phi - \phi_0. \tag{5.36}$$

The tachyon field T is holomorphic with respect to the trivial gauge connection and this should define an $\mathcal{N} = 2$ supersymmetric theory. The chiral tachyon superfield is given in this case by

$$\mathcal{T} = \Phi - \phi_0. \tag{5.37}$$

The boundary interaction is now

$$S_{boundary} = \frac{1}{2\pi} \int_{\partial\Sigma} \mathrm{d}x^0 \left[\frac{1}{2\lambda^2} \int \mathrm{d}\theta \mathrm{d}\bar{\theta}\, \overline{\Gamma}\Gamma + \mathrm{Re} \int \mathrm{d}\theta\, i\Gamma\, (\Phi - \phi_0) \right], \tag{5.38}$$

where we have introduced the coupling constant λ that has mass dimension 1/2 so that $\mathcal{T} = \Phi - \phi_0$ has dimension 0. The action, including the bulk one (5.2), is at most quadratic in all fields and the theory is renormalizable by itself. It is free and all the fields/parameters have their canonical dimension. In particular, the infra-red limit simply corresponds to $\lambda \to \infty$. Also the position parameter ϕ_0 is not renormalized, as we will show below in a more general context. In this way, we recover the interaction (5.4) that imposes the constraint $\Phi = \phi_0$. Thus, this system is identified as the D0-brane at $\phi = \phi_0$.

The above is compatible with some knowledge about D0-brane as a bound state of D2-anti-D2 system. For instance, it has a unit winding number at infinity [45, 42]. For this to be really identified as the D0-brane, $|T|$ should approach the vacuum value at infinity $|\phi| \to \infty$. As proposed/found in [29], the tachyon potential in this formulation is given by $\mathrm{e}^{-|T|^2/4}$ and the minimum is indeed at $|T| = \infty$. More importantly, following the proposal of [29], one can compute the open string field theory effective action as a function of λ (denoted by u in [29]) and it is minimized indeed at $\lambda \to \infty$. Furthermore, this computation gives the ratio of the tensions of the D2-brane and the D0-brane; We find $T_0/T_2 = (2\pi)^2$ which is the correct result (we are taking the unit $\alpha' = 1$: if we recover α' this is $T_0/T_2 = (2\pi)^2\alpha'$).

The Non-renormalization Theorem

Let us consider another configurations

$$A_\phi = A_{\overline{\phi}} = 0, \tag{5.39}$$

$$T = P(\phi, a_p) = a_0 + a_1\phi + \cdots a_k\phi^k. \tag{5.40}$$

This again preserves $\mathcal{N} = 2$ supersymmetry. The chiral tachyon superfield is given by

$$\mathcal{T} = P(\Phi, a_p), \tag{5.41}$$

and the boundary interaction is

$$S_{boundary} = \frac{1}{2\pi}\int_{\partial\Sigma} \mathrm{d}x^0 \left[\frac{1}{2\lambda^2}\int \mathrm{d}\theta\mathrm{d}\overline{\theta}\,\overline{\Gamma}\Gamma + \mathrm{Re}\int \mathrm{d}\theta\, i\Gamma P(\Phi, a_p) \right]. \tag{5.42}$$

This is no longer quadratic in fields and the system has a non-trivial interaction. It may appear hard to controle the quantum correction in this system. However, supersymmetry strongly constrains quantum corrections to the boundary F-term. We note that the system preserves the $U(1)$ R-symmetry under which Γ and Φ have charge 1 and 0 respectively. This shows that the boundary F-term is always linear in Γ, and the possible correction resides only in the boundary superpotential $P(\Phi, a_p)$. The effective boundary superpotential must be holomorphic in Φ and a_p's, and must respect the $U(1) \times U(1)$ global "symmetries" where Γ, Φ and a_p have charge $(-1, 0)$, $(0, 1)$ and $(1, -p)$ respectively. It is also required to approach the classical value $P(\Phi, a_p) = \sum_{p=0}^{k} a_p\Phi^p$ in the limit $a_p \to 0$. As in [35], these conditions are enough to constrain the boundary superpotential not to receive quantum correction at all. Of course, the boundary D-term $\int \mathrm{d}\theta\mathrm{d}\overline{\theta}\,\frac{1}{2\lambda^2}\overline{\Gamma}\Gamma$ can receive corrections.

Multiple D0-Branes

Let us consider the boundary superpotential

$$\mathcal{T} = \prod_{a=1}^{k}(\Phi - \phi_a). \tag{5.43}$$

We have seen that ϕ_a are not renormalized. When $\phi_a \neq \phi_b$ for $a \neq b$, the infra-red limit simply chooses one ϕ_a and there are in total k copies of the trivial fixed point we have considered above. In particular, the boundary entropy at the infra-red limit is k times that of the trivial one, and so is the tension

$$\text{tension} = kT_0. \tag{5.44}$$

Thus, this corresponds to k D0-branes located at $\phi_1, \ldots, \phi_k$.

Let us compute the open string Witten index — $\mathrm{Tr}(-1)^F$ of the theory formulated on the segment $0 \leq x^1 \leq \pi$. We first consider the case where one end of the string carries the above boundary interaction and the other end is free (pure Neumann corresponding to D2-brane). For the purpose of computing the index, we can take the zero mode approximattion where we ignore the x^1 dependence. Then, the supercharge Q is given by

$$\begin{aligned} Q &= \overline{\psi}\dot{\phi} - i\lambda\eta\prod_{a=1}^{k}(\phi - \phi_a) \\ &= -i\frac{\partial}{\partial\overline{\phi}}\begin{pmatrix} 0 & 0 \\ 1 & 0 \end{pmatrix} \otimes \begin{pmatrix} 1 & 0 \\ 0 & 1 \end{pmatrix} - i\lambda\prod_{a=1}^{k}(\phi - \phi_a)\begin{pmatrix} 1 & 0 \\ 0 & -1 \end{pmatrix} \otimes \begin{pmatrix} 0 & 1 \\ 0 & 0 \end{pmatrix} \end{aligned} \tag{5.45}$$

where the matrix representation is with respect to the basis $(|0\rangle, \overline{\psi}|0\rangle) \otimes (|0\rangle, \overline{\eta}|0\rangle)$. The supersymmetry equation $Q = Q^{\dagger} = 0$ is solved by the wave-function $f|0\rangle + g\overline{\psi}\overline{\eta}|0\rangle$ where f and g are functions of $\phi, \overline{\phi}$ that obey

$$\begin{pmatrix} -\partial/\partial\overline{\phi} & \lambda\prod_{a=1}^{k}(\phi - \phi_a) \\ \lambda\prod_{a=1}^{k}(\overline{\phi} - \overline{\phi}_a) & -\partial/\partial\phi \end{pmatrix}\begin{pmatrix} f \\ g \end{pmatrix} = 0. \tag{5.46}$$

This is identical to the Dirac equation for the fermion coupled to a k-vortex. As is well known, there are k normalizable solutions for any values of ϕ_a. Thus, we find that the Witten index is in this case

$$\mathrm{Tr}(-1)^F = k. \tag{5.47}$$

Next let us consider the case where both ends of the string carry the above boundary interaction. In that case, we have two boundary fermions η_0 and

η_π, one at $x^1 = 0$ and the other at $x^1 = \pi$, and the supercharge Q in the zero mode approximation is given by

$$\begin{aligned} Q &= \overline{\psi}\dot{\phi} - i\lambda(\eta_0 + \eta_\pi)\prod_{a=1}^{k}(\phi - \phi_a) \\ &= -i\frac{\partial}{\partial\overline{\phi}}\left(\begin{smallmatrix} 0 & 0 \\ 1 & 0 \end{smallmatrix}\right)\otimes\left(\begin{smallmatrix} 1 & 0 \\ 0 & 1 \end{smallmatrix}\right)\otimes\left(\begin{smallmatrix} 1 & 0 \\ 0 & 1 \end{smallmatrix}\right) \\ &\quad -i\lambda\prod_{a=1}^{k}(\phi - \phi_a)\left\{\left(\begin{smallmatrix} 1 & 0 \\ 0 & -1 \end{smallmatrix}\right)\otimes\left(\begin{smallmatrix} 0 & 1 \\ 0 & 0 \end{smallmatrix}\right)\otimes\left(\begin{smallmatrix} 1 & 0 \\ 0 & 1 \end{smallmatrix}\right) + \left(\begin{smallmatrix} 1 & 0 \\ 0 & -1 \end{smallmatrix}\right)\otimes\left(\begin{smallmatrix} 1 & 0 \\ 0 & -1 \end{smallmatrix}\right)\otimes\left(\begin{smallmatrix} 0 & 1 \\ 0 & 0 \end{smallmatrix}\right)\right\}. \end{aligned} \tag{5.48}$$

It is straightforward to show that the number of bosonic and fermionic supersymmetric ground states are the same. Thus the index in this case vanishes

$$\mathrm{Tr}(-1)^F = 0. \tag{5.49}$$

The results (5.47) and (5.49) are consistent with the interpretation of the boundary interaction as k D0-branes; As is well-known [46], the Witten index in this situation is the intersection number of the corresponding cycles. The complex plane $\mathbb{C}$ and k points in $\mathbb{C}$ have intersection number k, whereas the points in $\mathbb{C}$ have self intersection number zero, in agreement with (5.47) and (5.49).

D-Brane wrapped on a Divisor

The above construction generalizes straightforwardly to the case where the target space is an arbitrary Kahler manifold X. Let $\mathcal{L}$ be a holomorphic line bundle over X with a hermitian fibre metric h. We assume that it has a global holomorphic section F. Let us consider a boundary Fermi-superfield Γ with values in $\mathcal{L}^{-1}$ and the following boundary interaction

$$S_{boundary} = \frac{1}{2\pi}\int_{\partial\Sigma} \mathrm{d}x^0 \left[\int \mathrm{d}\theta \mathrm{d}\overline{\theta}\, h^{-1}(\Phi,\overline{\Phi})\overline{\Gamma}\Gamma + \mathrm{Re}\int \mathrm{d}\theta\, i\Gamma F(\Phi)\right]. \tag{5.50}$$

This corresponds to a configuration of the space filling D-brane and anti-D-brane. The D-brane and the anti-D-brane support the gauge bundle $\mathcal{L}$ (with the hermitian connection associated with h) and the trivial bundle $\mathcal{O}_X$ respectively, and the tachyon configuration is given by

$$\mathcal{O}_X \xrightarrow{F} \mathcal{L}. \tag{5.51}$$

We note that $h(\Phi, \overline{\Phi})$ that appears in the boundary D-term can receive a lot of quantum corrections, but $F(\Phi)$ is not renormalized at all.

In the case where the zero of F is simple and $F = 0$ is a smooth hypersurface D in X, we expect that $h(\Phi, \overline{\Phi})$ vanishes in the infra-red limit and we obtain a constraint

$$F(\Phi^i) = 0, \tag{5.52}$$

on the boundary. Then, the system can be identified as a D-brane wrapped on D. We note that the co-normal bundle (the normal cotangent bundle) of D in X is equal to $\mathcal{L}|_D$. Thus, the Fermi superfield Γ can be considered as taking values in the co-normal bundle of D. This is consistent with the interpretation of (the lowest component η of) Γ as the Gamma matrix in the normal bundle, which is the basic element of the Atiyah-Bott-Shapiro construction of lower-dimensional D-branes [42]. In a more general case where $F = 0$ does not define a smooth submanifold but a divisor D, the boundary interaction may flow to a non-trivial fixed point. For instance, if $F = f^k$ with f transversal, the system corresponds to k D-branes at $f = 0$. If $F = f_1 f_2$ where f_1 and f_2 have common zeroes, it corresponds to intersecting D-branes.

A remark is now in order. It is natural to expect that the data for the bundle $\mathcal{L}$ are chiral parameters. However, they do not appear in the boundary superpotential but in the transition function that relates Γ's in different patches. Thus, in the present description these parameters are not manifestly chiral. This is analogous to the similar drawback of the patchwise description of the non-linear sigma model: it is not manifest that Kahler class parameters are twisted chiral. As we will see shortly, if the bulk theory is realized as the linear sigma model, one can find a global descriptions where patch-wise definition is not necessary.

5.2 Multiple D-Branes and Anti-D-Branes

We next consider the system of m D-branes and $\bar{m}$ anti-D-branes. Here we do not construct the boundary interaction for general configurations, but provide constructions for a certain class of configurations. In particular, we consider the case where $m = \bar{m} = 2^{n-1}$ for some positive integer n. In such a case, the $2^{n-1} + 2^{n-1}$ dimensional Chan-Paton factor is realized on the irreducible representation S of the n-dimensional complex Clifford algebra

$$\{\eta_i, \overline{\eta}_i\} = \delta_{i,j}, \quad \{\eta_i, \eta_j\} = \{\overline{\eta}_i, \overline{\eta}_j\} = 0, \quad i, j = 1, \dots, n. \tag{5.53}$$

The representation S is constructed from a vector $|0\rangle$ annihilated by η_i by multiplying creation operators $\overline{\eta}_i$. It decomposes into two subspace S_+ and S_-, each with dimension 2^{n-1}, which consist of vectors $\overline{\eta}_{i_1}\overline{\eta}_{i_2}\cdots\overline{\eta}_{i_s}|0\rangle$ with even s and odd s respectively. The Chan Paton matrix takes the form (5.5) where the block decomposition corresponds to the decomposition $S = S_+ \oplus S_-$. The diagonal blocks A_1, A_2 are represented by even polynomials in $\eta_i, \overline{\eta}_i$ whereas the off-diagonal blocks $T, \overline{T}$ are represented by odd polynomials.

We consider the sigma model on $\mathbb{R}^{2n} = \mathbb{C}^n$ with the real coordinates x^μ or the complex coordinates $\phi^i = x^{2i-1} + ix^{2i}$. We take the following configuration

$$\begin{pmatrix} A^1 & T \\ \overline{T} & A^2 \end{pmatrix} = \sum_{\mu=1}^{2n} x^\mu \Gamma^\mu = \sum_{\mu=1}^{2n} x^\mu \begin{pmatrix} 0 & \sigma^\mu \\ \overline{\sigma}^\mu & 0 \end{pmatrix}, \tag{5.54}$$

where Γ^μ are the $2n$ dimensional Gamma matrices. This is motivated by the Atiyah-Bott-Shapiro construction [47] that has been proposed to be identified in [42] as the tachyon configuration for the condimension $2n$ D-brane. (A linear profile is also the one that is seen from the D-brane probe [49].) Since the Gamma matrices Γ^μ are the real and the imaginary parts of η_i, this configuration is represented by

$$\sum_{i=1}^{n} \eta_i \phi^i + \sum_{i=1}^{n} \overline{\eta}_i \overline{\phi}^i. \tag{5.55}$$

Repeating what we have done in the system of one D-brane and one anti-D-brane, we obtain the following boundary interaction corresponding to this configuration:

$$S_{boundary} = \frac{1}{2\pi} \sum_{i=1}^{n} \int_{\partial\Sigma} dx^0 \left[\frac{1}{2\lambda_i^2} \int d\theta d\bar{\theta}\, \overline{\Gamma}_i \Gamma_i + \mathrm{Re} \int d\theta\, i\Gamma_i \Phi^i \right]. \qquad (5.56)$$

Here Γ_i are the boundary Fermi superfields with the lowest component η_i.

This boundary interaction is quadratic in all fields and is renormalizable by itself. In fact, this is simply the sum of n copies of the system of a D0-brane in the complex plane. The parameters λ_i go to infinity in the infra-red limit and we obtain the constraint

$$\Phi^i = 0 \quad \text{at B-boundary.} \qquad (5.57)$$

Thus, this system is identified as a D0-brane at the origin of $\mathbf{C}^n$. The partition function of the system is simply given by the product $\prod_{i=1}^{n} Z(\lambda_i)$ where $Z(\lambda_i)$ is the partition function for the system corresponding to a D0-brane in the complex plane. Following the proposal of [29] on the open string field theory action, we can compute the ratio of the tensions, say of D9-brane and D$(9-2n)$-brane. We find $T_{9-2n}/T_9 = (2\pi)^{2n}$ which is again the correct result (in $\alpha' = 1$).

We note here another representation of the tachyon configuration (5.54) [48, 47]. The spinor representation S tensored with the trivial bundle $\mathcal{O}$ over $\mathbf{C}^n$ can be identified as the exterior algebra over $\mathcal{O}^{\oplus n}$ under which

$$\eta_{i_1}\eta_{i_2}\cdots\eta_{i_k}|\mathrm{top}\rangle \longleftrightarrow e_{i_1} \wedge e_{i_2} \wedge \cdots \wedge e_{i_k}, \qquad (5.58)$$

where $|\mathrm{top}\rangle := \overline{\eta}_1\overline{\eta}_2\cdots\overline{\eta}_n|0\rangle$ and $(e_1, \ldots, e_n)$ is a "basis" of $\mathcal{O}^{\oplus n}$. The operator $\sum_{i=1}^{n} \eta_i \phi^i$ that appears in the expression (5.55) is then identified as the wedge product by $\phi = \sum_{i=1}^{n} e_i \phi^i$. From this we see that the tachyon configuration is obtained by folding the complex

$$\mathcal{O} \xrightarrow{\phi\wedge} \bigwedge^1 \mathcal{O}^{\oplus n} \xrightarrow{\phi\wedge} \bigwedge^2 \mathcal{O}^{\oplus n} \xrightarrow{\phi\wedge} \cdots \xrightarrow{\phi\wedge} \bigwedge^n \mathcal{O}^{\oplus n} \qquad (5.59)$$

into maps between $\bigwedge^{even} \mathcal{O}^{\oplus n}$ and $\bigwedge^{odd} \mathcal{O}^{\oplus n}$. The complex (5.59) is called the Koszul complex. We note that the operator $\phi\wedge$ appears in the boundary F-term in (5.56).

5.3 D-Branes in Gauge Theory

Now, we generalize the above construction of B-type D-branes to supersymmetric gauge theory. This yields a global description of D-branes in toric manifolds.[1]

Space-filling D-Brane

We first provide the the boundary interaction corresponding to the space-filling D-brane in the non-linear sigma model limit. For simplicity we will mainly be talking about the $U(1)$ gauge theory with a single chiral matter field, but we will freely move to more genral cases as the generalization is obvious. For the bulk action S and other things, we use the notation fixed in Section 4.1.

The supersymmetry variation of the action S in (4.9) is a non-vanishing boundary term, as studied in [13]. If we modify the action as

$$\begin{aligned} S_B \;=\; S + \frac{1}{4\pi}\int_{\partial\Sigma} \mathrm{d}x^0 \Big[& i(\overline{\psi}_-\psi_+ - \overline{\psi}_+\psi_-) - i(\sigma - \overline{\sigma})|\phi|^2 \\ & + \frac{1}{2e^2}\left\{\partial_1|\sigma|^2 + 2\mathrm{Im}(\sigma(D + iv_{01}))\right\} + i(t\sigma - \bar{t}\overline{\sigma}) \Big], \end{aligned} \tag{5.60}$$

it is invariant under B-type supersymmetry (with the trivial phase $e^{i\beta} = 1$)

$$\delta S_B = 0. \tag{5.61}$$

The boundary condition derived by varying the action S_B contains $\psi_+ = \psi_-$ and $\sigma = \overline{\sigma}$ which are completed as

$$\begin{aligned} \mathcal{D}_+\Phi &= \mathcal{D}_-\Phi, \\ \Sigma &= \overline{\Sigma}, \end{aligned} \qquad \text{at B-boundary,} \tag{5.62}$$

where $\mathcal{D}_\pm = e^{-V} D_\pm e^V$. Note that the first condition is gauge invariant since $e^V\Phi$ transforms as $e^V\Phi \to e^{i\overline{A}} e^V \Phi$ with $\overline{A}$ anti-chiral. We also have another

[1] Boundary conditions in gauged linear sigma models for B-type D-branes were studied in [13, 15]. A construction similar to the one in this subsection has been presented in the talk [26].

boundary condition

$$v_{01} = -e^2\theta. \tag{5.63}$$

Under the boundary condition (5.62), the boundary terms in (5.60) simplifies so that the action becomes

$$S_B = S + \frac{\theta}{2\pi}\int\limits_{\partial\Sigma} dx^0\,\frac{\sigma+\overline{\sigma}}{2}. \tag{5.64}$$

In the sigma model limit $e\sqrt{r}\to\infty$, e.g. in the $\mathbb{C}\mathrm{P}^{N-1}$ model where there are N fields of charge 1, the field σ is frozen at

$$\sigma = \frac{\sum_{i=1}^{N}\psi_{i-}\overline{\psi}_{i+}}{\sum_{i=1}^{N}|\phi|^2}. \tag{5.65}$$

Thus, the boundary term in (5.64) is interpreted as a fermion bilinear in the non-linear sigma model. One can see that this is equal to the fermion bilinear boundary term in the supersymmetric B-field coupling

$$\frac{1}{2}\int\limits_{\Sigma} B_{IJ}d\phi^I\wedge d\phi^J + \frac{i}{4}\int\limits_{\partial\Sigma} dx^0 B_{IJ}\psi^I\psi^J. \tag{5.66}$$

Here the bulk B-field term comes from the Theta term $\frac{\theta}{2\pi}\int_\Sigma v_{01}d^2x$ where v_μ is given in the sigma model limit by

$$v_\mu = \frac{i}{2}\frac{\sum_{i=1}^{N}\overline{\phi_i}\overleftrightarrow{\partial}_\mu\phi_i}{\sum_{i=1}^{N}|\phi_i|^2}. \tag{5.67}$$

This is a gauge field of the line bundle $\mathcal{O}(1)$ over $\mathbb{C}\mathrm{P}^{N-1}$.

As remarked in [13], there are different formulations of the non-linear sigma model when the boundary is coupled to a $U(1)$ gauge field with non-vanishing field strength. This is true also when the B-field is non-vanishing. In one formulation we change the boundary condition of the bosonic fields from pure Neumann to mixed Dirichlet-Neumann condition, and accordingly the boundary condition of the fermons is changed as well. In the other formulation, we do not touch the boundary condition (for both bosons and fermions) but, for supersymmetry, we add a fermion-bilinear term on the boundary as (5.66).

As explained in [50] in the bosonic string theory, the two formulations lead to the same space-time theory. The consideration in [13] corresponds to the first formulation. Here we took the second formulation; (5.62) reduces to the pure Neumann boundary condition in the non-linear sigma model limit and the boundary term in (5.64) reduces to the boundary term in (5.66).

The Boundary Interaction for Lower-dimensional Branes

Now we construct a boundary interaction corresponding to brane-anti-brane system with tachyon condensation. To be specific, we consider the $U(1)$ gauge theory with N chiral superfields Φ_i of charge Q_i which reduces at low enough energies to the non-linear sigma model on a toric manifold $X = \mathbb{C}^N//\mathbb{C}^\times$. We denote by $\mathcal{O}_X(p)$ the line bundle $(\mathbb{C}^N \times \mathbb{C})//\mathbb{C}^\times$ over X where $\lambda \in \mathbb{C}^\times$ acts on the second factor $c \in \mathbb{C}$ by $c \mapsto \lambda^p c$.

We first recall that the bulk gauge symmetry and the vector superfield become, when restricted to B-boundary, a boundary (chiral) gauge symmetry and a boundary vector superfield. The bulk Wess-Zumino gauge reduceds to the Wess-Zumino gauge on the boundary where the vector superfield is expressed as

$$V = 2\theta\overline{\theta}\left(v_0 - \frac{\sigma + \overline{\sigma}}{2}\right). \tag{5.68}$$

In the sigma model limit (where v_μ and σ are given by (5.67) and (5.65) in the case $X = \mathbb{C}\mathrm{P}^{N-1}$), this combination $v_0 - (\sigma + \overline{\sigma})/2$ is precisely of the form $A_0 = A_I \partial_0 \phi^I - \frac{i}{4} F_{IJ} \psi^I \psi^J$ that appears in (5.15), where A_I is the gauge field of the line bundle $\mathcal{O}_X(1)$.

Let $F(\Phi)$ be a polynomial of Φ_i of charge q. For this we introduce a boundary Fermi superfield Γ of charge $-q$. We consider the following boundary interaction

$$S_{boundary} = \frac{1}{2\pi}\int_{\partial\Sigma} \mathrm{d}x^0 \left[\frac{1}{2\lambda^2}\int \mathrm{d}\theta \mathrm{d}\overline{\theta}\, \overline{\Gamma}\, \mathrm{e}^{-qV}\Gamma + \mathrm{Re}\int \mathrm{d}\theta\, \Gamma F(\Phi)\right] \tag{5.69}$$

This is manifestly gauge invariant and supersymmetric. At low enough energies where the bulk theory reduces to the non-linear sigma model on X, Γ reduces to a boundary Fermi superfield with values in $\mathcal{O}_X(-q)$ and $F(\Phi)$ determines a holomorphic section of $\mathcal{O}_X(q)$. In this limit, the above boundary interaction reduces to the one given in (5.50) where the hermitian metric h is the one coming from the standard Euclidean metric of $\mathbf{C}$. In particular, this boundary interaction corresponds to the tachyon configuration

$$\mathcal{O}_X \xrightarrow{F} \mathcal{O}_X(q). \tag{5.70}$$

If F has only simple zero, this corresponds to the D-brane wrapped on the hypersurface D

$$F = 0. \tag{5.71}$$

Of course, the charge $-q$ for Γ is compatible with the fact that the co-normal bundle of the hypersurface D is equal to $\mathcal{O}_X(-q)|_D$.

Intersection of Hypersurfaces

It is straightforward to generalize the construction for D-branes wrapped on the intersection of hypersurfaces $F_1 = 0, \ldots, F_l = 0$. If F_β is a charge q_β polynomial, the boundary interaction is just

$$S_{boundary} = \sum_{\beta=1}^{l} \frac{1}{2\pi} \int_{\partial\Sigma} \mathrm{d}x^0 \left[\frac{1}{2\lambda_\beta^2} \int \mathrm{d}\theta \mathrm{d}\overline{\theta}\, \overline{\Gamma}_\beta\, \mathrm{e}^{-q_\beta V} \Gamma_\beta + \mathrm{Re} \int \mathrm{d}\theta\, \Gamma_\beta F_\beta(\Phi) \right], \tag{5.72}$$

where Γ_β is a boundary Fermi superfield of charge $-q_\beta$. We note that the vector R-symmetry (that becomes the boundary R-symmetry at B-boundary) is always unbroken in the bulk theory. Thus, the boundary F-term is always linear in Γ_β, and the non-renormalization theorem applies to F_β, as before.

The tachyon configuration is the one obtained from the Koszul complex associated with $\mathcal{E} = \oplus_{\beta=1}^{l} \mathcal{O}_X(q_\beta)$ and $F = \sum_{\beta=1}^{l} e_\beta F_\beta$;

$$\mathcal{O}_X \xrightarrow{F\wedge} \bigwedge^1 \mathcal{E} \xrightarrow{F\wedge} \bigwedge^2 \mathcal{E} \xrightarrow{F\wedge} \cdots \xrightarrow{F\wedge} \bigwedge^l \mathcal{E}, \tag{5.73}$$

by folding into the maps between $\bigwedge^{even} \mathcal{E}$ and $\bigwedge^{odd} \mathcal{E}$.

6 B-Type D-Branes in Landau-Ginzburg Model

In this section, we consider B-type D-branes in a theory with bulk superpotential. This leads to the construction of D-branes in linear sigma models corresponding to hypersurfaces or complete intersections in toric manifolds. Another motivation is to study the mirror of the A-type D-branes identified in Section 4.

Let us consider a Landau-Ginzburg model with the superpotential $W(\Phi)$. The bulk action includes the F-term

$$S_W = \int_\Sigma \mathrm{d}^2x\, \mathrm{Re} \int \mathrm{d}^2\theta\, W(\Phi). \tag{6.1}$$

The B-type supersymmetry transformation of this F-term is

$$\delta S_W = \int_{\partial\Sigma} \mathrm{d}^2x\, \mathrm{Re} \int \mathrm{d}^2\theta \left[-\overline{\epsilon}(\overline{\mathcal{Q}}_+ + \overline{\mathcal{Q}}_-)W(\Phi) + \epsilon(\mathcal{Q}_+ + \mathcal{Q}_-)W(\Phi)\right]. \tag{6.2}$$

Using the relations $\mathcal{Q}_\pm W(\Phi)|_{\overline{\theta}_\pm=0} = \frac{\partial}{\partial\theta^\pm} W(\Phi)|_{\overline{\theta}^\pm=0}$ and $\overline{\mathcal{Q}}_\pm W(\Phi) = (\overline{D}_\pm - 2i\theta^\pm\partial_\pm)W(\Phi) = -2i\theta^\pm\partial_\pm W(\Phi)$, and performaing the partial integration, we obtain

$$\begin{aligned} \delta S_W &= \int_{\partial\Sigma} \mathrm{d}x^0\, \mathrm{Re} \int \mathrm{d}^2\theta\, i\overline{\epsilon}(\theta^+ - \theta^-)W(\Phi) \\ &= \int_{\partial\Sigma} \mathrm{d}x^0\, \mathrm{Re} \int_B \mathrm{d}\theta\, (-i\overline{\epsilon})W(\Phi). \end{aligned} \tag{6.3}$$

This vanishes if we require

$$W(\Phi) = \text{constant} \quad \text{at B-boundary.} \tag{6.4}$$

This is the case if we consider a D-brane on which W is a constant, which is the condition found in [13] by component analysis.

We can apply this argument to the linear sigma model which corresponds to sigma models on a complete intersection M in a toric manifold X. In this case, it is natural to choose zero as the constant value of the superpotential

(6.4). Combining this with the construction in the previous subsection, we can construct the boundary interaction that corresponds to D-branes in M wrapped on the holomorphic cycles defined as the intersection of M and $F_1 = \cdots = F_l = 0$.

6.1 D0-Branes in Massive Theory

In what follows, we focus our attention to D0-branes in massive LG models. By *massive*, we mean that all the critical points of the superpotential W are non-degenerate. In other words, all the critical points are isolated and the Hessian (the determinant of the second derivative matrix) is non-vanishing at each of them. More general cases such as higher dimensional D-branes in scale invariant models are also important, say, for string theory applications, but they will be discussed elsewhere.

Since D0-brane is a point, the condition (6.4) is vacuous. However, if the point is not one of the critical points, any classical configuration will not attain the zero energy. We expect that the worldsheet supersymmetry will be spontaneously broken.[1] To examine this, let us look at the expression of the supercharge

$$Q = \frac{1}{2\pi} \int \mathrm{d}x^1 \left\{ g_{i\bar{\jmath}}(\overline{\psi}_-^{\bar{\jmath}} + \overline{\psi}_+^{\bar{\jmath}})\partial_0\phi^i + g_{i\bar{\jmath}}(\overline{\psi}_-^{\bar{\jmath}} - \overline{\psi}_+^{\bar{\jmath}})\partial_1\phi^i + (\psi_-^i - \psi_+^i)\partial_i W \right\}. \tag{6.5}$$

Since the boundary point, say at $x^1 = \pi$ is locked at that point, we see that the supersymmetry is indeed broken for any configuration. Thus, we will not consider such a D-brane. In other words, *D0-branes must be located at one of the critical points of W*.

6.2 Supersymmetric Ground States

Let us compute the supersymmetry index or, if possible, determine the supersymmetric ground states for an open string stretched between two critical

[1]Non-zero energy does not necessarily mean supersymmetry breaking, as the example of A-type D-branes in LG model shows [13].

points p_a and p_b of W. Let us first consider the case $p_a \neq p_b$. Then, by the same reason as above, there will be no supersymmetric ground states. In particular, the supersymmetry index vanishes

$$\mathrm{Tr}_{ab}(-1)^F = 0 \quad a \neq b. \tag{6.6}$$

Let us next consider the case where $a = b$. One can always choose the variables so that $\phi_i = 0$ at p_a and $W = \sum_{i=1}^n m\Phi_i^2 + \cdots$ where $+\cdots$ are cubic or higher order terms. For the purpose of computing the index, one can deform the Kahler potential so that it takes the form $K = \sum_{i=1}^n |\Phi_i|^2 + \cdots$ and one can also neglect the higher order terms $+\cdots$ in K and W. Then, the computation reduces to that of the free massive theory.

The Free Massive Theory

We are thus led to consider the theory of a single chiral superfield Φ with the Kahler potential $|\Phi|^2$ and the superpotential

$$W = m\Phi^2. \tag{6.7}$$

Since this theory is free, not only the Witten index, but also the complete spectrum can be determined.

The action of the strip $\mathbf{R} \times [0, \pi]$ after elimination of the auxiliary field is given by

$$\begin{aligned} S = \frac{1}{2\pi} \int_{\mathbf{R}\times[0,\pi]} \mathrm{d}^2x \Big(& |\partial_0\phi|^2 - |\partial_1\phi|^2 - |m\phi|^2 + i\overline{\psi}_-(\partial_0 + \partial_1)\psi_- + i\overline{\psi}_+(\partial_0 - \partial_1)\psi_+ \\ & - m\psi_+\psi_- - \overline{m}\overline{\psi}_-\overline{\psi}_+ \Big). \end{aligned} \tag{6.8}$$

The boundary condition is that of the D-brane at the critical point $\phi = 0$:

$$\Phi = 0 \quad \text{at B-bounday}. \tag{6.9}$$

In components, this is

$$\left.\begin{aligned} &\phi = 0, \\ &\psi_- + \psi_+ = 0, \end{aligned}\right. \quad \text{at } x^1 = 0, \pi. \tag{6.10}$$

The B-type supersymmetry transformation is given by

$$\begin{aligned}\delta\phi &= \epsilon(\psi_- + \psi_+),\\ \delta(\psi_- + \psi_+) &= -2i\bar{\epsilon}\partial_0\phi,\\ \delta(\psi_- - \psi_+) &= 2i\bar{\epsilon}\partial_1\phi + 2\epsilon\overline{m}\overline{\phi}.\end{aligned} \tag{6.11}$$

This motivates us to change the variables as

$$b := \frac{\psi_- + \psi_+}{\sqrt{2}}, \quad c := \frac{\psi_- - \psi_+}{\sqrt{2}}, \tag{6.12}$$

so that the boundary condition is simply $b = 0$ at $x^1 = 0, \pi$. In terms of these variables, the fermionic part of the action can be written as

$$S_F = \frac{1}{2\pi}\int_{\mathbf{R}\times[0,\pi]} \mathrm{d}^2x\Big(i\bar{b}\partial_0 b + i\bar{c}\partial_0 c + \bar{b}(i\partial_1 c + \overline{mc}) + (-i\partial_1\bar{c} + mc)b \Big) \tag{6.13}$$

From this we see that we also need to impose the boundary condition $i\partial_1 c + \overline{mc} = 0$ at $x^1 = 0, \pi$. Thus, we are led to the following mode expansion:

$$\phi = \sum_{n=1}^{\infty} \phi_n(\mathrm{e}^{inx^1} - \mathrm{e}^{-inx^1}), \tag{6.14}$$

$$b = \sum_{n=1}^{\infty} b_n(\mathrm{e}^{inx^1} - \mathrm{e}^{-inx^1}), \tag{6.15}$$

$$i\partial_1 c + \overline{mc} = \sum_{n=1}^{\infty} \sqrt{n^2 + |m|^2}\, d_n(\mathrm{e}^{inx^1} - \mathrm{e}^{-inx^1}), \tag{6.16}$$

where $\sqrt{n^2 + |m|^2}$ is for later convenience. The last equation and its complex conjugate are solved by

$$c = c_0(x) - \sum_{n=1}^{\infty}\left\{ \frac{n}{\sqrt{n^2+|m|^2}} d_n(\mathrm{e}^{inx^1} + \mathrm{e}^{-inx^1}) + \frac{\overline{m}}{\sqrt{n^2+|m|^2}} d_n^{\dagger}(\mathrm{e}^{inx^1} - \mathrm{e}^{-inx^1})\right\}. \tag{6.17}$$

where $c_0(x)$ solves the equations $i\partial_1 c + \overline{mc} = 0$ and $-i\partial_1\bar{c} + mc = 0$. A general solution of the latter is given by

$$c_0(x) = c_0^+\,\mathrm{e}^{|m|x^1} + c_0^-\,\mathrm{e}^{-|m|x^1} \tag{6.18}$$

in which $c_0^\pm$ is "real" in the sense that

$$(c_0^\pm)^\dagger = \mp i \frac{m}{|m|} c_0^\pm . \tag{6.19}$$

In terms of these Fourier modes, the action can be written as $S = \int dt L$ where

$$L = i\bar{\xi}\dot{\xi} + \sum_{n=1}^{\infty} \left\{ |\dot{\phi}_n|^2 - (n^2 + |m|^2)|\phi_n|^2 + i b_n^\dagger \dot{b}_n + i d_n^\dagger \dot{d}_n + \sqrt{n^2 + |m|^2}(b_n^\dagger d_n + d_n^\dagger b_n) \right\}, \tag{6.20}$$

in which ξ is the following complex combination of c_0^+ and c_0^-:

$$\xi = \sqrt{\frac{e^{\pi|m|} - e^{-\pi|m|}}{4\pi|m|}} \left(\xi^+ + i\xi^- \right); \quad \xi^\pm = e^{\pm\pi|m|/2} \sqrt{\mp i \frac{m}{|m|}} c_0^\pm . \tag{6.21}$$

This system is indeed supersymmetric with respect to the variation

$$\delta\phi_n = \epsilon b_n, \quad \delta b_n = -i\bar{\epsilon}\dot{\phi}_n, \quad \delta d_n = \bar{\epsilon}\sqrt{n^2 + |m|^2}\phi_n, \quad \delta\xi = 0, \tag{6.22}$$

that follows from (6.11). The system of ϕ_n, b_n, d_n for each n is the (complexified) supersymmetric harmonic oscillator and the quantization is standard. In particular, it has a unique supersymmetric ground state $|0\rangle_n$. On the other hand, the zero mode system of ξ has vanishing Hamiltonian and the two states $|0\rangle_0$ and $\bar{\xi}|0\rangle_0$ are both supersymmetric ground states. Thus, we see that the total system has two supersymmetric ground states

$$|0\rangle, \quad \bar{\xi}|0\rangle, \tag{6.23}$$

where $|0\rangle = \otimes_{n=0}^{\infty} |0\rangle_n$. In particular the index vanishes

$$\mathrm{Tr}(-1)^F = 0. \tag{6.24}$$

The General Case

From the above analysis, we conclude in the general massive LG model that

$$\mathrm{Tr}_{ab}(-1)^F = 0 \quad \textit{for any } a \textit{ and } b. \tag{6.25}$$

Namely, the index vanishes not only for $a \neq b$ but also for $a = b$. Moreover, in the case where the quadratic approximation around the critical point is good enough, we see from the above analysis that (for $a = b$) there are 2^n supersymmetric ground states if there are n LG fields, half bosonic and half fermionic;

$$|0\rangle, \quad \overline{\xi}_i|0\rangle, \quad \overline{\xi}_i\overline{\xi}_j|0\rangle, \quad \ldots, \quad \overline{\xi}_1\overline{\xi}_2\cdots\overline{\xi}_n|0\rangle. \tag{6.26}$$

We claim that this is true for *any* critical point p_a of a massive LG theory if the manifold on which the superpotential is defined is Calabi-Yau (like $\mathbb{C}^n$ or $(\mathbf{C}^\times)^n$). Namely, in such a theory, quadratic approximation is always exact as long as determining the supersymmetric ground states is concerned. This can be seen by the correspondence of the supersymmetric ground states and the boundary chiral ring elements. To explain this it is best to perform topological twisting.

6.3 Open Topological Landau-Ginzburg Model

The twisting can be performed, as usual, by gauging the $U(1)$ R-symmetry by the worldsheet spin connection. In the present case, the vector $U(1)$ R-symmetry is broken by the massive superpotential. We assume here that the target space M on which the LG superpotential is defined is a non-compact Calabi-Yau manifold so that the axial $U(1)$ R-symmetry is unbroken and we can twist the theory (B-twist). Twisting changes the spin of the fields as shown by the new notation below

$$\begin{aligned} \overline{\psi}_-^{\bar{i}} &= \psi^{\bar{i}}, \quad \overline{\psi}_+^{\bar{i}} = \overline{\psi}^{\bar{i}}, \\ \psi_-^{i} &= \rho_z^i, \quad \psi_+^i = \rho_{\bar{z}}^i, \end{aligned} \tag{6.27}$$

($\psi, \overline{\psi}$ are scalars while ρ_z and $\rho_{\bar{z}}$ define a 1-form). Energy-momentum tensor is exact with respect to the operator

$$Q = \overline{Q}_+ + \overline{Q}_-, \tag{6.28}$$

which is scalar after twisting, and we define the space of "physical operators" as the Q-cohomology of the operators. The variation of the fields $\delta = -\bar{\epsilon}Q$ in the new notation is given by

$$\begin{array}{ll} \delta\phi^i = 0, & \\ \delta(\psi^{\bar{\imath}} - \overline{\psi}^{\bar{\imath}}) = \bar{\epsilon} g^{\bar{\imath}j}\partial_j W, & \delta\overline{\phi}^{\bar{\imath}} = \bar{\epsilon}(\psi^{\bar{\imath}} + \overline{\psi}^{\bar{\imath}}), \\ \delta\rho^i_\mu = -2\bar{\epsilon} J^\nu{}_\mu \partial_\nu \phi^i, & \delta(\psi^{\bar{\imath}} + \overline{\psi}^{\bar{\imath}}) = 0, \end{array} \tag{6.29}$$

where $J^\nu{}_\mu$ is the worldsheet complex structure. From this we see that the physical operators are the holomorphic functions of ϕ^i (i.e. holomorphic functions on M) modulo functions of the form $v^i\partial_i W$ where $v^i\partial_i$ is a holomorphic vector field on M. The physical operators are in one-to-one correspondense with the supersymmetric ground states of the original LG model on the periodic circle (which are identified as the Q-cohomology classes of states). The state corresponding to an operator O is the one that appears at the boundary circle of the semi-infinite cigar of the twisted model where O is inserted at the tip [51]. The correlation functions of operators $O_1, \ldots, O_s$ on a Riemann surface of genus g is given by

$$\langle O_1 \cdots O_s \rangle_g = \sum_{p_a:\text{critical point}} O_1(p_a) \cdots O_s(p_a)(\det \partial_i\partial_j W)^{g-1}(p_a). \tag{6.30}$$

Here the coordinates defining the derivatives $\partial_i\partial_j W$ are such that the holomorphic n-form Ω is expressed as $\mathrm{d}\phi^1 \wedge \cdots \wedge \mathrm{d}\phi^n$ (a choice of Ω is required because of the chiral fermionic determinant).

The above summarizes the topological LG model on a Riemann surface without a boundary [52]. One important thing to notice is that the operator Q in (6.28) is the one that is conserved when the theory is formulated on the strip with B-type boundary conditions/interactions. This suggests that one can also consider twisting B-type boundary theory. (B-boundary breaks the axial $U(1)_R$ but this is not a problem: the worldsheetboundary also breaks the local rotation symmetry.) For D0-branes in a massive LG model, we only have to translate the boundary condition. The Wick rotation to the Euclidean signature has to be made first on the strip [13]: we continue $x^0 \to -ix^2$ and the complex coordinate z that appears in (6.27) is defined by $z = x^1 + ix^2$. Then, the boundary condition $\phi^i = \text{const}$, $\psi^i_- + \psi^i_+ = 0$ and $\overline{\psi}^{\bar{\imath}}_- + \overline{\psi}^{\bar{\imath}}_+ = 0$ is translated as

$$\phi^i = const, \tag{6.31}$$

$$\rho^i_n = 0, \tag{6.32}$$
$$\psi^{\bar{\imath}} + \overline{\psi}^{\bar{\imath}} = 0, \tag{6.33}$$

where *const* is the coordintae value of a critical point, say p_a, and ρ^i_n is the normal component to the boundary. In particular, the fields remaining on the boundary are $\theta^{\bar{\imath}} := \psi^{\bar{\imath}} - \overline{\psi}^{\bar{\imath}}$, the tangent component ρ^i_τ of ρ^i, and the normal derivatives of all fields including $\eta^{\bar{\imath}} := \psi^{\bar{\imath}} + \overline{\psi}^{\bar{\imath}}$. The Q-variation of these fields can be read from (6.29) as

$$\begin{aligned} &\delta\partial_n\phi^i = 0, \\ &\delta\theta^{\bar{\imath}} = 0, \\ &\delta\rho^i_\tau = 2\bar{\epsilon}\partial_n\phi^i, \end{aligned} \qquad \begin{aligned} &\delta\partial_n\overline{\phi}^{\bar{\imath}} = \bar{\epsilon}\partial_n\eta^{\bar{\imath}}, \\ &\delta\partial_n\eta^{\bar{\imath}} = 0. \end{aligned} \tag{6.34}$$

From this we see that Q-cohomology classes are made of $\theta^{\bar{\imath}}$ and there are 2^n of them:

$$1, \;\; \theta^{\bar{\imath}}, \;\; \theta^{\bar{\imath}}\theta^{\bar{\jmath}}, \;\; \dots, \;\; \theta^{\bar{1}}\theta^{\bar{2}}\cdots\theta^{\bar{n}}. \tag{6.35}$$

As in the case without boundary, there is a one-to-one correspondence between the supersymmetric ground states of the original LG model on the segement $[0,\pi]$ and the Q-cohomology classes of the boundary operators; The state corresponding to an operator O is the one that appears at the back of a semi-infinite thin tongue of the twisted model where O is inserted at the tip [53]. Therefore, we have established from (6.35) that the spectrum (6.26) of the supersymmetric ground states is an exact result.

Let us compute some correlation functions. We first consider the correlation functions on the finite size cylinder, $\Sigma = S^1 \times [0,\pi]$. We impose the boundary condition corresponding to the D0-brane at p_a and p_b at the boundary circles $S^1 \times 0$ and $S^1 \times \pi$. If $p_a \neq p_b$, no configuration is Q-invariant and therefore all the topological correlation function vanishes. If $p_a = p_b$, the constant map to p_a is Q-invariant and the path-integral can be exactly performed by the quadtratic approximation around the constant map. Thus, we can compute the correlators using the free massive theory. (One can consider it as a sum of n decoupled free system; Since the deformation of D-term does not affect the topological correlation functions, one can choose the Kahler potential so that $K = \sum_{i=1}^n |\Phi_i|^2 + \cdots$ at the same time as $W = \sum_{i=1}^n m_i\Phi_i^2 + \cdots$.) First thing to notice is that there are $2n$ fermionic zero modes on the cylinder; for each i the functions $c_0(x)$ and $\bar{c}_0(x)$ as in (6.18) with $|m|\bar{c}_0^{\pm} = \mp imc_0^{\pm}$ define the zero mode. Thus, we must insert $2n$ fermionic operators for the amplitude to be non-vanishing. In particular, the partition function vanishes (this rederives $\mathrm{Tr}(-1)^F = 0$). Now, let us insert

$$(\theta)^n := \theta^{\bar{1}}\theta^{\bar{2}}\cdots\theta^{\bar{n}} \tag{6.36}$$

at a point of each boundary circle. The computation reduces to that of the $n = 1$ free massive theory which we have studied above. In fact we have already developed the machinery of computation. Since $\theta = \overline{\psi}_- - \overline{\psi}_+ = \sqrt{2}\bar{c}$, what we want to compute is

$$\mathrm{Tr}((-1)^F \, \mathrm{e}^{-\beta H} 2\bar{c}(0)\bar{c}(\pi)). \tag{6.37}$$

It is easy to see that only the zero mode $\bar{c}_0(0)\bar{c}_0(\pi)$ contributes in this computation. In terms of the normalized variables $\xi^{\pm}$ or ξ, $\bar{\xi}$, we have

$$\begin{aligned} \bar{c}_0(0)\bar{c}_0(\pi) &= \left(\sqrt{-i\frac{m}{|m|}}\,\mathrm{e}^{-\frac{\pi|m|}{2}}\xi^+ + \sqrt{i\frac{m}{|m|}}\,\mathrm{e}^{\frac{\pi|m|}{2}}\xi^-\right)\left(\sqrt{-i\frac{m}{|m|}}\,\mathrm{e}^{\frac{\pi|m|}{2}}\xi^+ + \sqrt{i\frac{m}{|m|}}\,\mathrm{e}^{-\frac{\pi|m|}{2}}\xi^-\right) \\ &= \frac{m}{|m|}(\mathrm{e}^{\pi|m|} - \mathrm{e}^{-\pi|m|})\xi^-\xi^+ = 2\pi im\,(\bar{\xi}\xi - 1). \end{aligned} \tag{6.38}$$

The constant term $2\pi im(-1)$ is irrelevant and hence (6.37) is

$$4\pi im\left(\langle 0|\bar{\xi}\xi|0\rangle - \langle 0|\xi\bar{\xi}\xi\bar{\xi}|0\rangle\right) = 4\pi im(0-1) = -4\pi im. \tag{6.39}$$

We note that m is the second derivative of the superpotential $W''(\phi = 0) = 2m$. Thus, the correlation function in the general case is given by (up to a numerical factor)

$$\langle(\theta)^n(0)(\theta)^n(\pi)\rangle^{aa}_{\text{cylinder}} = \det \partial_i\partial_j W(p_a). \tag{6.40}$$

On the other hand, by the factorization of the topological correlators, we have

$$\langle(\theta)^n(0)(\theta)^n(\pi)\rangle^{aa}_{\text{cylinder}} = \langle(\theta)^n O_c\rangle^a_{\text{disk}}\eta^{cd}\langle O_d(\theta)^n\rangle^a_{\text{disk}}, \tag{6.41}$$

where η^{ab} is the inverse matrix of $\eta_{ab} = \sum_c O_a(p_c)O_b(p_c)/\det\partial_i\partial_j W(p_c)$. If we choose as the basis of physical operators the functions ϵ_a such that $\epsilon_a(p_b) = \delta_{a,b}$, then we have $\eta^{ab} = \delta_{a,b}\det\partial_i\partial_j W(p_a)$. It thus follows from (6.40) and (6.41) that $\langle(\theta)^n\epsilon_a\rangle^a_{\text{disk}} = 1$. Also it is obvious that $\langle(\theta)^n\epsilon_b\rangle^a_{\text{disk}} = 0$ if $b \neq a$. To summarize, we have obtained

$$\langle(\theta)^n O\rangle^a_{\text{disk}} = O(p_a). \tag{6.42}$$

The Sine-Gordon Model

As an example, let us consider the $\mathcal{N}=2$ supersymmetric sine-Gordon model; the LG model of a priodic variable $Y \equiv Y + 2\pi i$ with the superpotential

$$W = \mathrm{e}^{-Y} + q\,\mathrm{e}^{Y}. \tag{6.43}$$

This superpotential has two critical points $\mathrm{e}^{-Y} = \pm\sqrt{q}$ at which the Hessian is $\partial_Y^2 W = \pm 2\sqrt{q}$. Some non-vanishing sphere amplitudes are

$$\langle \mathrm{e}^{-Y} \rangle_{S^2} = 1, \tag{6.44}$$

$$\langle \mathrm{e}^{-Y}\,\mathrm{e}^{-Y}\,\mathrm{e}^{-Y} \rangle_{S^2} = q. \tag{6.45}$$

On the other hand, the disk amplitudes are computed using (6.42) as

$$\langle \theta \rangle^{\pm}_{\mathrm{disk}} = 1, \tag{6.46}$$

$$\langle \theta\,\mathrm{e}^{-Y} \rangle^{\pm}_{\mathrm{disk}} = \pm\sqrt{q}, \tag{6.47}$$

where the superscript $\pm$ stands for the location of the D-brane, $\mathrm{e}^{-Y} = \pm\sqrt{q}$.

6.4 Comparison to the Sigma Models

The sigma model on an n-dimensional toric manifold is mirror to a LG model of n variables. As shown in Section 4.5, the D0-branes in such a LG model are the mirror of the D-branes wrapped on a certain Lagrangian torus. Thus, such D-branes in the toric sigma model must have the same properties as D0-branes in the LG model studied in this section when the theory is massive: Open string Witten index must be zero for any pair of D-branes; Space of supersymmetric ground states must be 2^n dimensional as in (6.26); the topologically twisted theory must have the same correlation functions. Here we check some of these properties directly in the non-linear sigma model (although it is not necessary because we have a *proof* of the mirror symmetry).

For our purpose, it is convenient to start with twisting the theory. We are now considering A-twist where $Q = \overline{Q}_+ + Q_-$ becomes the scalar operator that defines "physical operators". If we put A-type boundary condition/interaction, we can also consider twisting the theory with boundaries. We recall that our D-brane is wrapped on a real n-dimensional torus T embedded in the toric manifold X. The theory has t, the complexified Kahler class, and s_i, the parameters determing the location of T and the holonomy of flat $U(1)$ bundle on T (which are related so that (e^{-s_i}) is at the critical point of (4.45)). The twisted theory depends only on these parameters and independent of the detail of the metric. In particular, one can choose the metric of X so that a neighborhood of T is that of the n-torus T^n in the flat cylinder $\mathbb{C}^n/\mathbb{Z}^n$, where T^n is the real section $\mathbf{R}^n/\mathbf{Z}^n$. This is possible at the values of s_i we are choosing.

Now, the A-twist changes the spin of the fermions so that the following renaming is natural

$$\begin{aligned} \psi_-^i = \chi^i, \quad \overline{\psi}_+^{\bar{i}} = \overline{\chi}^{\bar{i}}, \\ \overline{\psi}_-^{\bar{i}} = \rho_z^{\bar{i}}, \quad \psi_+^i = \rho_{\bar{z}}^i. \end{aligned} \tag{6.48}$$

The boundary condition $\phi^i = \overline{\phi}^{\bar{i}}$, $\psi_-^i = \overline{\psi}_+^{\bar{i}}$ and $\overline{\psi}_-^{\bar{i}} = \psi_+^i$ is translated to

$$\phi^i = \overline{\phi}^{\bar{i}}, \tag{6.49}$$

$$\chi^i = \overline{\chi}^{\bar{i}}, \tag{6.50}$$

$$\rho_z^{\bar{i}} = \rho_{\bar{z}}^i, \tag{6.51}$$

where z is a worldsheet coordinate whose real part is normal to the boundary. The variation $\delta = \epsilon Q$ of the remaining variables at the boundary is given by

$$\delta(\phi^i + \overline{\phi}^{\bar{i}}) = \epsilon(\chi^i + \overline{\chi}^{\bar{i}}), \tag{6.52}$$

$$\delta(\chi^i + \overline{\chi}^{\bar{i}}) = 0, \tag{6.53}$$

$$\delta(\rho_z^{\bar{i}} + \rho_{\bar{z}}^i) = 2i\epsilon(\partial_{\bar{z}}\phi^i - \partial_z\overline{\phi}^{\bar{i}}). \tag{6.54}$$

From this we see that the Q-cohomology classes are in one-to-one correspondence with the de Rham cohomology classes of the torus T^n, or

$$Q\text{-cohomology group} = H^*_{DR}(T^n), \tag{6.55}$$

just as Q-cohomology group $= H^*_{DR}(X)$ in the bulk theory. By the state-operator correspondence as before, this is identified as the space of supersymmetric ground states of the theory on the segment with the same boundary condition at the two ends. Obviously, $H^*_{DR}(T^n)$ is 2^n-dimensional and has a basis like (6.26). In particular, Witten index, identified as the Euler number of T^n, vanishes.

The $\mathbb{C}P^1$ *Model*

Let us consider the $\mathbb{C}P^1$ model. This theory is mirror to the sine Gordon model with $q = e^{-t}$ and must reproduce the result obtained in the previous subsection. As we have seen in Section 4.6, the D-branes for the two values of (s_i), $s_1 = s_2 = t/2$ and $s_1 = s_2 = t/2 + \pi i$, are both wrapped on the equator of $\mathbb{C}P^1$ but differ in the Wilson line.

The topological $\mathbb{C}P^1$ sigma model for worldsheet without a boundary has been well-studied. It has two operators 1 and H where H is a second cohomology class represented by a delta-function 2-form at a point of $\mathbb{C}P^1$. Non-vanishing sphere correlation functions are

$$\langle H \rangle_{S^2} = 1, \tag{6.56}$$

$$\langle HHH \rangle_{S^2} = e^{-t}. \tag{6.57}$$

The first comes from the constant maps; one insertion of H require the insertion point to be mapped a given point of $\mathbb{C}P^1$ and there is only one such map. The second comes from the degree 1 maps; the three insertions of H requires three insertion points to be mapped to given three points (one for each) of $\mathbb{C}P^1$ and there is one such map. The factor e^{-t} comes from the classical action. The result (6.56) and (6.57) are in agreement with (6.44) and (6.45) under the identification $H = e^{-Y}$.

Now let us consider the amplitudes on the disk D^2. The non-trivial boundary operator is the first cohomology class of the equator T of $\mathbb{C}P^1$ which is represented as the delta function 1-form at a point of T. We denote it by ϑ. Let us count the number of deformations of maps in some classes. First, the constant maps. Since the boundary ∂D^2 must be mapped to the equator T, the whole disk must also be mapped to T. Obviously, there is a one dimensional modulus — the position in T. Next, degree one maps. As is well-known,

$SL(2,\mathbb{R})$ can be considered as the parameter space of such maps (it comes from the action on the upper-half plane as $\zeta \to (a\zeta + b)/(c\zeta + d)$). Thus, the parameter space is three-dimensional. The maps of higher degree have more moduli. For an amplitude to be non-vanishing, the number of moduli must match with the axial R-charge of the inserted operators.

Now let us consider an amplitude with just a ϑ insertion at a point of the boundary ∂D^2. Since one ϑ has axial R-charge 1 only the constant maps can contribute. The insertion of ϑ requires the insetion point to be mapped to a given point in T. There is only one such constant map. Thus, we obtain

$$\langle \vartheta \rangle_{D^2}^{\pm} = 1, \tag{6.58}$$

where the superscript $\pm$ distinguishes the $U(1)$ holonomy of the D-brane. Next let us consider the case where ϑ is inserted in ∂D^2 and H is inserted in the interior of D^2. The total axial R-charge is 3 and thus, degree one maps can contribute. The insertion of H requires the insertion point to be mapped to a given point in $\mathbb{CP}^1$. This reduces the 3 moduli to one. The insertion of ϑ further reduces the moduli and there is only one map obeying the requirement. The classical weight is $\mathrm{e}^{-t/2}$ or $\mathrm{e}^{-t/2+\pi i}$ depending on the Wilson line of the D-brane. Thus, we obtain

$$\langle \vartheta H \rangle_{D^2}^{\pm} = \pm\, \mathrm{e}^{-t/2}. \tag{6.59}$$

The results (6.58) and (6.59) reproduce the sine-Gordon result (6.58) and (6.59).

Remark. Open topological field theory has been studied from axiomatic point of view in [54, 55].

Acknowledgement

I would like to thank M. Douglas, T. Eguchi, M. Gutperle, A. Iqbal, H. Itoyama, S. Kachru, S. Katz, A. Lawrence, H. Liu, J. McGreevy, G. Moore, M. Naka, M. Nozaki, B. Pioline, R. Thomas, N. Warner, S.-K. Yang, E. Zaslow, and especially E. Martinec for valuable discussions. I also thank M. Aganagic and C. Vafa for explaining their work and for useful discussions. I am grateful to KIAS, Seoul; Aspen Center for Physics, Colorado; SI-2000, Yamanashi; and New High Energy Theory Center at Rutgers University where parts of this work were carried out, for their hospitality. This work is supported in part by NSF-DMS 9709694.

References

[1] E. Witten, "Phases Of N = 2 Theories In Two Dimensions", Nucl. Phys. **B403** (1993) 159 [hep-th/9301042].

[2] K. Hori and C. Vafa, "Mirror Symmetry", hep-th/0002222.

[3] K. Becker, M. Becker and A. Strominger, "Five-branes, membranes and non-perturbative string theory", Nucl. Phys. **B456** (1995) 130 [hep-th/9507158].

[4] A. Strominger, S.-T. Yau and E. Zaslow, "Mirror Symmetry is T-Duality", Nucl. Phys. **B479** (1996) 243-259 [hep-th/9606040].

[5] H. Ooguri, Y. Oz and Z. Yin, "D-branes on Calabi-Yau spaces and their mirrors", Nucl. Phys. **B477** (1996) 407 [hep-th/9606112].

[6] A. Recknagel and V. Schomerus, "D-Branes in Gepner Models", Nucl. Phys. **B531** (1998) 185 [hep-th/9712186];
M. Gutperle and Y. Satoh, "D-branes in Gepner models and supersymmetry", Nucl. Phys. **B543** (1999) 73 [hep-th/9808080]; "D0-branes in Gepner models and N=2 black holes", Nucl. Phys. **B555** (1999) 477 [hep-th/9902120].

[7] I. Brunner, M. R. Douglas, A. Lawrence and C. Romelsberger, "D-branes on the quintic", JHEP**0008** (2000) 015 [hep-th/9906200].

[8] E.-E. Diaconescu and J. Gomis, "Fractional branes and boundary states in orbifold theories", JHEP**0010** (2000) 001 [hep-th/9906242].

[9] S. Govindarajan, T. Jayaraman and T. Sarkar, "Worldsheet approaches to D-branes on supersymmetric cycles", Nucl. Phys. **B580** (2000) 519-547 [hep-th/9907131];
S. Govindarajan and T. Jayaraman, "On the Landau-Ginzburg description of Boundary CFTs and special Lagrangian submanifolds", JHEP**0007** (2000) 016 [hep-th/0003242].

[10] D.-E. Diaconescu and C. Romelsberger, "D-Branes and Bundles on Elliptic Fibrations", Nucl. Phys. **B574** (2000) 245-262 [hep-th/9910172]; P. Kaste, W. Lerche, C.A. Lutken, J. Walcher, "D-Branes on K3-Fibrations", Nucl. Phys. **B582** (2000) 203-215 [hep-th/9912147]; E. Scheidegger, "D-branes on some one- and two-parameter Calabi-Yau hypersurfaces", JHEP**0004** (2000) 003 [hep-th/9912188]; M. Naka and M. Nozaki, "Boundary States in Gepner Models", JHEP**0005** (2000) 027 [hep-th/0001037]; I. Brunner and V. Schomerus, "D-branes at Singular Curves of Calabi-Yau Compactifications", JHEP**0004** (2000) 020 [hep-th/0001132]; "On superpotentials for D-branes in Gepner models", JHEP**0010** (2000) 016 [hep-th/0008194]. J. Fuchs, C. Schweigert and J. Walcher, "Projections in string theory and boundary states for Gepner models", Nucl. Phys. **B588** (2000) 110 [hep-th/0003298]; J. Fuchs, P. Kaste, W. Lerche, C. Lutken, C. Schweigert and J. Walcher, "Boundary Fixed Points, Enhanced Gauge Symmetry and Singular Bundles on K3", hep-th/0007145.

[11] S. Kachru, S. Katz, A. Lawrence and J. McGreevy, "Open string instantons and superpotentials", Phys. Rev. D **62** (2000) 026001 [hep-th/9912151]; "Mirror symmetry for open strings", Phys. Rev. D **62** (2000) 126005 [hep-th/0006047].

[12] K. Sugiyama, "Comments on Central Charge of Topological Sigma Model with Calabi-Yau Target Space", Nucl. Phys. **B591** (2000) 701-737 [hep-th/0003166]; K. Sugiyama and S. Yamaguchi, "D-branes on a noncompact singular Calabi-Yau manifold", hep-th/0011091.

[13] K. Hori, A. Iqbal and C. Vafa, "D-Branes And Mirror Symmetry", hep-th/0005247.

[14] W. Lerche, "On a boundary CFT description of nonperturbative N = 2 Yang-Mills theory", hep-th/0006100; W. Lerche, C. A. Lutken and C. Schweigert, "D-branes on ALE spaces and the ADE classification of conformal field theories", hep-th/0006247; W. Lerche and J. Walcher, "Boundary rings and N = 2 coset models", hep-th/0011107.

[15] S. Govindarajan, T. Jayaraman and T. Sarkar, "On D-branes from gauged linear sigma models", hep-th/0007075.

[16] A. Kapustin and D. Orlov, "Vertex algebras, mirror symmetry, and D-branes: The case of complex tori", hep-th/0010293.

[17] M. R. Douglas, "D-branes, categories and N = 1 supersymmetry", hep-th/0011017.

[18] J. Hashiba and M. Naka, "Landau-Ginzburg description of D-branes on ALE spaces", hep-th/0011101.

[19] M. Aganagic and C. Vafa, "Mirror symmetry, D-branes and counting holomorphic discs", hep-th/0012041.

[20] E. Witten, "Chern-Simons gauge theory as a string theory", In: *The Floer memorial volume*, H. Hofer, C.H. Taubes, A. Weinstein and E. Zehender editors, Prog. Math. **133** (Birkhauser, 1995), hep-th/9207094.

[21] N.P. Warner, "Supersymmetry in Boundary Integrable Models", Nucl. Phys. **B450** (1995) 663-694 [hep-th/9506064].

[22] M. Douglas, B. Fiol and C. Römelsberger, "Stability and BPS branes", hep-th/0002037; "The spectrum of BPS branes on a noncompact Calabi-Yau", hep-th/0003263; B. Fiol and M. Marino, "BPS states and algebras from quivers", JHEP **0007** (2000) 031; B. Fiol, "The BPS Spectrum of N=2 SU(N) SYM and Parton Branes", hep-th/0012079.

[23] T. Hauer and A. Iqbal, "Del Pezzo Surfaces and Affine 7-brane Backgrounds", JHEP **0001** (2000) 043 [hep-th/9910054]; B.R. Greene and C.I. Lazaroiu, "Collapsing D-Branes in Calabi-Yau Moduli Space: I", hep-th/0001025; C.I. Lazaroiu,, "Collapsing D-branes in one-parameter models and small/large radius duality", hep-th/0002004; K. Mohri, Y. Ohtake and S.-K. Yang, "Duality Between String Junctions and D-Branes on Del Pezzo Surfaces", hep-th/0007243; K. Mohri, Y. Onjo and S.-K. Yang, "Closed Sub-Monodromy Problems, Local Mirror Symmetry and Branes on Orbifolds", hep-th/0009072.

[24] D.-E. Diaconescu and M. R. Douglas "D-branes on Stringy Calabi-Yau Manifolds", hep-th/0006224; S. Govindarajan and T. Jayaraman, "D-branes, Exceptional Sheaves and Quivers on Calabi-Yau manifolds: From Mukai to McKay", hep-th/0010196; A. Tomasiello, "D-branes on Calabi-Yau manifolds and helices", hep-th/0010217; P. Mayr, "Phases of Supersymmetric D-branes on Kahler Manifolds and the McKay correspondence", hep-th/0010223.

[25] Y. Oz, T. Pantev and D. Waldram, "Brane-antibrane systems on Calabi-Yau spaces", JHEP **0102** (2001) 045 [hep-th/0009112].

[26] Talk given by S. Kachru in Strings 2000, on the work by S. Kachru, A. Lawrence and J. McGreevy to appear, http://feynman.physics.lsa.umich.edu/cgi-bin/s2ktalk.cgi?kachru

[27] R. G. Leigh, "Dirac-Born-Infeld Action From Dirichlet Sigma Model", Mod. Phys. Lett. **A4** (1989) 2767.

[28] J. A. Harvey, D. Kutasov and E. J. Martinec, "On the relevance of tachyons", hep-th/0003101.

[29] D. Kutasov, M. Marino and G. Moore, "Remarks on tachyon condensation in superstring field theory", hep-th/0010108.

[30] J. R. David, "Tachyon condensation using the disc partition function", hep-th/0012089.

[31] A.A. Gerasimov and S.L. Shatashvili, "On Exact Tachyon Potential in Open String Field Theory", JHEP**0010** (2000) 034 [hep-th/0009103].

[32] D. Kutasov, M. Marino and G. Moore, "Some Exact Results on Tachyon Condensation in String Field Theory", JHEP**0010** (2000) 045 [hep-th/0009148].

[33] E. Witten, "On Background Independent Open-String Field Theory", Phys.Rev. **D46** (1992) 5467 [hep-th/9208027]; "Some Computations in Background Independent Open-String Field Theory", Phys.Rev. D47 (1993) 3405 [hep-th/9210065];

S.L. Shatashvili, "Comment on the Background Independent Open String Theory", Phys. Lett. **311** (1993) 83-86 [hep-th/9303143]; "On the Problems with Background Independence in String Theory", hep-th/9311177.

[34] L. Alvarez-Gaume and D. Z. Freedman, "Kahler Geometry And The Renormalization Of Supersymmetric Sigma Models", Phys. Rev. D **22** (1980) 846;

L. Alvarez-Gaume, D. Z. Freedman and S. Mukhi, "The Background Field Method And The Ultraviolet Structure Of The Supersymmetric Nonlinear Sigma Model", Annals Phys. **134** (1981) 85.

[35] N. Seiberg, "Naturalness versus supersymmetric nonrenormalization theorems", Phys. Lett. **B318** (1993) 469 [hep-ph/9309335].

[36] E. Martinec, "$\mathcal{N} = 2$ Superspace Boundary Conditions", a note sent to the author (June, 2000).

[37] H. Itoyama and P. Moxhay, "Multiparticle Superstring Tree Amplitudes", Nucl. Phys. **B293** (1987) 685.

[38] A. Sen, "Non-BPS states and branes in string theory", hep-th/9904207.

[39] J. Ishida and A. Hosoya, "Path Integral For A Color Spin And Path Ordered Phase Factor", Prog. Theor. Phys. **62** (1979) 544;
L. Alvarez-Gaumé, "Supersymmetry and the Atiyah-Singer Index Theorem", Commun. Math. Phys. **90** (1983) 161;
D. Friedan and P. Windey, "Supersymmetric Derivation of the Atiyah-Singer Index and the Chiral Anomaly", Nucl. Phys. **B235** (1984) 395;
N. Marcus and A. Sagnotti, "Group Theory From "Quarks" At The Ends Of Strings", Phys. Lett. **B188** (1987) 58.
E. D'Hoker and D.G. Gagné, "Worldline Path Integrals for Fermions with General Couplings", Section 5, Nucl. Phys. **B467** (1996) 297-312 [hep-th/9512080].

[40] A. Sen, "BPS D-branes on non-supersymmetric cycles", JHEP **9812** (1998) 021 [hep-th/9812031].

[41] P. Horava, "Type IIA D-branes, K-theory, and matrix theory", Adv. Theor. Math. Phys. **2** (1999) 1373 [hep-th/9812135].

[42] E. Witten, "D-branes and K-theory", JHEP **9812** (1998) 019 [hep-th/9810188].

[43] A. Sen, "Type I D-particle and its interactions", JHEP **9810** (1998) 021 [hep-th/9809111].

[44] A. Sen, "SO(32) spinors of type I and other solitons on brane-antibrane pair", JHEP **9809** (1998) 023 [hep-th/9808141].

[45] A. Sen, "Tachyon condensation on the brane antibrane system", JHEP **9808** (1998) 012 [hep-th/9805170].

[46] M. R. Douglas and B. Fiol, "D-branes and discrete torsion. II", hep-th/9903031.

[47] M. F. Atiyah, R. Bott and A. Shapiro, "Clifford Modules", Topology **3** (1964) 3.

[48] M. F. Atiyah, *K-Theory* (Benjamin, New York, 1969).

[49] K. Hori, "D-branes, T-duality, and index Theory", Adv. Theor. Math. Phys. **3** (1999) [hep-th/9902102].

[50] A. Abouelsaood, C. G. Callan, C. R. Nappi and S. A. Yost, "Open Strings In Background Gauge Fields", Nucl. Phys. **B280** (1987) 599.

[51] S. Cecotti and C. Vafa, "Topological antitopological fusion", Nucl. Phys. **B367** (1991) 359.

[52] C. Vafa, "Topological Landau-Ginzburg models", Mod. Phys. Lett. **A6** (1991) 337.

[53] M. Bershadsky, S. Cecotti, H. Ooguri and C. Vafa, "Kodaira-Spencer theory of gravity and exact results for quantum string amplitudes", Commun. Math. Phys. **165** (1994) 311 [hep-th/9309140].

[54] G. Moore and G. Segal, talk given by G. Moore in Strings 2000, http://feynman.physics.lsa.umich.edu/cgi-bin/s2ktalk.cgi?moore page 15-5; also in G. Moore, "Some comments on branes, G-flux, and K-theory", hep-th/0012007.

[55] C. I. Lazaroiu, "On the structure of open-closed topological field theory in two dimensions", hep-th/0010269.

THE CONNECTEDNESS OF THE MODULI SPACE OF MAPS TO HOMOGENEOUS SPACES

B. KIM

Pohang University of Science and Technology
Email: bumsig@postech.ac.kr

R. PANDHARIPANDE

California Institute of Technology
Email: rahulp@cco.caltech.edu

We prove the connectedness of the moduli space of maps (of fixed genus and homology class) to the homogeneous space G/P by degeneration via the maximal torus action. In the genus 0 case, the irreducibility of the moduli of maps is a direct consequence of connectedness. An analysis of a related Bialynicki-Birula stratification of the map space yields a rationality result: the (coarse) moduli space of genus 0 maps to G/P is a rational variety. The rationality argument depends essentially upon rationality results for quotients of SL2 representations proven by Katsylo and Bogomolov.

0 Introduction

Let X be a compact algebraic homogeneous space: $X = \mathbf{G}/\mathbf{P}$ where $\mathbf{G}$ is a connected complex semisimple algebraic group and $\mathbf{P}$ is a parabolic subgroup. Let $\beta \in H_2(X, \mathbb{Z})$. The (coarse) moduli space $\overline{M}_{g,n}(X, \beta)$ of n-pointed genus g stable maps parameterizes the data

$$[\mu : C \to X, p_1, \ldots, p_n]$$

satisfying:

(i) C is a complex, projective, connected, reduced, (at worst) nodal curve of arithmetic genus g.

(ii) The points $p_i \in C$ are distinct and lie in the nonsingular locus.

(iii) $\mu_*[C] = \beta$.

(iv) The pointed map μ has no infinitesimal automorphisms.

Since X is convex, the genus 0 moduli space $\overline{M}_{0,n}(X, \beta)$ is of pure dimension

$$\dim(X) + \int_\beta c_1(T_X) + n - 3.$$

Moreover, $\overline{M}_{0,n}(X,\beta)$ is locally the quotient of a nonsingular variety by a finite group. For general g, the space $\overline{M}_{g,n}(X,\beta)$ may have singular components of different dimensions. Stable maps in algebraic geometry were first defined in [Ko]. Basic properties of the moduli space $\overline{M}_{g,n}(X,\beta)$ can be found in [BM], [FP], and [KoM]. The following connectedness result is proven here.

Theorem 1 *$\overline{M}_{g,n}(\mathbf{G}/\mathbf{P},\beta)$ is a connected variety.*

This result may be viewed as analogous to the connectedness of the Hilbert scheme of projective space proven by Hartshorne. As in [Har], connectedness is obtained via maximal degenerations.

Since $\overline{M}_{0,n}(X,\beta)$ has quotient singularities, connectedness is equivalent to irreducibility.

Corollary 1 *$\overline{M}_{0,n}(\mathbf{G}/\mathbf{P},\beta)$ is an irreducible variety.*

Corollary 1 is easy to verify in case X is a projective space. When X is a Grassmannian, the irreducibility follows from Strømme's Quot scheme analysis [S]. A proof of Corollary 1 can be found in case $\mathbf{G} = \mathbf{SL}$ in [MM]. For the variety of partial flags in $\mathbb{C}^n$, a proof of irreducibility using flag-Quot schemes is established in [Ki]. Results of Harder closely related to Corollary 1 appear in [Ha]. There is an independent proof by J. Thomsen for the irreducibility of $\overline{M}_{0,n}(\mathbf{G}/\mathbf{P},\beta)$ in [T].

The moduli space $\overline{M}_{g,n}(X,\beta)$ has a natural locally closed decomposition indexed by stable, pointed, modular graphs τ (see [BM]). The strata correspond to maps with domain curves of a fixed topological type and a fixed distribution β_τ of β. The graph τ determines a complete moduli space of stable maps

$$\overline{M}_{\tau,n}(X,\beta_\tau)$$

together with a canonical morphism:

$$\pi_\tau : \overline{M}_{\tau,n}(X,\beta_\tau) \to \overline{M}_{g,n}(X,\beta). \tag{1}$$

A closed decomposition is determined by the images of these morphisms (1). Theorem 1 is a special case of the following result.

Theorem 2 *$\overline{M}_{\tau,n}(\mathbf{G}/\mathbf{P},\beta_\tau)$ is a connected variety.*

Since $\overline{M}_{\tau,n}(X,\beta_\tau)$ is normal in the genus 0 case, we obtain the corresponding corollary.

Corollary 2 *Let $g = 0$. $\overline{M}_{\tau,n}(\mathbf{G}/\mathbf{P},\beta_\tau)$ is an irreducible variety.*

In particular, all the boundary divisors of $\overline{M}_{0,n}(X,\beta)$ are irreducible.

Theorem 2 is proven by studying the maximal torus action on X. The method is to degenerate a general $\mathbf{G}$-translate of a map $\mu : C \to X$ onto a canonical 1-dimensional configuration of $\mathbb{P}^1$'s in X determined by the maximal torus and the Bialynicki-Birula stratification of X.

In the genus 0 case, we study the Bialynicki-Birula stratification of $\overline{M}_{0,n}(X,\beta)$. The following result is then deduced from the rationality of torus fixed components.

Theorem 3 *$\overline{M}_{0,n}(\mathbf{G}/\mathbf{P},\beta)$ is rational.*

The fixed component rationality is equivalent to a rationality result for certain quotients of $\mathbf{SL}_2$-representations proven by Katsylo and Bogomolov [Ka], [Bog]. It should be noted that the fixed components will in general be contained in the boundary of the moduli space of maps – the compactifaction by stable maps therefore plays an important role in the proof.

The rationality of the Hilbert schemes of rational curves in projective space (birational to $\overline{M}_{0,0}(\mathbb{P}^r,d)$) is a consequence of Katsylo's results [Ka] and was also studied by Hirschowitz in [Hi].

The main part of this paper was written in 1996 at the Mittag-Leffler Institute where the authors benefitted from discussions with many members. Thanks are especially due to I. Ciocan-Fontanine, B. Fantechi, W. Fulton, T. Graber, and B. Totaro. Conversations with F. Bogomolov were also helpful. B. K. was partially supported by KOSEF grant 1999-2-102-003-5, POSTECH grant 1999 and BK21. R. P. was partially supported by NSF grant DMS-9801574 and an A. P. Sloan foundation fellowship.

1 The torus action on G/P

Let $\mathbf{G}$ be a connected complex semisimple algebraic group. Let $\mathbf{P}$ be a parabolic subgroup. Select a maximal algebraic torus $\mathbf{T}$ and Borel subgroup $\mathbf{B}$ of $\mathbf{G}$ satisfying:

$$\mathbf{T} \subset \mathbf{B} \subset \mathbf{P} \subset \mathbf{G}.$$

Let $(\mathbf{G}/\mathbf{P})^{\mathbf{T}}$ denote the fixed point set of the left $\mathbf{T}$-action on $\mathbb{G}/\mathbb{P}$. Three special properties of this $\mathbf{T}$-action will be needed:

(i) The $\mathbf{T}$-action has isolated fixed points.

(ii) For every point $p \in (\mathbf{G}/\mathbf{P})^{\mathbf{T}}$, there exits a $\mathbf{T}$-invariant open set U_p containing p which is $\mathbf{T}$-equivalent to a vector space representation of $\mathbb{T}$.

(iii) Let $\mathbb{C}^* \subset \mathbf{T}$ correspond to an interior point of a Weyl chamber. Then, $(\mathbf{G}/\mathbf{P})^{\mathbb{C}^*} = (\mathbf{G}/\mathbf{P})^{\mathbf{T}}$, and the Bialynicki-Birula decomposition obtained from the $\mathbb{C}^*$-action is an affine *stratification* of $\mathbf{G}/\mathbf{P}$.

A stratification is a decomposition such that the closures of the strata are unions of strata. In general, the Bialynicki-Birula decomposition obtained from a $\mathbb{C}^*$-action on a nonsingular variety need not be a stratification.

The claims (i)-(iii) are well known. Only a brief summary of the arguments will be presented here. Let $\mathbf{W}$ be the Weyl group of $\mathbb{G}$ relative to $\mathbf{T}$.

Lemma 1 $|(\mathbf{G}/\mathbf{B})^{\mathbf{T}}| = |\mathbf{W}|$, *and* $\mathbf{W}$ *acts transitively on* $(\mathbf{G}/\mathbf{B})^{\mathbf{T}}$.

Proof. See, for example, [Bor].

In particular, $(\mathbf{G}/\mathbf{B})^{\mathbf{T}}$ is a finite set.

Lemma 2 *The natural map* $(\mathbf{G}/\mathbf{B})^{\mathbf{T}} \to (\mathbf{G}/\mathbf{P})^{\mathbf{T}}$ *is surjective.*

Proof. Let $p \in (\mathbf{G}/\mathbf{P})^{\mathbf{T}}$. The invariant fiber (isomorphic to $\mathbf{P}/\mathbf{B}$) over the fixed point p is a nonsingular projective variety, and hence contains a $\mathbf{T}$-fixed point by the Borel fixed point theorem (or, alternatively, this is a Hamiltonian action on a compact manifold).

Therefore, $\mathbf{W}$ acts transitively on the finite set $(\mathbf{G}/\mathbf{P})^{\mathbf{T}}$.

A representation $\psi : \mathbf{T} \to \mathbf{GL}(V)$ is *fully definite* if there exists a $\mathbb{C}^*$-basis of $\mathbf{T}$ for which all the weights of the representation are positive integers. Equivalently, a fully definite representation can be written

$$\psi(t_1, \ldots, t_r)v_j = \prod_{i=1}^{r} t_i^{\lambda_{ij}} \cdot v_j$$

where $\lambda_{ij} > 0$ for some choice of $\mathbb{C}^*$-basis of $\mathbb{T}$ and $\mathbb{C}$-basis $\{v_j\}$ of V.

The point $1 \in \mathbf{G}/\mathbf{B}$ corresponding to the identity element of $\mathbb{G}$ is a $\mathbb{T}$-fixed point. The $\mathbf{T}$-action induces a representation

$$\phi : \mathbf{T} \to \mathbf{GL}(\mathrm{Tan}_1 \mathbb{G}/\mathbb{B}).$$

Lemma 3 *The representation ϕ is fully definite.*

Proof. The natural quotient map $q : \mathbf{G} \to \mathbb{G}/\mathbf{B}$ is $\mathbb{T}$-equivariant for the conjugation action on $\mathbf{G}$ and the left action on $\mathbf{G}/\mathbb{B}$. The differential of q yields an isomorphism from the Adjoint representation of $\mathbf{T}$ on $\mathrm{Lie}(\mathbb{G})/\mathrm{Lie}(\mathbb{B})$ to ϕ. $\mathrm{Lie}(\mathbf{G})/\mathrm{Lie}(\mathbf{B})$ is the space of positive roots. This $\mathbf{T}$-representation space has n simple roots (where n is the rank of $\mathbb{G}$). All the 1-dimensional representations in $\mathrm{Lie}(\mathbf{G})/\mathrm{Lie}(\mathbf{B})$ are non-negative tensor products of these simple roots. Moreover, the n weight vectors of these simple roots are independent in the lattice of 1-dimensional representations of the torus $\mathbf{T}$. Lemma 3 now follows from Lemma 4 below.

Lemma 4 *Let $\psi : \mathbf{T} \to \mathbf{GL}(\mathbb{C}^n)$ be an n dimensional representation of a rank n torus $\mathbf{T}$. If the $n \times n$ matrix of weights is nonsingular, then the representation is fully definite.*

Proof. See [Bi].

Lemma 5 *The $\mathbf{T}$-representation $\mathrm{Tan}_1 \mathbf{G}/\mathbf{P}$ is fully definite.*

Proof. There is a surjection of $\mathbb{T}$-modules given by the differential $\mathrm{Tan}_1 \mathbf{G}/\mathbf{B} \to \mathrm{Tan}_1 \mathbf{G}/\mathbb{P}$.

Proposition 1 *For every $p \in (\mathbb{G}/\mathbb{P})^{\mathbf{T}}$, there exists a $\mathbb{T}$-invariant Zariski open set $U_p \subset \mathbf{G}/\mathbf{P}$ of p which is $\mathbf{T}$-equivalent to a vector space representation of $\mathbf{T}$.*

Proof. By a theorem of Bialynicki-Birula [Bi], it suffices to show the tangent representation of $\mathbf{T}$ is fully definite at p. This is a consequence of Lemma 5 and the transitivity of the $\mathbf{W}$-action on $(\mathbf{G}/\mathbf{P})^{\mathbf{T}}$. (In fact, only definiteness of the tangent representation is needed in [Bi].)

Let $\mathbb{C}^* \subset \mathbf{T}$ correspond to an interior point of a Weyl chamber. By the analysis of the tangent representation ϕ, every point of $(\mathbf{G}/\mathbf{B})^{\mathbf{T}}$ is an isolated fixed point of $\mathbb{C}^*$. The equality $(\mathbb{G}/\mathbb{B})^{\mathbb{C}^*} = (\mathbb{G}/\mathbb{B})^{\mathbb{T}}$ follows. Since the map $(\mathbb{G}/\mathbf{B})^{\mathbb{C}^*} \to (\mathbf{G}/\mathbf{P})^{\mathbb{C}^*}$ is surjective, $(\mathbf{G}/\mathbf{P})^{\mathbb{C}^*} = (\mathbf{G}/\mathbf{P})^{\mathbf{T}}$.

For each $p \in (\mathbf{G}/\mathbf{P})^{\mathbf{T}}$, let A_p be the set of points $x \in \mathbf{G}/\mathbf{P}$ such that

$$\lim_{t \to 0} tx = p.$$

By Proposition 1, A_p is isomorphic to the affine space $\mathbb{C}^{r_p}$ where r_p is the number of positive weights in the $\mathbb{C}^*$-representation $\mathrm{Tan}_p\mathbf{G}/\mathbf{P}$. The set $\{A_p\}$ is the Bialynicki-Birula affine decomposition of $\mathbf{G}/\mathbf{P}$. In fact, $\{A_p\}$ coincides (up to the Weyl group action) with the (open) Schubert cell stratification of $\mathbf{G}/\mathbf{P}$. This is essentially proven in [Bor] for the case $\mathbf{G}/\mathbf{B}$. The general case $\mathbf{G}/\mathbf{P}$ is proven in [A]. Therefore, $\{A_p\}$ is a stratification.

2 The $\mathbb{C}^*$-flow

Let $\mathbb{C}^* \subset \mathbf{T}$ correspond to an interior point of a Weyl chamber. Let $s, x_1, \ldots, x_l \in (\mathbf{G}/\mathbf{P})^{\mathbf{T}}$ be the fixed points corresponding to the unique maximal dimensional stratum A_s and the complete set of codimension 1 strata, $A_1, \ldots, A_l$, respectively. The points of A_s flow ($t \to 0$) to s, and the points of A_i flow ($t \to 0$) to x_i. Let $U = A_s \cup A_1 \cup \ldots \cup A_l$. Since the Bialynicki-Birula decomposition $\{A_p\}$ is a stratification, U is a Zariski open set with complement of codimension at least 2.

The inverse action of $\mathbb{C}^*$ on $\mathbf{G}/\mathbf{P}$ is also a torus action on $\mathbf{G}/\mathbf{P}$ with the same fixed point set. Let $A'_s, A'_1, \ldots, A'_l$ be the affine strata for the inverse action corresponding to the fixed points $s, x_1, \ldots, x_l$. Let $\dim(\mathbf{G}/\mathbf{P}) = m$. Since,

$$\dim(A_p) + \dim(A'_p) = m,$$

$A'_1, \cdots, A'_l$ are the complete set of 1-dimensional strata for the inverse action. Moreover, the closure $P_i = \overline{A'}_i$ can contain only the unique 0-dimensional stratum $A'_s = s$. We have shown the closures P_i are contained in U. Each P_i is isomorphic to $\mathbb{P}^1$ (Chevelley [C] proves the closed Schubert cells have singularities in codimension at least 2). The intersection pairing

$$P_i \cap \overline{A}_j = \delta(i - j)$$

follows from the above analysis. Since the closed strata of the inverse action freely generate the integral homology, the classes

$$[P_1], \ldots, [P_l] \in H_2(\mathbf{G}/\mathbf{P}, \mathbf{Z})$$

span an integral basis of $H_2(\mathbf{G}/\mathbf{P}, \mathbf{Z})$.

Let $f : C \to \mathbf{G}/\mathbf{P}$ be a non-constant stable map satisfying the following properties:

(i) The image $f(C)$ lies in U.

(ii) C intersects (via f) the divisors A_i transversely at nonsingular points of C.

(iii) All the markings of C have image in A_s.

If $[f]$ represents the class

$$\beta = \sum_{i=1}^{l} a_i[P_i] \in H_2(\mathbf{G}/\mathbf{P}, \mathbb{Z}),$$

then let C meet A_i at the a_i distinct points

$$\{x_{i,1}, \ldots, x_{i,a_i}\}.$$

We will study the induced $\mathbb{C}^*$-action on $\overline{M}_{g,n}(\mathbf{G}/\mathbf{P}, \beta)$ by translation of maps. Let $F : C_0 \to \mathbf{G}/\mathbf{P}$ be the limit in the space of stable maps,

$$F = \lim_{t \to 0} tf$$

where $t \in \mathbb{C}^*$.

Define a map $\tilde{F} : \tilde{C} \to \mathbf{G}/\mathbf{P}$ as follows. Let the domain $\tilde{C}$ be:

$$\tilde{C} = C \cup \bigcup_{i=1}^{l} (\cup_{j=1}^{a_i} \mathbf{P}^1_{i,j})$$

where $\mathbf{P}^1_{i,j}$ is a projective line attached to C at the point $x_{i,j}$. Let the markings of $\tilde{C}$ coincide with the markings of C (note the markings of C are disjoint from the set $\{x_{i,j}\}$ by condition (ii)). Define $\tilde{F}$ by $\tilde{F}(C \subset \tilde{C}) = s$ and

$$\tilde{F}|_{\mathbf{P}^1_{i,j}} : \mathbf{P}^1_{i,j} \cong P_i$$

for each i and j.

Proposition 2 *If f satisfies conditions (i-iii), then the $t \to 0$ limit F equals the stabilization of $\tilde{F}$.*

Proof. Let $\Delta^\circ \subset \Delta$ be the punctured holomorphic disk at the origin. Let

$$h : C \times \Delta^\circ \to \mathbf{G}/\mathbf{P}$$

be the map defined by $h(c,t) = tf(c)$. The $\mathbb{C}^*$-action on A_s extends to a map

$$\mathbb{C} \times A_s \to A_s$$

since the $\mathbb{C}^*$-action on A_s is a vector space representation with positive weights. The map h thus extends to a map

$$h : C \times \Delta \setminus \{x_{i,j} \times 0\} \to \mathbf{G/P}$$

since the f-image of $C \setminus \{x_{i,j}\}$ lies in A_s. Note,

$$h(C \setminus \{x_{i,j}\}, 0) = s. \tag{2}$$

After a suitable blow-up

$$\gamma : S \to C \times \Delta$$

supported along the isolated nonsingular points $\{x_{i,j} \times 0\}$ of $C \times \Delta$, there is a morphism $h' : S \to \mathbf{G/P}$.

The limit as $t \to 0$ of $tf(x_{i,j})$ equals x_i. Hence, the exceptional divisor $C_{i,j}$ of γ over $x_{i,j}$ connects the points x_i to s under the map h'. The image $h'(C_{i,j})$ thus represents an effective curve class containing the class $[P_i]$. By degree considerations over all the exceptional divisors $C_{i,j}$, we conclude $h'(C_{i,j})$ is of curve class exactly $[P_i]$. As P_i is the unique $\mathbb{C}^*$-fixed curve of class $[P_i]$ connecting the points x_i and s,

$$h'(C_{i,j}) = P_i.$$

We may assume S to be nonsingular (away from the original nodes of C) and each $C_{i,j}$ to be a normal crossings divisor – possibly after further blow-ups and base changes altering only the special fiber over $0 \in \Delta$. We then conclude each $C_{i,j}$ has a single component which is mapped to P_i isomorphically (and the other components of $C_{i,j}$ are contracted).

After blowing-down the h'-contracted components of each $C_{i,j}$, we obtain a map $h'' : S'' \to \mathbf{G/P}$ which is a family of nodal maps over Δ. The fiber of S'' over $t = 0$ is isomorphic to $\tilde{C}$. Moreover, the condition $\tilde{F}(C \subset \tilde{C}) = s$ follows directly from (2).

The limit stable map F is then simply obtained by stabilizing the map $\tilde{F}$. We have carried out the stable reduction of the family of maps tf (see [FP]).

3 Connectedness

Let $[\mu]$ denote the point $[\mu : C \to X, p_1, \ldots, p_n] \in \overline{M}_{g,n}(X, \beta)$. The stable, pointed, modular graph τ with $H_2(X, \mathbb{Z})$-structure canonically associated to $[\mu]$ consists of the following data:

(i) The pointed dual graph of C:

(a) The vertices V_τ correspond to the irreducible components of the curve C.

(b) The edges correspond to the nodes.

(c) The markings correspond to the marked points p_i.

(ii) The genus function, $g_\tau : V_\tau \to \mathbb{Z}^{\geq 0}$, where $g_\tau(v)$ is the geometric genus of the corresponding component of C.

(iii) The $H_2(X, \mathbb{Z})$-structure, $\beta_\tau : V_\tau \to H_2(X, \beta)$, where $\beta_\tau(v)$ equals the μ push-forward of the fundamental class of the corresponding component of C.

Following [BM], define $M_{\tau,n}(X, \beta_\tau)$ to be the moduli space of maps μ *together* with an isomorphism of τ_μ with a fixed stable graph τ. The space $\overline{M}_{\tau,n}(X, \beta_\tau)$ is the compactification via stable maps where the vertices of V_τ may correspond to nodal curves. Note $M_{\tau,n}(X, \beta_\tau)$ may not be dense in $\overline{M}_{\tau,n}(X, \beta_\tau)$.

There is a canonical morphism

$$\pi_\tau : \overline{M}_{\tau,n}(X, \beta_\tau) \to \overline{M}_{g,n}(X, \beta).$$

As τ varies over possible graphs, the images of π_τ determine a (closed) decomposition of the moduli space of maps.

Let τ be a stable, pointed, modular graph with $H_2(\mathbf{G}/\mathbf{P}, \mathbb{Z})$-structure. The connectedness of $\overline{M}_{\tau,n}(\mathbf{G}/\mathbf{P}, \beta_\tau)$ will now be established.

Proof of Theorem 2. If $\beta_\tau = 0$, the irreducibility of $\overline{M}_{\tau,n}(\mathbf{G}/\mathbf{P}, \beta_\tau)$ is a direct consequence of the irreducibility of the corresponding stratum in $\overline{M}_{g,n}$ and the irreducibility of $\mathbf{G}/\mathbf{P}$. We may thus assume $\beta_\tau \neq 0$.

Fix the $\mathbb{C}^*$-action on $\mathbf{G}/\mathbf{P}$ as studied in Section 2. Consider an arbitrary point

$$[\mu] \in \overline{M}_{\tau,n}(\mathbf{G}/\mathbf{P}, \beta_\tau).$$

By the Kleiman-Bertini Theorem, a general $\mathbf{G}$-translate f of μ satisfies conditions (i-iii) of Section 2. As $\mathbf{G}$ is connected, $[\mu]$ is connected to its general $\mathbf{G}$-translate $[f]$.

The point $[f]$ is connected to the limit:

$$[F] = \lim_{t \to 0}[tf].$$

To prove the connectedness of $\overline{M}_{\tau,n}(\mathbf{G}/\mathbf{P}, \beta_\tau)$, it suffices to prove the set of limits F lies in a connected locus of the moduli space. We will first construct the required connected locus of $\overline{M}_{\tau,n}(\mathbf{G}/\mathbf{P}, \beta_\tau)$.

The pair (τ, β_τ) canonically determines a family of maps γ_b with nodal domains over a base $b \in B$. For $v \in V_\tau$, let $\beta_\tau(v) = \sum_i a_i^v[P_i]$. Define the base space B as follows:

$$B = \prod_{v \in V_\tau} \overline{M}_{g(v),\mathrm{val}(v)+\sum_i a_i^v},$$

where val(v) is the valence of v in τ (including nodes and markings). The extra $\sum_i a_i^v$ markings each correspond to a basis homology element – with a_j^v of these markings corresponding to $[P_j]$. The degenerate cases $\overline{M}_{0,1}$ and $\overline{M}_{0,2}$ in the product B are taken to be points. B is irreducible and hence connected.

For $b = \prod_v [b_v] \in B$, let

$$\gamma_b : D_b \to \mathbf{G}/\mathbf{P}$$

be defined as follows:

(i) D_b is obtained by attaching the curves b_v by connecting nodes as specified by τ and further attaching $\mathbb{P}^1$'s to each of the extra points $\sum_i a_i^v$.

(ii) For each subcurve $b_v \subset D_b$, $\gamma_b(b_v) = s$.

(iii) For each $\mathbb{P}^1$ corresponding to $[P_j]$, $\gamma_b(\mathbb{P}^1) \cong P_j$.

The family of maps γ_b over B then defines a morphism (via stabilization):

$$\epsilon : B \to \overline{M}_{\tau,n}(\mathbf{G}/\mathbf{P}, \beta_\tau).$$

Certainly the image variety $\epsilon(B)$ is connected.

By Proposition 2, the limit F is simply the stabilization of $[\tilde{F}]$. Since $\tilde{F} = \gamma_b$ for some b, the set of limits F lies in a connected locus of $\overline{M}_{\tau,n}(\mathbf{G}/\mathbf{P}, \beta_\tau)$. This concludes the proof of Theorem 2.

Theorem 1 is a special case of Theorem 2 (where τ has a single vertex). Corollary 2 is a simple consequences of Theorem 2.

Proof of Corollary 2. In the genus 0 case, τ is a tree with genus function identically zero. The moduli stack

$$\overline{\mathcal{M}}_{\tau,n}(\mathbf{G}/\mathbf{P}, \beta_\tau) \tag{3}$$

is constructed as a fiber product over the evaluation maps obtained from the edges of τ. We will prove $\overline{\mathcal{M}}_{\tau,n}(\mathbf{G}/\mathbf{P}, \beta_\tau)$ is a nonsingular Deligne-Mumford stack by induction on the number of vertices of τ.

First, suppose τ has only 1 vertex v. Then, the moduli stack (3) is $\overline{\mathcal{M}}_{0,\mathrm{val}(v)}(\mathbf{G}/\mathbf{P}, \beta_\tau(v))$ – a nonsingular moduli stack by the convexity of $\mathbf{G}/\mathbf{P}$.

Next, let τ have m vertices and let v be an extremal vertex (v is incident to exactly 1 edge). Let $p \in \mathbf{G}/\mathbf{P}$ be a point. By the Kleiman-Bertini Theorem,

$$\mathrm{ev}_1^{-1}(p) \subset \overline{\mathcal{M}}_{0,\mathrm{val}(v)}(\mathbf{G}/\mathbf{P}, \beta_\tau(v)) \tag{4}$$

is a nonsingular Deligne-Mumford stack for the general point p (and hence *every* point p). Let τ' be the graph obtained by removing v from τ and adding an extra marking corresponding to the broken node. The moduli stack (3) is fibered over

$$\overline{\mathcal{M}}_{\tau',n'+1}(\mathbf{G}/\mathbf{P}, \beta'_\tau) \tag{5}$$

with fiber (4). As (5) is nonsingular by induction, the stack (3) is thus nonsingular. This completes the induction step.

Finally, since $\overline{\mathcal{M}}_{\tau,n}(\mathbf{G}/\mathbf{P}, \beta_\tau)$ is a nonsingular and connected Deligne-Mumford stack, it is irreducible.

4 Rationality

We first review a basic rationality result proven in a sequence papers by Katsylo and Bogomolov [Ka], [Bog]. Let $V = \mathbb{C}^2$ be a vector space. Let $a_1, a_2, \ldots, a_n$ be a sequence of positive integers with $\sum_i a_i \geq 3$. Then, the quotient

$$\mathbb{P}(\mathrm{Sym}^{a_1} V^*) \times \cdots \times \mathbb{P}(\mathrm{Sym}^{a_n} V^*) \; // \; \mathbf{PGL}(V) \tag{6}$$

is a rational variety – we may take any non-empty invariant theory quotient. Geometrically, the quotient (6) is birational to the moduli space quotient

$$M_{0,\sum_i a_i} \; / \; \Sigma_{a_1} \times \Sigma_{a_2} \times \cdots \times \Sigma_{a_n} \tag{7}$$

where Σ is the symmetric group. Essentially, the rationality of (6) is deduced from rationality in case $n = 1$ [Ka]. Proofs in the $n = 1$ case may be found in [Ka], [Bog].

We will also need the following simple Lemma.

Lemma 6 *Let W be any finite dimensional linear representation of $\mathbf{A}$ where $\mathbf{A} = \Sigma_2$ or $\mathbf{A} = \Sigma_3$. Then, $W/\mathbf{A}$ is rational.*

Proof. By the complete reducibility of representations and the fact that a $\mathbf{GL}$-bundle is locally trivial in the Zariski topology, it suffices to prove the

Lemma in case W is an irreducible representation. It is then easily checked by hand the two irreducible representation of Σ_2 and the three irreducible representations of Σ_3 have rational quotients.

Proof of Theorem 3. Fix the $\mathbb{C}^*$-action on $\mathbf{G}/\mathbf{P}$ as studied in Section 2. We first consider the moduli space

$$\overline{M} = \overline{M}_{0,n}(\mathbf{G}/\mathbf{P}, \beta = \sum_i a_i[P_i])$$

where the property

$$n + \sum_i a_i \geq 4 \tag{8}$$

is satisfied.

Let τ be the graph with a single vertex v with n markings, and let $\beta_\tau(v) = \sum_i a_i[P_i]$. Let γ_b over B be the family of maps constructed canonically from (τ, β_τ) in the proof of Theorem 2. The base B is simply:

$$B = \overline{M}_{0,n+\sum_i a_i}. \tag{9}$$

The map γ_b over a general point $b \in B$ has no map automorphisms (as $n + \sum_i a_i \geq 4$). Hence, the image $\epsilon(B)$ in $\overline{M}$ intersects the nonsingular (automorphism-free) locus of the moduli space $\overline{M}^0 \subset \overline{M}$. Let

$$\epsilon(B)^0 = \epsilon(B) \cap \overline{M}^0,$$

and let $B^0 = \epsilon^{-1}(\epsilon(B)^0)$. The map

$$B^0 \to \epsilon(B)^0$$

is simply a quotient of B^0 by the natural $\Sigma_{a_1} \times \cdots \times \Sigma_{a_n}$ action on (9). By the rationality result (6), $\epsilon(B)^0$ is rational.

Consider now the $\mathbb{C}^*$-action on $\overline{M}^0$ by translation. As $\overline{M}^0$ is a nonsingular, irreducible, quasi-projective variety, we may study the Bialynicki-Birula stratification of $\overline{M}^0$. By the proof of Theorem 2, $\epsilon(B)^0$ is a $\mathbb{C}^*$-fixed locus which contains the limit,

$$\lim_{t\to 0} t[f],$$

of the general point $[f] \in \overline{M}^0$. By [Bi], $\overline{M}^0$ is birational to an affine bundle over $\epsilon(B)^0$. Therefore, $\overline{M}$ is rational. The proof of Theorem 3 is complete in case (8) is satisfied.

Next, we will consider the case where the sum (8) is at most 3. In this case, the base B is a point. If $\epsilon(B)$ lies in the automorphism-free locus, the previous argument proving the rationality of $\overline{M}_{0,n}(\mathbf{G}/\mathbf{P},\beta)$ is still valid. There are exactly four cases in which the point $\epsilon(B)$ corresponds to a map with nontrivial automorphisms:

(i) $n=0$, $\beta=3[P_i]$.

(ii) $n=0$, $\beta=2[P_i]+[P_j]$, $i\neq j$.

(iii) $n=0$, $\beta=2[P_i]$.

(iv) $n=1$, $\beta=2[P_i]$.

Here, the Deligne-Mumford stack structure of these moduli spaces is important. The automorphism group in case (i) is Σ_3 and in cases (ii-iv) is Σ_2. In each case, we will show the coarse moduli space $\overline{M}_{0,n}(\mathbf{G}/\mathbf{P},\beta)$ is birational to a quotient of a linear representation of the corresponding automorphism group.

Consider first the case (i): $n=0$, $\beta=3[P_i]$. Let $\epsilon(B)=[\gamma]$. Let $[\mu]$ denote the unique 3-pointed stable map obtained from γ by marking each $\mathbb{P}^1\cong P_i$ by a point lying over x_i. Certainly, $[\mu]\in\overline{M}^0_{0,3}(\mathbf{G}/\mathbf{P},\beta)$

We will study:

$$N\subset\overline{M}^0_{0,3}(\mathbf{G}/\mathbf{P},\beta)$$

where N is the component of the locus of transverse intersection of the three divisors $\mathrm{ev}_1^{-1}(\overline{A}_i)$, $\mathrm{ev}_2^{-1}(\overline{A}_i)$, and $\mathrm{ev}_3^{-1}(\overline{A}_i)$ containing $[\mu]$. The torus $\mathbb{C}^*$ acts on N by translation. By an argument exactly parallel to the flow result of Proposition 2, we deduce

$$\lim_{t\to 0}t[f]=[\mu]$$

for a general element $[f]\in N$. As N is a nonsingular, quasi-projective scheme, Theorem 2.5 of [Bi] implies that N is $\mathbb{C}^*$-equivariantly birational to the tangent $\mathbb{C}^*$-representation at $[\mu]$.

There is a Σ_3-action on N by permutation of the markings. The $\mathbb{C}^*$ and Σ_3 actions commute. A slightly refined version of Theorem 2.5 of [Bi] shows N is $\mathbb{C}^*\times\Sigma_3$-equivariantly birational to the tangent $\mathbb{C}^*\times\Sigma_3$-representation at $[\mu]$. Lemma 7 below explains the refinements of the results of [Bi] needed here. N/Σ_3 is birational to $\overline{M}_{0,0}(\mathbf{G}/\mathbf{P},\beta)$. Hence, by Lemma 6, Theorem 3 is proven in case (i).

A similar strategy is used in cases (ii-iv). In each of these cases, let $\epsilon(B)=[\gamma]$ and let $[\mu]$ denote the rigidification by adding 2 new markings

$\bullet, \bullet'$ which lie over x_i. The locus N is chosen as the corresponding transverse intersection locus of $\mathrm{ev}_\bullet^{-1}(\overline{A}_i)$ and $\mathrm{ev}_{\bullet'}^{-2}(\overline{A}_i)$ in the maps space with the new markings. N is then $\mathbb{C}^* \times \Sigma_2$-equivariantly birational to the tangent $\mathbb{C}^* \times \Sigma_2$-representation of N at $[\mu]$ by the refined Lemma 7. Theorem 3 is then a consequence of Lemma 6 since N/Σ_2 is birational to the moduli space of maps considered in the case.

Lemma 7 *Let* $\mathbf{A}$ *be a finite group. Let* S *be a nonsingular, irreducible, quasi-projective scheme with a* $\mathbb{C}^* \times \mathbf{A}$*-action and a* $\mathbb{C}^* \times \mathbf{A}$*-fixed point* $s \in S$. *Let* T_s *denote the* $\mathbb{C}^* \times \mathbf{A}$*-representation on the tangent space at* s. *Suppose the* $\mathbb{C}^*$*-action is fully definite at* s. *Then, there is* $\mathbb{C}^* \times \mathbf{A}$*-equivariant isomorphism between an open set of* (S, s) *and* $(T_s, 0)$.

Proof. We note $\mathbb{C}^* \times \mathbf{A}$ is a linearly reductive group. By Theorem 2.4 of [Bi] for linearly reductive group actions, we may find a third nonsingular irreducible pointed space (Z, z) with a $\mathbb{C}^* \times \mathbf{A}$-action and equivariant, étale, morphisms:

$$\pi_1 : (Z, z) \to (S, s),$$

$$\pi_2 : (Z, z) \to (T_s, s).$$

In the proof of Theorem 2.5 of [Bi], such morphisms π_1 and π_2 are proven to be open immersions *by a study of only the* $\mathbb{C}^*$*-action.* Hence, the morphisms π_1 and π_2 are open immersions in our case. By the full definiteness of the $\mathbb{C}^*$-representation on T_s, the morphism π_2 is then an isomorphism.

References

[A] E. Akyildiz, *Bruhat decomposition via G_m-action,* (English. Russian summary) Bull. Acad. Polon. Sci. Sr. Sci. Math. 28 (1980), no. 11-12, 541–547 (1981).

[BM] K. Behrend and Yu. Manin, *Stacks of stable maps and Gromov-Witten invariants*, Duke Math. J. **85** (1996), no. 1, 1-60.

[Bi] Bialynicki-Birula, *Some theorems on actions of algebraic groups*, Ann. of Math. (2) **98** (1973), 480-497.

[Bog] F. Bogomolov, *Rationality of the moduli of hyperelliptic curves of arbitrary genus,* Proceedings of the 1984 Vancouver conference in algebraic geometry, 17-37, CMS Conf. Proc., 6, Amer. Math. Soc., Providence, R.I., 1986.

[Bor] A. Borel, *Linear algebraic groups,* Notes taken by Hyman Bass, W. A. Benjamin: New York-Amsterdam, 1969.

[C] C. Chevalley, *Séminaire C. Chevalley, 1956–1958*, Classification des groupes de Lie algbriques, Secrtariat mathmatique, 11 rue Pierre Curie, Paris 1958.

[DM] P. Deligne and D. Mumford *The irreducibility of the space of curves of given genus*, Inst. Hautes tudes Sci. Publ. Math. No. **36** 1969 75–109.

[FP] W. Fulton and R. Pandharipande, *Notes on stable maps and quantum cohomology*, Notes on stable maps and quantum cohomology. Algebraic geometry—Santa Cruz 1995, 45-96, Proc. Sympos. Pure Math., **62**, Part 2, Amer. Math. Soc., Providence, RI, 1997.

[Ha] G. Harder, *Chevalley groups over function fields and automorphic forms*, Ann. of Math. **100** (1974), 249-306.

[Har] R. Hartshorne, *Connectedness of the Hilbert scheme*, Inst. Hautes Etudes Sci. Publ. Math. **29** (1966), 5-48.

[Hi] A. Hirschowitz, *La rationalité des schémas de Hilbert de courbes gauches rationnelles suivant Katsylo,*. Algebraic curves and projective geometry (Trento, 1988), 87-90, Lecture Notes in Math., 1389, Springer: Berlin-New York, 1989.

[Ka] P. Katsylo, *Rationality of the field of invariants of reducible representations of the group* $\mathbf{SL}_2$, Moscow Univ. Math. Bull. **39** (1984), no. 5, 80-83.

[Ki] B. Kim, *Gromov-Witten invariants for flag manifolds,* Ph.D. Thesis, University of California at Berkeley, 1996.

[Ko] M. Kontsevich, *Enumeration of rational curves via torus actions*, in *The moduli space of curves*, R. Dijkgraaf, C. Faber, and G. van der Geer, eds., Birkhauser, 1995, pp 335-368.

[KoM] M. Kontsevich and Yu. Manin, *Gromov-Witten classes, quantum cohomology, and enumerative geometry*, Commun. Math. Phys. **164** (1994), 525-562.

[MM] B. Mann and M. Milgram, *On the moduli space of* $\mathbf{SU}(n)$ *monopoles and holomorphic maps to flag manifolds*, J. Diff. Geom. **38** (1993), 39-103.

[S] S. Strømme, *On parametrized rational curves in Grassmann varieties*, Space curves (Rocca di Papa, 1985), 251-272, Lecture Notes in Math., 1266, Springer: Berlin-New York, 1987.

[T] J. Thomsen, *Irreducibility of* $\overline{M}_{0,n}(G/P,\beta)$, Internat. J. Math. 9 (1998), no. 3, 367-376.

HOMOLOGICAL MIRROR SYMMETRY AND TORUS FIBRATIONS

MAXIM KONTSEVICH
IHES, 35 route de Chartres, F-91440, France
E-mail: maxim@ihes.fr

YAN SOIBELMAN
Department of Mathematics, KSU, Manhattan, KS 66506, USA
Email: soibel@math.ksu.edu

1 Introduction

1.1 Homological mirror symmetry and degenerations

Mathematically mirror symmetry can be interpreted in many ways. In this paper we will make a bridge between two approaches: the homological mirror symmetry ([Ko]) and the duality between torus fibrations (a version of Strominger-Yau-Zaslow conjecture, see [SYZ]).

The mirror symmetry is a duality between Calabi-Yau manifolds, i.e. complex manifolds which carry a Kähler metric with vanishing Ricci curvature. In fact, it is rather duality not between individual manifolds, but between manifolds in certain "degenerating" families ("large complex structure limit" and "large symplectic structure limit"). In this paper we propose a differential-geometric model of this degeneration. In particular, we conjecture that in the limit both dual manifolds X and $X^{\vee}$ become fiber bundles with toroidal fibers over the *same* base $\overline{Y}$ (see Section 3). Metric space $\overline{Y}$ is a compactification with some mild singularities of a (real) Riemannian manifold Y whose dimension is half of the dimension of X and $X^{\vee}$. Also, the manifold Y carries a rich geometric structure, including certain "combinatorial" data (so-called integral affine structure). This picture is partially motivated by the classical theory of collapsing Riemannian manifolds developed by M. Gromov and others (see for ex. [CG]). Another origin of our geometric conjectures is the [SYZ] version of mirror duality. We recast it in somewhat different terms in Section 2, devoted to the moduli space of conformal field theories and its natural compactification. In a recent preprint [GW] similar differential-geometric conjectures were suggested and verified in the case of degenerating K3-surfaces.

The Homological Mirror Conjecture proposed in [Ko] is a statement about equivalence of two A_{∞}-categories: the (derived) category of coherent sheaves on a Calabi-Yau manifold X and the Fukaya category of the dual Calabi-Yau manifold $X^\vee$. The former is defined in holomorphic (or algebraic) terms, the latter is defined in terms of symplectic geometry.

We apply the ideas of the theory of collapsing Riemannian manifolds to the Homological Mirror Conjecture. Let us call *differential-geometric* this model for degenerating Calabi-Yau manifolds. It gives a clear picture for the degeneration of the Fukaya category. The Fukaya category $F(X^\vee, \omega^\vee)$ of a symplectic manifold $(X^\vee, \omega^\vee)$, with $[\omega^\vee] \in H^2(X^\vee, \mathbf{Z})$, and its degeneration are defined as A_∞-categories over the field of Laurent formal power series $\mathbf{C}((q))$. The parameter q enters in the story when one writes higher compositions (Massey products), which have expressions $q^{\int_\beta \omega^\vee}$ as coefficients, $\beta \in H_2(X^\vee, \mathbf{Z})$. We can set $q = exp(-1/\varepsilon), \varepsilon \to 0$, where the parameter ε corresponds to the rescaling of the symplectic form: $\omega^\vee \mapsto \omega^\vee/\varepsilon$. If $[\omega^\vee]$ does not belong to $H^2(X, \mathbf{Z})$, one can work over the field $\mathbb{C}_\varepsilon := \{\sum_{i\geq 0} a_i e^{-\lambda_i/\varepsilon} | \, a_i \in \mathbf{C}, \lambda_i \in \mathbf{R}, \lambda_i \to +\infty\}$.

In the case of torus fibrations, a full subcategory of the limiting Fukaya category can be described in terms of the Morse theory on the base of the torus fibration. The higher products giving the A_∞-structure can be written as sums over sets of planar trees. In the case of cotangent bundles (instead of torus fibrations) this description was proposed earlier by Fukaya and Oh (see [FO]).

On the holomorphic side of mirror symmetry, the degeneration of the dual family X_q is described in non-archimedean terms: we have a Calabi-Yau manifold $\mathcal{X}_{mer}$ over the field $\mathbb{C}_q^{mer}$ of germs at $q = 0$ of meromorphic functions. Changing scalars, we get a Calabi-Yau manifold $\mathcal{X}_{form}$ over the local field of Laurent series $\mathbb{C}((q))$. Let us call *analytic* this degeneration picture. There is a description of a class of algebraic Calabi-Yau manifolds (over arbitrary local fields, complete with respect to discrete valuations) in terms of real C^∞-manifolds with integral affine structures. We expect that differential-geometric and (non-archimedean) analytic pictures of the degeneration are equivalent. This equivalence reflects two different ways of looking at Calabi-Yau manifolds: differential-geometric (via Kähler metrics with vanishing Ricci curvature) and algebro-geometric (via smooth projective varieties with vanishing canonical class).

The Homological Mirror Conjecture says that the Fukaya category $F(X^\vee, \omega^\vee)$ is equivalent (as an A_∞-category over $\mathbb{C}((q))$) to the derived category of coherent sheaves $D^b(\mathcal{X}_{form})$. We expect that it implies well-known numerical predictions for the number of rational curves on a Calabi-Yau manifold (genus zero Gromov-Witten invariants).

Using our conjectures about the collapse of Calabi-Yau manifolds, we offer a general approach to the proof of Homological Mirror Conjecture. We apply it in the case when the torus fibration has no singularities. This happens in the case of abelian varieties. In general, one should investigate the input of singularities of the base of torus fibration. In the same vein, we discuss the relationship between Riemannian manifolds with integral affine structures and varieties over non-archimedean fields in the simplest case of flat tori and abelian varieties. The general case will be discussed elsewhere.

It should be clear from the above discussion that the non-archimedean analysis plays an important role in the formulation and the proof of Homological Mirror Conjecture. Analytic picture of the degeneration seems to be related to the theory of rigid analytic spaces in the version of Berkovich (see [Be]). In particular, there is a striking similarity between the base of torus fibration and a certain canonically defined subset (see 3.3) of the Berkovich spectrum of an algebraic Calabi-Yau manifold over a local field. This subject definitely deserves further investigation.

1.2 Content of the paper

In Section 2 we discuss motivations from the Conformal Field Theory. In Section 3 we formulate the conjectures about analytic and differential-geometric pictures of the large complex structure limit. In Section 4 we describe a general framework of A_∞-pre-categories adapted to the transversality problem in the definition of the Fukaya category. Section 5 is devoted to the Fukaya category and its degeneration. The reader will notice an advantage of working over the field of Laurent power series: one can consider *all* local systems over Lagrangian submanifolds (in the conventional approach unitarity of the holonomy is required). Section 6 is devoted to the A_∞-category of smooth functions introduced by Fukaya (and then studied by Fukaya and Oh in [FuO]). We prove that this A_∞-category has a very simple de Rham model. This part of the paper can be read independently of the rest. On the other hand, the technique and the general scheme of the proof will be used later in Section 8, devoted to the Homological Mirror Conjecture. One important technical tool is an explicit A_∞-structure on a subcomplex of a differential-graded algebra (see [GS], [Me]). We restate the formulas from [Me] in term of sums over a set of planar trees. Our proof of the equivalence of Morse and de Rham A_∞-categories uses the approach to Morse theory from [HL]. Section 7 is devoted to the analytic side of the Homological Mirror Conjecture. We assign a rigid analytic space to the class of torus fibrations discussed in Section 3. We also construct a mirror symmetry functor for torus fibrations in terms of non-archimedean geometry. The use of non-archimedean analysis allows us to avoid problems with convergence of series in the definition of the Fukaya category. In Section 8 we construct the A_∞-pre-category which is equivalent to a full A_∞-subcategory of the A_∞-version of the derived category of coherent sheaves on a Calabi-Yau manifold over $\mathbf{C}_\varepsilon$. We prove that this category is equivalent to an A_∞-subcategory of the Fukaya category of the mirror dual torus fibration. In Appendix (Section 9) we describe the analogs of our constructions in the case of complex geometry.

Acknowledgements. Research of Y.S. was partially supported by IHES, the fellowship from the Clay Mathematics Institute and CRDF grant UM-2091. He also thanks the IHES for hospitality and excellent research conditions.

2 Degenerations of unitary Conformal Field Theories

In this section we will explain physical motivations for our picture of mirror symmetry. We assume that the reader is familiar to some extent with the basic notions of Conformal Field Theory. For example, the lectures [Gaw] contain most of what we need.

Unitary Conformal Field Theory (abbreviated by CFT below) is well-defined mathematically. It is described by the following data:

1) A real number $c \geq 0$ called central charge.

2) A bi-graded pre-Hilbert *space of states* $H = \oplus_{p,q \in \mathbf{R}_{\geq 0}} H^{p,q}, p - q \in \mathbf{Z}$ such that $dim(\oplus_{p+q \leq E} H^{p,q})$ is finite for every $E \in \mathbf{R}_{\geq 0}$. Equivalently, there is an action of the Lie group $\mathbf{C}^*$ on H, so that $z \in \mathbf{C}^*$ acts on $H^{p,q}$ as $z^p \bar{z}^q := (z\bar{z})^p \bar{z}^{q-p}$.

3) An action of the product of Virasoro and anti-Virasoro Lie algebras $Vir \times \overline{Vir}$ (with the same central charge c) on H, so that the space $H^{p,q}$ is an eigenspace of the generator L_0 (resp. $\overline{L}_0$) with the eigenvalue p (resp. q).

4) The space H carries some additional structures derived from the operator product expansion (OPE). The OPE is described by a linear map $H \otimes H \to H \widehat{\otimes} \mathbf{C}\{z, \bar{z}\}$. Here $\mathbf{C}\{z, \bar{z}\}$ is the topological ring of formal power series $f = \sum_{p,q} c_{p,q} z^p \bar{z}^q$ where $c_{p,q} \in \mathbf{C}$, $p, q \to +\infty$, $p, q \in \mathbf{R}$, $p - q \in \mathbf{Z}$. The OPE satisfies a list axioms, which we are not going to recall here (see [Gaw]).

Let $\phi \in H^{p,q}$. Then the number $p + q$ is called the *conformal dimension* of ϕ (or the *energy*), and $p - q$ is called the *spin* of ϕ. Notice that, since the spin of ϕ is an integer number, the condition $p + q < 1$ implies $p = q$.

The central charge c can be described by the formula $dim(\oplus_{p+q \leq E} H^{p,q}) = exp(\sqrt{4/3\pi^2 cE(1 + o(1))}$ as $E \to +\infty$. It is expected that all possible central charges form a countable well-ordered subset of $\mathbf{Q}_{\geq 0} \subset \mathbf{R}_{\geq 0}$. If $H^{0,0}$ is a one-dimensional vector space, the corresponding CFT is called irreducible. A general CFT is a sum of irreducible ones. The *trivial* CFT has $H = H^{0,0} = \mathbb{C}$ and it is the unique irreducible unitary CFT with $c = 0$.

Remark 1 *Geometric considerations of this paper are related to $N = 2$ Superconformal Field Theories (SCFT). There is a version of the above data and axioms for SCFT. In particular, each $H^{p,q}$ is a hermitian super vector space. There is an action of the super extension of the product of Virasoro and anti-Virasoro algebra on H. In the discussion of the moduli spaces below we will not distinguish between CFTs and SCFTs, because except of some minor details, main conclusions are true in both cases.*

2.1 Moduli space of Conformal Field Theories

For a given CFT one can consider its group of symmetries (i.e. automorphisms of the space $H = \oplus_{p,q} H^{p,q}$ preserving all the structures). It is expected that the group of symmetries is a compact Lie group of dimension less or equal than $dim\, H^{1,0}$.

Let us fix $c_0 \geq 0$ and $E_{min} > 0$, and consider the moduli space $\mathcal{M}_{c\leq c_0}^{E_{min}}$ of all irreducible CFTs with the central charge $c \leq c_0$ and

$$min\{p+q > 0 | H^{p,q} \neq 0\} \geq E_{min}$$

It is expected that $\mathcal{M}_{c\leq c_0}^{E_{min}}$ is a compact real analytic stack of finite local dimension. The dimension of the base of the minimal versal deformation of a given CFT is less or equal than $dim\, H^{1,1}$. We define $\mathcal{M}_{c\leq c_0} = \cup_{E_{min}>0} \mathcal{M}_{c\leq c_0}^{E_{min}}$. We would like to compactify this stack by adding boundary components corresponding to certain asymptotic descriptions of the theories with $E_{min} \to 0$. The compactified space is expected to be a compact stack $\overline{\mathcal{M}}_{c\leq c_0}$. In what follows we will loosely use the word "space" instead of the word "stack".

Remark 2 *There are basically only two classes of rigorously defined CFTs: the rational theories (RCFT) and the lattice CFTs. Considerations of this paper correspond to the case of sigma models which produce neither of these. The description of sigma models as path integrals corresponding to certain Lagrangians did not give yet a mathematically satisfactory construction. As we will explain below, there is an alternative way to speak about sigma models in terms of degenerations of CFTs.*

2.2 Physical picture of a simple collapse

In order to compactify $\mathcal{M}_{c\leq c_0}$ we consider degenerations of CFTs as $E_{min} \to 0$. A degeneration is given by a one-parameter (discrete or continuous) family $H_\varepsilon, \varepsilon \to 0$ of bi-graded spaces as above, where $(p,q) = (p(\varepsilon), q(\varepsilon))$. These spaces are equipped with OPEs. The subspace of fields with conformal dimensions vanishing as $\varepsilon \to 0$ gives rise to a commutative algebra $H^{small} = \oplus_{p(\varepsilon)\ll 1} H_\varepsilon^{p(\varepsilon),p(\varepsilon)}$ (the algebra structure is given by the leading terms in OPEs). The spectrum X of H^{small} is expected to be a compact space ("manifold with singularities") such that $dim\, X \le c_0$. It follows from the conformal invariance and the OPE, that the grading of H^{small} (rescaled as $\varepsilon \to 0$) is given by the eigenvalues of a second order differential operator defined on the smooth part of X. The operator has positive eigenvalues and is determined up to multiplication by a scalar. This implies that the smooth part of X carries a metric g_X, which is also defined up to multiplication by a scalar. Other terms in OPEs give rise to additional differential-geometric structures on X.

Thus, as a first approximation to the real picture, we assume the following description of a "simple collapse" of a family of CFTs. The degeneration of the family is described by the point of the boundary of $\overline{\mathcal{M}}_{c\le c_0}$ which is a triple $(X, \mathbf{R}_+^* \cdot g_X, \phi_X)$, where the metric g_X is defined up to a positive scalar factor, and $\phi_X : X \to \mathcal{M}_{c\le c_0 - dim\, X}$ is a map. One can have some extra conditions on the data. For example, the metric g_X can satisfy the Einstein equation.

Although the scalar factor for the metric is arbitrary, one should imagine that the curvature of g_X is "small", and the injectivity radius of g_X is "large". The map ϕ_X appears naturally from the point of view of the simple collapse of CFTs described above. Indeed, in the limit $\varepsilon \to 0$, the space H_ε becomes an H^{small}-module. It can be thought of as a space of sections of an infinite-dimensional vector bundle $W \to X$. One can argue that fibers of W generically are spaces of states of CFTs with central charges less or equal than $c_0 - dim\, X$. This is encoded in the map ϕ_X. In the case when CFTs from $\phi_X(X)$ have non-trivial symmetry groups, one expects a kind of a gauge theory on X as well.

Purely bosonic sigma-models correspond the case when $c_0 = c(\varepsilon) = dim\, X$ and the residual theories (CFTs in the image of ϕ_X) are all trivial. The target space X in this case should carry a Ricci flat metric. In the supersymmetric case the target space X is a Calabi-Yau manifold, and the residual bundle of CFTs is a bundle of free fermion theories.

Remark 3 *We expect that all compact Ricci flat manifolds (with the metric defined up to a constant scalar factor) appear as target spaces of degenerating CFTs. Thus, the construction of the compactification of the moduli space of CFTs should include as a part a compactification of the moduli spaces of Einstein manifolds. Notice that in differential geometry there is a fundamental* *result of Gromov (see [G]) about the precompactness of the moduli space of pointed connected complete Riemannian manifolds of a given dimension, with the Ricci curvature bounded from below. There is a deep relationship between the compactification of the moduli space of CFTs and the Gromov's compactification. It seems that one can deduce from certain physical arguments that all target spaces appearing as limits of CFTs have non-negative Ricci curvature.*

2.3 Multiple collapse and the structure of the boundary

In terms of the Virasoro operator L_0 the collapse is described by a subset (cluster) S_1 in the set of eigenvalues of L_0 which approach to zero "with the same speed", as $E_{min} \to 0$. The next level of the collapse is described by another subset S_2 of eigenvalues of L_0. Elements of S_2 approach to zero "modulo the first collapse" (i.e. at the same speed, but "much slower" than elements of S_1). One can continue to build a tower of degenerations. It leads to an hierarchy of boundary strata. Namely, if there are further degenerations of CFTs parameterized by X, one gets a fiber bundle over the space of triples $(X, \mathbf{R}_+^* \cdot g_X, \phi_X)$ with the fiber which is the space of triples of similar sort. Finally, we obtain the following qualitative geometric picture of the boundary $\partial\overline{\mathcal{M}}_{c \le c_0}$.

A boundary point is given by the following data:

1) A finite tower of maps of compact topological spaces $p_i : \overline{X}_i \to \overline{X}_{i-1}, 0 \le i \le k$, $\overline{X}_0 = \{pt\}$.

2) A sequence of smooth manifolds $(X_i, g_{X_i}), 0 \le i \le k$, such that X_i is a dense subspace of $\overline{X}_i$, and $dim\, X_i > dim\, X_{i-1}$, and p_i defines a fiber bundle $p_i : X_i \to X_{i-1}$.

3) Riemannian metrics on the fibers of the restrictions of p_i to X_i, such that the diameter of each fiber is finite. In particular the diameter of X_1 is finite, because it is the only fiber of the map $p_1 : X_1 \to \{pt\}$.

4) A map $X_k \to \mathcal{M}_{c \le c_0 - dim\, X_k}$.

The data above are considered up to the natural action of the group $(\mathbf{R}_+^*)^k$ (it rescales the metrics on fibers).

There are some additional data, like non-linear connections on the bundles $p_i : X_i \to X_{i-1}$. The set of data should satisfy some conditions, like differential equations on the metrics. We cannot formulate this portion of data more precisely in general case. It will be done below in the case of $N = 2$ SCFTs corresponding to sigma models with Calabi-Yau target spaces.

2.4 Example: Toroidal models

Non-supersymmetric toroidal model is described by the so-called Narain lattice, endowed with some additional data. More precisely, let us fix the central charge $c = n$ which is a positive integer number. What physicists call the Narain lattice $\Gamma^{n,n}$ is a unique unimodular lattice of rank $2n$ and the signature (n,n). It can be described as $\mathbf{Z}^{2n}$ equipped with the quadratic form $Q(x_1,...,x_n,y_1,...,y_n) = \sum_i x_i y_i$. The moduli space of toroidal CFTs is

$$\mathcal{M}^{tor}_{c=n} = O(n,n,\mathbf{Z})\backslash O(n,n,\mathbf{R})/O(n,\mathbf{R}) \times O(n,\mathbf{R}).$$

Equivalently, it is a quotient of the open part of the Grassmannian $\{V_+ \subset \mathbf{R}^{n,n} |\, dim\, V_+ = n, Q_{|V} > 0\}$ by the action of $O(n,n,\mathbf{Z}) = Aut(\Gamma^{n,n}, Q)$. Let V_- be the orthogonal complement to V_+. Then every vector of $\Gamma^{n,n}$ can be uniquely written as $\gamma = \gamma_+ + \gamma_-$, where $\gamma_\pm \in V_\pm$. For the corresponding CFT one has

$$\sum_{p,q} dim(H^{p,q}) z^p \bar{z}^q = \Big|\prod_{k\geq 1}(1-z^k)\Big|^{-2n} \sum_{\gamma\in\Gamma^{n,n}} z^{Q(\gamma_+)} \bar{z}^{-Q(\gamma_-)}$$

Let us try to compactify the moduli space $\mathcal{M}^{tor}_{c=n}$. Suppose that we have a one-parameter family of toroidal theories such that $E_{min}(\varepsilon)$ approaches zero. Then for corresponding vectors in H_ε one gets $p(\varepsilon) = q(\varepsilon) \to 0$. It implies that $Q(\gamma(\varepsilon)) = 0, Q(\gamma_+(\varepsilon)) \ll 1$. It is easy to see that one can add vectors $\gamma(\varepsilon)$ satisfying these conditions. Thus one gets a (part of) lattice of the rank less or equal than n. In the case of "maximal" simple collapse the rank will be equal to n. One can see that the corresponding points of the boundary give rise to the following data: $(X, \mathbf{R}^*_+ \cdot g_X, \phi^{triv}_X; B)$, where (X, g_X) is a flat n-dimensional torus, $B \in H^2(X, \mathbf{R}/\mathbf{Z})$ and ϕ^{triv}_X is the constant map form X to the trivial theory point in the moduli space of CFTs. These data in turn give rise to a toroidal CFT, which can be realized as a sigma model with the target space (X, g_X) and given B-field B. The residual bundle of CFTs on X is trivial.

Let us consider a 1-parameter family of CFTs defined by the family $(X, \lambda g_X, \phi^{triv}_X; B = 0)$, where $\lambda \in (0, +\infty)$. There are two degenerations of this family, which define two points of the boundary $\partial\overline{\mathcal{M}}^{tor}_{c=n}$. As $\lambda \to +\infty$, we get a toroidal CFT defined by $(X, \mathbf{R}^*_+ \cdot g_X, \phi^{triv}_X; B = 0)$. As $\lambda \to 0$ we get $(X^\vee, \mathbf{R}^*_+ \cdot g_{X^\vee}, \phi^{triv}_X; B = 0)$, where $(X^\vee, g_{X^\vee})$ is the dual flat torus.

There might be further degenerations of the lattice. Thus one obtains a stratification of the compactified moduli space of lattices (and hence CFTs). Points of the compactification are described by flags of vector spaces $0 =$

$V_0 \subset V_1 \subset V_2 \subset ... \subset V_k \subset \mathbf{R}^n$. In addition one has a lattice $\Gamma_{i+1} \subset V_{i+1}/V_i$, considered up to a scalar factor. These data give rise to a tower of torus bundles $X_k \to X_{k-1} \to ... \to X_1 \to \{pt\}$ over tori with fibers $(V_{i+1}/V_i)/\Gamma_{i+1}$. If $V_k \simeq \mathbf{R}^{n-l}, l \geq 1$, then one has also a map from the total space X_k of the last torus bundle to the point $[H_k]$ in the moduli space of toroidal theories of smaller central charge: $\phi_n : X_k \to \mathcal{M}^{tor}_{c=l}$, $\phi_k(X_k) = [H_k]$.

2.5 *Example: WZW model for SU(2)*

In this case we have a discrete family with $c = \frac{3k}{k+2}$, where $k \geq 1$ is an integer number called *level*. In the limit $k \to +\infty$ one gets $X = SU(2) = S^3$ equipped with the standard metric. The corresponding bundle is the trivial bundle of trivial CFTs (with $c = 0$ and $H = H^{0,0} = \mathbf{C}$). Analogous picture holds for an arbitrary compact simply connected simple group G.

2.6 *A-model and B-model of N = 2 SCFT as boundary strata*

The boundary of the compactified moduli space $\overline{\mathcal{M}}^{N=2}$ of $N = 2$ SCFTs with a given central charge contains an open stratum given by sigma models with Calabi-Yau targets. Each stratum is parameterized by the classes of equivalence of quadruples $(X, J_X, \mathbf{R}^*_+ \cdot g_X, B)$ where X is a compact real manifold, J_X a complex structure, g_X is a Calabi-Yau metric, and $B \in H^2(X, i\mathbf{R}/\mathbf{Z})$ is a B-field. The residual bundle of CFTs is a bundle of free fermion theories.

As a consequence of supersymmetry, the moduli space $\mathcal{M}^{N=2}$ of superconformal field theories is a complex manifold which is locally isomorphic to the product of two complex manifolds. [a] It is believed that this decomposition (up to certain corrections) is global. Also, there are two types of sigma models with Calabi-Yau targets: A-models and B-models. Hence, the traditional picture of the compactified moduli space looks as follows:

Here the boundary consists of two open strata (A-stratum and B-stratum) and a mysterious meeting point. This point corresponds, in general, to a submanifold of codimension one in the closure of A-stratum and of B-stratum.

We argue that this picture should be modified. There is another open stratum of $\partial\overline{\mathcal{M}}^{N=2}$ (we call it T-stratum). It consists of toroidal models (i.e. CFTs associated with Narain lattices), parameterized by a manifold Y with a Riemannian metric defined up to a scalar factor. This subvariety meets both A and B strata along the codimension one stratum corresponding to the double collapse.

---

[a] Strictly speaking, one should exclude models with chiral fields of conformal dimension

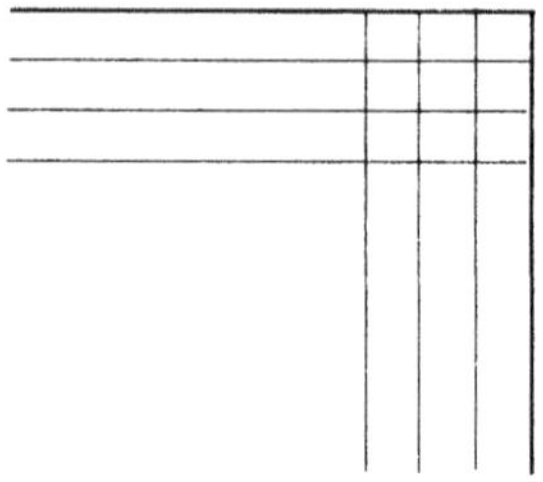

Figure 1. The traditional picture of the compactified moduli space

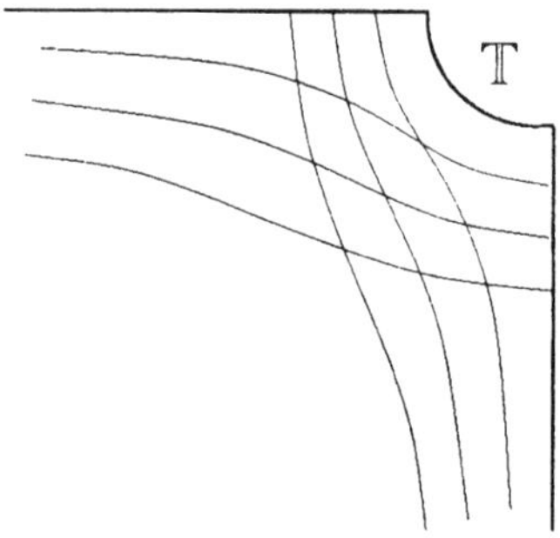

Figure 2. The "true" picture is obtained from the traditional one by the real blow-up at the corner

2.7 Mirror symmetry and the collapse

Mirror symmetry is related to the existence of two different strata of the boundary $\partial\overline{\mathcal{M}}^{N=2}$ which we called A-stratum and B-stratum. As a corollary, same quantities admit different geometric descriptions near different strata. In the traditional picture, one can introduce natural coordinates in a small neighborhood of a boundary point corresponding to $(X, J_X, \mathbb{R}_+^* \cdot g_X, B)$. Skipping X from the notation, one can say that the coordinates are (J, g, B) (complex structure, Calabi-Yau metric and the B-field). Geometrically, the pairs (g, B) belong to the preimage of the Kähler cone under the natural map $Re : H^2(X, \mathbb{C}) \to H^2(X, \mathbb{R})$ (more precisely, one should consider B

$(2,0)$, e.g. sigma models on hyperkähler manifolds, see [AM].

as an element of $H^2(X, i\mathbb{R}/\mathbb{Z})$). It is usually said, that one considers an open domain in the complexified Kähler cone with the property that with the class of metric $[g]$ it contains also the ray $t[g], t \gg 1$. The mirror symmetry gives rise to an identification of neighborhoods of $(X, J_X, \mathbb{R}_+^* \cdot g_X, B_X)$ and $(X^\vee, J_{X^\vee}, \mathbb{R}_+^* \cdot g_{X^\vee}, B_{X^\vee})$ such that J_X is interchanged with $[g_{X^\vee}] + iB_{X^\vee})$ and vice versa.

We can describe this picture in a different way. Using the identification of complex and Kähler moduli, one can choose $([g_X], B_X, [g_{X^\vee}], B_{X^\vee})$ as local coordinates near the meeting point of A-stratum and B-stratum. There is an action of the additive semigroup $\mathbb{R}_{\geq 0} \times \mathbb{R}_{\geq 0}$ in this neighborhood. It is given explicitly by the formula $([g_X], B_X, [g_{X^\vee}], B_{X^\vee}) \mapsto (e^{t_1}[g_X], B_X, e^{t_2}[g_{X^\vee}], B_{X^\vee})$ where $(t_1, t_2) \in \mathbb{R}_{\geq 0} \times \mathbb{R}_{\geq 0}$. As $t_1 \to +\infty$, a point of the moduli space approaches the B-stratum, where the metric is defined up to a positive scalar only. The action of the second semigroup $\mathbb{R}_{\geq 0}$ extends by continuity to the non-trivial action on the B-stratum. Similarly, in the limit $t_2 \to +\infty$ the flow retracts the point to the A-stratum.

This picture should be modified, if one makes a real blow-up at the corner, as discussed before. Again, the action of the semigroup $\mathbb{R}_{\geq 0} \times \mathbb{R}_{\geq 0}$ extends continuously to the boundary. Contractions to A-stratum and B-stratum carry non-trivial actions of the corresponding semigroups isomorphic to $\mathbb{R}_{\geq 0}$. Now, let us choose a point in, say, A-stratum. Then the semigroup flow takes it along the boundary to the new stratum, corresponding to the double collapse. The semigroup $\mathbb{R}_{\geq 0} \times \mathbb{R}_{\geq 0}$ acts trivially on this stratum. A point of the double collapse is also a limiting point of a 1-dimensional orbit of $\mathbb{R}_{\geq 0} \times \mathbb{R}_{\geq 0}$ acting on the T-stratum. Explicitly, the element (t_1, t_2) changes the size of the tori defined by the Narain lattices, rescaling them with the coefficient $e^{t_1 - t_2}$. This flow carries the point of T-stratum to another point of the double collapse, which can be moved then inside of the B-stratum. The whole path, which is the intersection of $\partial \overline{\mathcal{M}}^{N=2}$ and the $\mathbb{R}_{\geq 0} \times \mathbb{R}_{\geq 0}$-orbit, connects an A-model with the corresponding B-model through the stratum of toroidal models.

The T-portion of the path (we call it T-path) connects dual torus fibrations over the same Riemannian base. This is mirror symmetry in our picture.

This description is inspired by [SYZ]. The reader notices however, that in our picture, the mirror symmetry phenomenon is explained entirely in terms of the ***boundary*** of the compactified moduli space. In order to explain the mirror symmetry phenomenon it is not necessary to build full SCFTs. It is sufficient to work with simple toroidal models on the boundary of the compactified moduli space $\overline{\mathcal{M}}^{N=2}$. Also, in contrast with [SYZ], we do not use supersymmetric cycles (D-branes) in our description.

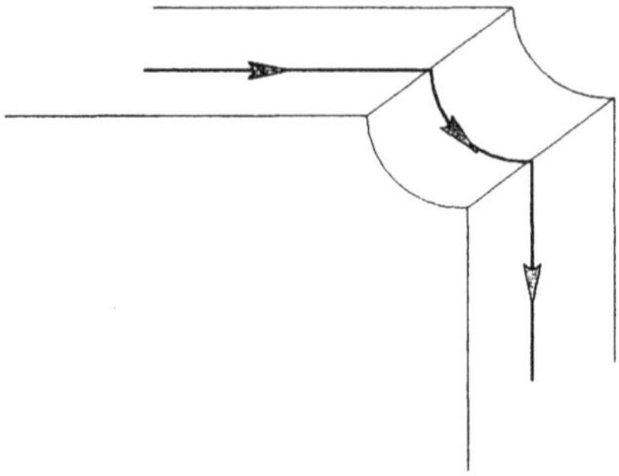

Figure 3. The path on the boundary from A-stratum to B-stratum via T-stratum

3 Calabi-Yau manifolds in the large complex structure limit

3.1 Maximal degenerations of Calabi-Yau manifolds

Let $\mathbf{C}_q^{mer} = \{f = \sum_{n \geq n_0} a_n q^n\}$ be the field of germs at $q = 0$ of meromorphic functions in one complex variable.

Let $\mathcal{X}_{mer}$ be an algebraic n-dimensional Calabi-Yau manifold over $\mathbb{C}_q^{mer}$ (i.e. $\mathcal{X}_{mer}$ is a smooth projective manifold over $\mathbf{C}_q^{mer}$ with the trivial canonical class: $K_{\mathcal{X}} = 0$). We fix an algebraic non-vanishing volume element $vol \in \Gamma(\mathcal{X}_{mer}, K_{\mathcal{X}})$. The pair $(\mathcal{X}_{mer}, vol)$ defines a 1-parameter analytic family of complex Calabi-Yau manifolds $(X_q, vol_q), 0 < |q| < r_0$, for some $r_0 > 0$.

Let $[\omega] \in H^2_{DR}(\mathcal{X}_{mer})$ be the cohomology class in the ample cone. Then for every q, such that $0 < |q| < r_0$ it defines a Kähler class ω_q on X_q. By the Yau theorem, there exists a unique Calabi-Yau metric g_{X_q} on X_q with the Kähler class $[\omega_q]$.

It follows from the resolution of singularities, that as $q \to 0$ one has the following formula:

$$\int_{X_q} vol_q \wedge \overline{vol}_q = C(log|q|)^m |q|^{2k}(1 + o(1))$$

for some $C \in \mathbf{C}^*, k \in \mathbf{Z}, 0 \leq m \leq n = dim\,(\mathcal{X}_{mer})$.

Definition 1 *We say that $\mathcal{X}_{mer}$ has maximal degeneration at $q = 0$ if in the formula above we have $m = n$.*

Let us show that this definition is equivalent to the usual one, given in terms of variations of Hodge structures (see [Mo]).

Proposition 1 *The Calabi-Yau manifold $\mathcal{X}_{mer}$ has maximal degeneration iff for all sufficiently small q there exists a vector $v \in H^n(X_q, \mathbb{C})$ such that $(T - Id)^{n+1} v = 0$ and $(T - id)^n v \neq 0$ where T is the monodromy operator.*

Proof. First of all, notice that the volume $\int_{X_q} vol_q \wedge \overline{vol_q}$ can be calculated cohomologically as the Poincaré pairing $\langle [vol_q], [\overline{vol_q}] \rangle$ in the (primitive part of) middle cohomology $H^n(X_q, \mathbb{C})$.

We can assume (after passing to a cover by adding a root of q) that the operator T is unipotent. Let us trivialize the bundle $H^n(X_q, \mathbb{C})$ over the punctured disc by multiplication by

$$q^{-log(T)/2\pi i} = \sum_{k=0}^{n} \left(\frac{log(q)}{-2\pi i} \right)^k \frac{(log(T))^k}{k!}$$

The nilpotent orbit theorem says that the Hodge filtration extends to a holomorphic filtration on the trivialized bundle over the whole disc, including the point $q = 0$. Thus, the bundle $q^{-log(T)/2\pi i}(H^{n,0})$ extends to a line bundle over the disc, and the section $q^{-log(T)/2\pi i}([vol_q])$ is a non-zero meromorphic section of this bundle. After the multiplication by an appropriate power of q we may assume further that this section is holomorphic and non-vanishing at $q = 0$. We denote this holomorphic section by $a(q)$, $a(0) \neq 0$.

Now let us calculate the volume:

$$\langle [vol_q], [\overline{vol_q}] \rangle = \sum_{k,l \geq 0} \left\langle \left(\frac{log(q)}{2\pi i} \right)^k \frac{(log(T))^k}{k!} a(q), \left(\frac{log(\overline{q})}{-2\pi i} \right)^l \frac{(log(T))^l}{l!} \overline{a(q)} \right\rangle =$$

$$= \sum_{k,l \geq 0} \left(\frac{log(q)}{2\pi i} \right)^k \left(\frac{log(\overline{q})}{-2\pi i} \right)^l \frac{(-1)^l}{k!l!} \left\langle (log(T))^{k+l} a(q), \overline{a(q)} \right\rangle$$

Here we use the fact that operator $log(T)$ is real and also skew-symmetric with respect to the Poincaré pairing. It follows from the equality $log(T)^{n+1} = 0$ (which holds automatically) that in the sum above all terms with $k + l > n$ vanish. The contribution of terms with $k + l = n$ is equal to

$$\frac{(log(|q|)^n}{(\pi i)^n n!} \langle (log(T)^n a(q), \overline{a(q)} \rangle.$$

It follows easily from the standard properties of variations of polarized Hodge structures (see e.g. [De]) that if $log(T)^n \neq 0$ then in fact $log(T)^n(a(0)) \neq 0$, and moreover $\langle (log(T)^n a(0), \overline{a(0)} \rangle \neq 0$. Now the statement of the Proposition becomes obvious. ■

Notice that in [De] a slightly stronger condition of maximal degeneration was imposed: the weight filtration on $H^*(X_q, \mathbb{C})$ associated with the monodromy operator should be complementary to the Hodge filtration.

Let us recall the definition of the Gromov-Hausdorff metric ρ_{GH}. It is a metric on the space of isometry classes of compact metric spaces. We say that two metric spaces M_1 and M_2 are ε-close in ρ_{GH} if there exists a metric space M containing both M_1 and M_2 as metric subspaces, such that M_1 belongs to the ε-neighborhood of M_2 and vice versa. Then $\rho_{GH}(M_1, M_2)$ is given by the minimum of such ε.

Let us rescale the Calabi-Yau metric: $g_{X_q}^{new} = g_{X_q}/diam(X_q, g_{X_q})^2$. Thus we obtain a 1-parameter family of Riemannian manifolds $X_q^{new} = (X_q, g_{X_q}^{new})$ of the diameter 1.

Conjecture 1 *If $\mathcal{X}_{mer}$ has maximal degeneration at $q = 0$ then there is a limit $(\overline{Y}, g_{\overline{Y}})$ of X_q^{new} in the Gromov-Hausdorff metric, such that:*

a) $(\overline{Y}, g_{\overline{Y}})$ is a compact metric space, which contains a smooth oriented Riemannian manifold (Y, g_Y) of dimension n as a dense open metric subspace. The Hausdorff dimension of $Y^{sing} = \overline{Y} \setminus Y$ is less or equal than $n - 2$.

b) Y carries an integral affine structure. This means that it carries a torsion-free flat connection ∇ with the holonomy contained in $SL(n, \mathbb{Z})$.

c) The metric g_Y has a potential. This means that it is locally given in affine coordinates by a symmetric matrix $(g_{ij}) = (\partial^2 K/\partial x_i \partial x_j)$, where K is a smooth function (defined modulo adding an affine function, i.e. the sum of a linear function and a constant).

d) In affine coordinates the metric volume element is constant, $det(g_{ij}) = det(\partial^2 K/\partial x_i \partial x_j) = const$ (real Monge-Ampère equation).

At the end of this section we propose a non-rigorous explanation of our conjecture based on differential-geometric considerations.

Remark 4 *1) Since the matrix (g_{ij}) defined by the metric g_Y is positive, the function K is convex. In particular, there is locally well-defined Legendre transform of K. This fact will be used later, when we will discuss the duality of Monge-Ampère manifolds.*

2) It seems plausible that in the case when all X_q are simply-connected, and $h^{k,0}(X_q) = 0$ for $0 < k < n$, the metric space $\overline{Y}$ is a homological sphere of dimension n. In all examples it is in fact homeomorphic to S^n.

The conjecture opens the way for compactification of the moduli space of Calabi-Yau metrics on a given Calabi-Yau manifold M, by adding as a boundary component the set of pairs $(\overline{Y}, \mathbb{R}_+^* \cdot g_{\overline{Y}})$ for all 1-parameter maximal degenerations X_q, such that $X_{q'} = M$ for some q'. This corresponds to a choice of a "cusp" in the moduli space of Calabi-Yau manifolds. This choice is usually described in terms of certain algebro-geometric data: the action of the monodromy operator, variation of Hodge structures, mixed Hodge structure of the special fiber, etc. The previous conjecture offers a pure "metric" description of a cusp.

It follows from part b) of the conjecture that one can choose a ∇-covariant lattice $T^{\mathbf{Z}}_{Y,y} \subset T_{Y,y}, y \in Y$. Suppose we are given a triple (Y, g_Y, ∇), satisfying the properties a)-c) of the conjecture, and we have fixed a covariant lattice $T^{\mathbf{Z}}_Y$ in the tangent bundle T_Y. Then we can construct a 1-parameter family of *non-compact* complex Calabi-Yau manifolds, endowed with Ricci flat Kähler metrics. Namely, let X^ε be the total space of the torus bundle $p_\varepsilon : X^\varepsilon \to Y$ with fibers $T_{Y,y}/\varepsilon T^{\mathbf{Z}}_{Y,y}, y \in Y, 0 < \varepsilon \leq \varepsilon_0$. The total space TY of the tangent bundle T_Y carries a canonical complex structure coming from the isomorphism $T_{TY} \simeq \pi^* T_Y \oplus \pi^* T_Y \simeq \pi^* T_Y \otimes \mathbf{C}$ where $\pi : TY \longrightarrow Y$ is the canonical projection (here we use the affine structure on Y). Using the same identification, we introduce a metric on TY, namely $g_{TY} = \pi^* g_Y \oplus \pi^* g_Y$. It is easy to see, that g_{TY} is a Kähler metric with the potential $\pi^* K$. It follows from the Monge-Ampère equation that the metric g_{TY} is Ricci flat. Passing to the quotient, we obtain on X^ε a complex structure J_{X^ε} and a Ricci flat Kähler metric g_{X^ε}.

Let $U \subset Y$ be an open simply-connected subset. Then there is an action of the torus $T^n \simeq T_{Y,y}/T^{\mathbf{Z}}_{Y,y}$ on $p_\varepsilon^{-1}(U), y \in U$ (different tori are identified for different points $y \in U$ by means of the connection ∇). It implies that for any $t \in H^1(Y, (T_Y/T^{\mathbf{Z}}_Y)^{discr})$ (cohomology with coefficients in the local system of tori considered as abstract groups) one can define a twisted manifold $X^{\varepsilon,t}$, which is the total space of the torus fibration $p_{\varepsilon,t} : X^{\varepsilon,t} \to Y$.

Roughly speaking, the next conjecture says that the "leading asymptotic term" of the family of Calabi-Yau manifolds $X^{new}_q, q = e^{-1/\varepsilon}$ near the point of maximal degeneration $\varepsilon = 0$, is isomorphic up to a twist to the family $(X^\varepsilon, J_{X^\varepsilon})$ associated with the torus bundle described above.

More precisely, we formulate it as follows.

Conjecture 2 *Let $(\mathcal{X}_{mer}, vol) = (X_q, vol_q)$ be a 1-parameter family of maximally degenerate Calabi-Yau manifolds, and X^{new}_q be the family with rescaled metrics, as before. There exist a constant $C > 0$ and a function $t(q)$ such that Kähler manifolds X^{new}_q and $X^{\varepsilon(q),t(q)}$ with $\varepsilon(q) = C(log|q|)^{-1}$ are close to each other (as $q \to 0$) in the following sense:*

for any $\delta > 0$ there exist a decomposition $X_q = X^{sm}_q \sqcup X^{sing}_q$ and an embedding of smooth manifolds $j_q : X_q \to p^{-1}_{\varepsilon(q),t(q)}(Y \setminus (Y^{sing})^\delta)$, where $(Y^{sing})^\delta$ is a δ-neighborhood of Y^{sing}, such that:

a) (X_q, X^{sing}_q) converges in the Gromov-Hausdorff metric to the pair

$(\overline{Y}, Y^{sing})$.

b) j_q *identifies up to* $o(1)$ *terms, uniformly in* $x \in X_q^{sm}$*, the scalar products and complex structures on the tangent spaces* $T_x X_q$ *and* $T_{j_q(x)} X^{\varepsilon(q),t(q)}$.

There is the following motivation for the Conjectures 1 and 2. In general, for a degenerating family of Riemannian metrics with non-negative Ricci curvature, one expects a description in terms of a tower of fibrations (collapses) with singularities (compare with 2.3). [b] In the case of Kähler manifolds there are two basic pictures of a simple collapse. The first case is when both the base and the fiber are Kähler manifolds. In the second case fibers are flat totally real tori of dimension m and the base looks locally as a product of a domain in $\mathbb{R}^m$ with a Kähler manifold. The logarithmic factor in the asymptotic behavior of the volume should come only from torus fibers. Thus, the largest possible power of the logarithm can appear only when we have a tower of purely torus fibrations. It seems that the fixing (up to a scalar) of the Kähler class forbids the multiple collapse. These considerations give an intuitive "explanation" of our conjectures.

Remark 5 *During the preparation of this text we learned that conjectures similar to ours were proposed independently by M. Gross and P. Wilson (see [GW]). A remarkable achievement in [GW] consists of the verification of conjectures in the case of degenerating* $K3$ *surfaces, together with a precise description of the behavior of metrics near singular fibers. Also, in a recent preprint [Le] mirror symmetry was discussed from a similar point of view. In the main body of the present paper we will consider degenerations of complex abelian varieties. In this case the conjectures obviously hold.*

3.2 Monge-Ampère manifolds and duality of torus fibrations

In this section we propose a mathematical language for the geometric mirror symmetry, understood as a duality of torus fibrations.

Definition 2 *A Monge-Ampère manifold is a triple* (Y, g, ∇)*, where* (Y, g) *is a smooth Riemannian manifold with the metric* g*, and* ∇ *is a flat connection on* T_Y *such that:*

a) ∇ *defines an affine structure on* Y.

b) Locally in affine coordinates $(x_1, ..., x_n)$ *the matrix* $((g_{ij}))$ *of* g *is given by* $((g_{ij})) = ((\partial^2 K / \partial x_i \partial x_j))$ *for some smooth real-valued function* K.

c) The Monge-Ampère equation $det((\partial^2 K / \partial x_i \partial x_j)) = const$ *is satisfied.*

[b] Some steps in the program of compactification of the space of metrics are accomplished now (see e.g. [CC]), but still there are many non-clarified issues.

Monge-Ampère manifolds were studied (under a different name) in [CY] where is was proven that if Y is *compact* then its finite cover is a torus.

Let us consider a (non-compact) example motivated by the mirror symmetry for K3 surfaces (see also [GW]). Let S be a complex surface endowed with a holomorphic non-vanishing volume form vol_S, and $\pi : S \to C$ be a holomorphic fibration over a complex curve C, such that fibers of π are non-singular elliptic curves.

We define a metric g_C on C as the Kähler metric associated with the $(1,1)$-form $\pi_*(vol_S \wedge \overline{vol_S})$. Let us choose (locally on C) a basis (γ_1, γ_2) in $H_1(\pi^{-1}(x), \mathbf{Z}), x \in C$. We define two closed 1-forms on C by the formulas

$$\alpha_i = Re(\int_{\gamma_i} vol_S), i = 1, 2.$$

It follows that $\alpha_i = dx_i$ for some functions $x_i, i = 1, 2$. We define an affine structure on C, and the corresponding connection ∇, by saying that (x_1, x_2) are affine coordinates. One can check directly that (C, g_C, ∇) is a Monge-Ampère manifold. In a typical example of elliptic fibration of a K3 surface, one gets $C = \mathbf{CP}^1 \setminus \{z_1, ..., z_{24}\}$, where $\{z_1, ..., z_{24}\}$ is a set of distinct 24 points in $\mathbf{CP}^1$.

Returning to the general case, we can restate a portion of our conjectures by saying that the smooth part of the Gromov-Hausdorff limit of a maximally degenerate family of Calabi-Yau manifolds is a Monge-Ampère manifold with an integral affine structure.

There is a well-known duality on local solutions of the Monge-Ampère equation.

Lemma 1 *Let $U \subset \mathbf{R}^n$ be a convex open domain in $\mathbf{R}^n$ equipped with the standard affine coordinates $(x_1, ..., x_n)$, and $K : U \to \mathbf{R}$ be a convex function satisfying the Monge-Ampère equation. Then the Legendre transform $\widehat{K}(y_1, ..., y_n) = max_{x \in U}(\sum_i x_i y_i - K(x_1, ..., x_n))$ also satisfies the Monge-Ampère equation.*

Proof. The graph of $L = dK$ is a Lagrangian submanifold in $T^*\mathbf{R}^n = \mathbf{R}^n \oplus (\mathbf{R}^n)^*$. Let p_1 and p_2 be the natural projections to the direct summands. They are local diffeomorphisms. Since K is defined up to the adding of an affine function, the graph itself is defined up to translations. The Monge-Ampère equation corresponds to the condition $p_1^*(vol_{\mathbf{R}^n}) = p_2^*(vol_{\mathbf{R}^{*n}})$, where $vol_{\mathbf{R}^n}$ (resp. $vol_{\mathbf{R}^{*n}}$) denotes the standard volume form on $\mathbf{R}^n$ (resp. $\mathbf{R}^{*n}$). The manifold L can be considered as a graph of $d\widehat{K}$. Thus $\widehat{K}$ satisfies the Monge-Ampère equation as well. The Lemma is proved. ∎

The manifold L carries a Riemannian metric g_L induced by the indefinite metric $\sum_i dx_i\, dy_i$ on $\mathbf{R}^n \oplus (\mathbf{R}^n)^*$. This metric is given by the matrix $(\partial^2 K/\partial x_i \partial x_j)$ in coordinates $(x_1, ..., x_n)$, and by the matrix $(\partial^2 \widehat{K}/\partial y_i \partial y_j)$ in the dual coordinates. Thus on L we have a metric, and two affine structures (pullbacks of the standard affine structures on the coordinate spaces). Hence we have two structures of the Monge-Ampère manifold on L. It is easy to see that the local pictures can be glued together. This leads to the following result.

Proposition 2 *For a given Monge-Ampère manifold (Y, g_Y, ∇_Y) there is a canonically defined dual Monge-Ampère manifold $(Y^\vee, g_Y^\vee, \nabla_Y^\vee)$ such that (Y, g_Y) is identified with $(Y^\vee, g_Y^\vee)$ as Riemannian manifolds, and the local system $(T_{Y^\vee}, \nabla_Y^\vee)$ is naturally isomorphic to the local system dual to (T_Y, ∇_Y).*

Corollary 1 *If ∇_Y defines an integral affine structure on Y (i.e. the holonomy of ∇_Y belongs to $GL(n, \mathbf{Z})$), then $\nabla_Y^\vee$ defines an integral affine structure on $Y^\vee$. As thc dual covariant lattice one takes the lattice $(T_Y^{\mathbf{Z}})^\vee$, which is dual to $T_Y^{\mathbf{Z}}$ with respect to the metric g_Y.*

Now we can state the geometric counterpart of the mirror symmetry conjecture.

Conjecture 3 *Smooth parts of maximal degenerations of dual families of Calabi-Yau manifolds are dual Monge-Ampère manifolds with dual integral affine structures.*

Monge-Ampère manifolds with integral affine structures are real analogs of Calabi-Yau manifolds. In fact the mirror duality in the sense of this section holds for a larger class of manifolds. We define an *AK-manifold* (AK stands for *affine* and *Kähler*) as in the Definition 2, but dropping the condition c) (Monge-Ampère equation), see also [CY]. The reader can check easily that all constructions of this section, including the duality of torus fibrations hold for AK-manifolds as well.

Remark 6 *The idea to use the Legendre transform for the purposes of mirror symmetry was around for some time (see for example [H], [Le]).*

Remark 7 *In our description of geometric mirror symmetry we ignore the B-fields. In what follows we will always assume that $B = 0$.*

3.3 Speculations about relations with non-archimedean geometry

Considerations from CFT and from differential geometry indicate that the integral affine structure on Y does not depend on the choice of the Kähler class of Calabi-Yau metrics. Thus, we obtain a "combinatorial" invariant $(Y, T_Y^{\mathbf{Z}})$ of (maximally degenerating) Calabi-Yau variety over the local field

$K = \mathbb{C}((q))$. One can argue that in this case there will be a canonical atlas of coordinate charts such that the transition maps belong to the group $SAff(n,\mathbf{Z}) := SL(n,\mathbf{Z}) \ltimes \mathbf{Z}^n$. The natural question arises whether one can define and calculate it purely algebraically, without the use of transcendental methods and Calabi-Yau metrics. We expect that the answer to this question is positive. In other words there exists a canonical way to associate the data $(Y, T_Y^{\mathbf{Z}})$ with arbitrary smooth projective variety X, $c_1(T_X) = 0$ having "maximal degeneration" over an arbitrary field K with a discrete valuation.

The conjectural answer (only for the compactification $\overline{Y}$ of Y) is the following: let us choose (after an extension of the field K) a model with stable reduction. Call an irreducible component D of the special fiber X_0 *essential* if the order of pole at D of the global volume element on X is maximal among all components of X_0. We define topological space $\overline{Y}(X_0)$ as the Clemens complex spanned by essential divisors (see [LTY]). Roughly speaking, k-cells of $\overline{Y}$ correspond to irreducible components of $(k+1)$-fold intersections of essential divisors.

Using ideas from motivic integration and from Berkovich theory of non-archimedean analytic spaces (see [Be]), one can prove the following result.

Proposition 3 *For different choices of models with stable reduction, the spaces $\overline{Y}(X_0)$ can be canonically identified.*

We will sketch the idea of the proof. [c] Following [Be], one can associate to an algebraic variety X over a local field K, a K-analytic space X^{an}. For an affine X, points of X^{an} are $\mathbb{R}$-valued multiplicative seminorms on the algebra of functions $\mathcal{O}(X)$, extending a given norm on K. In our case $K = \mathbb{C}((q))$.

Step 1. One defines the set of *divisorial points* $D(X,K)$ as the direct limit of the sets of irreducible components of special fibers, taken over all models of X over $\mathbf{C}[[q]]$ (not necessarily smooth). One has a natural embedding $D(X,K) \subset X^{an}$ (the seminorm corresponding to D is the exponent of the order of the pole of the volume form on the divisor D).

Similarly one defines $D(X,K') = D(X \otimes K')/Gal(K'/K)$, where K'/K is a Galois extension of finite order. Taking direct limit of $D(X,K')$ over all K' one gets the set $D(X)$.

Step 2. One checks that $D(X) \subset X^{an}$, and in fact $D(X)$ is a dense subset in X^{an}. For local fields with finite or countable residue field set $D(X)$ is countable, while X^{an} is connected.

Step 3. Having a divisor $D \in D(X)$ one defines the number

$$v(vol, D) = \frac{ord_{(X\otimes K', D)}(vol) + 1}{ram(K'/K)},$$

[c] After the lecture of M.K. at Rutgers University on October 30, 2000.

where *ord* means the order of zero of the volume form *vol* on $D \subset X \otimes K'$, and *ram* means the ramification index. One checks that $v(vol, D)$ is well-defined. In p-adic case this number can be derived from the contribution of D to the integral $\int_{X(K)} |vol|_K$, where $|vol|_K$ is the associated measure.

Step 4. One checks that $\overline{Y}(X_0)$ is the closure in X^{an} of the set of divisors $D \in D(X)$ such that $v(vol, D)$ is minimal. Hence it is independent of a choice of a model. This explains the statement of the theorem. ∎

In examples coming from toric geometry the space $\overline{Y} = \overline{Y}(X_0)$ is always a manifold. It is not clear yet what is the origin of the smooth part $Y \subset \overline{Y}$, and of the affine structure on it. Conjecturally, all this comes from a map $\pi : X(\overline{K}) \to \overline{Y}$ where $\overline{K}$ is the algebraic closure of K. In the differential-geometric picture of torus fibrations (when $K = \mathbf{C}_{mer}$) the map π is obvious: it associates with a meromorphic (finitely ramified) family of points $x_q \in X_q$ the limit point $lim_{q \to 0}\, x_q \in Y$ in the metric sense. Also, the differential-geometric picture suggests that the closure of the image $\pi(Z(\overline{K}))$ where $Z \subset K$ is an algebraic subvariety, should be a piecewise linear closed subset of Y, and linear pieces of it have rational directions. In particular, if Z is a curve then $\pi(Z(\overline{K}))$ is a graph in Y. This opens a way to express Gromov-Witten invariants of X in terms of the Feynman expansion for certain quantum field theory on Y.

Also, we expect that the choice of an ample class in $NS(X) \otimes \mathbf{R}$ on X gives rise to the ***dual*** integral affine structure on Y defined again in some purely algebro-geometric way. If the ample class is the first Chern class of a line bundle, then there should be also a canonical reduction of the dual integral affine structure to a $SAff(n, \mathbb{Z})$ -structure.

4 A_∞-algebras and A_∞-categories

4.1 Two problems with the general definition

The purpose of this section is to describe the framework in which the results concerning A_∞-categories will be formulated. We would like to make few comments even before recalling a definition of the Fukaya category. These comments are informal. Precise definitions will be given later in this section.

There are two main problems with the definition of the Fukaya category. First, morphisms can be defined only for transversal Lagrangian submanifolds (in particular, the identity morphism is never defined). Second, since there are pseudo-holomorphic discs with the boundary on a given Lagrangian submanifold, one has to add a composition m_0 to the set of compositions $m_n, n \geq 1$. As a result, the spaces of morphisms are not complexes: $m_1^2 \neq 0$.

On the other hand, the derived category of coherent sheaves arises from an A_∞-category without m_0 and with the condition $m_1^2 = 0$. Hence one should explain in which sense two A_∞-categories in question are equivalent. The above-mentioned problems can be resolved by an appropriate generalization of the notion of A_∞-category. We offer such a generalization below (we call it A_∞-pre-category). We remark that there exists a better generalization. It involves numerous preparations and will be given elsewhere (see [KoS]). On the other hand, the problem with m_0 does not appear in the version of the Fukaya category for abelian varieties considered in this paper. Hence, for the purposes of present paper it is sufficient to work with A_∞-pre-categories (or A_∞-categories with transversal structure, cf. [P1]). This gives a partial solution to the transversality problem, and provides a solution to the problem with the identity morphisms.

Using A_∞-pre-categories we are going to formulate and prove in Section 8 a variant of the Homological Mirror Conjecture for torus fibrations. It can be applied to the case of abelian varieties. In particular, one can obtain certain formulas for Massey products for abelian varieties in terms of partial theta-sums similar to those considered in [P1].

4.2 Non-unital A_∞-algebras and A_∞-categories

Let $A = \oplus_{i\in\mathbf{Z}} A^i$ be a $\mathbf{Z}$-graded module over a field k. [d] As usual, we will denote by $A[n]$ the graded k-module such that $(A[n])^i = A^{i+n}$ for all i.

Definition 3 *A structure of non-unital A_∞-algebra on A is given by a codifferential d of degree $+1$ on the cofree tensor coalgebra $T_+(A[1]) = \oplus_{n\geq 1}(A[1])^{\otimes n}$.*

The codifferential d is by definition a coderivation, such that $d^2 = 0$. It is uniquely determined by its "Taylor coefficients" $m_n : A^{\otimes n} \to A[2-n], n \geq 1$. The condition $d^2 = 0$ can be rewritten as a sequence of quadratic equations

$$\sum_{i+j=n+1}\sum_{0\leq l\leq i} \epsilon(l,j) m_i(a_0, ..., a_{l-1}, m_j(a_l, ..., a_{l+j-1}), a_{l+j}, ..., a_n) = 0$$

where $a_m \in A$, and $\epsilon(l,j) = (-1)^{j\sum_{0\leq s\leq l-1} deg(a_s))+l(j-1)+j(i-1)}$. In particular, $m_1^2 = 0$.

Definition 4 *A morphism of non-unital A_∞-algebras (A_∞-morphism for short) $(V, d_V) \to (W, d_W)$ is a morphism of tensor coalgebras $T_+(V[1]) \to T_+(W[1])$ of degree zero, which commutes with the codifferentials.*

[d] In what follows one can replace k by a $\mathbf{Z}$-graded commutative associative algebra and assume that all k-modules are projective.

A morphism f of non-unital A_∞-algebras is determined by its "Taylor coefficients" $f_n : V^{\otimes n} \to W[1-n], n \geq 1$ satisfying the system of equations

$$\sum_{1\leq l_1<...,<l_i=n}(-1)^{\gamma_i} m_i^W(f_{l_1}(a_1,...,a_{l_1}),$$
$$f_{l_2-l_1}(a_{l_1+1},...,a_{l_2}),...,f_{n-l_{i-1}}(a_{n-l_{i-1}+1},...,a_n)) =$$
$$\sum_{s+r=n+1}\sum_{1\leq j\leq s}(-1)^{\epsilon_s} f_s(a_1,...,a_{j-1},m_r^V(a_j,...,a_{j+r-1}),a_{j+r},...,a_n).$$

Here

$$\epsilon_s = r\sum_{1\leq p\leq j-1} deg(a_p) + j - 1 + r(s-j),$$

$$\gamma_i = \sum_{1\leq p\leq i-1}(i-p)(l_p - l_{p-1} - 1) + \sum_{1\leq p\leq i-1}\nu(l_p)\sum_{l_{p-1}+1\leq q\leq l_p} deg(a_q)$$

where we use the notation $\nu(l_p) = \sum_{p+1\leq m\leq i}(1 - l_m + l_{m-1})$ and set $l_0 = 0$.

Definition 5 *A non-unital A_∞-category C over k is given by the following data:*

1) A class of objects $Ob(C)$.

2) For any two objects X_1 and X_2 a $\mathbb{Z}$-graded k-module of morphisms $Hom(X_1, X_2)$.

3) For any sequence of objects $X_0, ..., X_n$, $n \geq 1$, a morphism of k-modules (called a composition map) $m_n : \otimes_{0\leq i\leq n-1} Hom(X_i, X_{i+1}) \to Hom(X_0, X_n)[2-n]$.

It is required that for any sequence of objects $X_0, ..., X_N$, $N \geq 0$ the graded k-module $A = A(X_0, ..., X_N) := \oplus_{i,j} Hom(X_i, X_j)$, equipped with the direct sum of the compositions $m_n, n \geq 1$, is a non-unital A_∞-algebra.

The class of objects $Ob(C)$ will be often denoted by C. We hope it will not lead to a confusion.

Remark 8 *A non-unital A_∞-algebra A can be considered as a non-unital A_∞-category with one object X such that $Hom(X, X) = A$.*

Definition 6 *A functor $F : C_1 \to C_2$ between non-unital A_∞-categories is given by the following data:*

1) A map of classes of objects $\phi : C_1 \to C_2$.

2) For any finite sequence of objects $X_0, ..., X_n$, $n \geq 0$, a morphism of graded k-modules $f_n : \otimes_{0\leq i\leq n-1} Hom_{C_1}(X_i, X_{i+1}) \to Hom_{C_2}(\phi(X_0), \phi(X_n))[1-n]$.

The following condition holds for any $X_1, ..., X_N \in C_1$: the sequence $f_n, n \geq 1$ defines an A_∞-morphism

$$\oplus_{i,j} Hom_{C_1}(X_i, X_j) \to \oplus_{i,j} Hom_{C_2}(\phi(X_i), \phi(X_j)).$$

Remark 9 *Let C be a non-unital A_∞-category. Let us replace spaces of morphisms by their cohomology with respect to m_1. In other words, we define $Hom_{H(C)}(X,Y) := \{Ker\, m_1\}/\{Im\, m_1\}$, where $m_1 : Hom_C(X,Y) \to Hom_C(X,Y)[1]$ is the composition map. Then $H(C) = (C, Hom_{H(C)}(\cdot,\cdot))$ gives rise to a "non-unital" category structure with the class of objects C and composition of morphisms induced by m_2. We write "non-unital" because there are no identity morphisms $id_X \in Hom_{H(C)}(X,X)$.*

4.3 A_∞-pre-categories

We start with the notion of *non-unital* A_∞-pre-category. It allows us to work with "transversal" sequences of objects. [e] Then we will introduce the notion of A_∞-pre-category. It provides us with a replacement of the identity morphisms. Roughly speaking, we will have the identity morphism up to homotopy.

Definition 7 *Let k be a $\mathbf{Z}$-graded commutative associative ring as before. A non-unital A_∞-pre-category over k is defined by the following data:*

a) A class of objects C.

b) For any $n \geq 1$ a subclass C_n^{tr} of C^n, $C_1^{tr} = C$, called the class of transversal sequences.

c) For $(X_1, X_2) \in C_2^{tr}$ a $\mathbf{Z}$-graded k-module of morphisms $Hom(X_1, X_2)$.

d) For a transversal sequence of objects $(X_0, ..., X_n)$, $n \geq 0$, a morphism of k-modules (composition map) $m_n : \otimes_{0\leq i \leq n-1} Hom(X_i, X_{i+1}) \to Hom(X_0, X_n)[2-n]$.

It is required that a subsequence $(X_{i_1}, ..., X_{i_l}), i_1 < i_2 < ... < i_l$ of a transversal sequence $(X_1, ..., X_n)$ is transversal, and that the composition maps satisfy the same system of equations as for non-unital A_∞-categories. Explicitly:

$$\sum_{i+j=n+1} \sum_{0\leq l \leq i} \epsilon(l,j) m_i(a_0, ..., a_{l-1}, m_j(a_l, ..., a_{l+j}), a_{l+j+1}, ..., a_n) = 0,$$

where $a_m \in Hom(X_m, X_{m+1})$, $\epsilon(l,j) = (-1)^{j \sum_{0\leq s \leq l-1} deg(a_s) + \ell(j-1) + j(i-1)}$.

Definition 8 *A functor $F : C \to \mathcal{D}$ between non-unital A_∞-pre-categories is given by the following data:*

1) A map of classes of objects $\phi : C \to \mathcal{D}$, such that $\phi^n(C_n^{tr}) \subset \mathcal{D}_n^{tr}$.

2) For any transversal sequence of objects $(X_0, ..., X_n)$, $n \geq 1$ in C, a morphism of graded k-modules

$$f_n : \otimes_{0\leq i\leq n-1} Hom_C(X_i, X_{i+1}) \to Hom_{\mathcal{D}}(\phi(X_0), \phi(X_n))[1-n].$$

[e] The notion of "transversality" is purely formal in this section. The choice of the name will become clear after concrete applications in the geometric context, see next sections.

These data satisfy the following property: the sequence $f_n, n \geq 1$ defines an A_∞-morphism $\oplus_{i<j} Hom_{\mathcal{C}}(X_i, X_j) \to \oplus_{i<j} Hom_{\mathcal{D}}(\phi(X_i), \phi(X_j))$.

The reader have noticed that we use the summation only over the increasing pairs of indices $i < j$. It differs from the case of non-unital A_∞-pre-categories. The reason is that we do not require the transversality to be a symmetric relation on objects. It is possible that $Hom(X_0, X_1)$ exists, but $Hom(X_1, X_0)$ does not. In the case when all Hom's are defined, two discussed definitions agree. In particular, a non-unital A_∞-category is the same as a non-unital A_∞-pre-category such that $\mathcal{C}_n^{tr} = \mathcal{C}^n$ for any $n \geq 1$.

Definition 9 *Let $\mathcal{C}$ be a non-unital A_∞-pre-category, $(X_1, X_2) \in \mathcal{C}_2^{tr}$. We say that $f \in Hom^0(X_1, X_2)$ (zero stands for degree) is a quasi-isomorphism if $m_1(f) = 0$, and for any objects X_0 and X_3 such that $(X_0, X_1, X_2) \in \mathcal{C}_3^{tr}$ and $(X_1, X_2, X_3) \in \mathcal{C}_3^{tr}$ one has: $m_2(f, \cdot) : Hom(X_0, X_1) \to Hom(X_0, X_2)$ and $m_2(\cdot, f) : Hom(X_2, X_3) \to Hom(X_1, X_3)$ are quasi-isomorphisms of complexes.*

Definition 10 *An A_∞-pre-category is a non-unital A_∞-pre-category $\mathcal{C}$, satisfying the following extension property:*

For any finite collection of transversal sequences $(S_i)_{i \in I}$ in $\mathcal{C}$ and an object X there exist objects X_+ and X_- and quasi-isomorphisms $f_- : X_- \to X$, $f_+ : X \to X_+$ such that extended sequences (X_-, S_i, X_+) are transversal for any $i \in I$.

Remark 10 *Let $\mathcal{C}$ be an A_∞-pre-category. Then partially defined on $H(\mathcal{C}) = (\mathcal{C}, Hom_{H(\mathcal{C})}(\cdot,\cdot))$ composition m_2 extends uniquely, so that it defines a structure of a category on $H(\mathcal{C})$.*

Definition 11 *Let $\mathcal{C}$ and $\mathcal{D}$ be A_∞-pre-categories over k. An A_∞-functor $F : \mathcal{C} \to \mathcal{D}$ is a functor between the corresponding non-unital A_∞-pre-categories such that F takes quasi-isomorphisms in $\mathcal{C}$ to quasi-isomorphisms in $\mathcal{D}$.*

There is an important notion of equivalence of A_∞-pre-categories (and A_∞-categories). We are planning to provide all the details elsewhere (see [KoS]). For the purposes of present paper we will be using the following definition (which is in fact a theorem in the more general framework).

Definition 12 *An A_∞-functor $F : \mathcal{C} \to \mathcal{D}$ between A_∞-pre-categories is called an A_∞-equivalence functor if:*

a) Every object $Y \in \mathcal{D}$ is quasi-isomorphic to an object $\phi(X), X \in \mathcal{C}$.

b) The functor induces quasi-isomorphisms of non-unital A_∞-algebras of morphisms, corresponding to all transversal sequences of objects.

Definition 13 *Two A_∞-pre-categories $\mathcal{C}$ and $\mathcal{D}$ are called equivalent if there exists a finite sequence of A_∞-pre-categories $(\mathcal{C}_0, \dots, \mathcal{C}_n)$, $\mathcal{C}_0 = \mathcal{C}$, $\mathcal{C}_0 = \mathcal{D}$ such that for every i, $0 \leq i \leq k-1$ there exists an A_∞-equivalence functor* *from C_i to C_{i+1} or vice versa.*

We suggest the language of A_∞-pre-categories in order to replace more conventional A_∞-categories with strict identity morphisms.

Definition 14 *An A_∞-category with strict identity morphisms is a non-unital A_∞-category C, such that for any object X there exists an element $1 = 1_X \in Hom^0(X,X)$ (identity morphism) such that $m_2(1,f) = m_2(f,1) = f$ and $m_n(f_1,...,1,...,f_n) = 0, n \neq 2$ for any morphisms $f, f_1, ..., f_n$.*

An A_∞-category C with strict identity morphisms is an A_∞-pre-category, because (in the previous notation) we can extend a transversal sequence S to (X,S,X), and set $X_+ = X_- = X$, $f_\pm = 1_X$. Another remark is that if C has only one object, it is an A_∞-algebra with the strict unit. One can try to develop the deformation theory of such algebras along the lines of [KoS1]. The problem is that the corresponding operad is not free, and the standard theory becomes complicated. We hope that the framework of A_∞-pre-categories is appropriate for the purposes of deformation theory of A_∞-categories. The following conjecture gives another evidence in favor of such a generalization of A_∞-categories.

Conjecture 4 *Let us define the notion of equivalent A_∞-categories with strict identity morphisms similarly to the case of A_∞-pre-categories (see above). Then the equivalence classes of A_∞-pre-categories are in one-to-one correspondence with the equivalence classes of A_∞-categories with strict identity morphisms.*

4.4 Example: directed A_∞-pre-categories

There is a useful special case of the notion of A_∞-pre-category (independently a similar notion was suggested in [Se]).

Definition 15 *A directed A_∞-pre-category is an A_∞-pre-category such that*

a) There is bijection of the class of objects and the set integer numbers: $C \simeq \mathbf{Z}$. We denote by X_i the object corresponding to $i \in \mathbf{Z}$.

b) Transversal sequences are $(X_{i_1}, ..., X_{i_n}), i_1 < i_2 < ... < i_n$.

The extension property is equivalent to the following one: for any object X_i there are exist objects $X_j, j < i$ and $X_m, m > i$ which are quasi-isomorphic to X_i. Then one can formulate the following version of the previous conjecture.

Conjecture 5 *Equivalence classes of directed A_∞-pre-categories are in one-to-one correspondence with the equivalences classes of A_∞-categories with strict identity morphisms and countable class of objects.*

Having an A_∞-category C with strict identity morphisms, and countable class of objects, one can construct an infinite sequence of objects $(X_i)_{i\in\mathbf{Z}}$ such that each objects appears infinitely many times for positive and negative i. Then a directed A_∞-pre-category C' is defined by setting $Hom_{C'}(X_i, X_j) = Hom_C(X_i, X_j)$ for $i < j$. All other $Hom's$ are not defined.

5 Fukaya category and its degeneration

5.1 Fukaya category

Fukaya category (of a compact symplectic manifold) in the approach presented here will be in fact an A_∞-pre-category. Our definition is not given in the maximal generality, but it will be sufficient for the main application to abelian varieties. For more elaborated definitions see [Fu1], [FuOOO].

Let (V, ω) be a compact symplectic manifold of dimension $2n$, such that $c_1(T_V) = 0 \in H^2(V, \mathbf{Z})$. The Fukaya category (with the trivial B-field) associated with (V, ω) depends on some additional data, which we are going to describe below.

We fix an almost complex structure J compatible with ω and a smooth everywhere non-vanishing differential form Ω, which is $(n, 0)$-form with respect to J. Let L be an oriented Lagrangian submanifold. Then one has a map $Arg_L := Arg_{\Omega_{|L}} : L \to \mathbf{R}/2\pi\mathbf{Z}$, where $Arg_{\Omega_{|L}}(x)$ is the argument of the non-zero complex number $\Omega(e_1 \wedge ... \wedge e_n)$, and $e_1, ..., e_n$ is an oriented basis of $T_xL, x \in L$.

Definition 16 *Objects of the Fukaya category $F(V, \omega, J, \Omega)$ are triples $(L, \rho, \widetilde{Arg}_L)$, where L is a compact oriented Lagrangian submanifold of V with a spin structure (called the support of the object), ρ is a local system on L (i.e. a complex vector bundle with flat connection), and $\widetilde{Arg}_L : L \to \mathbb{R}$ a continuous lift of Arg_L.*

We require that for any element $\beta \in \pi_2^{free}(V, L) := \pi_0(Maps((D^2, \partial D^2), (V, L)))$, the pairing $([\omega], \beta)$ is equal to zero.

We will sometimes denote the Fukaya category by $F(V, \omega)$, or simply by $F(V)$. We will also often omit from the notation the lifted argument function, thus denoting an object simply by (L, ρ). The choice of a spin structure is essential for signs, and the choice of the lift $\widetilde{Arg}_L$ is necessary for $\mathbb{Z}$-grading.

Let $\mathbf{C}_\varepsilon$ be the field consisting of formal series $f = \sum_{i \geq 0} c_i e^{-\lambda_i/\varepsilon}$, such that $c_i \in \mathbf{C}, \lambda_i \in \mathbf{R}, \lambda_0 < \lambda_1 < ..., \lambda_i \to +\infty$. In the case when $[\omega] \in H^2(V, \mathbb{Z})$, one can in fact work over the field $\mathbf{C}((q))$, where $q = exp(-\frac{1}{\varepsilon})$. In general we equip $\mathbf{C}_\varepsilon$ with the adic topology: a fundamental system of neighborhoods of zero consists of sets $U_x = \{f = \sum_{i \geq 0} c_i e^{-\lambda_i/\varepsilon} | \lambda_i \geq x, i \geq 0\}, x \in \mathbf{R}$.

Definition 17 *For two objects with transversal supports we define the space* *of morphisms such as follows*

$$Hom_{F(V,\omega)}((L_1,\rho_1,\widetilde{Arg}_1),(L_2,\rho_2,\widetilde{Arg}_2)) := (\oplus_{x\in L_1\cap L_2} Hom(\rho_{1x},\rho_{2x}))\otimes \mathbf{C}_\varepsilon.$$

Thus morphisms form a finite-dimensional vector space over the field $\mathbf{C}_\varepsilon$. There is a $\mathbf{Z}$-grading of the space of morphisms given in terms of Maslov index $deg : L_1 \cap L_2 \to \mathbf{Z}$ (see [Fu2], [Ko], [Se]). Maslov index depends on a choice of the lift $\widetilde{Arg}_L$ (it corresponds to a choice of a point in the universal covering of the bundle of the Lagrangian Grassmannians).

Remark 11 *The condition* $([\omega],\beta) = 0$ *is introduced for convenience only. It helps to avoid the problem with the composition* m_0 *we mentioned before. The condition holds in the case when* V *is a torus with the constant symplectic form, and* L *is a Lagrangian subtorus. This is our main application in present paper. In general there is a way to work with non-trivial* m_0, *if it is small in the adic topology.*

Now we are going to describe the A_∞-structure. It is defined by means of a collection of maps (higher compositions) of graded vector spaces $m_k^{F(V)}$: $\otimes_{0\le i\le k-1} Hom_{F(V)}((L_i,\rho_i),(L_{i+1},\rho_{i+1})) \to Hom_{F(V)}((L_0,\rho_0),(L_k,\rho_k))[2-k]$, where $k \ge 1$ and the sequence $(L_0,...,L_k)$ corresponds to a transversal sequence of objects (the latter notion will be defined below).

In the case, when all local systems are trivial of rank one, the map m_k is defined such as follows. Let D be a standard disc $D = \{z \in \mathbf{C} \,|\, |z| \le 1\}$. Let us fix a sequence $(L_0,...,L_k)$ of supports of objects with pairwise transversal intersections, intersection points $x_i \in L_i \cap L_{i+1}, 0 \le i \le k-1$, $x_k \in L_0 \cap L_k$, and $\beta \in \pi_2^{free}(V, \cup_{0\le i\le k} L_i)$. We denote by $\mathcal{M}(L_0,...,L_k;x_0,...,x_k;\beta)$ the set of collections $(y_0,...,y_k;\psi)$, where $y_i, 0 \le i \le k$ are cyclically ordered pairwise distinct points on the boundary ∂D, and $\psi : D \to (V,J)$ a pseudo-holomorphic map such that $\psi(y_i) = x_i, \psi(\overline{y_iy_{i+1}}) \subset L_i, 0 \le i \le k, y_0 = y_k$, $[\phi] = \beta$. Here $\overline{y_iy_{i+1}}$ denotes the arc between y_i and y_{i+1}. There is a natural action of $PSL(2,\mathbf{R})$ on $\mathcal{M}(L_0,...,L_k;x_0,...,x_k;\beta)$ arising from the holomorphic action on D by fractional linear transformations. The action is free except of the case $k = 1, x_0 = x_1, \beta = 0$, which is not relevant for our purposes.

Let $x_i \in L_i \cap L_{i+1}, 0 \le i \le k-1, x_k \in L_0 \cap L_k$ satisfy the condition $deg\, x_k = \sum_{0\le i\le k-1} deg\, x_i + 2 - k$. Then the matrix element $(m_k(x_0,x_1,...,x_{k-1}),x_k)$ is given by the formula $(m_k(x_0,x_1,...,x_{k-1}),x_k) = \sum \pm q^{(\beta,[\omega])}$, where sum is taken over all $PSL(2,\mathbf{R})$-orbits of points in $\mathcal{M}(L_0,...,L_k;x_0,...,x_k;\beta)$. Signs are derived from orientations of certain cycles in the moduli space $\mathcal{M} = \mathcal{M}(L_0,...,L_k;x_0,...,x_k;\beta)/PSL(2,\mathbf{R})$. We will comment on them below (see [Fu1], [FuOOO] for more details). In the case of non-trivial local systems there is an additional factor for each summand. It corresponds to the holonomies of local system along the arcs.

Now we will describe the transversality condition. Assume that we are given a sequence of objects $(L_i, \rho_i), 0 \leq i \leq k$ of the Fukaya category. We say that they are transversal if the following conditions hold:

1) There are only pairwise intersections $L_i \cap L_j$, and they are transversal.

2) For any subsequence $(L_{i_0}, ..., L_{i_m})$, $m \geq 1$, $i_0 < i_1 < ... < i_m$, any choice of intersection points $x_{i_m} \in L_{i_0} \cap L_{i_m}$, $x_{i_p} \in L_{i_p} \cap L_{i_{p+1}}, 0 \leq p \leq m-1$ such that $deg\, x_{i_m} - (\sum_{0 \leq i \leq m-1} deg\, x_{i_p} + 2 - m) = 0$, and any $\beta \in \pi_2^{free}(V, \cup_{0 \leq p \leq m} L_{i_p})$, the corresponding component of the moduli space $\mathcal{M}(L_{i_0}, ..., L_{i_m}; x_{i_0}, ..., x_{i_m}; \beta)/PSL(2, \mathbf{R})$ contains only smooth points, and is zero-dimensional.

3) If $deg\, x_{i_m} - (\sum_{0 \leq i \leq m-1} deg\, x_{i_p} + 2 - m) < 0$ then the corresponding component is empty.

Let us comment on these conditions (for more details see [FuOOO]). The first one is needed to define morphisms. The quotient set

$$\mathcal{M} = \mathcal{M}(L_{i_0}, ..., L_{i_m}; x_{i_0}, ..., x_{i_m}; \beta)/PSL(2, \mathbf{R})$$

which appears in the second condition locally can be identified with the space of solutions of a non-linear elliptic problem. For the linearized problem the corresponding Fredholm operator has index $deg\, x_{i_m} - (\sum_{0 \leq i \leq m-1} deg\, x_{i_p} + 2 - m)$. We define smooth points $\mathcal{M}^{sm}$ of $\mathcal{M}$ as such points where the cokernel of the Fredholm operator is trivial. Then $\mathcal{M}^{sm}$ is a smooth manifold of the dimension equal to the index. Moreover, one checks that the spaces $\mathcal{M}^{sm}$ carry natural orientations given by the determinants of the corresponding Fredholm operators. It is here where the choice of spin structures on L_i enters into the game. It follows that in the zero-dimensional case what we get is a set of points with multiplicities ± 1 (in particular, the multiplicities are integer numbers). Multiple covers and stable maps which appear in the definition of Gromov-Witten invariants and produce non-trivial denominators, do not appear in our framework for the Fukaya category (in our case all "stable maps" in components of the virtual degree zero are embeddings. Ramified coverings appear in higher codimensions). Therefore one can define the Fukaya category over the ring $\mathbf{Z}_\varepsilon$ (the integral version of $\mathbf{C}_\varepsilon$). The number of points counted with signs gives a tensor coefficient of m_k.

Composition maps satisfy a system of quadratic equations, thus making $F(V, \omega)$ into a non-unital A_∞-pre-category. One can check that it is in fact an A_∞-pre-category. Proof of the extension property is based on the following result of Fukaya (see [Fu2], [Se]).

Proposition 4 *Let (L_t, ρ_t) be an object obtained by a small Hamiltonian de-*

formation of an object (L,ρ) of $F(V)$. Then (L_t,ρ_t) and (L,ρ) are quasi-isomorphic.

For example, a sequence consisting of one object (L,ρ) can be extended to a transversal sequence $((L_{t_1},\rho_{t_1}),(L,\rho))$. Similarly, one can extend any finite set of transversal sequences.

It is easy to see that the set of connected components of the space of pairs (J,Ω) (equipped with the natural topology) is a principal homogeneous space over the lattice $H^1(V,\mathbf{Z})$. Namely, $f:V\to U(1)$ acts on (J,Ω) such as follows: $(J,\Omega)\mapsto(J,f\Omega)$. The following theorem can be derived from [Fu2].

Theorem 1 *There exists a set Σ of the second category (in the sense of Baire) in the space of almost complex structures compatible with ω such that Fukaya categories $F(V,\omega,J_1,\Omega_1)$ and $F(V,\omega,J_2,\Omega_2)$ are equivalent as long as $J_1,J_2\in\Sigma$, and (J_1,Ω_1) is homotopic to (J_2,Ω_2).*

Therefore the equivalence class of the Fukaya category depends on the connected component of the space of pairs.

Remark 12 *Definitions of this section (and other sections of the paper) can be modified in order to accommodate the case of non-zero B-field. We are not going to do that in order to avoid more complicate notations.*

5.2 Fukaya-Oh category for torus fibration

Let (Y,g_Y,∇) be an AK-manifold with integral affine structure. The covariant lattice is denoted by $T_Y^{\mathbf{Z}}$, as before. From now on we will assume that Y is *compact.* This is a severe restriction. It was proven in [CY] that in this case a finite cover of space Y is a torus with the standard affine structure. It appears in the collapse of complex abelian varieties.

The manifold $X^\vee=T_Y^*/(T_Y^{\mathbf{Z}})^\vee$ is the total space of the torus bundle $p^\vee:X^\vee\to Y$. It carries a natural symplectic form $\omega=\omega_{X^\vee}$ induced from the standard one on T^*Y. We endow $X^\vee$ with a 1-parameter family of complex structures $J_\eta,\eta\to 0$ compatible with ω. Indeed, the manifold $X_\eta^\vee:=T_Y^*/\eta(T_Y^{\mathbf{Z}})^\vee$ carries a canonical complex structure described before. We identify $X^\vee$ and $X_\eta^\vee$ by the map $(y,v)\mapsto(y,\eta v)$, where $y\in Y,v\in T_{Y,y}^*$. Using this identification, we pull back to $X^\vee$ the complex structure and the metric. The fibers of $p^\vee:X^\vee\to Y$ are flat Lagrangian tori for all values of η.

We define on $(X^\vee,J_\eta)$ a nowhere vanishing $(n,0)$-form Ω_η such as follows. Let us fix an oriented orthonormal basis $e_1,...,e_n$ in $T_{Y,y}^*,y\in Y$. We define Ω_η as the n-form on $X^\vee$, which is invariant with respect to the $T_{Y,y}^*/(T_{Y,y}^{\mathbf{Z}})^\vee$-action, and is equal to $\bigwedge_{1\le j\le n}((p^\vee)^*e_j+\sqrt{-1}J_\eta(p^\vee)^*e_j)$.

Let L be a compact oriented Lagrangian submanifold of $X^\vee$ such that

$p^{\vee}_{|L}$ is an unramified covering, and the orientation of L is induced from the orientation of Y. We claim that there is a canonical choice $\widetilde{Arg}_L^{can} : L \to \mathbb{R}$ for the function $\widetilde{Arg}_L : L \to \mathbf{R}$. Indeed, for any point $x \in X^{\vee}$ the space of Lagrangian subspaces in $T_{X^{\vee},x}$, which are transversal to the vertical tangent space $T_x^{vert} = Ker(p^{\vee})_*$ is contractible. Let us consider the space $\mathcal{L}$ of pairs (x, l) such that $x \in X^{\vee}$ and $l \subset T_{X^{\vee},x}$ is a Lagrangian subspace, which is transversal to T_x^{vert}, and endowed with the orientation induced from Y. Then the function $(x, l) \mapsto Arg_{(\Omega_\eta)_{|l}}(x) \in \mathbb{R}/2\pi\mathbb{Z}$ admits a unique continuous lifting $\widetilde{Arg} : \mathcal{L} \to \mathbf{R}$, vanishing at $(x, J_\eta(T_x^{vert})), x \in X^{\vee}$. Restricting this function to L we obtain $\widetilde{Arg}_L^{can}$.

We will denote by $F^\eta(X^{\vee})$ the Fukaya category $F(X^{\vee}, \omega, J_\eta, \Omega_\eta)$, and by $F^\eta_{unram}(X^{\vee})$ its full A_∞-pre-subcategory with objects $(L, \rho, \widetilde{Arg}_L^{can})$ such that L is a compact Lagrangian submanifold with the orientation induced from Y, $p^{\vee}_{|L}$ is an unramified covering, and $\widetilde{Arg}_L^{can}$ was described above. To simplify the notations we will denote objects of these categories by (L, ρ).

Remark 13 *One can check that for transversal Lagrangian submanifolds L_1 and L_2 as above, the Maslov index at any $x \in L_1 \cap L_2$ is equal to the Morse index at $p^{\vee}(x)$ of the smooth Morse function $f_1 - f_2 : Y \to \mathbb{R}$ such that locally near x one has $L_i = graph\,(df_i)\,(mod(T_Y^{\mathbf{Z}})^{\vee}), i = 1, 2$.*

It follows from the results of [FuO] that there exists a limit of the family of A_∞-pre-categories $F^\eta_{unram}(X^{\vee})$, $\eta \to 0$ in the following sense. Objects and morphisms of $F^\eta_{unram}(X^{\vee})$ do not depend on η and remain the same in the limit. The compositions $m_k^{F^\eta_{unram}(X^{\vee})}$ have limits as $\eta \to 0$ in the adic topology of $\mathbf{C}_\varepsilon$. They will be explicitly described below.

The following result can be derived from [FuO].

Proposition 5 *The limiting A_∞-pre-category is equivalent to $F^\eta_{unram}(X^{\vee})$ for all sufficiently small η.*

We will denote this A_∞-pre-category by $FO(X^{\vee})$ and call it the *Fukaya-Oh category* of $X^{\vee}$ (or *degenerate Fukaya category* of $X^{\vee}$).

Remark 14 *In what follows we will assume that $dim Y > 1$. The case $dim Y = 1$ is somewhat different, but also it is much more simple (see for example [P1]). In particular, $F^\eta_{unram}(X^{\vee})$ does not depend on η in this case.*

As we said before, the objects and morphisms for $FO(X^{\vee})$ are the same as for $F^\eta_{unram}(X^{\vee})$. In order to define the composition map

$$m_k : \otimes_{0 \le i \le k-1} Hom((L_i, \rho_i), (L_{i+1}, \rho_{i+1})) \to Hom((L_0, \rho_0), (L_k, \rho_k))[2-k]$$

one uses the standard formulas, but the sum runs over certain two-dimensional surfaces in $X^{\vee}$ described below. For a sequence $((L_0, \rho_0), ..., (L_k, \rho_k)), k \ge 1$ of objects in $FO(X^\vee)$ we consider immersed two-dimensional surfaces $S \to X^\vee$ such that:

a) Boundary of S belongs to $L_0 \cup ... \cup L_k$.

b) $S = (\cup_\alpha T_\alpha) \cup (\cup_\beta S_\beta)$ where and T_α are geodesic triangles in fibers of $p^\vee$, hence they are projected to points in Y.

c) Each S_β is a union of 1-parameter families of geodesic intervals contained in fibers of $p^\vee$ (i.e. a "strip"). Moreover, $p^\vee_{|S_\beta} : S_\beta \to Y$ is a fibration over a connected interval $I_\beta := p^\vee(S_\beta)$ immersed in Y. Fibers of S_β over the interior points of I_β are geodesic intervals of strictly positive length. Fibers of S_β over the boundary points of I_β are either edges of triangles T_α or intersection points $x_i \in L_i \cap L_{i+1}, 0 \leq i \leq k-1$, $x_k \in L_0 \cap L_k$.

d) Intervals I_β are edges of an immersed planar trivalent tree $\Gamma \subset Y$. Points $p^\vee(T_\alpha)$ are internal vertices of Γ. Tail vertices of Γ are projections of the intersection points $x_0, \ldots, x_k$.

e) Let $r : T^*_Y \to X^\vee$ be the natural fiberwise universal covering. If the Lagrangian manifolds $r^{-1}(L_i), i = 0, ..., k$ are locally given by differentials of smooth functions $f_i, i = 0, ..., k$ on Y, then the edges of Γ must be gradient lines of $f_i - f_j$. Intersection points of $r^{-1}(L_i)$ and $r^{-1}(L_j)$ correspond to critical points of $f_i - f_j$.

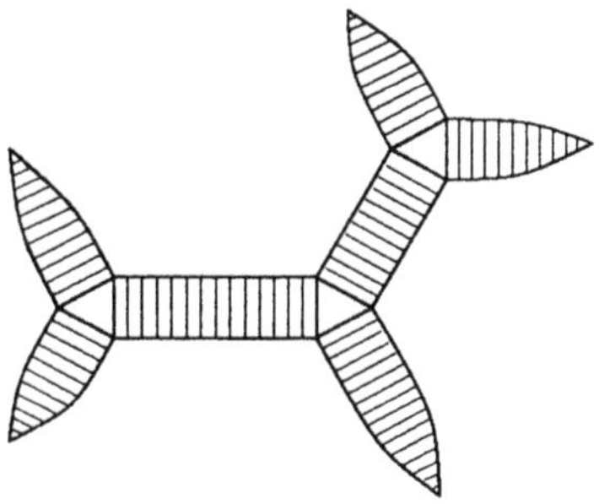

Figure 4. Typical surface S

The projection of surface S to Y is a gradient tree, with tail vertices being critical points of $f_i - f_{i+1}$ or of $f_0 - f_k$, and edges $p^\vee(S_\beta)$ being the gradient lines of functions $f_i - f_j$, where $i = i(\beta), j = j(\beta), i < j$. The triangles are mapped into the internal vertices of the tree.

Compositions $m_k = m_k^{FO(X^\vee,\omega)}$ are given by the standard formulas, but now we are counting surfaces S described in a)-d). The weight $q^{\langle [S],[\omega]\rangle}$ can be written as $exp(-\frac{1}{\varepsilon}\sum_\beta var_{p^\vee(S_\beta)} f_\beta)$, where $f_\beta = f_{i(\beta)} - f_{j(\beta)}$, and var is

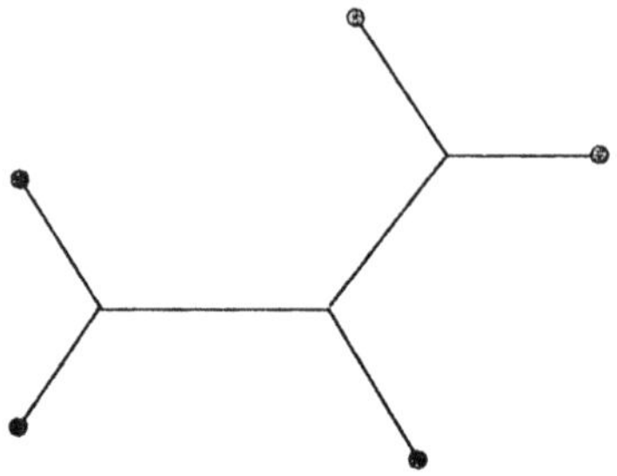

Figure 5. Tree $\Gamma = p^\vee(S)$ for surface S as above

the (positive) variation of the function along the gradient line.

The transversality condition for a sequence of objects of Fukaya-Oh category can be formulated similarly to the case of Fukaya category.

The reader can compare our considerations with those from [FuO]. The fibers of $p^\vee$ are "small" tori (of the size $O(\eta)$). The base Y is "large" (of the size of $O(1)$). Hence, the Lagrangian manifolds are close to the zero section of $p^\vee$. This is similar to the situation considered in [FuO]. Indeed, in [FuO] the authors study the A_∞-subcategory of $F(T_Y^*)$ (where Y is an arbitrary smooth compact manifold), with the objects (L,ρ) such that $L = \eta\, graph(df)$, $f : Y \to \mathbf{R}$ is a smooth function. In other words, they considered Lagrangian sections of the natural projection $T_Y^* \to Y$, which are close to the zero section. When $\eta \to 0$, pseudo-holomorphic discs get "stretched" along the fibers of $p^\vee$. Thus they look like the surfaces S described above. Then the higher compositions of the Fukaya category "approach" the compositions $m_k^{FO(X^\vee,\omega)}$. This was proved in [FuO] in the case when $X^\vee$ was replaced by T_Y^*. Considerations from [FuO] apply in our case as well.

Remark 15 *One can extend the Fukaya-Oh category considering Lagrangian submanifolds in $X^\vee$ which are not necessarily unramified coverings of Y. For example, one can try to add to $FO(X^\vee)$ new objects which are local systems on Lagrangian tori which are fibers of the projection $p^\vee : X^\vee \to Y$. It seems that with these objects one can go much further than with transversal ones. For example, in the general case of torus fibrations with singular fibers, one can argue that for almost any $y \in Y$ there are no limiting holomorphic discs with the boundary in the torus $(p^\vee)^{-1}(y)$. The set of such points y is the complement to a countable union Z of hypersurfaces in Y (this follows from the fact that $dim(Y^{sing}) = dim(Y) - 2$). Thus, we get a large collection of honest objects without the parasitic composition m_0. The total picture seems* *to be quite intricate, as examples show that the subset Z is everywhere dense. Presumably, it is related with some mysterious non-abelian 1-cocycle which we will discuss later in the remark in section 7.1.*

6 Morse-Smale complex and the category of Morse functions

6.1 Notations from Morse theory

Let (Y, g_Y) be a compact oriented Riemannian manifold of dimension n, $f : Y \to \mathbf{R}$ be a smooth Morse function. We will denote the set of critical points of f by $Cr(f)$. If $x \in Cr(f)$, we will denote by U_x (resp. S_x) the unstable (resp. stable) submanifolds associated with x. Namely, $U_x = \{y \in Y \,|\, lim_{t \to +\infty} e^{-t\, grad(f)} y = x\}$, and $S_x = \{y \in Y \,|\, lim_{t \to +\infty} e^{t\, grad(f)} y = x\}$.

Let $ind(x)$ be the Morse index of x, i.e. the negative rank of the quadratic form $(\partial^2 f)_{|T_x Y}, x \in Cr(f)$. The manifolds S_x and U_x are diffeomorphic to open balls of dimensions $ind(x)$ and $n - ind(x)$ respectively. It follows that the cohomology of S_x with compact support is a graded vector space with the only non-zero 1-dimensional component in degree $ind(x)$: $H^*_c(S_x) \simeq \mathbb{Z}[-ind(x)]$. A choice of generator defines an orientation of S_x. If the function f satisfies Morse-Smale transversality condition, i.e. for any $x, y \in Cr(f)$ the manifolds S_x and U_y intersect transversely, then $Y = \sqcup_{x \in Cr(f)} S_x$ is a cell decomposition of Y. The cohomology $H^*(Y, \mathbb{Z})$ can be computed as the cohomology of the Morse complex $(M^*(Y,f), \partial)$, with the components $M^i(Y,f) = \sum_{x \in Cr(f), ind(x)=i} H^i_c(S_x)$. Let us choose orientations of manifolds S_x for all $x \in Cr(f)$. We endow U_x with the dual orientations. The graded module $M^i(Y,f)$ can be identified with $\oplus_{0 \le i \le n} \mathbb{Z}^{Cr_i}[-i]$ where Cr_i is the set of critical points of f of index i. The choice of orientations gives a basis $([x])_{x \in Cr(f)}$ of $M^*(Y,f)$.

The differential ∂ is the standard Morse differential:

$$\partial([x]) = \sum_{y \in Cr(f), ind(y) = ind(x)+1} deg((U_x \cap S_y)/\mathbf{R}) \cdot [y],$$

where $(U_x \cap S_y)/\mathbf{R}$ an oriented 0-dimensional manifold (a set of points with signs), and $deg(\cdot) \in \mathbf{Z}$ denotes the total number of points counted with signs. The action of $\mathbf{R}$ arises from the natural reparametrization $x \mapsto x + t$ of the gradient trajectories.

There is also a generalization $M^*(Y, f, \rho)$ of the Morse complex for a flat vector bundle ρ (see [BZ], [HL]).

6.2 Morse A_∞-category of smooth functions

Here we will define the Morse category of smooth functions $M(Y)$ following [FuO]. It will be an A_∞-pre-category over $\mathbb{C}$. Objects of $M(Y)$ are pairs (f, ρ), where $f : Y \to \mathbf{R}$ is a smooth function, and ρ is a local system of finite-dimensional complex vector spaces on Y. Before defining the transversality of objects, we will define the transversality of functions.

Suppose we are given a sequence of smooth functions $(f_0, ..., f_k), k \geq 2$ such that all $f_i - f_j, i \neq j$ are Morse functions, and a sequence of critical points $x_i \in Cr(f_i - f_{i+1}), 0 \leq i \leq k-1, x_k \in Cr(f_0 - f_k)$. We will use oriented binary planar trees in order to describe certain moduli spaces associated with such sequences. Let us fix a planar trivalent tree T with $k+1$ tails vertices. Among the tail vertices we choose one and call it the *root* vertex. Let us orient edges of T along the shortest paths towards the root. Thus, T becomes a binary tree considered as an *oriented* tree . We depict T inside of the standard unit disc $D \subset \mathbf{R}^2$ in such a way that tail vertices of T belong to ∂D, and connected components of $D \setminus T$ are cyclically numbered from 0 to k in the clockwise order. We assume that numbers attached to two regions near the root vertex are 0 and k.

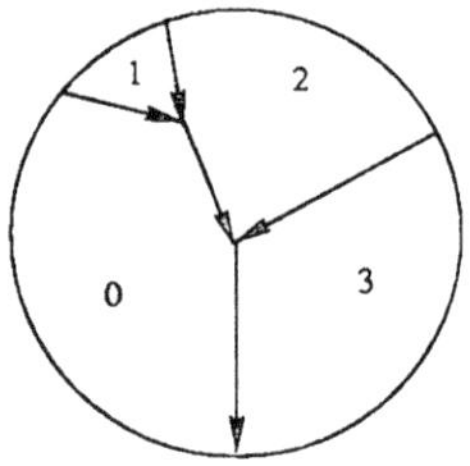

Figure 6. Planar binary tree embedded into the circle

We define a gradient immersion of T into Y as a continuous map $j : T \to Y$ such that:

1) The restriction $j|_e$ is an orientation preserving homeomorphism of the edge e onto an interval in the gradient line of $f_{l(e)} - f_{r(e)}$, where the label $l(e)$ (resp. $r(e)$) corresponds to the region of $D \setminus T$ which is left (resp. right) to e.

2) Each tail vertex v is mapped to the point $x_v \in Cr(f_{l_v} - f_{r_v})$, where l_v (resp. r_v) is the label of the region which is left (resp. right) to the only tail edge containing v.

We will need immersed binary trees (let us call them *gradient* trees) in order to define compositions and transversal sequences in the Morse A_∞-pre-category. These structures can be defined in terms of certain varieties, which we are going to describe now.

Suppose that we are given a sequence of functions $(f_0, ..., f_k), k \geq 2$ and critical points $(x_0, \ldots, x_k)$ as above, and a binary planar tree T. Let us consider the manifold $Y(T) = Y^{V_i(T)}$, where $V_i(T)$ is the set of internal vertices of T. We are going to define several submanifolds in $Y(T)$. For each tail vertex $v_m, 0 \leq m \leq k-1$ we define $Z_{v_m} = \pi_{\hat{v}_m}^{-1}(U_{x_m})$ and for $m = k$ we define $Z_{v_k} = \pi_{\hat{v}_k}^{-1}(S_{x_k})$. Here $\hat{v}_l$ denotes the second endpoint of the edge of T containing v_l, and $\pi_v : Y(T) \to Y$ is the canonical projection on the factor corresponding to $v \in V_i(T)$.

For pair (f_i, f_j) we define a subset $Z_{i,j} \subset Y \times Y$, consisting of pairs (y_1, y_2) such that $y_1 \neq y_2$ and $y_2 = e^{t\, grad(f_i - f_j)} y_1$ for some $t > 0$. Then $Z_{i,j}$ is a non-compact submanifold of $Y \times Y$.

An edge e of T we call internal if both endpoints of it are internal vertices. The set of internal edges we denote by $E_i(T)$. For each internal edge $e \in E_i(T)$, which separates two regions labeled by $l(e)$ (left) and $r(e)$ (right), we define a submanifold $Z_e = \pi_e^{-1}(Z_{l(e),r(e)})$, where $\pi_e : Y(T) \to Y \times Y$ is the natural projection.

It follows from the definitions that the space of gradient immersions of a given T as above, up to homeomorphisms preserving tails, can be identified with $\mathcal{M}(T; f_0, ..., f_k; x_0, ..., x_k) := (\cap_{0 \leq m \leq k} Z_{v_m}) \cap (\cap_{e \in E_i(T)} Z_e) \subset Y^{V_i(T)}$.

Definition 18 *We say that a sequence $(f_0, \ldots, f_k)$, $k \geq 2$ is T-transversal for a given tree T, if for any sequence of intersection points $(x_0, \ldots, x_k)$ such that*

$$\sum_{i=0}^{k-1} ind(x_i) - ind(x_k) \leq k-2$$

the collection of submanifolds $\left((Z_{v_m})_{0 \leq m \leq k}, (Z_e)_{e \in E_i(T)}\right)$ is transversal in $Y(T)$ (i.e. intersection of any subcollection is transversal). For $k = 1$, we say that (f_0, f_1) is T-transversal (there is only one tree T in this case) if $f_0 - f_1$ is a Morse function, satisfying the Morse-Smale transversality condition.

Remark 16 *As in the case of Fukaya category we consider here only spaces $\mathcal{M}(T; f_0, ..., f_k; x_0, ..., x_k)$ of (virtual) dimensions less or equal than zero. Our condition in the case of strictly negative dimension means that the moduli space is empty.*

It can be proven (see [Fu1]) that there exists a subset of second Baire category in $(C^\infty(Y))^{\mathbf{Z}}$ such that for any element $(f_i)_{i \in \mathbf{Z}}$ of this set and for any strictly increasing sequence of integers $i_0 < \cdots < i_k)$ and for any planar tree T with $k+1$ tails, the sequence $(f_{i_0}, \ldots, f_{i_k})$ is T-transversal.

Definition 19 *A sequence of objects $(f_0,\rho_0),...,(f_k,\rho_k)$ is called transversal if for any $m \geq 1$ and any binary tree T with $m+1$ tails, an arbitrary subsequence $(f_{i_0},...,f_{i_m}), i_0 < ... < i_m$ is T-transversal.*

For any two transversal objects $W_0 = (f_0,\rho_0)$ and $W_1 = (f_1,\rho_1)$ we define the space of morphisms $Hom_{M(Y)}(W_0,W_1)$ as the Morse complex $M^*(Y, f_0 - f_1, \rho_0^* \otimes \rho_1)$. Now we define the A_∞-structure on $M(Y)$.

The map $m_1 : Hom((f_0,\rho_0),(f_1,\rho_1)) \to Hom((f_0,\rho_0),(f_1,\rho_1))[1]$ is the standard differential in the Morse-Smale complex. Higher compositions m_k where $k \geq 2$ for transversal sequences of objects are linear maps

$$m_k : \otimes_{0\leq i\leq k-1} Hom((f_i,\rho_i),(f_{i+1},\rho_{i+1})) \to Hom((f_0,\rho_0),(f_k,\rho_k))[2-k]$$

Each m_k is defined as a sum $m_k = \sum \pm m_{k,T}$ where T runs through the set of isomorphism classes of oriented binary planar trees with $(k+1)$ tails. Let us describe the summands $m_{k,T}$. For simplicity we will give the formulas in the case when all local systems are trivial of rank one.

Let us fix critical points $y_i \in Cr(f_i - f_{i+1}), 0 \leq i \leq k-1, y_k \in Cr(f_0 - f_k)$, such that $\sum_{0\leq i\leq k-1} ind(y_i) = ind(y_k) + 2 - k$, and orientations of manifolds $S_{y_i}, 0 \leq i \leq k$. It follows from the definition of a transversal sequence that the moduli space of gradient trees $\mathcal{M}(T; f_0,...,f_k; y_0,...,y_k)$ is an oriented compact zero-dimensional manifold.

Definition 20 *We define compositions $m_k, k \geq 2$ by the formula*

$$m_k([y_0],...,[y_{k-1}]) = \sum_{[T]} \sum_{y_k \in Cr(f_0 - f_k)} deg(\mathcal{M}(T; f_0,...,f_k; y_0,...,y_k)) \cdot [y_k]$$

where $[T]$ is the equivalence class of T as an abstract oriented planar tree, and $deg(\cdot) \in \mathbf{Z}$ is the total number of points counted with signs, as before.

For local systems of higher ranks one proceeds as in the case of Fukaya categories, using flat connections in order to define an analog of the holonomy of local systems.

One can obtain slightly different formulas for m_k in the following way. For any point $\gamma \in \mathcal{M}(T; f_0,...,f_k; y_0,...,y_k)$ we define the weight

$$w_\gamma = exp(-\frac{1}{\varepsilon} \sum_{e \in E(T)} var_\gamma(f_{l(e)} - f_{r(e)})) \in \mathbf{C}_\varepsilon.$$

Here $var_\gamma(f_{l(e)} - f_{r(e)}) > 0$ is a variation of $f_{l(e)} - f_{r(e)}$ along the gradient line $\gamma(e)$, which is defined such as follows: $var_\gamma(f_{l(e)} - f_{r(e)}) = (f_{l(e)} - f_{r(e)})(y_{max}) - (f_{l(e)} - f_{r(e)})(y_{min})$, where y_{max} and y_{min} are the end-points of $\gamma(e)$, such that $(f_{l(e)} - f_{r(e)})(y_{max}) - (f_{l(e)} - f_{r(e)})(y_{min}) > 0$. After extension of scalars to $\mathbf{C}_\varepsilon$ one can choose another basis in $Hom_{M(Y)}(W_0, W_1)$, namely $[y]_{new} = [y]exp(\frac{(f_0(y)-f_1(y))}{\varepsilon})$ for $y \in Cr(f_0 - f_1)$. Then the formulas for m_k will be modified. The contribution of each γ will be multiplied by w_γ. The formulas will be similar to those for the Fukaya-Oh category (see Section 5.2).

6.3 De Rham A_∞-category of smooth functions

The other A_∞-pre-category we are interested in will be a differential-graded category (dg-category for short). In other words, it is an A_∞-category with strict identity morphisms and vanishing compositions $m_n, n \geq 3$. We will call it *de Rham category* of Y and denote by $DR(Y)$. Objects of $DR(Y)$ are same as for $M(Y)$. They are pairs (f, ρ), where $f : Y \to \mathbf{R}$ is a smooth function and ρ is a local system on Y. Morphisms are complexes defined by the formula

$$Hom_{DR(Y)}((f_0, \rho_0), (f_1, \rho_1)) = \Gamma(Y, \wedge^* T^*_Y \otimes Hom(\rho_0, \rho_1)).$$

Notice that the space of morphisms does not depend on f_0 and f_1. The composition of morphisms is defined in the obvious way: in a local trivialization of ρ_0 and ρ_1 it is given by the product of matrices with the coefficients in $\Omega^*(Y)$.

Now we can formulate the main result of this section.

Theorem 2 *A_∞-pre-categories $M(Y)$ and $DR(Y)$ are equivalent.*

The proof of the theorem will occupy the rest of the section. First, we will discuss a version of formulas from "homological perturbation theory" (see [GS], [Me]). They will give an A_∞-structure on a subcomplex of a dg-algebra. Then we will discuss an approach to the proof based on the ideas of [HL]. It seems plausible that an alternative proof (but, presumably, much more difficult) can be obtained within the framework of Witten complex, using methods of [BZ].

6.4 A_∞-structure on a subcomplex

In this section we are going to restate in a convenient form some results from [GS] and [Me].

Let $(A, m_n), n \geq 1$ be a non-unital A_∞-algebra, $\Pi : A \to A$ be an idempotent which commutes with the differential $d = m_1$. In other words, Π is a linear map of degree zero such that $d\Pi = \Pi d, \Pi^2 = \Pi$. Assume that we are given an homotopy $H : A \to A[-1]$, $1 - \Pi = dH + Hd$. Let us denote the image of Π by B. Then we have an embedding $i : B \to A$ and a projection $p : A \to B$, such that $\Pi = i \circ p$.

Let us introduce a sequence of linear operations $m_n^B : B^{\otimes n} \to B[2-n]$ in the following way [f]:

a) $m_1^B := d^B = p \circ m_1 \circ i$;

b) $m_2^B = p \circ m_2 \circ (i \otimes i)$;

c) $m_n^B = \sum_T \pm m_{n,T}, n \geq 3$.

Here the summation is taken over all oriented planar trees T with $n+1$ tails vertices (including the root vertex), such that the (oriented) valency $|v|$ (the number of incoming edges) of every internal vertex of T is at least 2. In order to describe the linear map $m_{n,T} : B^{\otimes n} \to B[2-n]$ we need to make some preparations. Let us consider another tree $\bar{T}$ which is obtained from T by the insertion of a new vertex into every internal edge. As a result, there will be two types of internal vertices in $\bar{T}$: the "old" vertices, which coincide with the internal vertices of T, and the "new" ones, which can be thought geometrically as the midpoints of the internal edges of T.

To every tail vertex of $\bar{T}$ we assign the embedding i. To every "old" vertex v we assign m_k with $k = |v|$. To every "new" vertex we assign the homotopy operator H. To the root we assign the projector p. Then moving along the tree down to the root one reads off the map $m_{n,T}$ as the composition of maps assigned to vertices of $\bar{T}$.

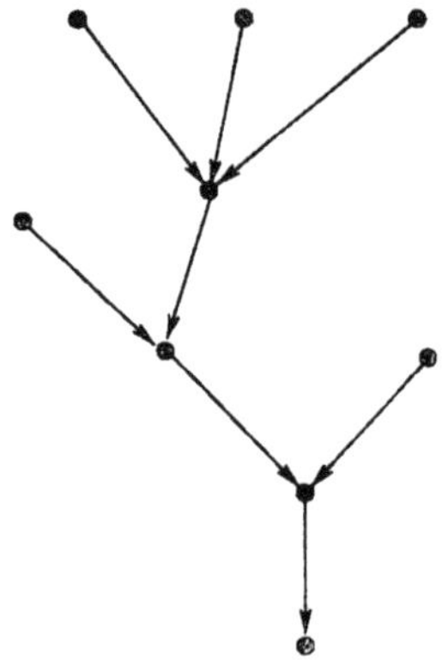

Figure 7. Tree T

Proposition 6 *The linear map m_1^B defines a differential in B.*

Proof. Clear. ■

[f] In a special case similar formulas appeared in [Go]

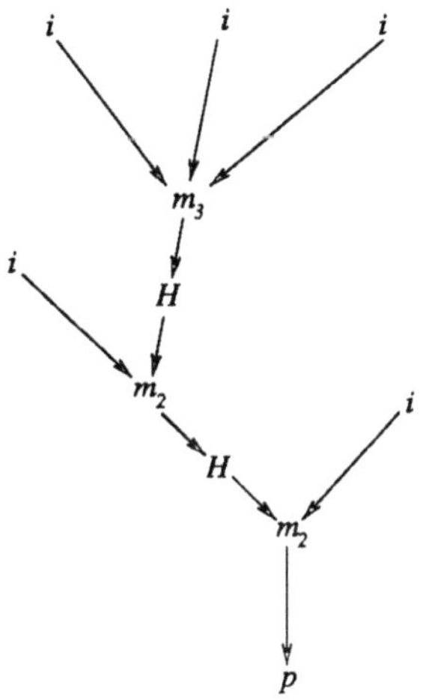

Figure 8. Tree $\bar{T}$

Theorem 3 *The sequence $m_n^B, n \geq 1$ gives rise to a structure of an A_∞-algebra on B.*

Sketch of the proof. The proof is quite straightforward, so we just briefly show main steps of computations.

First, one observes that p and i are homomorphisms of complexes. In order to prove the theorem we will replace for a given $n \geq 2$ each summand $m_{n,T}$ by a different one, and then compute the result in two different ways. Let us consider a collection of trees $\{\bar{T}_e\}_{e \in E(\bar{T})}$ such that $\bar{T}_e$ is obtained from $\bar{T}$ in the following way:

a) we split the edge e into two edges by inserting a new vertex w_e inside e;

b) the remaining part of $\bar{T}$ is unchanged.

We assign $d = m_1$ to the vertex w_e, and keep all other assignments untouched. In this way we obtain a map $m_{n,\bar{T}_e} : B^{\otimes n} \to B[3-n]$.

Let us consider the following sum (with appropriate signs):

$$\hat{m}_n^B = \sum_T \sum_{e \in E(\bar{T})} \pm m_{n,\bar{T}_e}.$$

We can compute it in two different ways: using the relation $1 - \Pi = dH + Hd$, and using the formulas for $d(m_j)$, $j \geq 2$ given by the A_∞-structure on A. The case of the relation $1 - \Pi = dH + Hd =: d(H)$ gives

$$\hat{m}_n^B = d(m_n^B) - m_n^{B,\Pi} + m_n^{B,1}$$

where $m_n^{B,\Pi}$ is defined analogously to m_n^B, with the only difference that we assign to a new vertex operator Π instead of H for some edge $e \in E_i(T)$. Similarly, the summand $m_n^{B,1}$ is defined if we assign to a new vertex operator $1 = id_A$ instead of H. Formulas for $d(m_j)$ are quadratic expressions in m_l, $l < j$. This gives us another identity

$$\hat{m}_n^B = m_n^{B,1}$$

Thus we have $d(m_n^B) = m_n^{B,\Pi}$, and it is exactly the A_∞-constraint for the collection $(m_n^B)_{n\geq 1}$. ∎

Moreover, using similar technique, one can prove the following result.

Proposition 7 *There is a canonical A_∞-morphism $g : B \to A$, which defines a quasi-isomorphism of A_∞-algebras.*

For the convenience of the reader we give an explicit formula for a canonical choice of g. The operator $g_1 : B \to A$ is defined as the inclusion i. For $n \geq 2$ we define g_n as the sum of terms $g_{n,T}$ over all planar trees T with $n+1$ tails. Each term $g_{n,T}$ is similar to the term $m_{n,T}$ defined above, the only difference is that we insert operator H instead of p into the root vertex.

One can also construct an explicit A_∞-quasi-isomorphism $A \to B$.

Remark 17 *a) Similar construction works in the case of an arbitrary non-unital A_∞-category. In that case one needs projectors $\Pi_{X,Y}$ and homotopies $H_{X,Y}$ for every graded space of morphisms $Hom(X,Y)$. All formulas remain the same as in the case of A_∞-algebras. The resulting A_∞-category with the spaces of morphisms given by $\Pi_{X,Y}(Hom(X,Y))$ is equivalent to the original one. We will use this fact later.*

b) Propositions 4 and 5 should hold in a much more general case of algebras over operads (see e.g. [M]).

6.5 Projectors and homotopies in Morse theory

We would like to apply formulas for the A_∞-structure on a subcomplex to the proof of the Theorem 2. In order to do that we need to identify the Morse complex with a direct summand of the de Rham complex. Our approach is based on the ideas of Harvey and Lawson (see [HL]).

Let Y be a compact oriented smooth manifold, $dim\, Y = n$. The space of currents $D'(Y)$ we will identify with the space of distribution-valued differential forms. Continuous linear operators $\Omega^*(Y) \to D'(Y)$ are given by their

Schwartz kernels, which are elements of $D'(Y \times Y)$. Smoothing operators $D'(Y) \to \Omega^*(Y)$ have kernels in $\Omega^*(Y \times Y) \subset D'(Y \times Y)$.

With any oriented submanifold $Z \subset Y$, $\dim Z = k$ of finite volume we associate a canonical current $[Z]$ of degree $n-k$ (namely, we can integrate smooth k-forms over Z).

Let g_Y be a Riemannian metric on Y, and f be a Morse-Smale function. The gradient flow $exp(t\,grad(f)), t \geq 0$ gives rise to a 1-parameter semigroup acting on $\Omega^*(Y)$: $\psi^t(\alpha) = exp(t\,grad(f))_*(\alpha)$. Schwartz kernel of ψ^t is $[G_t]$ where manifold $G_t \subset Y \times Y$ is given by $G_t := graph(exp(t\,grad(f)))$. We also have the identity

$$id - \psi^t = dH^t + H^t d,$$

where $H^t : \Omega^*(Y) \to \Omega^*(Y) \subset D'(Y)$ is a linear operator of degree -1 defined by the distributional kernel $[Z_t]$, $Z_t := \cup_{0 \leq t' \leq t} graph(exp(t'\,grad(f)))$.

It is checked in [HL] that this picture has a limit as $t \to +\infty$. Namely, there exist limits of currents $[G_t]$ and $[Z_t]$:

$$[G_\infty] = lim_{t \to +\infty}[G_t] = \sum_{x \in Cr(f)} [S_x] \times [U_x]$$

$$[Z_\infty] = lim_{t \to +\infty}[Z_t] = [\cup_{0 \leq t < +\infty} G_t]$$

Linear operators ψ^∞ (of degree zero) and H^∞ (of degree -1), corresponding to these kernels, map $\Omega^*(Y)$ to $D'(Y)$ and satisfy the identity

$$i - \psi^\infty = dH^\infty + H^\infty d,$$

where $i : \Omega^*(Y) \to D'(Y)$ is the natural inclusion. According to the de Rham theorem this inclusion is a quasi-isomorphism of complexes, therefore ψ^∞ is. Morally, $\Pi_\infty := \psi^\infty$ should be thought of as a projector. The image $\Pi_\infty(\Omega^*(Y)) \subset D'(Y)$ coincides with $\oplus_{x \in Cr(f)} \mathbf{R} \cdot [U_x]$. We have

$$\Pi_\infty(\alpha) = \sum_{x \in Cr(f)} \left(\int_{S_x} \alpha\right) \cdot [U_x] = \sum_{x \in Cr(f)} \int_Y (\alpha \wedge [S_x]) \cdot [U_x].$$

Moreover, the operator Π_∞ commutes with the differentials. Hence the complex $\Pi_\infty(\Omega^*(Y))$ is a finite-dimensional subcomplex of $D'(Y)$ isomorphic to the Morse complex $M^*(Y, f)$. In fact it is quasi-isomorphic to both complexes $\Omega^*(Y)$ and $D'(Y)$. In this way Harvey and Lawson prove that the de Rham cohomology is isomorphic to the cohomology of Morse complex.

In order to construct actual projectors and homotopies we will proceed as follows. Let $\rho_\delta, \delta \to 0$ be a family of smooth closed differential n-forms on $Y \times Y$ such that $supp(\rho_\delta)$ belongs to the open δ-neighborhood N_δ of the diagonal $diag \subset Y \times Y$, and the cohomology class of ρ_δ in $H^n_c(N_\delta, \mathbf{R})$ is the same as of $[diag]$.

We define $R_\delta : D'(Y) \to \Omega^*(Y)$ as the integral operator given by the kernel ρ_δ.

Lemma 2 *1) The operator R_δ is a homomorphism of complexes.*

2) If $Z_1, Z_2 \in Y$ are two oriented submanifolds of finite volume such that they intersect transversely at finitely many points, and $dim\, Z_1 + dim\, Z_2 = dim\, Y$, $\overline{Z}_1 \cap \overline{Z}_2 = Z_1 \cap Z_2$, then for sufficiently small δ one has:

$$\int_Y R_\delta([Z_1]) \wedge R_\delta([Z_2]) = deg(Z_1 \cap Z_2) \in \mathbf{Z}$$

3) There exists a linear operator $h_\delta : \Omega^*(Y) \to \Omega^*(Y)$ such that its kernel has support in N_δ, the wave front $WF(h_\delta)$ is the conormal bundle of $diag \subset Y \times Y$, and

$$dh_\delta + h_\delta d = id - (R_\delta)_{|\Omega^*(Y)}.$$

Proof . Part 1) follows from the fact that ρ_δ is a closed current. Part 2) follows from the fact that R_δ changes the supports of $Z_i, i = 1, 2$ by $O(\delta)$. To prove part 3) one observes that the operators id and $(R_\delta)_{|\Omega^*(Y)}$ preserve the space of smooth forms $\Omega^*(Y)$, and ρ_δ is cohomologous to $[diag]$. ■

Let $x, y \in Cr(f)$ be two critical points of the same Morse index. Then $deg(S_x \cap U_y) = \delta_{xy}$ (the Kronecker symbol). By the part 2) of the Lemma, for sufficiently small δ we obtain the identity

$$\int_Y R_\delta([S_x]) \wedge R_\delta([U_y]) = \delta_{xy}$$

This implies the following result.

Proposition 8 *Let us define for a sufficiently small δ a linear operator $D'(Y) \to \Omega^*(Y)$ by the formula $\Pi_\delta(\alpha) = \sum_{x \in Cr(f)} (\int_Y \alpha \wedge R_\delta([S_x])) \cdot R_\delta([U_x])$. Then*

1) $\Pi_\delta^2(\alpha) = \Pi_\delta(\alpha)$ if $\alpha \in \Omega^*(Y)$, and $\Pi_\delta d = d\Pi_\delta$.

2) The image $\Pi_\delta(M^*(Y, f))$ is a subcomplex in $\Omega^*(Y)$ which is canonically isomorphic to the Morse complex $M^*(Y, f)$.

We define a homotopy operator $H_\delta : \Omega^*(Y) \to \Omega^*(Y)[-1]$ as an integral operator given by the kernel $(R_\delta \boxtimes R_\delta)[Z_\infty] + (h_\delta \boxtimes h_\delta)([diag])$. (The last summand is well-defined because of the condition on the wave front of h_δ). It is easy to check that the following identity holds:

$$id - \Pi_\delta = dH_\delta + H_\delta d.$$

Thus we have a family of homotopies and projectors parameterized by δ.

Remark 18 *One can define the projector Π_δ using another canonical element $\sum_{x \in Cr(f)} [S_x] \otimes R_\delta([U_x])$, instead of $\sum_{x\in Cr(f)} R_\delta([S_x]) \otimes R_\delta([U_x])$, as we did. The above Proposition holds for the new canonical element as well.*

There is a version of the previous construction, which will be useful in the next subsection. Namely, we start with a differential $n+1$-form ρ on $Y \times Y \times (0,1)$ such that for the support of $supp(\rho)$ belongs to $\sqcup_{\delta>0}(N_\delta, \delta)$ for all sufficiently small $\delta \in (0,1)$, and ρ defines the same cohomology class in $H^n_c(Y \times Y \times (0,1))$ as $[diag] \times (0,1)$.

Let us consider now the spaces $\Omega^*_0(Y) := \varinjlim_{\delta\to 0} \Omega^*(Y \times (0,\delta))$ and $D'_0(Y) := \varinjlim_{\delta \to 0} \Omega^*(0,\delta)\widehat{\otimes} D'(Y)$. It is easy to see that both complexes $\Omega^*_0(Y)$ and $D'_0(Y)$ are quasi-isomorphic to $\Omega^*(Y)$.

We define a linear operator $R : D'_0(Y) \to \Omega^*_0(Y)$ similarly to the definition of R_δ. Then the Lemma and the Proposition hold with obvious changes. We will denote the corresponding objects by the same letters as before, skipping the subscript δ (like H for the homotopy and Π for the projector). Morally, they are obtained from the old objects by extending them as differential forms "in the direction of δ".

6.6 Proof of the theorem from 6.3

For simplicity we will assume that all local systems are trivial and have rank one. The general case is completely similar.

We are going to construct the following chain of A_∞-equivalences connecting $DR(Y)$ and $M(Y)$:

$$DR(Y) \hookrightarrow DR_0(Y) \hookleftarrow DR_0^{tr}(Y) \hookleftarrow DR_0^{tr,\Pi}(Y)) \leftarrow M(Y)$$

Classes of objects of all these categories will be the same, and all functors will be identical on objects.

The A_∞-pre-category $DR_0(Y)$ is in fact a dg-category, i.e. all sequences of objects are transversal, compositions m_k vanish for $k \geq 3$ and it has strict identity morphisms. The space $Hom_{DR_0(Y)}(f_0, f_1)$ is defined as

$\varinjlim_{\delta\to 0}\Omega^*(Y\times(0,\delta)) = \Omega_0^*(Y)$. Clearly the space of morphisms does not depend on objects. Using the wedge product of differential forms we make $DR_0(Y)$ into a dg-category over the field $\mathbb{C}$. There is a natural functor $DR(Y)\to DR_0(Y)$, which is the identity map on objects. On morphisms it is the natural embedding of $\Omega^*(Y)$ as the subspace of forms on $Y\times(0,\delta)$, which are pullbacks of forms on Y. Clearly it establishes an equivalence of A_∞-categories.

The A_∞-pre-category $DR_0^{tr}(Y)$ is defined as the full subcategory of $DR_0(Y)$, and it differs from the latter only by the choice of transversal sequences. Namely, we use the same notion of transversality in $DR_0^{tr}(Y)$ as in the Morse category.

The next A_∞-pre-category $DR_0^{tr,\Pi}(Y)$ is obtained from $DR_0^{tr}(Y)$ by applying homological perturbation theory. For any two transversal objects f_0, f_1 of $DR_0^{tr}(Y)$ we define $Hom_{DR_0^{tr,\Pi}(Y)}(f_0,f_1)$ as $\Pi_{f_0,f_1}(\Omega_0^*(Y))$. Here Π_{f_0,f_1} is the projector Π corresponding to the Morse function f_0-f_1, it was described at the end of the previous subsection. We also have homotopies H_{f_0,f_1} associated with f_0-f_1. Then formulas of homological perturbation theory (summation over trees) give rise to an A_∞-pre-category $DR_0^{tr,\Pi}(Y)$ and an equivalence $DR_0^{tr,\Pi}(Y)\to DR_0^{tr}(Y)$.

The last functor $\Psi : M(Y)\to DR_0^{tr,\Pi}(Y))$ will have no non-trivial higher components Ψ_n for $n\geq 2$. The first component Ψ_1 of it is a linear map

$$\Psi_1 : Hom_{M(Y)}(f_0,f_1)\to Hom_{DR_0^{tr,\Pi}(Y)}(f_0,f_1)$$

for every transversal pair (f_0,f_1). Recall that $Hom_{M(Y)}(f_0,f_1)$ has a basis $\{[x]\}$ labeled by critical points $x\in Cr(f_0-f_1)$. We define $\Psi_1([x])$ as $R([S_x])$. It is clear that Ψ_1 gives a quasi-isomorphism of complexes for every transversal pair (f_0,f_1).

Now, we claim that Ψ is an A_∞-functor. This means that Ψ_1 maps all higher compositions in $M(Y)$ to higher compositions in $DR_0^{tr,\Pi}(Y)$. This follows directly from the descriptions of higher compositions in both categories in terms of planar trees, see Sections 6.2, 6.4. Notice that the number of functions in a given sequence of objects is finite. For all sufficiently small δ every summand in the formula for $m_k^{M(Y)}$, corresponding to a binary tree T, coincides with the summand for $m_k^{M(Y)}$ corresponding to the same T (we can assume that δ is so small that the part 2) of the Lemma 2 can be applied). The theorem is proved. ■

7 A_∞-structure for the derived category of coherent sheaves

7.1 Rigid analytic space

It will be helpful (although not necessary) for the reader of this section to be familiar with basic facts of non-archimedean analysis (see [BGR]). For any smooth manifold Y with integral affine structure we will construct a sheaf $\mathcal{O}_Y$ of $\mathbf{C}_\varepsilon$-algebras on Y. Stalks $\mathcal{O}_{Y,y}$ of this sheaf are noetherian algebras, and one can define the notion of coherent sheaves of $\mathcal{O}_Y$-modules. If $Y = \mathbb{R}^n/\mathbb{Z}^n$ is the torus with the standard integral affine structure then the category of coherent $\mathcal{O}_Y$-modules will be equivalent (by a non-archimedean version of GAGA) to the category of coherent sheaves on an abelian variety over the field $\mathbf{C}_\varepsilon$.

We start with the local picture. We denote by $v : \mathbf{C}_\varepsilon \to \mathbf{R} \cup \{+\infty\}$ a (non-discrete) valuation defined by $v(\sum_{\lambda_1<\lambda_2<\dots} c_i e^{-\lambda_i/\varepsilon}) = \lambda_1$ if $c_1 \neq 0$ and $v(0) = +\infty$.

Definition 21 *Let $U \subset \mathbf{R}^n$ be an open subset of the standard vector space $\mathbf{R}^n$. We define $\mathcal{O}_{\mathbf{R}^n}(U)$ as the vector space over $\mathbf{C}_\varepsilon$ consisting of formal Laurent series*

$$f = \sum_{k_1,\dots,k_n \in \mathbf{Z}^n} a_{k_1\dots k_n} z_1^{k_1}\dots z_n^{k_n},$$

where $z_1, \dots, z_n$ are formal variables, $a_{k_1\dots k_n} \in \mathbf{C}_\varepsilon$, and for any $(y_1, \dots, y_n) \in U$ we have: $lim_{\sum_i |k_i| \to \infty}(v(a_{k_1\dots k_n}) + \sum_i k_i y_i) = +\infty$.

It follows from the definition that if $f \in \mathcal{O}_{\mathbf{R}^n}(U)$ and $(z_1, \dots, z_n) \in (\mathbf{C}_\varepsilon^*)^n$ then the series $\sum_{k_1,\dots,k_n} a_{k_1\dots k_n} z_1^{k_1}\dots z_n^{k_n}$ converges in the adic topology as long as $(v(z_1), \dots, v(z_n)) \in U$.

We introduce an action of the group $GL(n, \mathbf{Z}) \ltimes \mathbf{R}^n$ on $(\mathbf{R}^n, \mathcal{O}_{\mathbf{R}^n})$ such as follows:

a) $GL(n, \mathbf{Z})$ acts simultaneously by the linear change of coordinates and linear transformation of indices $(k_1, \dots, k_n)$ in the series;

b) translations $(t_1, \dots, t_n) \in \mathbf{R}^n$ act on the coordinates $(y_1, \dots, y_n)$ by the shift $(y_1, \dots, y_n) \mapsto (y_1 + t_1, \dots, y_n + t_n)$, and on the series by the rescaling of coefficients

$$\sum_{k_1,\dots,k_n} a_{k_1\dots k_n} z_1^{k_1}\dots z_n^{k_n} \mapsto \sum_{k_1,\dots,k_n} (a_{k_1\dots k_n} e^{-\sum_i t_i k_i/\varepsilon}) z_1^{k_1}\dots z_n^{k_n}.$$

Using this action we define the sheaf $\mathcal{O}_Y$ for an arbitrary smooth manifold Y with integral affine structure.

We claim that there is a canonically associated to Y a rigid analytic space Y^{an} defined over $\mathbf{C}_\varepsilon$. Here is the construction. Let us consider a covering of Y by open subsets U_i such that all non-empty intersections $U_{i_1 i_2 \dots i_k} := U_{i_1} \cap \dots \cap U_{i_k}$ in some local affine coordinates are convex polyhedra whose faces have rational slopes. Every $U_{i_1 i_2 \dots i_k}$ can be identified with the intersection of finitely many half-spaces, such that their pre-images under the map $v^n : (\mathbf{C}_\varepsilon^*)^n \to \mathbf{R}^n$ are sets of the type $\{(z_1, \dots, z_n) | \, v(z_1^{k_1} \dots z_n^{k_n}) \geq C\}$ for some rational $C > 0$. It is known after Tate that such a system of inequalities defines an affinoid domain (i.e. a local model for a rigid analytic space over $\mathbf{C}_\varepsilon$).

Definition 22 *We define Y^{an} as the rigid analytic space over $\mathbb{C}_\varepsilon$ obtained by gluing the local data $(U_i, \mathcal{O}_{U_i})$ by means of the action of $GL(n, \mathbb{Z}) \ltimes \mathbb{R}^n$.*

It is easy to see that Y^{an} is canonically defined, and that the category $Coh(Y^{an})$ of coherent analytic sheaves on Y^{an} (in the sense on analytic geometry) is equivalent to the category of coherent $\mathcal{O}_Y$-modules (i.e. locally finitely generated $\mathcal{O}_Y$-modules).

To every algebraic variety $\mathcal{Y}$ over $\mathbb{C}_\varepsilon$ one can associate canonically a rigid analytic space $\mathcal{Y}^{an}$. If $\mathcal{Y}$ is projective then the category $Coh(\mathcal{Y}^{an})$ is equivalent to the category $Coh(\mathcal{Y})$ of algebraic coherent sheaves on $\mathcal{Y}$ (GAGA theorem).

Assume that $Y = \mathbf{R}^n/\Lambda$ is an n-dimensional torus equipped with the standard integral affine structure induced by $\mathbb{Z}^n \subset \mathbf{R}^n$, and Λ is a lattice commensurable with $\mathbf{Z}^n$. The following result can be derived from [Mum].

Proposition 9 *In the previous notation one has $Y^{an} \simeq \mathcal{Y}^{an}$ where $\mathcal{Y}$ is an abelian variety over $\mathbf{C}_\varepsilon$.*

Let us return to the picture of metric collapse in the case of abelian varieties. Since the collapse was defined by rescaling of the lattice (see Section 2) one can prove that $\mathcal{Y}$ is isomorphic to the original abelian variety over $\mathbb{C}_\varepsilon$. Therefore in the case of abelian varieties we have two equivalent descriptions of the collapse: the one in terms of Riemannian geometry and the one in terms of analytic non-archimedean geometry.

Remark 19 *For the case of collapse with singular fibers, the rigid analytic space Y^{an} constructed as above, seems to be a "wrong" one. First of all, it is not compact because Y is not compact. But there is also a more fundamental problem. It seems that Y^{an} can not be embedded into a compact analytic space associated with a projective algebraic variety. There are several indications that there exists another sheaf of algebras $\mathcal{O}'_Y$ which is (locally on Y) isomorphic to $\mathcal{O}_Y$, and the rigid analytic space $(Y^{an})'$ associated with $(Y, \mathcal{O}'_Y)$ admits an algebraic compactification. In general, sheaves $\mathcal{O}'_Y$ which are twisted versions of $\mathcal{O}_Y$ are classified by the first non-abelian cohomology $H^1(Y, \underline{Aut}(\mathcal{O}_Y))$ where $\underline{Aut}(\mathcal{O}_Y)$ is the sheaf of groups of automorphisms of $\mathcal{O}_Y$. Thus, in the mirror symmetry for Calabi-Yau manifolds which are not* *abelian varieties, we expect a new ingredient, the cohomology class* $[\mathcal{O}'_Y]$.

7.2 A_∞-structure on coherent sheaves

There is a sheaf of abelian groups Af_Y on Y given by locally affine functions with integral slopes (such functions locally are given by $l = c + \sum_{1\le i\le n} m_i y_i$ where $m_i \in \mathbf{Z}, c \in \mathbf{R}$). There is a morphism of sheaves $exp : Af_Y \to \mathcal{O}_Y^*$ given by $l \mapsto exp(l) := e^{-c/\varepsilon} \prod_{1\le i\le n} z_i^{m_i}$.

Let (Y, g) be an AK-manifold (see Section 3.2). We are going to define a characteristic class $[g]$ of the metric, which will be an analog of the cohomology class of a Kähler form in complex geometry. Let $Af_Y \otimes \mathbf{R}$ be a sheaf of all real-valued locally affine functions. For a cover by convex sets $(U_i)_{i\in I}$ one can choose smooth functions K_i such that $g_{|U_i} = \partial^2 K_i$. Then $K_i - K_j \in Af_Y \otimes \mathbf{R}(U_i \cap U_j)$, defines a 1-cocycle whose cohomology class we denote by $[g] \in H^1(Y, Af_Y \otimes \mathbf{R})$. If the *dual* affine structure (see Section 3) is integral, we get a class $[g]$ in the subgroup $H^1(Y, Af_Y)/torsion \subset H^1(Y, Af_Y \otimes \mathbb{R})$. We will call such classes *integral*. In this case $exp([g]) \in H^1(Y^{an}, \mathcal{O}_Y^*)$ is the first Chern class of a line bundle on Y^{an}. By analogy with the Kähler geometry we expect that this line bundle is ample. In the case when (Y, g) is a flat torus, the ampleness can be proven directly (see [BL]).

From now on we assume that $[g]$ is integral. Then by GAGA (see [Be], Prop. 3.14) the category of analytic coherent sheaves on Y^{an} is equivalent to the category of algebraic coherent sheaves on the corresponding algebraic projective variety $\mathcal{Y}$.

The sheaf $\mathcal{O}_Y$ admits a resolution $\widehat{\Omega}_Y^*$ by a soft sheaf of dg-algebras. Locally, for a small open $U \subset Y$, sections of $\widehat{\Omega}_Y^*$ are given by sums $\alpha = \sum_{i_1,\dots,i_n} c_{i_1\dots i_n} z_1^{i_1} \dots z_n^{i_n}$ where $c_{i_1\dots i_n} = \sum_j c_{j,i_1\dots i_n} e^{-\lambda_{j,i_1\dots i_n}/\varepsilon}, c_{j,i_1\dots i_n} \in \Omega^*(U)$ with the same convergence conditions as for the sheaf $\mathcal{O}_Y$. Differential is given by the de Rham differential acting on the coefficients $c_{j,i_1\dots i_n}$.

We define a dg-category $\mathcal{C}(Y)$ such as follows. Objects are finite complexes of locally free $\mathcal{O}_Y$-modules of finite rank. For any two such complexes E_1 and E_2 we define the space of morphisms as

$$Hom_{\mathcal{C}(Y)}(E_1, E_2) = \Gamma(Y, Hom_{\mathcal{O}_Y}(E_1, E_2) \widehat{\otimes}_{\mathcal{O}_Y} \widehat{\Omega}_Y^*),$$

where we use the completed tensor product in the r.h.s. Differential and grading on the spaces of morphisms are induced by those on $E_1, E_2, \widehat{\Omega}_Y^*$. We will treat $\mathcal{C}(Y)$ as an A_∞-pre-category in which all sequences of objects are transversal and there are no higher compositions except m_1 and m_2.

For a given projective algebraic variety V over a field, one can define canonically an equivalence class of A_∞-categories $D^b_{A_\infty}(V)$. It is obtained by the following enhancement of the bounded derived category of coherent sheaves on V. Objects of this A_∞-category are the same as of the derived category of coherent sheaves. In order to define the space of morphisms between two objects, one replaces them by arbitrary chosen acyclic resolutions by locally free sheaves (e.g. the Godement resolutions) and then takes the global sections of the space of morphisms between resolutions in the category of complexes of sheaves. In this way one obtains a dg-category. In the case of projective varieties over complex numbers, there is an alternative construction in terms of complexes of holomorphic vector bundles and Dolbeault forms. Different choices of resolutions lead to A_∞-equivalent categories. We will denote the (A_∞-equivalence) class of these categories by $D^b_{A_\infty}(V)$.

By definition, spaces of morphisms of $\mathcal{C}(Y)$ are resolutions of the corresponding spaces of sheaves of $\mathcal{O}_Y$-modules. Then using GAGA theorem from [Be], one concludes that the following result holds.

Proposition 10 *The category $\mathcal{C}(Y)$ is A_∞-equivalent to $D^b_{A_\infty}(\mathcal{Y})$, where $\mathcal{Y}$ is the projective algebraic variety corresponding to the analytic space Y^{an} assigned to Y.*

8 Homological mirror conjecture

In the previous section we constructed an A_∞-category $\mathcal{C}(Y)$ which is A_∞-equivalent to the derived category of coherent sheaves on a Calabi-Yau manifold over the field $\mathbb{C}_\varepsilon$. In this section we are going to construct a chain of A_∞-pre-categories and A_∞-equivalences (cf. with Section 6.6)

$$\mathcal{C}_{unram}(Y) \hookrightarrow \mathcal{C}_{unram,0}(Y) \hookleftarrow \mathcal{C}^{tr}_{unram,0}(Y) \hookleftarrow \mathcal{C}^{tr,\Pi}_{unram,0}(Y) \leftarrow FO(X^\vee)$$

and a functor $F : \mathcal{C}_{unram}(Y) \to \mathcal{C}(Y)$ which establishes an equivalence between $\mathcal{C}_{unram}(Y)$ and a full subcategory of $\mathcal{C}(Y)$. Recall that the Fukaya-Oh category $FO(X^\vee)$, as defined in this paper, is also equivalent to a full subcategory of the Fukaya category $F(X^\vee)$. Thus, we establish an A_∞-equivalence between full subcategories of the Fukaya category $F(X^\vee)$ and of $\mathcal{C}(Y)$.

The approach we are going to use is completely similar to the one we used in the case of Morse theory in Section 6. All the categories in our chain of A_∞-equivalence functors from above will have the same class of objects, i.e. the same as the Fukaya-Oh category.

8.1 Mirror symmetry functor on objects

Here we will define dg-category $\mathcal{C}_{unram}(Y)$ and the fully faithful embedding F of this category to $\mathcal{C}(Y)$.

In the Appendix we will explain the conventional picture for the mirror symmetry functor in case of complex numbers. There we will use a kind of Fourier-Mukai transform along fibers of the torus fibration. The kernel of this transform is an analog of Poincaré bundle. If one starts with a local system on a Lagrangian section of $p^\vee : X^\vee \to Y$ then the transform makes from it a smooth bundle on X with the connection which is flat in the anti-holomorphic directions. In other words, one gets a holomorphic bundle on X.

These considerations cannot be literally repeated in the non-archimedean case. We are going to construct the functor F in the following way. Let (L,ρ) be an object of the category $FO(X^\vee)$ such that $rank(\rho) = 1$ and the projection $L \to Y$ is one-to-one map. The manifold L is locally given by the graph of $df\ (mod\,(T_Y^{\mathbf{Z}})^\vee)$, where f is a smooth function on Y. To such an object we assign a sheaf of rank one $\mathcal{O}_Y$-modules $F(L,\rho)$. For sufficiently small open $U \subset Y$ and chosen $f \in C^\infty(U)$ the sheaf $F(L,\rho)_{|U}$ is identified with $\mathcal{O}_{Y|U}$. Change $f \mapsto f + l$, where $l \in Af_Y(U)$ leads to the change of the trivialization of $F(L,\rho)_{|U}$ as $1_U \mapsto exp(l)\,1_U$ (here $1_U \in \mathcal{O}_Y(U)$ is the identity function). If $rank(\rho)$ is greater than one, we decompose $\rho_{|U}$ for small $U \subset Y$ into the sum of rank one local systems and then apply the construction. Analogously, if the covering $L \to Y$ has more than one leaf, we apply the previous construction to each leaf of the covering and then take the direct sum.

We will call F the mirror symmetry functor on objects. The category $\mathcal{C}_{unram}(Y)$ is defined as the dg-category whose class of objects is $Ob(FO(X^\vee))$, and the spaces of morphisms are

$$Hom_{\mathcal{C}_{unram}(Y)}((L_1,\rho_1),(L_2,\rho_2)) := Hom_{\mathcal{C}(Y)}(F(L_1,\rho_1),F(L_2,\rho_2))$$

8.2 Spectrum of a morphism and the semigroup

Let $E_i = F(L_i,\rho_i), i = 1,2$ be locally free $\mathcal{O}_Y$-modules (i.e. vector bundles) corresponding to objects $(L_i,\rho_i) \in FO(X^\vee), i = 1,2$. For any $\alpha \in Hom_{\mathcal{C}(Y)}(E_1,E_2)$ and a point $y \in Y$ we will define the spectrum of α at y as a certain (at most countable) discrete set of real numbers with finite multiplicities.

Let us assume first that $\rho_i, i = 1,2$ are trivial rank one local systems on $L_i, i = 1,2$, and $L_i, i = 1,2$ are unramified coverings of Y. For a sufficiently small open set U containing y we can write in local coordinates $L_i = graph(df_i)\ (mod\,(T_Y^{\mathbf{Z}})^\vee), i = 1,2$ for smooth functions $f_i : Y \to \mathbb{R}, i = 1,2$. Restriction to a small open set U of a morphism $\alpha \in Hom_{\mathcal{C}(Y)}(E_1,E_2)(U) = \widehat{\Omega}_Y^*(U)$ can be identified with the infinite series $\alpha = \sum_{i_1,\dots,i_n} c_{i_1\dots i_n} z_1^{i_1}\dots z_n^{i_n}$, where $c_{i_1 \dots i_n} = \sum_j c_{j,i_1 \dots i_n} e^{-\lambda_{j,i_1 \dots i_n}/\varepsilon}$ and $c_{j,i_1 \dots i_n} \in \Omega_Y^*(U)$.

We define the *spectrum of* α *at* $y \in U$ as the set of real numbers (with multiplicities)

$$Sp_y(\alpha) = \{-\lambda_{j,i_1 \dots i_n} + \sum_{1 \le k \le n} i_k y_k + f_1(y) - f_2(y)\},$$

where the germ of $c_{j,i_1 \dots i_n}$ at y is not equal to zero. One can check that $Sp_y(\alpha)$ is well-defined (i.e. does not depend on the local trivialization), and has the only limiting point at $s = -\infty$.

In the general case of higher rank local systems and Lagrangian manifolds which are unramified coverings of Y, we decompose $E_i, i = 1, 2$ locally near $y \in Y$ into the direct sum of trivial rank one $\mathcal{O}_Y$-modules. The spectrum of a morphism at the point y is then defined as the union of the spectra of morphisms between corresponding line bundles.

Remark 20 *One can use instead of the spectrum an* $\mathbb{R}$*-filtration* $Hom_{C(Y)}(E_1, E_2)^{\le s}$ *on the space of morphisms. It comes from the filtration on the stalks of sheaves of morphisms* $\underline{Hom}_{\mathcal{O}_Y}(E_1, E_2) \widehat{\otimes} \widehat{\Omega}_Y^*$ *(completed tensor product) defined by the condition* $\{-\lambda_{j,i_1 \dots i_n} + \sum_{1 \le k \le n} i_k y_k + f_1(y) - f_2(y)\} \le s$*. It is easy to see that* α *belongs to* $Hom_{C(Y)}(E_1, E_2)^{\le s}$ *iff for all* $y \in Y$ *one has* $Sp_y(\alpha) \subset (-\infty, s]$.

Let us consider a subspace $Hom_{C(Y)}^{alg}(E_1, E_2) \subset Hom_{C(Y)}(E_1, E_2)$ of algebraic morphisms. It consists of finite sums (both in z_i and $e^{-\lambda_{j,i_1 \dots i_n}/\varepsilon}$). It is dense in the space of all morphisms (analytic functions can be approximated by Laurent polynomials). Moreover, the space $Hom_{C(Y)}(E_1, E_2)$ coincides with the completion of $Hom_{C(Y)}^{alg}(E_1, E_2)$ with respect to the $\mathbb{R}$-filtration introduced above.

There is a 1-parameter semigroup $\phi^t, t \ge 0$ acting on $Hom_{C(Y)}^{alg}(E_1, E_2)$. In local coordinates ϕ^t acts on the coefficients $c_{j,i_1 \dots i_n}$ by moving them along the gradient flow of $f_1 - f_2$. In order to define it globally we need to describe the space $Hom_{C(Y)}^{alg}(E_1, E_2)$ in geometric terms. It will be done below.

Given two Lagrangian submanifolds $L_i \subset X^\vee, i = 1, 2$ as above, a point $y \in Y$, two points $x_i \in L_i, i = 1, 2$ such that $p^\vee(x_i) = y$, we define a set $P(L_1, L_2, y)$ of homotopy classes of paths $\gamma \in (p^\vee)^{-1}(y)$ starting at x_1 and ending at x_2. Each homotopy class contains a unique geodesic in the flat metric on the torus. We define the space $P(L_1, L_2) = \sqcup_{y \in Y} P(L_1, L_2, y)$. It carries an obvious topology such that the natural projection $\pi : P(L_1, L_2) \to Y$ is an unramified covering with countable fibers. Using the symplectic form ω on $X^\vee$ we define a closed 1-form μ on $P(L_1, L_2)$ by the formula $\mu = \int_\gamma \omega$.

Locally on Y we have: $L_i = df_i(mod\,(T_Y^{\mathbf{Z}})^\vee), i = 1,2$ where $f_i : Y \to \mathbf{R}$ are smooth functions. Then locally on $P(L_1, L_2)$ we have: $\mu = d(f_1 - f_2 + l)$, where l is a local section of the pullback of the sheaf Aff_Y. Clearly the function l is defined up the adding of a real constant. Thus obtain an $\mathbf{R}$-torsor on $P(L_1, L_2)$. Using the embedding $\mathbf{R} \to \mathbf{C}_\varepsilon^*$, $\lambda \mapsto exp(\lambda/\varepsilon)$ we get a $\mathbf{C}_\varepsilon^*$-torsor, which defines a local system $\mathbf{C}_\varepsilon^{tw}$ of 1-dimensional $\mathbf{C}_\varepsilon$-modules over $P(L_1, L_2)$. Fibers of $\mathbf{C}_\varepsilon^{tw}$ carry natural filtrations. Indeed, in a neighborhood of a point $(x_1, x_2, \gamma, y) \in P(L_1, L_2)$ we can choose a smooth function $f = f_1 - f_2 + l$ such that $\mu = df$. It defines a local trivialization of $\mathbf{C}_\varepsilon^{tw}$. In this trivialization the filtration is defined for $h \in \mathbf{C}_\varepsilon$ by the condition $v(h)(y) + f(y) \leq s, s \in \mathbf{R}$, where v is the valuation. We define a subsheaf $\mathbf{C}_\varepsilon^{tw,alg}$ of $\mathbf{C}_\varepsilon^{tw}$ by the requirement that in a local trivialization it is a subsheaf of finite sums of exponents.

Notice that there are natural projections $pr_i : P(L_1, L_2) \to L_i, i = 1,2$. Having local systems ρ_i on $L_i, i = 1,2$ we define local systems $\widehat{\rho}_i, i = 1,2$ on $P(L_1, L_2)$ as pullbacks with respect to $pr_i, i = 1,2$.

On $P(L_1, L_2)$ we define a sheaf $\underline{Hom}^{alg}(E_1, E_2)$ ($E_i, i = 1,2$ were defined previously) such as follows: $\underline{Hom}^{alg}(E_1, E_2) = \mathbf{C}_\varepsilon^{tw,alg} \otimes (\widehat{\rho}_1)^* \otimes \widehat{\rho}_2 \otimes \underline{\Omega}^*_{P(L_1,L_2)}$, where $\underline{\Omega}^*_{P(L_1,L_2)}$ is the sheaf of differential forms. We endow stalks of $\underline{Hom}^{alg}(E_1, E_2)$ with $\mathbf{R}$-filtrations induced by the filtration on $\mathbf{C}_\varepsilon^{tw}$ and trivial filtrations on the other tensor factors.

Let $\pi_!$ denotes the functor of direct image with compact support. Then $\pi_!(\underline{Hom}^{alg}(E_1, E_2)) = \pi_!(\mathbf{C}_\varepsilon^{tw} \otimes \widehat{\rho}_1^* \otimes \widehat{\rho}_2) \otimes \underline{\Omega}_Y^*$, where the last tensor factor is the sheaf of de Rham differential forms on Y.

We can identify $\mathbf{Z}^n$ with $H_1(T^n, \mathbf{Z})$, and the latter group naturally acts on homotopy classes of paths γ. On the other hand, the group ring of $\mathbf{Z}^n$ over $\mathbf{C}_\varepsilon$ can be identified with the ring of Laurent polynomials $\mathbf{C}_\varepsilon[z_1^{\pm 1}, ..., z_n^{\pm 1}]$. Let $\mathbf{C}_\varepsilon^{alg} \subset \mathbf{C}_\varepsilon$ be the subring of finite sums of exponents. It is easy to see that the structure of $\mathbf{C}_\varepsilon^{alg}[\mathbf{Z}^n]$-module on the sections of $\underline{Hom}^{alg}(E_1, E_2)$ corresponds to the structure of $\mathbf{C}_\varepsilon^{alg}[z_1^{\pm 1}, ..., z_n^{\pm 1}]$-module on its image under $\pi_!$. Using this observation one can prove that

$$Hom^{alg}_{C(Y)}(E_1, E_2) \simeq \Gamma(Y, \pi_!(\underline{Hom}^{alg}(E_1, E_2))) = \Gamma_c(P(L_1, L_2), \underline{Hom}^{alg}(E_1, E_2)),$$

where the isomorphism is induced by the natural morphism of sheaves

$$\pi_!(\underline{Hom}^{alg}(E_1, E_2)) \to \underline{Hom}^{alg}_{C(Y)}(E_1, E_2).$$

Here Γ_c refers to the functor of sections with compact support.

Using the metric on Y we assign to the 1-form μ a vector field ξ on $P(L_1, L_2)$. Locally ξ is the generator of the gradient flow of $f_1 - f_2 + l$. It is not difficult to show that there is no trajectory of the flow which goes to infinity for a finite time. Therefore the vector field ξ generates a 1-parameter semigroup ψ^t acting on $P(L_1, L_2)$. The following result is easy to prove.

Proposition 11 *The 1-parameter semigroup ψ^t decreases the filtration on stalks of points which do not belong to $L_1 \cap L_2$. More precisely,*

$$\psi^t(\underline{Hom}_p^{alg}(E_1, E_2)^s) \subset \underline{Hom}_{\psi^t(p)}^{alg}(E_1, E_2)^{s-\int_0^t \mu},$$

where $p \in P(L_1, L_2)$ is an arbitrary point.

Functor $\pi_!$ is compatible with the filtrations on the stalks of sheaves $\underline{Hom}_{C(Y)}^{alg}(E_1, E_2)$ and $\underline{Hom}^{alg}(E_1, E_2)$. It is easy to see that the completion of stalks of the former with respect to the filtration induced from the one on $\underline{Hom}^{alg}(E_1, E_2)$ coincides with $Hom_{C(Y)}(E_1, E_2)$. Since the semigroup ψ^t decreases the filtration, the semigroup ϕ^t extends continuously to the completion with respect to the filtration. Thus the following proposition holds.

Proposition 12 *The action of ϕ^t extends continuously from $Hom_{C(Y)}^{alg}(E_1, E_2)$ to $Hom_{C(Y)}(E_1, E_2)$.*

8.3 Homological mirror symmetry for abelian varieties

The approach and proofs are completely similar to those from Section 6, so we will omit the details. One has to change scalars to $\mathbb{C}_{varepsilon}$ in Section 6

In the previous subsection we defined the semigroup ψ^t acting on the sections with compact support $\Gamma_c(P(L_1, L_2), \underline{Hom}^{alg}(E_1, E_2))$. This action corresponds to the action of the semigroup ϕ^t on the space of morphisms $Hom_{C(Y)}(E_1, E_2)$. Notice that the sheaf $\underline{Hom}^{alg}(E_1, E_2) = \mathbb{C}_\varepsilon^{tw,alg} \otimes (\widehat{\rho}_1)^* \otimes \widehat{\rho}_2 \otimes \underline{\Omega}^*_{P(L_1,L_2)}$ is a subsheaf of $\mathbb{C}_\varepsilon^{tw} \otimes (\widehat{\rho}_1^* \otimes \widehat{\rho}_2) \otimes \underline{D}'_{P(L_1,L_2)}$, where $\underline{D}'_{P(L_1,L_2)}$ is the sheaf of distribution-valued differential forms on $P(L_1, L_2)$. Sections of the latter sheaf carry the natural topology: series in $exp(-\lambda_i/\varepsilon)$ with distributional coefficients converge, if they converge in the adic sense when paired with a test differential form. Similarly to the case of Morse theory (Section 6.5) one proves the following result.

Proposition 13 *For any $\beta \in \Gamma_c(P(L_1, L_2), \underline{Hom}^{alg}(E_1, E_2))$ there exists limit*

$$\psi^\infty(\beta) = lim_{t\to+\infty}\psi^t(\beta) \in \Gamma(P(L_1, L_2), \mathbb{C}_\varepsilon^{tw} \otimes (\widehat{\rho}_1^* \otimes \widehat{\rho}_2) \otimes \underline{D}'_{P(L_1,L_2)}).$$

The limit is not difficult to describe in terms of the gradient flow generating ψ^t. Using the fact that ψ^t moves the spectrum of a morphism to $-\infty$, one can prove similarly to the Section 6.5 that the limit $\psi^\infty(\beta)$ belongs to a finite-dimensional $\mathbb{C}_\varepsilon$-vector space generated by the distributions corresponding the unstable manifolds $U_x \subset P(L_1, L_2), x \in L_1 \cap L_2$. Clearly, the map $\beta \mapsto \psi^\infty(\beta)$ extends to the completion with respect to the filtration. It descends to the map $\alpha \mapsto \phi^\infty(\alpha)$, where $\alpha \in Hom_{\mathcal{C}(Y)}(E_1, E_2)$. The image of ϕ^∞ belongs to the space isomorphic to $Hom_{FO(X^\vee)}((L_1,\rho_1),(L_2,\rho_2))$.

Then we repeat the arguments from Section 6.6. Namely, we define an A_∞-pre-category $\mathcal{C}_{unram,0}(Y)$ in the same way as we defined the category $DR_0(Y)$ in Section 6.6. Objects of $\mathcal{C}_{unram,0}(Y)$ are the same as of $\mathcal{C}_{unram}(Y)$. The spaces of morphisms of $\mathcal{C}_{unram,0}(Y)$ are dg-modules over the dg-algebra $\mathbf{C}_\varepsilon \widehat{\otimes} \Omega_0^*$, where Ω_0^* is the dg-algebra of germs of differential forms in the auxiliary parameter δ at $\delta = 0 \in \mathbf{R}_{\geq 0}$ (cf. Section 6.6). Compositions of morphisms in $\mathcal{C}_{unram,0}(Y)$ are linear with respect to the dg-module structure. The transversality conditions in $\mathcal{C}_{unram,0}(Y)$ and $\mathcal{C}_{unram}(Y)$ by definition are the same as in $FO(X^\vee)$. Thus we obtain an A_∞-pre-category $\mathcal{C}^{tr}_{unram,0}(Y)$ which is A_∞-equivalent to $\mathcal{C}_{unram}(Y)$.

Using homological perturbation theory in the same way as in Section 6.6 (projectors and homotopies are now defined by means of the semigroup ϕ^t), we construct the analog of the category $DR_0^{tr,\Pi}(Y)$. We denote this A_∞-pre-category by $\mathcal{C}^{tr,\Pi}_{unrum,0}(Y)$. The spaces of morphisms of this category are completed tensor products of Ω_0^* with finite-dimensional $\mathbf{C}_\varepsilon$-vector spaces, spanned by the "smoothings" of the unstable currents $[U_x]$. These smoothings are defined by means of the operators R_δ in the same way as in Section 6.5. By definition, the category $\mathcal{C}^{tr,\Pi}_{unrum,0}(Y)$ has the same transversality conditions as the category $FO(X^\vee)$. The spaces of morphisms are naturally quasi-isomorphic to the corresponding spaces of morphisms in $FO(X^\vee)$. Repeating the arguments from Section 6.6 we will see that the A_∞-structure on $\mathcal{C}^{tr,\Pi}_{unram,0}(Y)$ is equivalent to the one on $FO(X^\vee)$. Indeed, we have a natural map from the space $Hom_{FO(X^\vee)}((L_1,\rho_1),(L_2,\rho_2))$ to the space $Hom_{\mathcal{C}^{tr,\Pi}_{unram,0}(Y)}(F(L_1,\rho_1),F(L_2,\rho_2))$. Let us recall that all the categories $\mathcal{C}_{unram}(Y)$, $\mathcal{C}_{unram,0}(Y)$, $\mathcal{C}^{tr}_{unram,0}(Y)$ and $\mathcal{C}^{tr,\Pi}_{unrum,0}(Y)$ have the same class of objects. Therefore we have the mirror symmetry functor on objects $F : FO(X^\vee) \to \mathcal{C}^{tr,\Pi}_{unrum,0}(Y)$ (see Section 8.1). On the other hand, the considerations above give rise to the linear maps of the spaces of morphisms. Let ν_{X_1,X_2} be this linear map for two objects $X_i = (L_i,\rho_i), i = 1,2$. The proof of the following proposition is completely similar to the corresponding one in Section 6.6. One works over the field $\mathbf{C}_\varepsilon$ and uses the formulas for m_k, $k \geq 1$ from the last paragraph of Section 6.2.

Proposition 14 *Let $E_i = F(X_i), 0 \leq i \leq k, k \geq 1$ be locally free rank one $\mathcal{O}_Y$-modules (vector bundles) corresponding to objects $X_i = (L_i,\rho_i) \in FO(X^\vee), 0 \leq i \leq k$. Then the formulas for*

$$m_k^{FO(X^\vee)} : \otimes_{0 \leq i \leq k} Hom(E_i, E_{i+1}) \to Hom(E_0, E_k)[2-k]$$

coincide (after the extension of scalars from $\mathbb{C}_\varepsilon$ to $\mathbb{C}_\varepsilon\widehat{\otimes}\Omega_0^*$) with the formulas for

$$m_k^{\mathcal{C}^{tr,\Pi}_{unram,0}(Y)} : \otimes_{0\leq i\leq k} Hom(X_i, X_{i+1}) \to Hom(X_0, X_k)[2-k]$$

when the spaces of morphisms are identified via the maps $\nu(X_i, X_j)$.

Therefore A_∞-pre-categories $\mathcal{C}^{tr,\Pi}_{unram,0}(Y)$ and $FO(X^\vee)$ are equivalent. On the other hand, it follows from Sections 6.4, 6.6 that $\mathcal{C}_{unram}(Y)$ and $\mathcal{C}^{tr,\Pi}_{unram,0}(Y)$ are also equivalent. Finally, applying the functor F, we obtain the following theorem.

Theorem 4 *The full subcategory $F(\mathcal{C}_{unram}(Y))$ of $C(Y)$ is A_∞-equivalent to $FO(X^\vee)$.*

This is our version of Homological Mirror Conjecture for abelian varieties.

Remark 21 *If we endow the torus $Y = \mathbb{R}^n/\mathbb{Z}^n$ with a flat metric and consider only flat Lagrangian subtori in $X^\vee$ then all higher compositions in the A_∞-pre-category $FO(X^\vee)$ can be written in terms of explicit "truncated theta series" analogous to those considered in [Ko] and [P1] in the case of elliptic curves.*

9 Appendix: constructions in the case of complex numbers

In the previous section we considered algebraic and analytic varieties over the complete local non-archimedean field $\mathbb{C}_\varepsilon$. In this section we explain our approach in the case of complex numbers (i.e. we will assume that ε is a fixed positive number). We should warn the reader that it is not yet clear how to obtain rigorous proof of the Homological Mirror Conjecture in this case. In particular, it is not known how to prove convergence of the series defining compositions in the Fukaya category. Nevertheless we will discuss the complex case because the geometry is more transparent. For example, one can construct the mirror symmetry functor on objects by means of the real version of Fourier-Mukai transform (see Section 9.1). From the point of view of main body of present paper, the Appendix can be treated as a geometric motivation. For this reason we will not stress that X is an abelian variety, but will be using our conjectures about the collapse, and the assumption that the base Y of the torus fibration is a smooth manifold with integral affine structure and Kähler potential. We will be also using the notation from Section 3.

9.1 *Mirror symmetry functor on objects over $\mathbb{C}$*

In the case of complex numbers the mirror symmetry functor assigns a holomorphic vector bundle $F(L,\rho)$ on $X = X_\varepsilon$ to a pair (L,ρ), where $L \subset X^\vee$

is a Lagrangian submanifold, such that the projection $p^{\vee}_{|L} : L \to Y$ is an unramified covering, and ρ is a local system on L. If L is a section of $p^{\vee}$, and $rank(\rho) = 1$, then $E = F(L, \rho)$ is a line bundle. In general, E can be locally represented as a sum $E \simeq \oplus_{\alpha \in A} E_{\alpha}$ where A is the set of leaves (i.e. connected components) of the covering $L \to Y$, and E_{α} is a holomorphic vector bundle of the rank equal to the rank of ρ on the leaf α.

The following explicit construction of the mirror symmetry functor on objects is not new, see e.g. [AP]. We start with the remark that there is a canonical $U(1)$-bundle on $X \times_Y X^{\vee}$ (Poincaré line bundle). It will be denoted by P. It admits a canonical connection, which will be described below . Let us fix $y \in Y$. Then $p^{-1}(y) \simeq T_{Y,y}/\varepsilon T^{\mathbb{Z}}_{Y,y}$ and $(p^{\vee})^{-1}(y) \simeq T^{*}_{Y,y}/(T^{\mathbb{Z}}_{Y,y})^{\vee}$. We identify torus $(p^{\vee})^{-1}(y)$ with the moduli space of $U(1)$-local systems on the torus $p^{-1}(y)$ trivialized over a point $0 \in p^{-1}(y)$. We define $U(1)$-bundle P to be the tautological bundle on $X \times_Y X^{\vee}$ corresponding to this description.

In order to describe the connection on P let us consider the fiberwise universal coverings $r : T_Y \to T_Y/T^{\mathbb{Z}}_Y$ and $r^{\vee} : T^{*}_Y \to T^{*}_Y/(T^{\mathbb{Z}}_Y)^{\vee}$. Then the pullback $\bar{P}$ of P to $T_Y \times_Y T^{*}_Y$ is canonically trivialized. Thus we can work in coordinates. Let $y = (y_1, ..., y_n)$ be coordinates on Y, $x = (x_1, ..., x_n)$ and $x^{\vee} = (x^{\vee}_1, ..., x^{\vee}_n)$ be coordinates on the fibers of $T_Y \to Y$ and $T^{*}_Y \to Y$ respectively. Deck transformations $x_j \mapsto x_j + \varepsilon n_j$, $n_j \in \mathbb{Z}$ act on $\bar{P}$ preserving the trivialization, and transformations $x^{\vee}_j \mapsto x^{\vee}_j + n^{\vee}_j$, $n^{\vee}_j \in \mathbb{Z}$ act on $\bar{P}$ by the multiplication by $exp(2\pi i/\varepsilon \sum_j n^{\vee}_j x_j)$.

Let ∇_0 be the trivial connection on $\bar{P}$. We consider the connection $\bar{\nabla}$ on $\bar{P}$ which is given by the following formula

$$\bar{\nabla} = \nabla_0 + 2\pi i/\varepsilon \sum_{1 \le j \le n} x^{\vee}_j dx_j.$$

Lemma 3 *The connection $\bar{\nabla}$ gives rise to a connection on P.*

Proof. Obviously, connection $\bar{\nabla}$ does not change under the transformation $x_j \mapsto x_j + \varepsilon n_j, n_j \in \mathbb{Z}$. The transformation $x^{\vee}_j \mapsto x^{\vee}_j + n^{\vee}_j, n^{\vee}_j \in \mathbb{Z}$ together with the gauge transformation of $\bar{\nabla}$ by $h = exp(2\pi i/\varepsilon \sum_j n^{\vee}_j x_j)$ also preserves $\bar{\nabla}$. This proves the Lemma. ∎

Let (L, ρ) be as above. The mirror symmetry functor assigns to it a holomorphic vector bundle $E = F(L, \rho)$ such that (in coordinates) its fiber over a point (y, x) is given by the formula $E(y, x) = \oplus_{\{x^{\vee} \in L, p^{\vee}(x^{\vee}) = y\}} \rho(x^{\vee}) \otimes P(x, x^{\vee})$. This vector bundle carries the induced connection ∇_E. In the case of unitary ρ the bundle E carries also a natural hermitean metric.

Proposition 15 *The $(0,2)$-part of the curvature $curv(\nabla_E)$ is trivial. In par-*

ticular, ∇_E *is a holomorphic connection.*

Proof. It follows from the fact that L is Lagrangian. Indeed, let us lift L to T^*_Y. Then locally in a neighborhood of a connected component of L, one can find a smooth real function $f = f(y)$ such that $L = df$. We can write the local equation for L: $x^\vee_j = \partial f/\partial y_j, 1 \leq j \leq n$. The connection ∇_E can be locally written as $\nabla_{E,0} + id_E \otimes (2\pi i/\varepsilon \sum_j \partial f/\partial y_j dx_j)$, where $\nabla_{E,0}$ is the trivial flat connection on the vector bundle E. Since the holomorphic coordinates on T_Y are given by $z_j = y_j + ix_j, i = \sqrt{-1}$, one sees that the $(0,2)$-part of the curvature is equal to $curv(\nabla_E)^{(0,2)} = const \times (\sum_{j,k} \partial^2 f/\partial y_j \partial y_k d\bar{z}_j d\bar{z}_k) = 0$. The Proposition is proved. ■

Definition 23 *For any two holomorphic vector bundles* E_1 *and* E_2 *on* X, *we define* $Hom_{Dolb}(E_1, E_2) = \Omega^{0,*}(X, Hom(E_1, E_2))$.

We consider the space of Dolbeault differential forms with values in the vector bundle $Hom(E_1, E_2)$ as a dg-algebra with respect to the $\bar{\partial}$-differential. In this way one gets a structure of A_∞-category (in fact a dg-category) on the derived category of coherent sheaves on X. One can show that this A_∞-structure is equivalent to the one mentioned in the main text.

9.2 Sectors in the space of Dolbeault forms

Let $E_i = F(L_i, \rho_i), i = 1, 2$ be holomorphic vector bundles as above. There is an analog of the dg-category $\mathcal{C}(Y)$ in the case of complex numbers. We will denote it by $\mathcal{A}(Y)$. Objects of $\mathcal{A}(Y)$ are holomorphic vector bundles on X of the type $E = F(L, \rho)$. Morphisms are sections of soft sheaves on Y. Namely, we define the sheaf $\underline{Hom}_{\mathcal{A}(Y)}(E_1, E_2)$ on Y as the direct image $p_*(\underline{Hom}_{Dolb}(E_1, E_2))$ (in the self-explained notation). Then $Hom_{\mathcal{A}(Y)}(E_1, E_2)$ are global sections of this sheaf. This sheaf corresponds to the sheaf
$\underline{Hom}_{\mathcal{C}(Y)}(E_1, E_2)$ in the non-archimedean geometry. Let us choose an open affine chart $U \subset Y, U \simeq \mathbf{R}^n$. Then $\Gamma(U, \underline{Hom}_{\mathcal{A}(Y)}(E_1, E_2))$ contains a subsheaf of finite Fourier sums with respect to the natural action of the torus T^n on $\Gamma(U \times T^n, \underline{Hom}_{Dolb}(E_1, E_2))$. Thus we have the sheaf $\underline{Hom}^{alg}_{\mathcal{A}(Y)}(E_1, E_2)$ which is an analog of the sheaf $\underline{Hom}^{alg}_{\mathcal{C}(Y)}(E_1, E_2)$ considered in the non-archimedean case. Notice that there exists a natural homomorphism of sheaves $j : p^*(\underline{\Omega}^*_Y) \to \underline{\Omega}^{0,*}_X$. The image of j consists of Dolbeault forms on X which have coefficients locally constant along fibers of p. In local coordinates j is given by the formula $f_{i_1,\ldots,i_n}(y) dy_{i_1} \wedge \ldots \wedge dy_{i_n} \mapsto f_{i_1,\ldots,i_n}(y) d\bar{z}_{i_1} \wedge \ldots \wedge d\bar{z}_{i_n}$, where $\bar{z}_k = y_k - \sqrt{-1} x_k, 1 \leq k \leq n$. It is easy to see that j is compatible with the structure of dg-algebras on de Rham and Dolbeault forms. Thus for

a pair of holomorphic vector bundles E_1 and E_2 on X we have a canonical structure of dg-module over $\underline{\Omega}_Y^*$ on the sheaf $\underline{Hom}_{\mathcal{A}(Y)}(E_1, E_2)$. In the case when $E_i = F(L_i, \rho_i), i = 1, 2$ the subsheaf $\underline{Hom}^{alg}_{\mathcal{A}(Y)}(E_1, E_2)$ is also a sheaf of dg-modules over $\underline{\Omega}_Y^*$.

As in the non-archimedean case there is a canonical decomposition of the stalk $\underline{Hom}^{alg}_{\mathcal{A}(Y)}(E_1, E_2)_y, y \in Y$ into the direct sum of dg-modules of finite rank over $\underline{\Omega}_Y^*$. Summands are labeled by the homotopy classes $[\gamma] \in P(L_1, L_2, y)$ and called *sectors*. We will denote them by $\underline{Hom}^{alg,[\gamma]}_{\mathcal{A}(Y)}(E_1, E_2)_y$. Informally, sectors correspond to "Fourier components" of Dolbeault forms in $\underline{Hom}^{alg}_{\mathcal{A}(Y)}(E_1, E_2)_y$ in the direction of torus fibers. Let us describe them more explicitly. For simplicity we will assume that $\rho_i, i = 1, 2$ are rank one trivial local systems, and $L_i, i = 1, 2$ intersect with each fiber of p at exactly one point. Then near $p^{-1}(y)$ we can write $L_i = graph(df_i)\,(mod(T_Y^{\mathbf{Z}})^{\vee}), i = 1, 2$, where f_i are germs at y of smooth functions on Y. ¿From the description of the Poincaré bundle P we deduce that $\underline{Hom}^{alg}_{\mathcal{A}(Y)}(E_1, E_2)_y$ is canonically identified with the space of germs of $\overline{\partial}$-forms near $T^n = p^{-1}(y)$, endowed with the twisted differential $\overline{\partial}'\alpha = \overline{\partial}\alpha + \frac{i}{\varepsilon}\sum \partial f/\partial y_i d\overline{z}_i \wedge \alpha$, where $f = f_1 - f_2$. Then the sector corresponding to a path γ consists of Dolbeault forms $\alpha = \sum_{i_1,\dots,i_n} exp(i\langle m, x\rangle/\varepsilon) f_{i_1\dots i_n}(y) d\overline{z}_1 \wedge \dots \wedge d\overline{z}_n$. Here vector $m = m(\gamma)$ is the homotopy class of the loop in $T^n = p^{-1}(y)$ which is the composition of three paths:

1) the path $[0,1] \to T^n, t \mapsto t(df_1)_y\, mod\,(T_Y^{\mathbf{Z}})^{\vee}$;
2) the path γ;
3) the path $[0,1] \to T^n, t \mapsto (1-t)(df_2)_y\, mod\,(T_Y^{\mathbf{Z}})^{\vee}$.

A choice of sector corresponds to the choice of monomial $z_1^{i_1}\dots z_n^{i_n}$ in the non-archimedean case. Homotopy classes of paths in non-archimedean approach correspond to summands of Fourier series. Locally each sector can be identified with the de Rham complex on Y. Namely, to a form $\alpha = \sum_{i_1,\dots,i_n} f_{i_1\dots i_n}(y) exp(\langle m, x\rangle) d\overline{z}_{i_1} \wedge \dots \wedge d\overline{z}_{i_n}$ we assign the form $\alpha_m = \sum_{i_1,\dots,i_n} f_{i_1\dots i_n}(y) exp(\frac{1}{\varepsilon}\langle m, x\rangle) dy_{i_1} \wedge \dots dy_{i_n}$, where $m = m(\gamma)$ defines the sector. It is easy to see that the differential $\overline{\partial}'$ on Dolbeault forms on X corresponds to the de Rham differential d on $\Omega^*(Y)$. In this way we obtain an isomorphism of complexes $\underline{Hom}^{alg,[\gamma]}_{\mathcal{A}(Y)}(E_1, E_2)_y \simeq \underline{\Omega}^*_{Y,y} \otimes \mathbf{C}$.

Remark 22 *When ε is not a fixed number, but a parameter $\varepsilon \to 0$, the coefficients $f_{i_1\dots i_n}(y)$ are asymptotic series in ε of the type $f_{i_1\dots i_n}(y, \varepsilon) = \sum_{j\geq 1} exp(-\lambda_j/\varepsilon) f_{j,i_1\dots i_n}$ where $\lambda_j \in \mathbf{R}$, $\lambda_1 < \dots < \lambda_j < \dots$, and $\lambda_j \to +\infty$.*

The set of exponents appearing in the expansion of α_m at y corresponds to the spectrum $Sp_y(\alpha)$ considered in the non-archimedean case.

9.3 Semigroup φ^t

Now we can define a semigroup $\varphi^t : Hom_{Dolb}(E_1, E_2) \to Hom_{Dolb}(E_1, E_2), 0 \leq t < +\infty$. This is an analog of the semigroup ϕ^t in the non-archimedean case. First, we identify the sector $\underline{Hom}^{alg,[\gamma]}_{\mathcal{A}(Y)}(E_1, E_2)_y$ with $\underline{\Omega}^*(Y, Hom(\rho_1, \rho_2))_y$ as above. Let us recall from the non-archimedean part, that to the homotopy class of a path γ we canonically associated a closed 1-form $\mu_\gamma = \int_\gamma \omega$, where ω is the symplectic form on $X^\vee$. Using the Riemannian metric g_Y on Y we assign to μ a vector field ξ_γ on Y. In a local trivialization it is given by $grad((f_1 - f_2 + \langle m(\gamma), \cdot \rangle)/\varepsilon)$. Then the infinitesimal action of φ^t is defined as the Lie derivative Lie_{ξ_γ}. Different Fourier components (sectors) move on Y with different speeds in different directions. Hence the picture is more complicated than in the case of Morse theory.

One can show that the generator $\Delta = \frac{d}{dt}|_{t=0}\varphi^t$ is a second order differential operator on $Hom_{Dolb}(E_1, E_2)$. When g_Y is a flat metric and $f_1 = f_2 = 0$ one can find the following explicit formula for Δ:

$$\Delta = i \sum_j \frac{\partial^2}{\partial x_j \partial y_j} - \frac{1}{\varepsilon} \sum_j \frac{\partial^2}{\partial x_j^2}.$$

It seems plausible that there is an extension of the semigroup $\varphi^t = e^{t\Delta}$, $t \geq 0$, from $Hom^{alg}_{\mathcal{A}(Y)}(E_1, E_2)$ to the whole space of morphisms $Hom_{\mathcal{A}(Y)}(E_1, E_2)$. Notice that Δ is not self-adjoint, and its real part is not elliptic. Nevertheless, we expect that the semigroup operator φ^t converges as $t \to +\infty$ to a "projector" as in the case of Morse theory.

References

[AM] P.S. Aspinwall, D. Morrison, String Theory on K3 surfaces, hep-th/9404151.

[AP] D. Arinkin, A. Polishchuk, Fukaya category and Fourier transform, math.AG/9811023.

[Be] V. Berkovich, Spectral theory and analytic geometry over non-Archimedean fields. Mathematical Surveys and Monographs, 33. Amer. Math. Soc., 1990.

[BGR] S. Bosch, U. Günter, R. Remmert, Non-archimedean analysis. Springer-Verlag, 1984.

[BL] S. Bosch, W. Lütkebohmert, Degenerating abelian varieties, Topology 30 (1991), no. 4, 653-698.

[BZ] J-M. Bismut, W. Zhang, An extension of a theorem by Cheeger and Müller, Asterisque 205 (1992).

[CC] J. Cheeger, T.H. Colding, On the structure of spaces with Ricci curvature bounded below. I. J. Differential Geom. 46 (1997), no. 3, 406-480.

[CG] J. Cheeger, M. Gromov, Collapsing Riemannian manifolds while keeping their curvature bounded, Parts I and II, J. Diff. Geom., 23 (1986), 309-346 and 32 (1990), 269-298.

[CY] S.-Y. Cheng, S.-T. Yau, The real Monge-Ampere equation and affine flat structures, Proceedings of the 1980 Beijing Symposium on Differential Geometry and Differential Equations, vol.1, Science Press, Beijing 1982, 339-370.

[De] P. Deligne, Local behavior of Hodge structures at infinity. Mirror symmetry, II, 683-699, AMS/IP Stud. Adv. Math., 1, Amer. Math. Soc., Providence, RI, 1997

[Fu 1] K. Fukaya, Morse homotopy and its quantization, AMS/IP Studies in Adv. Math., 2:1 (1997), 409-440.

[Fu2] K. Fukaya, A_∞-category and Floer homologies. Proc. of GARC Workshop on Geometry and Topology'93 (Seoul 1993), p. 1-102.

[FuO] K. Fukaya, Y.G. Oh, Zero-loop open string in the cotangent bundle and Morse homotopy, Asian Journ. of Math., vol. 1(1998), p. 96-180).

[FuOOO] K. Fukaya, Y.G. Oh, H. Ohta, K. Ono, Lagrangian intersection Floer theory-anomaly and obstruction. Preprint, 2000.

[G] M. Gromov, Metric structures for Riemannian manifolds, (J. Lafontaine and P. Pansu, editors), Birkhäuser, 1999.

[Gaw] K. Gawedzki, Lectures on Conformal Field Theory, in Mathematical Aspects of String Theory, AMS, 2000.

[Go] A. Goncharov, Multiple zeta-values, Galois groups, and geometry of modular varieties, math.AG/0005069

[GS] V. Gugenheim, J. Stasheff, On perturbations and A_∞-structures, Bul. Soc. Math. Belg. A38(1987), 237-246.

[GW] M. Gross, P. Wilson, Large complex structure limits of K3 surfaces, math.DG/0008018.

[H] N. Hitchin, The moduli space of special Lagrangian submanifolds, math.DG/9711002.

[HL] F. Harvey, B. Lawson, Finite volume flows and Morse theory, Ann. Math. 153 (2001), 1-25.

[Ko] M. Kontsevich, Homological algebra of mirror symmetry. Proc. ICM Zuerich, 1994, alg-geom/9411018.

[KoS] M. Kontsevich, Y. Soibelman, Deformation theory, (book in preparation).

[KoS1] M. Kontsevich, Y. Soibelman, Deformations of algebras over operads and Deligne conjecture, math.QA/0001151.

[Le] N. Leung, Mirror symmetry without corrections, math.DG/0009235.

[LTY] B. Lian, A. Todorov, S-T. Yau, Maximal Unipotent Monodromy for Complete Intersection CY Manifolds, math.AG/0008061.

[M] M. Markl, Homotopy Algebras are Homotopy Algebras, math.AT/9907138.

[Me] S. Merkulov, Strong homotopy algebras of a Kähler manifold, math. AG/9809172.

[Mo] D. Morrison, Compactifications of moduli spaces inspired by mirror symmetry, Journées de Géométrie Algébrique d'Orsay (Juillet 1992), Astérisque, vol. 218, 1993, pp. 243-271, also alg-geom/9304007.

[Mum] D. Mumford, An analytic construction of degenerating abelian varieties over complete rings. Compositio Math. 24:3 (1972), 239-272.

[P1] A. Polishchuk, A_∞-structures on an elliptic curve, math.AG/0001048.

[Se] P. Seidel, Vanishing cycles and mutations. math.SG/0007115.

[SYZ] A. Strominger, S-T. Yau, E. Zaslow, Mirror symmetry is T-duality, hep-th/9606040.

[W1] E. Witten, Supersymmetry and Morse theory, J. Diff. Geom., v. 17 (1982), 661-692.

Addresses:
M.K.: IHES, 35 route de Chartres, F-91440, France
maxim@ihes.fr

Y.S.: Department of Mathematics, KSU, Manhattan, KS 66506, USA
soibel@math.ksu.edu

GENUS-1 VIRASORO CONJECTURE ON THE SMALL PHASE SPACE

XIAOBO LIU

Department of Mathematics, University of Notre Dame, Notre Dame, IN 46556, USA

E-mail address: xliu3@nd.edu

The Virasoro conjecture proposed by Eguchi-Hori-Xiong and S. Katz predicts that the generating function of Gromov-Witten invariants is annihilated by infinitely many differential operators which form a half branch of the Virasoro algebra (cf. [7], [8], and [2]). As proved in [16], the genus-0 part of this conjecture follows from the genus-0 topological recursion relation (See [5] and [10] for alternative proofs). However the genus-1 part of this conjecture seems to be more complicated. The topological recursion relation is no longer sufficient for this case. Using an equation proved in [9], the genus-1 Virasoro conjecture for manifolds with semisimple quantum cohomology was proved in [5]. For arbitrary projective varieties, the genus-1 Virasoro conjecture is reduced to a single equation on the small phase space in [14]. Despite these progresses, a complete solution to the genus-1 Virasoro conjecture is still not available. We will discuss some open problems about the genus-1 Virasoro conjecture in this paper. Some of the results in this paper were announced in [14].

The study of the genus-1 Virasoro conjecture will be important to the understanding of the scope of the entire Virasoro conjecture. Although the genus-0 Virasoro conjecture holds for all compact symplectic manifolds, there are some non-trivial compatibility conditions which must be satisfied for the genus-1 Virasoro conjecture. Such compatibility conditions can be phrased in terms of the quantum cohomology of the underlying projective variety (see Section 2). There are strong evidence that similar compatibility conditions also exist for the higher genera Virasoro conjecture [15]. Because of such non-trivial compatibility conditions, at this moment, it is not clear whether the Virasoro conjecture holds for all projective varieties. A complete answer to this problem would largely depend on what could be discovered for the genus-1 case.

We would also like to make a comparison between the method used in [5] and the one used in this paper and [14]. The method used in [5] was based on the work in [4] where an explicit solution of an equation in [9] was given under the assumption that the quantum cohomology of the underlying manifold is semisimple. If the quantum cohomology is not semisimple or if the genus is bigger than 1, it seems very hard to find such kind of explicit forms of the potential functions. The method used in this paper and in [14] is based on the global tensor analysis on the phase spaces. No explicit solutions of the potential functions are needed. Therefore it could be easily adapted to study the higher genera Virasoro conjecture.

1 Genus-1 Virasoro conjecture on the small phase space

Using the genus-1 topological recursion relation, one can prove that the genus-1 Virasoro conjecture can be completely reduced to a problem on the small phase space, which is canonically identified with the space of cohomology classes of the underlying manifold (cf. [14]). The quantum cohomology defines a Frobenius manifold structure on the small phase space (cf. [3]). Let E be the Euler vector field on this Frobenius manifold and E^k the k-th quantum power of E, i.e. $E^k := \underbrace{E \bullet \cdots \bullet E}_{k}$, where $\bullet$ denotes the quantum product. Here we use the convention that E^0 is the constant vector field corresponding to the identity of the cohomology ring and $E^1 = E$. On the small phase space, the genus-1 Virasoro conjecture has the following simple form:

$$E^k F_1 = \phi_k \tag{1}$$

for all $k \geq 0$, where F_1 is the generating function of the genus-1 primitive Gromov-Witten invariants (this is the F_1^s in [14]) and ϕ_k is a function defined by genus-0 data (see formula (2)). For each k, equation (1) is the restriction of the genus-1 L_{k-1} constraint to the small phase space.

To define ϕ_k explicitly, we first choose a basis $\{\gamma_\alpha \mid \alpha = 1, \ldots, N\}$ of the space of cohomology classes $H^*(V;\mathbb{C})$ of the underlying manifold V. For simplicity, we assume $H^{\mathrm{odd}}(V;\mathbb{C}) = 0$. We choose the basis in such a way that γ_1 is the identity of the cohomology ring and $\gamma_\alpha \in H^{p_\alpha, q_\alpha}(V)$ for some integers p_α and q_α. Let $\{t^1, \ldots, t^N\}$ be the coordinates on $H^*(V;\mathbb{C})$ with respect to this basis. We can identify each γ_α with the vector field $\frac{\partial}{\partial t^\alpha}$ and further identify each cohomology class with a constant vector field. Let $\mu_\alpha = p_\alpha - d/2$ where d is the complex dimension of V. Then

$$E = c_1(V) + \sum_\alpha (\mu_1 + 1 - \mu_\alpha) t^\alpha \gamma_\alpha.$$

Let U be the matrix representing the quantum product by the Euler vector field, i.e. its entries are defined by $E \bullet \gamma_\alpha = \sum_\beta U_\alpha^\beta \gamma_\beta$. Let H be the column vector representing the vector field $\sum_{\alpha,\beta} \eta^{\alpha\beta} \gamma_\alpha \bullet \gamma_\beta$ where $(\eta^{\alpha\beta})$ is the inverse matrix of the intersection pairing η, i.e. $\eta_{\alpha\beta} = \int_V \gamma_\alpha \cup \gamma_\beta$. Let $\mu = \mathrm{diag}(\mu_1, \mu_2, \ldots, \mu_N)$. Then ϕ_k is defined by

$$\begin{aligned}
\phi_0 &:= 0.\\
\phi_1 &:= -\frac{1}{24}\int_V c_1(V) \cup c_{d-1}(V).\\
\phi_k &:= -\frac{1}{24}\sum_{m=0}^{k-1}(1,0,\ldots,0)U^m \mu U^{k-1-m}\eta H\\
&\quad -\frac{1}{4}\sum_{m=0}^{k-1}\mathrm{trace}\left(U^m \mu U^{k-1-m}\mu\right) \qquad (2)
\end{aligned}$$

for $k \geq 2$.

For $k = 0$ and 1, equation (1) follows from the puncture equation and divisor equation respectively. The following result was proved in [14]

Theorem 1.1 *The following statements are equivalent:*

(a) The genus-1 Virasoro conjecture is true.

(b) $E^k F_1 = \phi_k$ *for all* k.

(c) $E^k F_1 = \phi_k$ *for one* k *with* $k \geq 2$.

In particular, the genus-1 Virasoro conjecture can be reduced to $E^2 F_1 = \phi_2$. Besides the genus-1 topological recursion relation, in the proof of this theorem, we also used an equation discovered in [9]. In [14], we showed that this equation can be written in the following global form:

$$\begin{aligned}
&\sum_{g \in S_4} \{v_{g(1)} \bullet v_{g(2)}\}\{v_{g(3)} \bullet v_{g(4)}\}F_1 - 4v_{g(4)}\{v_{g(1)} \bullet v_{g(2)} \bullet v_{g(3)}\}F_1\\
&\qquad -6\left\{[\{v_{g(1)} \bullet v_{g(2)}\}, v_{g(3)}] \bullet v_{g(4)}\right\} F_1\\
&= \text{genus } 0 \text{ data} \qquad (3)
\end{aligned}$$

where v_i are arbitrary vector fields on the small phase space and S_4 is the permutation group of 4 elements. It is apparent that the power of this equation depends strongly on the structure of the quantum cohomology. For example, when the quantum cohomology is semisimple, there exists a canonical coordinate system whose coordinate vector fields behave extremely well under both Lie bracket (they are commutative) and quantum product (they are idempotents). Applying the above formula for such vector fields, one can express all second order partial derivatives of F_1 with respect to the canonical coordinates in terms of genus-0 data. Using the fact EF_1 is constant, one can further obtain all the first order partial derivatives of F_1. In fact, it was proved in [4] that when the quantum cohomology is semisimple, F_1 can be solved in terms of the tau function of the isomonodromy deformation associated to the

Frobenius structure defined by the quantum product. This leads to the proof of the genus-1 Virasoro conjecture under the semisimplicity assumption in [5]. For general cases, the canonical coordinate system does not exist. However, we still have a sequence of vector fields, i.e. $\{E^k \mid k \geq 0\}$ which behave reasonably well under quantum product and Lie bracket, i.e.

$$[E^k, E^m] = (m-k)E^{k+m-1}. \tag{4}$$

Applying equation (3) to these vector fields, we obtain (cf. [14]):

$$E^k F_1 = \frac{k}{2} E^{k-1}(E^2 F_1) + \text{genus } 0 \text{ data} \tag{5}$$

for $k \geq 0$. In other words, $E^k F_1$ can be computed in terms $E^2 F_1$ and genus-0 data. This leads to the proof of Theorem 1.1. However at this moment, we don't know whether equation (3) is sufficient to solve $E^2 F_1$ from genus-0 data for an arbitrary projective variety. If it is not, does it mean that there exist other universal relations for genus-1 Gromov-Witten invariants or the genus-1 Virasoro conjecture may not hold for some projective varieties? In our opinion, this is the most interesting open problem related to the genus-1 Virasoro conjecture.

One can also consider to apply equation (3) to E^k with $k < 0$. In fact, by the quasi-homogeneity equation,

$$E^k U = U^k + [\mu, U^k]$$

for all $k \geq 0$ (cf. [14]). In particular, $E^0 U$ is the identity matrix. This implies that at generic points, U is invertible and therefore E^k with $k < 0$ are well defined near such points and the above formula holds for all k. Then using product formula for derivation, one obtains that

$$E^k U^m - E^m U^k = (m-k)U^{m+k-1}$$

for all integers m and k. This implies that equation (4) also holds for negative integers m and k. Applying equation (3) to this larger sequence of vector fields and using the method similar to the one in [14], we can prove that for all integers m and k

$$E^k E^m F_1 - \frac{m(m-1)}{m+k-1} E^{m+k-1} F_1 = \text{genus } 0 \text{ data}$$

if $m + k \neq 1$, and

$$E^{m+1} E^{-m} F_1 - \frac{m(m+1)}{2} E^{-1} E^2 F_1 = \text{genus } 0 \text{ data.}$$

However, it seems that we still can not solve $E^2 F_1$ from these equations.

2 Compatibility of functions ϕ_k

We now turn to some necessary conditions for genus-1 Virasoro conjecture. As discussed before, genus-1 Virasoro conjecture implies $\phi_k = E^k F_1$ for all $k \geq 0$. For these equations to be true, ϕ_k must be compatible with the behavior of vector fields E_k. This observation produces two constraints for ϕ_k which are not obvious from the definition of these functions. The first constraint is due to the bracket relation (4):

$$E^k \phi_m - E^m \phi_k = \phi_{m+k-1} \tag{6}$$

for all nonnegative integers k and m. We call this constraint the **differential compatibility** of ϕ_k. The second constraint is due to the fact that the small phase space is finite dimensional. There exists $n \geq 0$ such that $\{E^k \mid 0 \leq k \leq n\}$ are linearly independent and there are functions f_i with $i = 0, \ldots, n$ such that

$$E^{n+1} = \sum_{i=0}^{n} f_i E^i. \tag{7}$$

If the genus-1 Virasoro conjecture is true, then we must have

$$\phi_{n+1} = \sum_{i=0}^{n} f_i \phi_i. \tag{8}$$

We call this constraint the **algebraic compatibility** of ϕ_k. Since ϕ_k are defined by purely genus-0 data. These two constraints for ϕ_k are actually constraints on the quantum cohomology of the underlying manifolds. In [14], it was proved that the differential compatibility holds for all projective varieties. However, at this moment, it is not clear whether the algebraic compatibility also holds for all projective varieties. If there exists a projective variety where the algebraic compatibility does not hold, this would give a counter example to the genus-1 Virasoro conjecture.

It is easy to prove the algebraic compatibility for some special cases, e.g. semisimple Frobenius manifolds, quantum cohomology of all curves and K3 surfaces.

Semisimple Frobenius manifolds: Although this follows from the result of [5] since algebraic compatibility is a necessary condition for genus-1 Virasoro conjecture, it is also simple to prove it directly. Let $A_k = \sum_{m=0}^{k-1} U^m \mu U^{k-1-m} \eta$. Then ϕ_k depends linearly on A_k. Therefore we only need to show the algebraic compatibility of A_k. Since $U\eta = \eta U^t$, $A_k = \sum_{m=0}^{k-1} U^m \mu \eta (U^t)^{k-1-m}$. In the semisimple case, the matrix U has distinct eigenvalues. So there exist matrices T and $D = \mathrm{diag}(\lambda_1, \ldots, \lambda_N)$ such that $U = TDT^{-1}$. Therefore

$$A_k = T\left(\sum_{m=0}^{k-1} D^k \Omega D^{k-1-m}\right) T^t$$

where $\Omega = (\Omega_{\alpha\beta}) := T^{-1}\mu\eta(T^{-1})^t$. Matrix Ω is antisymmetric since $\mu\eta$ is antisymmetric. Therefore the matrix $\sum_{m=0}^{k-1} D^k \Omega D^{k-1-m}$ has zeroes on the diagonal and its (α, β) entry for $\alpha \neq \beta$ is given by

$$\sum_{m=0}^{k-1} \Omega_{\alpha\beta} \lambda_\alpha^m \lambda_\beta^{k-1-m} = \Omega_{\alpha\beta} \frac{\lambda_\alpha^k - \lambda_\beta^k}{\lambda_\alpha - \lambda_\beta},$$

which depends linearly on D^k when D is fixed. At each fixed point, any linear relation on $\{E^k \mid k \geq 0\}$ gives linear relation among $\{D^k \mid k \geq 0\}$, which in turn produces linear relation among $\{A_k \mid k \geq 0\}$. This proves the algebraic compatibility of $\{\phi_k \mid k \geq 0\}$.

Curves and K3 surfaces: Rational curves have semisimple quantum cohomology and therefore are included in the above case. For elliptic curves, $E = t^1 E^0$ and $\phi_0 = \phi_1 = 0$. So algebraic compatibility is trivial. All curves with genus bigger than 1 and K3 surfaces have trivial quantum cohomology because of dimension reasons. It is straightforward to check that

$$E^2 = -(t^1)^2 E^0 + 2t^1 E.$$

Moreover for K3 surfaces $\phi_2 = \phi_1 = \phi_0 = 0$ and for a curve of genus-g, $\phi_0 = 0$, $\phi_1 = \frac{g-1}{12}$, and $\phi_2 = \frac{g-1}{6} t^1$. Therefore the algebraic compatibilities are satisfied. We should remark that for curves of genus bigger than 1, we must consider the full cohomology space (including odd dimensional cohomology classes). If we only consider the space of even dimensional cohomology classes, then the algebraic compatibility is no longer true. Actually this is also necessary for the differential compatibility. For all projective varieties, we must consider the full cohomology spaces for the differential compatibility to be true. The reason is that in the proof of differential compatibility (cf. [14]), we have used a formula in [1]. If we do not consider all cohomology classes, this formula would not be true.

Now come back to general cases. For later applications, we need to compute $E^k f_i$. We first observe that, since $E^{k+n+1} = E^k \bullet E^{n+1}$, we have

$$E^{k+n+1} = \sum_{i=0}^{n} f_i E^{k+i}, \tag{9}$$

for every $k \geq 0$. We have the following

Lemma 2.1 (i) $E^0 f_i = -(i+1)f_{i+1}$ for $0 \leq i \leq n-1$, and $E^0 f_n = n+1$.
(ii) $Ef_i = (n+1-i)f_i$ for $0 \leq i \leq n$.
(iii) $E^2 f_0 = f_n f_0$ and $E^2 f_i = (n-i+2)f_{i-1} + f_n f_i$ for $1 \leq i \leq n$.
(iv) *For* $k > 0$,

$$\begin{cases} E^k f_0 = f_0 E^{k-1} f_n, \\ E^k f_i = f_i E^{k-1} f_n + E^{k-1} f_{i-1} \text{ for } 1 \leq i \leq n. \end{cases}$$

Remark 2.2 Lemma 2.1 (i) and (iv) tell us that at each point, $E^m f_j$ is completely determined by the values of $f_0, \ldots, f_n$ at that point.

Proof of Lemma 2.1: We first prove formula (iv). By formulas (4) and (9), for $0 \leq m \leq k$,

$$\begin{aligned} (n+1+2m-k)E^{n+k} &= \left[E^{k-m},\ E^{n+m+1}\right] = \left[E^{k-m},\ \sum_{i=0}^{n} f_i E^{i+m}\right] \\ &= \sum_{i=0}^{n} \left(E^{k-m} f_i\right) E^{i+m} + \sum_{i=0}^{n} f_i \left[E^{k-m}, E^{i+m}\right] \qquad (10) \\ &= \sum_{i=0}^{n} \left(E^{k-m} f_i\right) E^{i+m} + \sum_{i=0}^{n} f_i (i+2m-k) E^{i+k-1}. \end{aligned}$$

Using the fact that $2mE^{k+n} = \sum_{i=0}^{n} 2m f_i E^{i+k-1}$, we obtain

$$\sum_{i=0}^{n} \left(E^{k-m} f_i\right) E^{i+m} = (n+1-k)E^{n+k} - \sum_{i=0}^{n} f_i (i-k) E^{i+k-1}.$$

Since the right hand side of this equation does not depend on m, so does the left hand side. Therefore we have

$$\sum_{i=0}^{n} \left(E^k f_i\right) E^i = \sum_{i=0}^{n} \left(E^{k-m} f_i\right) E^{i+m}, \qquad (11)$$

for all $0 \leq m \leq k$. In the special case $m = 1$, we have

$$\sum_{i=0}^{n} \left(E^k f_i\right) E^i = \sum_{i=0}^{n} \left(E^{k-1} f_i\right) E^{i+1}.$$

Replacing E^{n+1} on the right hand side of this equality by $\sum_{i=0}^{n} f_i E^i$ and using the fact that $\{E^0, \ldots, E^n\}$ are linearly independent, we obtain formula (iv).

Formula (i) is obtained from (10) by setting $k = m = 0$. Formula (ii) and (iii) are obtained by using (i) and the recursion formula (iv). □

Define

$$\psi_k := \phi_{n+1+k} - \sum_{i=0}^{n} f_i\phi_{i+k}.$$

The algebraic compatibility is equivalent to $\psi_0 = 0$. We will show that it also implies $\psi_k = 0$ for all $k \geq 0$. We need the following result

Lemma 2.3 (i) $E^0\psi_k = k\psi_{k-1}$ for $k \geq 1$.
(ii) $E\psi_k = (n+k)\psi_k$ for $k \geq 0$.
(iii) $E^m\psi_k = (n+k-m+1)\psi_{k+m-1} + \sum_{i=1}^{m-1}(E^i f_n)\psi_{k+m-1-i}$ for $k \geq 0$ and $m \geq 2$.

Proof: Since ϕ_0 and ϕ_1 are constant functions, the differential compatibility (6) implies that $E^0\phi_k = \phi_{k-1}$ and $E\phi_k = \phi_k$. The first two formulas then follow from Lemma 2.1 (i) and (ii) respectively.

To prove (iii) for the case $m = 2$, we first apply equation (6) to get

$$E^2\psi_k = E^{k+n+1}\phi_2 + (k+n-1)\phi_{k+n+2} - \sum_{j=0}^{n}\left(E^2 f_j\right)\phi_{k+j}$$
$$-\sum_{j=0}^{n} f_j\left\{E^{k+j}\phi_2 + (k+j-2)\phi_{k+j+1}\right\}.$$

Using Lemma 2.1 (iii) and formula (9), we obtain

$$E^2\psi_k = (k+n-1)\psi_{k+1} + f_n\psi_k.$$

Since $Ef_n = f_n$, this proves (iii) for the case $m = 2$.

On the other hand, since

$$E^2E^kf_n = [E^2, E^k]f_n + E^kE^2f_n = (k-2)E^{k+1}f_n + E^k(2f_{n-1} + f_n^2)$$

and $E^{k+1}f_n = f_nE^kf_n + E^kf_{n-1}$, we have

$$E^2E^kf_n = kE^{k+1}f_n$$

for all k. Using this formula and the fact $E^{m+1} = [E^2, E^m]/(m-2)$, a straightforward induction on m proves (iii). □

Two immediate consequences of Lemma 2.3 are the following

Corollary 2.4 *The following statements are equivalent:*
(a) $\{\phi_k \mid k \geq 0\}$ *are algebraically compatible.*
(b) $\psi_k = 0$ *for one* k.
(c) $\psi_k = 0$ *for all* $k \geq 0$.

Proof: This follows from the fact

$$\psi_k = \frac{1}{k+1} E^0 \psi_{k+1} = \frac{1}{n+k-2}(E^2 \psi_{k-1} - f_n \psi_{k-1}).$$

□

Corollary 2.5 *For all $k \geq 0$,*

$$\psi_{n+1+k} = \sum_{i=0}^{n} f_i \psi_{i+k}.$$

Proof: Since $E^{n+1} = \sum_{i=0}^{n} f_i E^i$, the corollary follows by applying Lemma 2.3 to the obvious relation

$$0 = E^{n+1}\psi_{k+1} - \sum_{i=0}^{n} f_i E^i \psi_{k+1}$$

and using the formula

$$f_j E^{j-i} f_n = E^{j-i+1} f_j - E^{j-i} f_{j-1},$$

which in turn follows from Lemma 2.1 (iv). □

Remark: This corollary gives non-trivial linear relations among functions ϕ_k. But it seems that this is not enough to prove the algebraic compatibility.

3 Some sufficient conditions for genus-1 Virasoro conjecture

Now we come back to the Virasoro conjecture. Let

$$h_k := E^k F_1 - \phi_k.$$

Then $h_0 = h_1 = 0$ and, as proved in [14],

$$E^k \frac{h_m}{m} = (m-1)\frac{h_{m+k-1}}{m+k-1} \tag{12}$$

for all $k \geq 0$ and $m > 0$. By Theorem 1.1, to prove the genus-1 Virasoro conjecture we only need to show that $h_2 = 0$. We first prove the following

Proposition 3.1 *Let*

$$Z_k := \sum_{i=0}^{n} \left(E^k f_i\right) E^i.$$

If the algebraic compatibility condition (8) holds, then $Z_k h_2 = 0$ for all $k \geq 0$.

Remark:

(a) Without assuming the algebraic compatibility, we can still obtain that $Z_k(E^2 F_1)$ can be computed in terms of genus-0 data.

(b) Formula (11) implies that $Z_k = E^k \bullet Z_0$. Therefore at each point, Z_k is determined by the values of f_i at this point.

Proof of Proposition 3.1: Setting $m = k$ in formula (10) and using formula (11), we obtain

$$Z_k = (n + k + 1)E^{n+k} - \sum_{i=0}^{n}(i + k) f_i E^{i+k-1}.$$

Therefore by equation (12),

$$Z_k h_2 = 2h_{n+k+1} - \sum_{i=0}^{n} 2 f_i h_{i+k}.$$

The right hand side of this equality is equal to 0 because of formula (9) and Corollary 2.4. □

Definition 3.2 *We say that a manifold V has* **non-degenerate** *quantum cohomology if at generic points of the small phase space, there exists an integer $m \geq 1$ such that E^m is contained in the linear span of $\{E^0, Z_k \mid k \geq 0\}$.*

An immediate consequence of Proposition 3.1 is the following

Theorem 3.3 *For any manifold V with non-degenerate quantum cohomology, if the algebraic compatibility condition (8) is satisfied , then the genus-1 Virasoro conjecture holds.*

Remark: Without assuming the algebraic compatibility, we can obtain that if the quantum cohomology of V is non-degenerate, then $E^k F_1$ can be computed in terms of genus-0 data for all $k \geq 0$.

Proof of Theorem 3.3: By equation (12), $E^0 h_2 = 0$. If for some positive integer m, E^m is contained in the span of $\{E^0, Z_k \mid k \geq 0\}$, then by Proposition 3.1, $E^m h_2 = 0$. By equation (12),

$$h_{m+1} = \frac{m+1}{2} E^m h_2 = 0.$$

If $m \geq 1$, then repeatedly taking derivatives (by $m - 1$ times) of h_{m+1} along the direction E^0 and using equation (12), we obtain that $h_2 = 0$. The theorem then follows from Theorem 1.1. □

To apply Theorem 3.3, we need to know which manifolds have non-degenerate quantum cohomology. In the rest of this paper, we discuss some sufficient conditions for the non-degeneracy of the quantum cohomology. To this end, it is interesting to know how large is the vector space spanned by $\{Z_k \mid k \geq 0\}$. We first notice that by the remark following Proposition 3.1 and formula (9),

$$Z_{n+1+k} = \sum_{i=0}^{n} f_i Z_{i+k}.$$

Therefore, at each point, $\{Z_k \mid k \geq 0\}$ and $\{Z_k \mid 0 \leq k \leq n\}$ span the same vector space. The following lemma gives us a sense on how large this vector space might be.

Lemma 3.4 *At each point t,*

$$\text{span}\{Z_k(t) \mid 0 \leq k \leq n\} = \text{span}\{E^k(t) \mid 0 \leq k \leq n\}$$

if and only if the polynomial in x

$$p_t(x) = x^{n+1} - \sum_{i=0}^{n} f_i(t) x^i$$

has no multiple roots.

Proof: The derivative of $p_t(x)$ with respect to x is

$$p_t'(x) = (n+1)x^n - \sum_{i=0}^{n-1} (i+1) f_{i+1}(t) x^i.$$

Note that the coefficients of $p_t'(x)$ are the same as the coefficients of $Z_0(t)$. The resultant of polynomials $p_t(x)$ and $p_t'(x)$ is the determinant of the following $(2n+1) \times (2n+1)$ matrix

$$\begin{pmatrix}
1, & -f_n, & -f_{n-1}, & \cdots\cdots, & -f_1 & -f_0, & & & \\
 & 1, & -f_n, & -f_{n-1}, & \cdots\cdots, & -f_1 & -f_0, & & \\
 & & \cdots\cdots & \cdots\cdots & \cdots & \cdots & \cdots\cdots & \cdots & \\
 & & & 1, & -f_n, & -f_{n-1}, & \cdots\cdots, & -f_1 & -f_0 \\
n+1, & -nf_n, & -(n-1)f_{n-1}, & \cdots\cdots, & -f_1, & & & & \\
 & n+1, & -nf_n, & -(n-1)f_{n-1}, & \cdots\cdots, & -f_1, & & & \\
 & & \cdots\cdots & \cdots\cdots & \cdots & \cdots & \cdots\cdots & & \\
 & & & \cdots\cdots & \cdots & \cdots & \cdots\cdots & \cdots & \\
 & & & & n+1, & -nf_n, & -(n-1)f_{n-1}, & \cdots\cdots, & -f_1
\end{pmatrix},$$

where non-zero entries of the first n rows are coefficients of $p_t(x)$ and non-zero entries of the last $n+1$ rows are coefficients of $p_t'(x)$. Performing elementary row transformations, we can transform this matrix to the following form

$$\begin{pmatrix} B & C \\ 0 & A \end{pmatrix},$$

where B is an $n \times n$ upper triangular matrix whose diagonal entries are 1, and $A = (a_{i,j})$, $0 \le i, j \le n$, is an $(n+1) \times (n+1)$ matrix whose entries are given by the recursion formula

$$
\begin{aligned}
&a_{n,0} = n+1, \quad a_{n,j} = -(n-j+1)f_{n-j+1} \text{ for } 1 \le j \le n; \\
&\text{For } 1 \le i \le n, \\
&\quad \begin{cases} a_{n-i,n} = f_0 a_{n-i+1,0}, \\ a_{n-i,j} = f_{n-j} a_{n-i+1,0} + a_{n-i+1,j+1}, \text{ for } 0 \le j \le n-1. \end{cases}
\end{aligned}
$$

Comparing this recursion formula with the recursion formula in Lemma 2.1, we obtain that $a_{i,j} = E^{n-i} f_{n-j}$ for all i and j. Therefore A is the coefficient matrix of representing $\{Z_n, Z_{n-1}, \ldots, Z_0\}$ in terms of $\{E^n, E^{n-1}, \ldots, E^0\}$. Since the determinant of A is equal to the resultant of $p_t(x)$ and $p_t'(x)$, A is invertible if and only if $p_t(x)$ has no multiple roots. This proves the lemma. □

Recall that a manifold has non-degenerate quantum cohomology if there exists one $m > 0$ such that at generic points, E^m is contained in the span of $\{E^0, Z_0, \ldots, Z_n\}$. Observe that if the first n columns of the matrix A in the proof of Lemma 3.4 has rank n, than E^m is contained in the span of $\{E^0, Z_0, \ldots, Z_n\}$ for all $m \ge 0$. Therefore such manifolds have non-degenerate quantum cohomology. However, to compare non-degeneracy with semisimplicity, we only need the following weaker result which corresponds to the case where A has rank $n+1$.

Corollary 3.5 *If at generic points of the small phase space of a manifold V, the polynomial*

$$p_t(x) = x^{n+1} - \sum_{i=0}^{n} f_i(t) x^i$$

has no multiple roots, then the quantum cohomology of V is non-degenerate.

In the case that the quantum cohomology of V is semisimple, at generic points of the small phase space, $\{E^k \mid 0 \le k \le n\}$ form a basis of the tangent space of the small phase space. With respect to this basis, the quantum multiplication by E has the following matrix representation

$$
\begin{pmatrix}
0 & 0 & 0 & \cdots & 0 & f_0 \\
1 & 0 & 0 & \cdots & 0 & f_1 \\
0 & 1 & 0 & \cdots & 0 & f_2 \\
0 & 0 & 1 & \cdots & 0 & f_3 \\
\cdots & \cdots & \cdots & \cdots & \cdots & \cdots \\
0 & 0 & 0 & \cdots & 1 & f_n
\end{pmatrix}.
$$

The polynomial $p_t(x)$ in Corollary 3.5 is precisely the characteristic polynomial of this matrix, and therefore has no multiple roots at semisimple points. Hence we have

Corollary 3.6 *If the quantum cohomology of a manifold V is semisimple, then it is also non-degenerate.*

Remark: This corollary implies that the genus-1 Virasoro conjecture is true for all manifolds with semisimple quantum cohomology (which has been proved in [5]). We also notice that for such manifolds, $\{E^k \mid k \geq 0\}$ span the entire tangent spaces of the small phase space. However for the quantum cohomology to be non-degenerate, the space spanned by $\{E^k \mid k \geq 0\}$ could be much smaller than tangent space. For example, for curves of high genera and K3 surfaces, the dimension of the space spanned by $\{E^k \mid k \geq 0\}$ is at most two. However the small phase spaces could have very large dimensions. From this point of view, non-degeneracy is much weaker than semisimplicity.

Another sufficient condition for the non-degeneracy is the following

Lemma 3.7 *If at every point of the small phase space, the dimension of the vector space spanned by $\{E^k \mid k \geq 0\}$ is less than or equal to 2, then the quantum cohomology is non-degenerate.*

Proof: The case where the dimension of the vector space spanned by $\{E^k \mid k \geq 0\}$ is 1 is trivial since E^k is proportional to E^0 for every $k > 1$. If the dimension of the vector space spanned by $\{E^k \mid k \geq 0\}$ is 2, then $E^2 = f_0 E^0 + f_1 E$ with $E^0 f_1 = 2$ (c.f. Lemma 2.1 (i)). Hence

$$E = \frac{1}{2}\left\{Z_0 - \left(E^0 f_0\right) E^0\right\}.$$

By definition, the quantum cohomology is non-degenerate. □

Remark: Using this simple lemma, we can get examples of quantum cohomologies which are non-degenerate but not semisimple. In fact, as we have seen in Section 2, for all curves and K3 surfaces, the dimension of the span of $\{E^k \mid k \geq 0\}$ are less than or equal to 2. Therefore the quantum cohomologies of these spaces are non-degenerate. However, except rational curves, the quantum cohomologies of these spaces are not semisimple. As an application, we obtain that the genus-1 Virasoro conjecture holds for all curves and K3 surfaces. The genus-1 Virasoro conjecture for elliptic curves is not known before. After this work has been finished, the author was informed that F. Zahariev found a combinatorial proof to the genus-1 Virasoro conjecture for elliptic curves. It seems that there is a gap in the proof of Virasoro conjecture for K3 surfaces in [10] since it omits to verify the genus-1 degree-0 case. Moreover, the proof contained in this paper is technically simpler since it doesn't use deformation invariance of GW invariants.

It would be interesting to find more geometric criteria for a manifold to have non-degenerate quantum cohomology. This is related to the problem: which manifolds have semisimple quantum cohomology (cf. [18] and [19])? So far, we do not have an example of projective variety whose quantum cohomology is not non-degenerate. However, the non-degeneracy is indeed a nontrivial constraint on the Frobenius manifold structure as we can see from the following example.

Example 3.8 As for the semisimplicity, the non-degeneracy can also be defined for an abstract Frobenius manifold in the same way. We consider the Frobenius manifold $M_n := H^*(\mathbb{C}P^n)$ where the Frobenius algebra structure is given by the ordinary cohomology ring structure at every point of M_n (c.f. Example 1.5 in [3]). It is not semisimple since it has nilpotent elements at each point. Let γ be a non-zero element of $H^2(\mathbb{C}P^n)$. Then $\{\gamma^k \mid 0 \leq k \leq n\}$ form a basis of M_n, where $\gamma^k = \underbrace{\gamma \cup \cdots \cup \gamma}_{k}$. We denote the corresponding coordinates by t_k, $0 \leq k \leq n$. (This notation is different from our convention before where superscripts were used instead of subscripts.) The Euler vector field on M_n is given by $E = \sum_{k=0}^{n}(1-k)t_k\gamma^k$. It is straightforward to verify that for $n \leq 3$, the dimension of the vector space spanned by $\{E^k \mid k \geq 0\}$ is less than or equal to 2. Therefore, in this case, M_n is non-degenerate. For $n = 4$ or 5, $E^3 = t_0^3 E^0 - 3t_0^2 E + 3t_0 E^2$. Therefore $Z_0 = 3t_0^2 E^0 - 6t_0 E + 3E^2$ and $Z_k = t_0^k Z_0$ for $k \geq 1$. Therefore M_4 and M_5 are degenerate. Notice that in this example, the polynomial in Corollary 3.5 is of the form $p_t(x) = (x - t_0)^3$. In fact, it is not hard to show that in general, if the dimension of the vector space spanned by $\{E^k \mid k \geq 0\}$ is equal to 3, then the Frobenius manifold is non-degenerate unless $p_t(x) = (x - g(t))^3$ for some function g. For M_6, we have $E^4 = -t_0^4 E^0 + 4t_0^3 E - 6t_0^2 E^2 + 4t_0 E^3$. Therefore $Z_0 = -4t_0^3 E^0 + 12t_0^2 E - 12t_0 E^2 + 4E^3$ and $Z_k = t_0^k Z_0$ for $k \geq 1$. Therefore M_6 is also degenerate.

References

[1] Borisov, L., *On betti numbers and Chern classes of varieties with trivial odd cohomology groups*, (alg-geom/9703023).

[2] Cox, D. and Katz, S., *Algebraic geometry and mirror symmetry*, 1999.

[3] Dubrovin, B., *Geometry of 2D topological field theories*, Integrable systems and quantum groups, Lecture Notes in Math. 1620, Springer, Berlin, 1996, 120-348.

[4] Dubrovin, B., Zhang, Y., *Bihamiltonian hierarchies in 2D topological field theory at one-loop approximation*, to appear in Comm. Math. Phys.,

(hep-th/9712232)

[5] Dubrovin, B., Zhang, Y., *Frobenius manifolds and Virasoro constraints*, (math.AG/9808048)

[6] Eguchi, T., Hori, K., and Xiong, C., *Gravitational Quantum Cohomology*, Int. J. Mod. Phys. A12 (1997) 1743-1782, (hep-th/9605225)

[7] Eguchi, T., Hori, K., and Xiong, C., *Quantum Cohomology and Virasoro Algebra*, Phys. Lett. B402 (1997) 71-80, (hep-th/9703086)

[8] Eguchi, T., Jinzinji, M., and Xiong, C., *Quantum Cohomology and Free Field Representation*, Nucl. Phys. B510 (1998) 608-622, (hep-th/9709152)

[9] Getzler, E., *Intersection theory on $\bar{M}_{1,4}$ and elliptic Gromov-Witten Invariants*, J. Amer. Math. Soc. 10 (1997) 973-998 (alg-geom/9612004)

[10] Getzler, E., *The Virasoro conjecture for Gromov-Witten invariants*, Algebraic Geometry: Hirzebruch 70, 1999, AMS, 147 - 176.

[11] Hertling, C., Manin, Y., *Weak Frobenius manifolds*, (math.QA/9810132)

[12] Hori, K., *Constraints For Topological Strings In $D \geq 1$*, Nucl. Phys. B439 (1995) 395, (hep-th/9411135)

[13] Li, J. and Tian, G., *Virtual moduli cycles and Gromov-Witten invariants of algebraic varieties*, J. Amer. Math. Soc., 11 (1998), 119-174.

[14] Liu, X., *Elliptic Gromov-Witten invariants and Virasoro conjecture*, to appear in Comm. Math. Phys.

[15] Liu, X., in preparation.

[16] Liu, X. and Tian, G., *Virasoro constraints for quantum cohomology*, J. Diff. Geom. 50 (1998), 537 - 591. (math.AG/9806028)

[17] Ruan, Y. and Tian, G., , *A mathematical theory of quantum cohomology*, J. Diff. Geom. 42 (1995), 259 - 367

[18] Tian, G., *Quantum cohomology and its associativity*, Current developments in mathematics, 1995 (Cambridge, MA.), 360 -401, Internat. Press.

[19] Tian, G., Xu, G., *On the semi-simplicity of the quantum cohomology algebras of complete intersections*, Math. Res. Lett. 4 (1997) 481-488. (9611035)

[20] Witten, E., *Two dimensional gravity and intersection theory on Moduli space*, Surveys in Diff. Geom., 1 (1991), 243-310.

OBSTRUCTION TO AND DEFORMATION OF LAGRANGIAN INTERSECTION FLOER COHOMOLOGY

HIROSHI OHTA *
Graduate School of Mathematics, Nagoya University,
Chikusa, Nagoya, 464-8602, Japan,
E-mail: ohta@math.nagoya-u.ac.jp

This is a short survey about obstruction theory to and deformation theory of Lagrangian intersection Floer cohomology, developed in our joint paper [FOOO]. The obstruction to define Lagrangian intersection Floer cohomology is systematically investigated and a system of the obstruction classes are constructed. Also, a filtered A_∞ algebra associated to Lagrangian submanifold is constructed and by using it, deformation of the Lagrangian submanifold and Floer cohomology is described in terms of the notion of filtered A_∞ algebra. Moreover, some applications of our theory to concrete problems in symplectic geometry are discussed.

CONTENTS

*Partially supported by Grant in Aid for Scientific Research no 12640066, the Ministry of Education, Science, Sports and Culture, Japan.

§0. Introduction

In symplectic geometry, there are two kinds of Floer cohomologies. One is the absolute version and the other is the relative version. The absolute version is related to the periodic Hamiltonian systems and the Arnold conjecture for the fixed point sets of the Hamiltonian diffeomorphisms of symplectic manifold. The relative version is related to Lagrangian intersection theory. Our Floer cohomology we will discuss here is the relative one. Roughly speaking, from the point of view of Morse theory, the generators of the Morse cochain complex in the absolute case are the set of fixed points of a Hamiltonian diffeomorphism. The spaces of the gradient trajectories, which are needed to define the coboundary operators, are moduli spaces of J-holomorphic maps from infinite cylinder such that two end points converge to corresponding two fixed points. By the removable singularity theorem for J-holomorphic maps, the space can be regarded as moduli space of J-holomorphic 2-spheres. The fundamental theory of moduli space of J-holomorphic curves *without boundary* is now established and we can define the Floer cohomology in absolute case for general symplectic manifolds. See [FO], [LT], [B] etc. However, in the relative case, we have new problems and difficulties which do not appear in the absolute case. The generators of the cochain complex correspond to intersection points of two Lagrangian submanifolds. To define coboundary operators, we have to study moduli spaces of J-holomorphic maps from infinite strip or disc with Lagrangian boundary condition. In particular, we have to study moduli space of J-holomorphic curves *with boundary*. If we define, as in a usual way, the "coboundary operator" δ in Lagrangian intersection Floer theory by counting the number of certain components of moduli spaces of J-holomorphic discs, δ does *not* satisfy $\delta \circ \delta = 0$ in general. This is essentially because the phenomena that holomorphic disc bubbles off at a point of the boundary of holomorphic disc happens. This is real codimension one phenomena. (See §1). This is the main trouble to overcome. We will study the obstruction to $\delta \circ \delta = 0$ systematically. Moreover, in the case of our obstructions vanish, we will develop a deformation theory of Lagrangian intersection Floer cohomologies. The obstruction and the deformation are described in terms of certain homological algebra, so called A_∞-algebra. Strictly speaking, we introduce and use a notion of *filtered* A_∞-algebra.

§1. Preliminaries and problems to overcome.

Let (M, ω) be a smooth symplectic manifold with real dimension $2n$ and L_0, L_1 closed Lagrangian submanifolds of M. We assume that our Lagrangian submanifold is always orientable. Although it is enough that we assume that L_0 and L_1 intersect cleanly in Bott's sense, we assume here that L_0 and L_1

intersect transversally for simplicity.

First of all, we briefly explain our setting. Consider the path space

$$\Omega(L_0, L_1) = \{\ell : [0,1] \to M \mid \ell(0) \in L_0, \ell(1) \in L_1\}.$$

We choose and fix a base point $\ell_0 \in \Omega(L_0, L_1)$ on each connected component of $\Omega(L_0, L_1)$. We now describe a covering space of the component $\Omega_{\ell_0}(L_0, L_1)$ of $\Omega(L_0, L_1)$ that contains ℓ_0. Consider the set of all pairs (ℓ, w) satisfying:

(1.1.1) $w(0,\cdot) = \ell_0$,
(1.1.2) $w(\tau, 0) \in L_0$, $w(\tau, 1) \in L_1$ for all $0 \leq \tau \leq 1$,
(1.1.3) $w(1,\cdot) = \ell$,

where $w : [0,1] \times [0,1] \to M$. We define an equivalence relation on this set as follows: First, we consider any closed loop

$$c : S^1 \to \Omega_{\ell_0}(L_0, L_1)$$

which will also define a pair of closed loops in L_0 and L_1 for $t = 0, 1$ respectively. Noting that every symplectic vector bundle over S^1 is trivial, the bundle c^*TM over $S^1 \times [0,1]$ is symplectically trivial. Therefore any such trivialization defines two closed loops of Lagrangian subspaces

$$\alpha_0, \alpha_1 : S^1 \to \Lambda(\mathbb{C}^n)$$

by

$$\alpha_0(\tau) = T_{c(\tau,0)}L_0, \qquad \alpha_1(\tau) = T_{c(\tau,1)}L_1,$$

in the trivialization. Here $\Lambda(\mathbf{C}^n)$ denotes the space of all Lagrangian subspaces in $\mathbf{C}^n$. We fix any such trivialization

$$\Psi : c^*TM \to S^1 \times [0,1] \times \mathbf{C}^n$$

and denote by $\mu_\Psi(\alpha_i)$ the Maslov index of the loop α_i in the trivialization Ψ. One can find that the difference

$$\mu_\Psi(\alpha_1) - \mu_\Psi(\alpha_0)$$

is independent of the choice of trivialization Ψ but depends only on the loop c. We denote this common number by $\mu(c)$ and call it the *Maslov index* of the loop c in $\Omega(L_0, L_1)$. It defines an integer valued homomorphism

$$\mu : \pi_1(\Omega_{\ell_0}(L_0, L_1), \ell_0) \to \mathbf{Z}. \tag{1.2}$$

Using (1.2) and the symplectic form ω, we define an equivalence relation $\sim$ on the set of all pairs (ℓ, w) satisfying (1.1). We denote by $w\#w'$ the concatenation of w and w' along ℓ, which will define a loop in $\Omega_{\ell_0}(L_0, L_1)$ based at ℓ_0.

Definition 1.3. We say that (ℓ, w) is equivalent to (ℓ, w') and write $(\ell, w) \sim (\ell, w')$ if the following conditions are satisfied

$$\int_w \omega = \int_{w'} \omega \quad \text{i.e.} \quad \int_{\overline{w}\#w'} \omega = 0 \tag{1.3.1}$$

$$\mu(\overline{w}\#w') = 0 \tag{1.3.2}$$

where $\overline{w}$ is the disc w with the opposite orientation.

We define a covering space of $\Omega_{\ell_0}(L_0, L_1)$ by

$$\widetilde{\Omega}_{\ell_0}(L_0, L_1) = \{(\ell, w) \mid \text{ satisfying (1.1)}\}/\sim.$$

We denote by $[\ell, w]$ the equivalence class of (ℓ, w). Now we define a functional $\mathcal{A} : \widetilde{\Omega}_{\ell_0}(L_0, L_1) \to \mathbf{R}$ by

$$\mathcal{A}([\ell, w]) = \int w^*\omega. \tag{1.4}$$

A simple standard calculation shows that the set of critical points of $\mathcal{A}$ on $\widetilde{\Omega}_{\ell_0}(L_0, L_1)$ are those $[\ell_p, w]$ where $\ell_p : [0, 1] \to M$ is the constant path corresponding to an intersection point $p \in L_0 \cap L_1$. We denote by $Cr_{\ell_0}(L_0, L_1)$ the set of all critical points of

$$\mathcal{A} : \widetilde{\Omega}_{\ell_0}(L_0, L_1) \to \mathbf{R},$$

and put $Cr(L_0, L_1) = \cup_{\ell_0} Cr_{\ell_0}(L_0, L_1)$. We next study the gradient lines of $\mathcal{A}$. As usual, we fix a compatible almost complex structure J on M and consider the induced Riemannian metric $g_J := \omega(\cdot, J\cdot)$. This will in turn induce an L^2-metric on $\widetilde{\Omega}_{\ell_0}(L_0, L_1)$. We now define the moduli space $\widetilde{\mathcal{M}}_J([\ell_p, w], [\ell_q, w'])$ as follows: $\widetilde{\mathcal{M}}_J([\ell_p, w], [\ell_q, w'])$ is the set of maps

$$u : \mathbf{R} \times [0, 1] \to M$$

with

$$u(\mathbf{R} \times \{0\}) \subset L_0, \quad u(\mathbf{R} \times \{1\}) \subset L_1, \tag{1.5.1}$$

(1.5.2) u satisfies

$$\begin{cases} \dfrac{\partial u}{\partial \tau} + J\dfrac{\partial u}{\partial t} = 0 \\ \lim\limits_{\tau \to -\infty} u(\tau, t) = p, \qquad \lim\limits_{\tau \to +\infty} u(\tau, t) = q \end{cases}$$

(1.5.3) $w \# u \sim w'$.

Here $w \# u$ is the obvious concatenation of w and u along the constant path ℓ_p.

From now on, we will suppress J from various notations whenever possible. Then we have the following:

Proposition 1.6. *There exists a map $\mu : Cr(L_0, L_1) \to \mathbf{Z}$ such that the space $\widetilde{\mathcal{M}}([\ell_p, w], [\ell_q, w'])$ has a Kuranish structure of dimension $\mu([\ell_q, w']) - \mu([\ell_p, w])$. We also assume that the pair (L_0, L_1) is relatively spin. Then the space will carry an orientation in the sense of Kuranish structure.*

Remark 1.7. (1) The space $\widetilde{\mathcal{M}}([\ell_p, w], [\ell_q, w'])$ is not a smooth manifold, in general. This trouble comes from the transversality problem. In order to overcome this problem, we have now an established machinery, so called *Kuranish structure* introduced in [FO]. We do not explain the notion of Kuranishi structure here. See [FO] and [FOOO]. When we use Kuranishi structure, the "(virtual) fundamental class" is defined only over $\mathbf{Q}$, not $\mathbf{Z}$. So we can not work over $\mathbf{Z}/2\mathbf{Z}$ coefficient in general. In this sense, we can not avoid the orientation problem. In this note, we do not mention about the transversality problem no more.

(2) The definition of relatively spin will be given in §2. We should note that this space is not always orientable, in general. In the absolute version of Floer cohomology for Hamiltonian diffeomorphism, the corresponding spaces of gradient trajectories (or connecting orbits) which are used to define the coboundary operator are always orientable and have a canonical orientation induced by an almost complex structure. The reason why we can not expect the space of connecting orbits for the Lagrangian intersection Floer cohomology is basically that the almost complex structure does not preserve the Lagrangian boundary condition.

We have an $\mathbf{R}$-action on $\widetilde{\mathcal{M}}([\ell_p, w], [\ell_q, w'])$ defined by the translation along the τ-direction, and put

$$\mathcal{M}([\ell_p, w], [\ell_q, w']) = \widetilde{\mathcal{M}}([\ell_p, w'], [\ell_p, w])/\mathbf{R}.$$

The standard Floer's cochain "complex" $(CF(L_0, L_1), \delta_0)$ (actually this is not a complex, in general) is defined as

Definition 1.8. We assume that the pair (L_0, L_1) is relatively spin.

$$CF^k(L_0, L_1) = \widehat{\bigoplus_{\substack{[\ell_p, w] \in Cr(L_0, L_1), \\ \mu([\ell_p, w]) = k}}} \mathbb{Q}[\ell_p, w] \tag{1.8.1}$$

$$\delta_0[\ell_p, w] = \sum_{\mu[\ell_q, w] = \mu[\ell_p, w]+1} \#(\mathcal{M}([\ell_p, w], [\ell_q, w']))[\ell_q, w']. \tag{1.8.2}$$

Here $\widehat{\bigoplus}$ means an appropriate completion. Since we use the Kuranishi structure of $\mathcal{M}([\ell_p, w], [\ell_q, w'])$, the number in the right hand side of (1.8.2) is a rational number.

For the absolute case of Floer cohomology, similar constructions have been used. However, there is a crucial difference for the case of Lagrangian intersections from the absolute case: The boundary $\partial\mathcal{M}([\ell_p, w], [\ell_q, w']))$ consists of more than

$$\bigcup_{\mu([\ell_r, w'']) = \mu([\ell_p, w])+1} \mathcal{M}([\ell_p, w], [\ell_r, w'']) \times \mathcal{M}([\ell_r, w''], [\ell_q, w']).$$

More precisely, the compactification of $\mathcal{M}([\ell_p, w], [\ell_q, w'])$ has extra codimension one components other than those of "split connecting orbits". The extra components come from bubbling-off discs. From the index formula, we know that bubbling-off spheres are phenomena of real codimension at least two (complex codimension at least one), while bubbling-off discs is of real codimension one in general. Therefore $\delta_0 \circ \delta_0 \neq 0$, in general. Thus we have an obstruction to define Floer cohomology in the relative case.

Thus we can summarize our problems (modulo transversality problems) to overcome as follows.

- **Obstruction problem:** We have to study the obstruction to $\delta_0 \circ \delta_0 = 0$ systematically.
- **Orientation problem:** Find a condition for the moduli space of J holomorphic curves with boundary (Lagrangian boundary condition) to be orientable. Moreover we have to discuss the problems about the orientations on various moduli spaces carefully.

§2. Orientation and obstruction classes.

To state our results, let us introduce some notations. We have two important group homomorphisms from $\pi_2(M, L)$:

$$\mathcal{A} : \pi_2(M, L) \to \mathbf{R}, \quad \text{and} \quad \mu_L : \pi_2(M, L) \to \mathbb{Z}. \tag{2.0}$$

Here $\mathcal{A}$ is defined by $\mathcal{A}(\beta) = \omega(\beta)$ for $\beta \in \pi_2(M, L)$, which is called symplectic area (or energy) and μ_L is called the Maslov index. We can define μ_L in a way similar to μ in §1. We omit the precise definition of μ_L. See [Oh], for example. We note that μ_L is always even when L is orientable.

Definition 2.1. For $\beta \in \pi_2(M, L)$, we denote by $\mathcal{M}_{k+1}(L, \beta)$ the set of all isomorphism classes of genus zero stable J holomorphic maps $w : D^2 \to M$ with $k + 1$ marked points on the boundary ∂D^2 such that $w(\partial D^2) \subset L$ and $[w] = \beta$. We denote by $\mathcal{M}_{k+1}^{\text{main}}(L, \beta)$ the component which corresponds to that the ordering of the marked points are cyclic. We call it a *main component.*

Remark 2.2. As usual, the stability means that the automorphism group of $((D; z_0, \ldots, z_{k+1}), w)$ is finite. Here the automorphism $\varphi : D \to D$ is the biholomorphic map such that $w \circ \varphi = w$ and $\varphi(z_i) = z_i$. Strictly speaking, we need impose extra interior marked points, (for example to handle sphere bubbles), but we omit these points here.

Then we have

Proposition 2.3. *$\mathcal{M}_{k+1}(L, \beta)$ has a Kuranishi structure of real dimension $n + \mu_L(\beta) - 3 + k + 1$. Here $n = \dim L$ and 3 is dimension of $\mathrm{Aut}(D^2) \cong PSL_2(\mathbf{R})$.*

As for the orientability of this space, we have to introduce a notion of relative spin Lagrangian submanifold.

Definition 2.4. (1) An orientable Lagrangian submanifold L in (M, ω) is called *relative spin* if there exists a class $st \in H^2(M; \mathbb{Z}/2\mathbb{Z})$ such that $w_2(TL) \equiv st|_L$ in $H^2(L; \mathbb{Z}/2\mathbb{Z})$. In particular, when L is spin, it is relative spin.

(2) A pair of Lagrangian submanifolds (L_0, L_1) is *relative spin* if there exits a class $st \in H^2(M; \mathbf{Z}/2\mathbf{Z})$ such that $w_2(TL_0) \equiv st|_{L_0}$ and $w_2(TL_1) \equiv st|_{L_1}$ simultaneously.

Now let us state our results. Firstly, we have the following theorem about the orientation problem.

Theorem 2.5. *We denote by $\widetilde{\mathcal{M}}$ the space of all J holomorphic maps $w : D^2 \to M$ with $w(\partial D) \subset L$. Assume that L is relative spin. Then $\widetilde{\mathcal{M}}$ is orientable. The orientation is given by the choice of an orientation of L, $st \in H^2(M; \mathbf{Z}/2\mathbf{Z})$ and a spin structure on $TL \oplus V|_{L^{(2)}}$, where V is the vector bundle on 3-skeleton of M determined by st and $L^{(2)}$ stands for 2-skeleton of L. Moreover, if a pair of Lagrangian submanifold (L_0, L_1) is relative spin, then $\widetilde{\mathcal{M}}([\ell_p, w], [\ell_q, w'])$ is orientable. The orientation is given by the choice of orientations of L_0 and L_1, $st \in H^2(M; \mathbb{Z}/2\mathbb{Z})$ and spin structures on $TL_0 \oplus V|_{L_0^{(2)}}$ and $TL_1 \oplus V|_{L_1^{(2)}}$*

This theorem can be proved by some gluing argument on the indices of families of linearized Dolbeault operators and an elementary topological argument.

As for the obstruction problem, we can show the following theorem.

Theorem 2.6. *Let L be an oriented Lagrangian submanifold of (M,ω). Assume that L is relative spin. Then we have the series of homology classes $\{o_k(L)\}_{k=1,2,\ldots}$ of L which satisfy the following significances:*

(1) $o_k(L) \in H_{n+\mu_L(\beta_k)-2}(L;\mathbb{Q})$. More precisely, $o_k(L)$ is in

$$\mathrm{Ker}\ (H_{n+\mu_L(\beta_k)-2}(L;\mathbb{Q}) \longrightarrow H_{n+\mu_L(\beta_k)-2}(M;\mathbb{Q})).$$

(2) $o_k(L)$ is defined if $o_j(L)=0$ in $H_*(L;\mathbb{Q})$ for every $j<k$.

(3) If all $o_k(L)$ vanish, then we can define the Floer cohomology $HF(L,L)$ by deforming the coboundary operators.

(4) Assume that a pair of Lagrangian submanifolds (L_0,L_1) is relative spin. (Then we can define the series $\{o_k(L_0)\}$ and $\{o_k(L_1)\}$.) If all $o_k(L_0)$ and $o_k(L_1)$ vanish, then we can define the Floer cohomology $HF(L_0,L_1)$ by deforming coboundary operators.

We call $o_k(L)$ an ***obstruction class***.

Remark 2.7. (1) Let us explain what the β_k's are. These are elements of $\pi_2(M,L)$ such that β_k is represented by a J-holomorphic disc. Then Gromov's compactness theorem implies that for each $C \geq 0$ the number of the set

$$\{\beta \in \pi_2(M,L) \mid \beta \text{ is represented by a } J \text{ holomorphic disc and } \mathcal{A}(\beta) \leq C\}$$

is finite. Therefore we have a partial order on the set of all $\beta \in \pi_2(M,L)$ which are represented by J holomorphic discs by the energy $\mathcal{A}$. Namely, we have

$$0=\mathcal{A}(\beta_0)<\mathcal{A}(\beta_1)\leq\mathcal{A}(\beta_2)\leq\cdots\leq\mathcal{A}(\beta_{k-1})\leq\mathcal{A}(\beta_k)\leq\cdots.$$

Here $\beta_0=0$ corresponds to constant maps.

(2) Taking (1) in Theorem 2.6 into account, we find that if $\mu_L(\beta_k)\geq 3$ for all k, then the obstruction classes automatically vanish. This condition was essentially used in the earlier work by Y-G Oh [Oh]. Here he defined Floer cohomology over $\mathbf{Z}/2\mathbf{Z}$ under some additional assumption which guarantees the trouble about the transversality problem does not happen, so the Kuranishi structure is not necessary into account in his case.

(3) We do not specify the coefficient ring of Floer cohomology here. See §4.

§3. Construction of the obstruction classes.

3.A) On orientations.

Before we explain the idea of construction of the obstruction classes, we like to mention a little bit about the orientations on various moduli spaces which will be used. From now on, we always assume that L is relative spin. We have an evaluation map at the marked point z_j

$$ev_j : \mathcal{M}_{k+1}(L,\beta) \longrightarrow L$$

for each $j = 0, 1, \ldots, k$, defined by $ev_j((w; z_0, \ldots, z_k)) = w(z_j)$. Let P_j be an oriented chain in L. We put $\deg P_j = n - \dim P_j$. We take a fibre product (in the sense of Kuranishi structure)

$$\mathcal{M}_{k+1}(L,\beta)_{(ev_1,\ldots,ev_k)} \times (P_1 \times \cdots \times P_k).$$

Then, by using the orientations on $\mathcal{M}_{k+1}(L,\beta)$ (defined by Theorem 2.5), P_j's and L, we can define the fibre product orientation on it. But we use a different orientation from the fibre product orientation.

Definition 3.1. We put

$$\mathcal{M}_1(\beta; P_1, \ldots, P_k) = (-1)^{\epsilon} \mathcal{M}_{k+1}(L,\beta)_{(ev_1,\ldots,ev_k)} \times (P_1 \times \cdots \times P_k)$$

Here ϵ is given by

$$\epsilon = (n+1) \sum_{j=1}^{k-1} \sum_{i=1}^{j} \deg P_i.$$

If we take the fibre product iteratively, we can rewrite the right hand side as

$$\mathcal{M}_1(\beta; P_1, \ldots, P_\ell) = (-1)^{\sum_{k=1}^{\ell-1} \sum_{j=1}^{k} \deg P_j}$$
$$\Big(\cdots ((\mathcal{M}_{\ell+1}(\ \beta\ )_{ev_1} \times_{f_1} P_1)_{ev_2} \times_{f_2} P_2) \times \cdots \Big)_{ev_\ell} \times_{f_\ell} P_\ell.$$

The "feeling" of the sign is an effect from the marked points. Roughly speaking, there might be two conventions when we consider the effect of the marked points. One is that we put all the parameters which describe the marked points on a "one side" in the fibre product. But we use another convention. We put the one dimensional parameters which describe the each marked point "one by one" in the fibre product. We call our convention *BARAMAKI* way. (We call the first one *HAKIYOSE* way.) If we change the ordering of the marked points, then we have another connected component (which are homeomorphic to each other). But the orientation might be changed. Under our orientation in Definition 3.1, we can find that the change is given by the following.

Proposition 3.2. *Let σ be the transposition element $(i, i+1)$ in the k-th symmetric group S_k. $(i = 1, \ldots, k-1)$. Then the action of σ on*

$$\mathcal{M}_1(\beta; P_1, \ldots, P_i, P_{i+1}, \ldots, P_k)$$

by changing the order of marked points is described by following.

$$\begin{aligned}&\sigma(\mathcal{M}_1(\beta; P_1, \ldots, P_i, P_{i+1}, \ldots, P_k))\\ &= (-1)^{(\deg P_i + 1)(\deg P_{i+1}+1)}\mathcal{M}_1(\beta; P_1, \ldots, P_{i+1}, P_i, \ldots, P_k).\end{aligned}$$

Now we explain the idea of the construction of our obstruction classes. We construct $o_k(L)$ inductively.

3.B) The first obstruction class.

We consider the space $\mathcal{M}_1(\beta_1)$ with the evaluation map ev_0,

$$ev_0 : \mathcal{M}_1(\beta_1) \longrightarrow L.$$

Note that $\mathcal{A}(\beta_1)$ is the minimal (non zero) area. Hence for an element in $\mathcal{M}_1(\beta_1)$, the bubbling off phenomena does not happen. Therefore $\partial\mathcal{M}_1(\beta_1) = \emptyset$. Thus $ev_0(\mathcal{M}_1(\beta_1))$ is a cycle in L, so defines a homology class. We define the first obstruction class $o_1(L)$ by the homology class:

$$o_1(L) = [ev_0(\mathcal{M}_1(\beta_1))].$$

The degree is given by $n + \mu_L(\beta_1) - 2$ because of Proposition 2.3.

3.C) The higher obstruction classes.

We suppose that $o_j(L) = 0$ for all $1 \le j \le k-1$. Under this situation, we are going to construct the k-th obstruction class $o_k(L)$. By the assumption we have bounding chains $\mathcal{B}_j = \mathcal{B}_j(L) \subset L$ such that

$$\partial\mathcal{B}_j(L) = (-1)^{n+1} o_j(L).$$

Here the orientation on $o_j(L)$ is given by the orientation on $\mathcal{M}_1(\beta_*; \mathcal{B}_*, \ldots, \mathcal{B}_*)$ defined in Definition 3.1 and the orientation on $\mathcal{B}_j(L)$ is given by changing the boundary orientation by $(-1)^{n+1}$. This sign plays an important role in the later argument. We put

$$\begin{aligned}&\mathcal{M}_1(\beta_k; \mathcal{B}_{i_1}, \ldots, \mathcal{B}_{i_m})\\ &= (-1)^{\epsilon_1}\mathcal{M}_{m+1}(L; \beta_k - \beta_{i_1} - \cdots - \beta_{i_m})_{(ev_{i_1}, \ldots, ev_{i_m})} \times (\prod_{\ell=1}^{m} \mathcal{B}_{i_\ell})\end{aligned}$$

for $i_1, \ldots, i_m < k$. Here ϵ_1 is given by the rule in Definition 3.1, that is

$$\epsilon_1 = (n+1)\frac{m(m-1)}{2},$$

because $\deg \mathcal{B}_{i_\ell} \equiv \mu_L(\beta_{i_\ell}) + 1 \equiv 1 \pmod 2$. (Note that since L is orientable, μ_L is always even.) It is easy to see that the dimension of $\mathcal{M}_1(\beta_k; \mathcal{B}_{i_1}, \ldots, \mathcal{B}_{i_m})$ is given by $n + \mu_L(\beta_k) - 2$. (Recall μ_L is a group homomorphism.) Then we define

Definition 3.3.

$$o_k(L) = \sum_{\substack{m=0,1,2,\ldots \\ i_1,\ldots,i_m<k \\ \beta_k - \sum_{j=1}^m \beta_{i_j} \in G_+(L)}} \frac{1}{m!}(ev_0(\mathcal{M}_1(\beta_k; \mathcal{B}_{i_1}, \ldots, \mathcal{B}_{i_m})),$$

where $G_+(L)$ stands for the subset of $\pi_2(M, L)$ whose elements are represented by J holomorphic discs. Note that the right hand side is a finite sum.

Then we have a chain in L defined by $o_k(L)$. What we have to show is that $o_k(L)$ defines a *cycle* . We can show the following.

Proposition 3.4. $\partial o_k(L) = \emptyset$.

If we ignore the sign problems, the proof is, in a sense, easy. That is, we have two kinds of boundaries of $\mathcal{M}_1(\beta_k; \mathcal{B}_{i_1}, \ldots, \mathcal{B}_{i_m})$ like as

$$\begin{aligned}
&\partial \mathcal{M}_1(\beta_k; \mathcal{B}_{i_1}, \ldots, \mathcal{B}_{i_m}) \\
&= (-1)^{\epsilon_1}\Big(\partial \mathcal{M}_{m+1}(L; \beta_k - \beta_{i_1} - \cdots - \beta_{i_m})_{(ev_{i_1}, \ldots, ev_{i_m})} \times (\prod_{\ell=1}^{m} \mathcal{B}_{i_\ell}) \\
&\quad \coprod \coprod_{\ell=1}^{m} (-1)^{n+m+nm} \mathcal{M}_{m+1}(L; \beta_k - \beta_{i_1} - \cdots - \beta_{i_m})_{(ev_{i_1}, \ldots, ev_{i_m})} \times \\
&\qquad (\mathcal{B}_{i_1} \times \cdots \times \partial \mathcal{B}_{i_\ell} \times \cdots \times \mathcal{B}_{i_m})\Big).
\end{aligned}$$

The first type boundaries correspond to the bubbling off J holomorphic dics $\partial \mathcal{M}_{m+1}(L; \beta_k - \beta_{i_1} - \cdots - \beta_{i_m})$ and the second type boundaries correspond to the case when the bounding chain $\mathcal{B}_{i_\ell}$ goes to $\partial \mathcal{B}_{i_\ell} = (-1)^{n+1} o_{i_\ell}(L)$. We take the summation in Definition 3.3 over all "lower" strata of moduli spaces. Therefore these two kinds of boundaries cancel each other. The non trivial issue is that they cancel each other with ***sign***. That is, we have to show that the orientations on these two kinds of boundaries are opposite. But in this note, we omit the proof.

Now let (L_0, L_1) be a pair of relative spin Lagrangian submanifolds. If all obstructions $o_k(L_0)$ and $o_k(L_1)$ vanish, then, as we state in Theorem 2.6.(4), we can define Floer cohomology $HF(L_0, L_1)$ by deforming the coboundary operators as follows. By assumption, we have bounding chains $\mathcal{B}_{0,*} = \mathcal{B}_*(L_0) \subset L_0$ and $\mathcal{B}_{1,*} = \mathcal{B}_*(L_1) \subset L_1$. By imposing marked points on the boundaries of the strip $[0,1] \times \mathbf{R}$ in (1.5) (ℓ marked points on $\{0\} \times \mathbb{R}$ and m marked points on $\{1\} \times \mathbb{R}$), we can define the fibre product like as

$$\begin{aligned}&\mathcal{M}([\ell_p, w], [\ell_q, w']; \mathcal{B}_{0,i_1}, \ldots, \mathcal{B}_{0,i_\ell}; \mathcal{B}_{1,j_1}, \ldots, \mathcal{B}_{1.j_m})\\ &= (-1)^{\epsilon_2} \mathcal{M}_{\ell,m}([\ell_p, w], [\ell_q, w'])_{(ev_1^0,\ldots,ev_\ell^0,ev_1^1,\ldots,ev_m^1)} \times (\prod_{k=1}^{\ell} \mathcal{B}_{0,i_k} \times \prod_{k=1}^{m} \mathcal{B}_{1,j_k}),\end{aligned}$$

where ϵ_1 is given by

$$\epsilon_1 = (n+1)(\sum_{k=1}^{\ell+m-1} \sum_{j=1}^{k} 1) = \frac{(n+1)(\ell+m-1)(\ell+m)}{2},$$

which is consistent with the rule in Definition 3.1. Now we define an operator $\delta_{\beta_{0,i_1},\ldots,\beta_{0,i_\ell};\beta_{1,j_1},\ldots,\beta_{1,j_m}}$ by

$$\begin{aligned}&\langle \delta_{\beta_{0,i_1},\ldots,\beta_{0,i_\ell};\beta_{1,j_1},\ldots,\beta_{1,j_m}}[\ell_p, w], [\ell_q, w'] \rangle\\ &:= \#\mathcal{M}([\ell_p, w], [\ell_q, w']; \mathcal{B}_{0,i_1}, \ldots, \mathcal{B}_{0,i_\ell}; \mathcal{B}_{1,j_1}, \ldots, \mathcal{B}_{1,j_m}).\end{aligned}$$

Then we can show

Proposition 3.5. *We define our modified coboundary operator δ by*

$$\delta := \sum_{\ell,m} \sum_{\beta_{0,i_1},\ldots,\beta_{0,i_\ell};\beta_{1,j_1},\ldots,\beta_{1,j_m}} \frac{1}{\ell! m!} \delta_{\beta_{0,i_1},\ldots,\beta_{0,i_\ell};\beta_{1,j_1},\ldots,\beta_{1,j_m}}.$$

Then it satisfies $\delta \circ \delta = 0$.

§4. A_∞-deformation of Lagrangian submanifold.

We should note that the constructions in the previous section depend, *a priori*, on various choices of the bounding chains, the almost complex structures, and the Kuranishi structures. In this note, we only discuss dependence on the bounding chains. (As for our conclusions about "independence", see Theorem 5.19 and Theorem 5.20.) To do this we use language of certain homological algebras. The key point is that we have to work at *chain level*, not homological level. Firstly, we construct a *filtered A_∞ algebra* associated to a relative spin Lagrangian submanifold L. Here we note that we do *not* assume the obstruction classes of L vanish.

4.A) A filtered A_∞ algebra.

First of all, we introduce the notion of *filtered A_∞ algebra* [FOOO]. Let us introduce the universal Novikov ring.

Definition 4.1. ([FOOO]). Let T and e be two formal variables. *The universal Novikov ring* Λ_{nov} is the totality of all formal sums $\sum a_i T^{\lambda_i} e^{n_i}$ such that

(4.1.1) $a_i \in \mathbf{Q}$, $\lambda_i \in \mathbf{R}$ and $n_i \in \mathbf{Z}$.

(4.1.2) $\lim_{i\to\infty} \lambda_i = \infty$.

We define its subset $\Lambda_{0,nov}$ by

$$\Lambda_{0,nov} = \left\{\sum a_i T^{\lambda_i} e^{n_i} \mid \lambda_i \geq 0, \text{ and } n_i = 0 \text{ if } \lambda_i = 0\right\}.$$

We define the product of elements of Λ_{nov} in an obvious way. Then Λ_{nov} is a commutative ring with the unit 1, and $\Lambda_{0,nov}$ is its subring. We define the grading by

$$\deg T^{\lambda} e^{n} = 2n.$$

Roughly speaking, λ_i stands for a filtration and n_i for a grading. Geometrically, λ_i corresponds to an energy $\mathcal{A}$ in §1 or §2, and n_i corresponds to the Maslov index. When we consider a pair of Lagrangian submanifolds, we use Λ_{nov} as well. (It might be helpful to keep the geometric back ground in your mind.) We remark the $\Lambda_{0,nov}$ is a local ring with the maximal ideal

$$\Lambda_{0,nov}^{+} = \{\sum a_i T^{\lambda_i} e^{n_i} \in \Lambda_{nov} \mid \lambda_i > 0\}$$

such that $\Lambda_{0,nov}/\Lambda_{0,nov}^{+} \cong \mathbf{Q}$. So when we reduce the coefficient ring to $\Lambda_{0,nov}/\Lambda_{0,nov}^{+} \cong \mathbf{Q}$, then we do not have filtrations. See (4.0) below.

Let $\oplus_{m\in\mathbf{Z}} C^m$ be a free graded $\Lambda_{0,nov}$ module. There is a filtration $F^{\lambda}C^m$ on C^m ($\lambda \in \mathbf{R}_{\geq 0}$), such that

(4.2.1) $F^{\lambda}C^m \subset F^{\lambda'}C^m$ if $\lambda > \lambda'$.

(4.2.2) $T^{\lambda_0} \cdot F^{\lambda}C^m \subset F^{\lambda+\lambda_0}C^m$.

(4.2.3) $e^k C^m \subset C^{m+2k}$.

(4.2.4) C^m is complete with respect to the filtration.

(4.2.5) C^m has a basis $\mathbf{e}_i$ such that $\mathbf{e}_i \in F^0 C^m$ and $\mathbf{e}_i \notin F^{\lambda}C^m$ for $\lambda > 0$.

We denote by C the completion of $\oplus_{m\in\mathbf{Z}} C^m$ with respect to the filtration. (4.2.3) means that the degree of e is 2. We put $(C[1])^m = C^{m+1}$ and

$$B_k(C[1]) = \bigoplus_{m_1,\cdots,m_k} (C[1])^{m_1} \otimes \cdots \otimes (C[1])^{m_k}.$$

Suppose that we have a sequence of maps $\mathfrak{m} = \{\mathfrak{m}_k\}_{k\geq 0}$ of degree $+1$

$$\mathfrak{m}_k : B_k(C[1]) \to C[1], \quad \text{for } k = 0, 1, \cdots.$$

We note that $\mathfrak{m}_0 : \Lambda_{0,nov} \to C[1]$. We assume that

$$\mathfrak{m}_k \left(F^{\lambda_1}C^{m_1} \otimes \cdots \otimes F^{\lambda_k}C^{m_k}\right) \subseteq F^{\lambda_1+\cdots+\lambda_k}C^{m_1+\cdots+m_k-k+2} \tag{4.2.6}$$

and

$$\mathfrak{m}_0(1) \in F^{\lambda'}C[1] \quad \text{for some } \lambda' > 0. \tag{4.2.7}$$

When we put

$$\overline{\mathfrak{m}}_k(x_1, \ldots, x_k) \equiv \mathfrak{m}_k(x_1, \ldots, x_k) \mod \Lambda_{0,nov}^+ \tag{4.0}$$

for $k = 0, 1, 2, \ldots$, then $\{\overline{\mathfrak{m}}_k\}$ defines an A_∞ algebra structure on $\overline{C} = C/\Lambda_{0,nov}^+C$ over $\mathbb{Q}$ introduced by Stasheff [St]. (Strictly speaking, he did not treat the map $\mathfrak{m}_0$. But this map is important when we discuss the obstruction theory. See Remark 4.9. Note that the filtration is defined by λ which is the power of T. Thus on the $\overline{C}$, we do not have filtrations. In this way $(\overline{C}, \overline{\mathfrak{m}} = \{\overline{\mathfrak{m}}_k\}_{k\geq 0})$ becomes an A_∞ algebra over $\Lambda_{0,nov}/\Lambda_{0,nov}^+ \cong \mathbb{Q}$.) We also assume that there exists a constant $\lambda'' > 0$ such that

$$\mathfrak{m}_k(\mathrm{e}_{i_1}, \ldots, \mathrm{e}_{i_k}) - \overline{\mathfrak{m}}_k(\mathrm{e}_{i_1}, \ldots, \mathrm{e}_{i_k}) \in F^{\lambda''}C[1]. \tag{4.2.8}$$

Here λ'' is independent of k and $\mathrm{e}_{i_1}, \ldots, \mathrm{e}_{i_k}$. The condition (4.2.8) is used when we construct a spectral sequence. See [FOOO] for details. We call (4.2.8) the *gap condition.*

Now the direct sum $B(C[1]) := \oplus_k B_k(C[1])$ has a structure of graded coalgebra. We regard $B(C[1])$ as a coalgebra and will construct a coderivation on it. The coproduct Δ is defined by :

$$\Delta(x_1 \otimes \cdots \otimes x_n) = \sum_{k=0}^{n} (x_1 \otimes \cdots \otimes x_k) \otimes (x_{k+1} \otimes \cdots \otimes x_n). \tag{4.3}$$

We can extend $\mathfrak{m}_k$ uniquely to a coderivation

$$d_k : \bigoplus_n B_n(C[1]) \to \bigoplus_n B_{n-k+1}(C[1]),$$

by

$$d_k(x_1 \otimes \cdots \otimes x_n) = \sum_{\ell=1}^{n-k+1} (-1)^{\deg x_1+\cdots+\deg x_{\ell-1}+\ell-1} x_1 \otimes \cdots \otimes \mathfrak{m}_k(x_\ell, \cdots, x_{\ell+k-1}) \otimes \cdots \otimes x_n \tag{4.4}$$

for $k \le n$ and $d_k = 0$ for $k > n$. Here and hereafter deg x means the degree of x *before* we shift it. When $k = 0$, we put $\mathfrak{m}_0(1)$ in the right hand side. Namely we define d_0 by

$$d_0(x_1 \otimes \cdots \otimes x_n) = \sum_{\ell=1}^{n+1} (-1)^{\deg x_1 + \cdots + \deg x_{\ell-1} + \ell - 1} x_1 \otimes \cdots \otimes x_{\ell-1} \otimes \mathfrak{m}_0(1) \otimes x_\ell \otimes \cdots \otimes x_n.$$

We want to consider the infinite sum $\widehat{d} = \sum d_k$. Therefore we need to consider a completion $\widehat{B}(C[1])$ of $B(C[1])$. We define a filtration $F^\lambda B_k(C[1])$ on $B_k(C[1])$ by

$$F^\lambda B_k(C[1]) = \bigcup_{\lambda_1 + \cdots + \lambda_k \ge \lambda} (F^{\lambda_1} C^{m_1} \otimes \cdots \otimes F^{\lambda_k} C^{m_k})$$

Let $\widehat{B}_k(C[1])$ be the completion with respect to the filtration.

Definition 4.5. $\widehat{B}(C[1])$ is the set of all formal sum $\sum_k \mathbf{x}_k$ where $\mathbf{x}_k \in \widehat{B}_k(C[1])$ such that

$$\mathbf{x}_k \in F^{\lambda_k} \widehat{B}_k(C[1])$$

with $\lim_{k \to \infty} \lambda_k \to \infty$.

Lemma 4.6. *If* (4.2) *is satisfied, then $\widehat{d}$ is well-defined as a map from $\widehat{B}(C[1])$ to $\widehat{B}(C[1])$.*

The proof is easy.

Now we introduce the following condition for an element $\mathbf{e}$ of C.

Condition 4.7.

(4.7.1) $\mathbf{m}_{k+1}(x_1, \cdots, \mathbf{e}, \cdots, x_k) = 0$, $k \ge 2$, $k = 0$.

(4.7.2) $\mathbf{m}_2(\mathbf{e}, x) = (-1)^{\deg x} \mathbf{m}_2(x, \mathbf{e}) = x$.

Definition 4.8. ([FOOO]) (1) $\mathfrak{m} = \{\mathbf{m}_k\}_{k \ge 0}$ defines a structure of *filtered A_∞ algebra* on C if (4.2) are satisfied and if $\widehat{d} \circ \widehat{d} = 0$. We call $\widehat{B}(C[1])$ the (completed) *bar complex* associated to the A_∞ algebra $(C, \mathfrak{m})$. If a filtered A_∞ algebra has an element $\mathbf{e}$ which satisfies Condition 4.7, the we call it *an A_∞ algebra with unit* and $\mathbf{e}$ a *unit*.

(2) For a filtered A_∞ algebra $(C, \mathfrak{m})$, we say that a filtered A_∞ algebra $(C', \mathbf{m}')$ is *an A_∞-deformation* of $(C, \mathbf{m})$, if $(\overline{C}', \overline{\mathfrak{m}}') = (\overline{C}, \overline{\mathfrak{m}})$. Here $(\overline{C}', \overline{\mathfrak{m}}')$ and $(\overline{C}, \overline{\mathfrak{m}})$ are defined by reducing the coefficient ring $\Lambda_{0,nov}$ to

$\Lambda_{0,nov}/\Lambda_{0,nov}^{+} \cong \mathbf{Q}$. We also say that a filtered A_∞ algebra $(C, \mathfrak{m})$ is *an A_∞-deformation* of an A_∞ algebra $(\overline{C}'', \overline{\mathfrak{m}}'')$ if $(\overline{C}, \overline{\mathfrak{m}}) = (\overline{C}'', \overline{\mathfrak{m}}'')$.

Remark 4.9. (1) The equation $\widehat{d} \circ \widehat{d} = 0$ produces infinitely many relations among $\mathfrak{m}_k$'s. For example, we have

$$\begin{aligned}
&\mathfrak{m}_1(\mathfrak{m}_0(1)) = 0,\\
&\mathfrak{m}_2(\mathfrak{m}_0(1), x) + (-1)^{\deg x+1}\mathfrak{m}_2(x, \mathfrak{m}_0(1)) + \mathfrak{m}_1(\mathfrak{m}_1(x)) = 0,\\
&\mathfrak{m}_3(\mathfrak{m}_0(1), x, y) + (-1)^{\deg x+1}\mathfrak{m}_3(x, \mathfrak{m}_0(1), y)\\
&+(-1)^{\deg x+\deg y+2}\mathfrak{m}_3(x, y, \mathfrak{m}_0(1))\\
&+\mathfrak{m}_2(\mathfrak{m}_1(x), y) + (-1)^{\deg x+1}\mathfrak{m}_2(x, \mathfrak{m}_1(y)) + \mathfrak{m}_1(\mathfrak{m}_2(x, y)) = 0,\\
&\cdots\cdots\cdots
\end{aligned}$$

In general, it is easy to show that $\widehat{d} \circ \widehat{d} = 0$ is equivalent to that for each k

$$\sum_{k_1+k_2=k+1} \sum_i (-1)^{\deg x_1+\cdots+\deg x_{i-1}+i-1} \mathfrak{m}_{k_1}(x_1, \cdots, \mathfrak{m}_{k_2}(x_i, \cdots, x_{i+k_2-1}), \cdots, x_k) = 0.$$

If $\mathfrak{m}_0 = 0$, then $\mathfrak{m}_1\mathfrak{m}_1 = 0$. So in this case $\mathfrak{m}_1$ plays a role of a (co)boundary operator. In this sense, $\mathfrak{m}_0$ describes an obstruction to that $\mathfrak{m}_1\mathfrak{m}_1 = 0$.

(2) In addition to $\mathfrak{m}_0 = 0$, suppose that $\mathfrak{m}_k = 0$ for $k \geq 3$. We put

$$\mathfrak{m}_1(x) = (-1)^{\deg x} dx, \quad \text{and} \quad \mathfrak{m}_2(x, y) = (-1)^{\deg x(\deg y+1)} x \wedge y,$$

where $\deg x$ denotes the degree of x as *cochain*. Then this is nothing but a DGA (differential graded algebra) and $\widehat{d} \circ \widehat{d} = 0$ implies that the usual Leibnitz rule and the associativity of the product structure. We note that the signs here are slightly different from those in [G-J]. (See also Remark 4.13 (2) below.)

4.B) A filtered A_∞ algebra associated to Lagrangian submanifold L.

Let $\overline{S}^k(L; \mathbf{Q})$ be a free $\mathbf{Q}$ module generated by all integral k-currents on L which are represented by singular chains. We denote by $C^k(L; \mathbf{Q})$ a countably generated submodule of $\overline{S}^k(L; \mathbf{Q})$. (We will use a method of "smooth correspondence". To do this we need and use the transversality argument and the Baire category theorem. This is why we introduce $C^k(L; \mathbf{Q})$. But the details are omitted here, see [FOOO].) Since an element in $C^k(L; \mathbf{Q})$ is represented by a singular chain, we sometimes write it as a singular chain representative (P, f). (But when we consider the orientation problems, we have to notice the difference of signs of boundary orientation and product (intersection) as *chain* or *cochain*, see Remark 4.13 (2) below.) We define $C^{\bullet}(L;\Lambda_{0,nov})$ by the completion of $C^{\bullet}(L;\mathbf{Q})\otimes\Lambda_{0,nov}$. For the convenience of notation, we put

$$C^{\bullet}=C^{\bullet}(L;\Lambda_{0,nov}):=\left(C^{\bullet}(L;\mathbf{Q})\otimes\Lambda_{0,nov}\right)^{\widehat{}}.$$

The degree in $C^{\bullet}(L;\Lambda_{0,nov})$ is the sum of the degree in $C^{\bullet}(L;\mathbf{Q})$ and the degree of the coefficients in $\Lambda_{0,nov}$. Using the filtration on $\Lambda_{0,nov}$, we can uniquely define the filtration on $C^k(L;\Lambda_{0,nov})$ which satisfies the following conditions;

$$C^k(L;\mathbf{Q})\subset F^0C^k(L;\Lambda_{0,nov})$$

and

$$C^k(L;\mathbf{Q})\not\subset F^{\lambda}C^k(L;\Lambda_{0,nov})\qquad\text{for }\lambda>0.$$

We now define the maps

$$\mathbf{m}_k:B_k(C[1](L;\Lambda_{0,nov}))\to C[1](L;\Lambda_{0,nov}).$$

of degree $+1$ for $k\geq 0$. To do this, we recall that $\mathcal{M}_{k+1}(\beta)$ is the set of pairs $((\Sigma,\vec{z}),w)$ where $w:(\Sigma,\partial\Sigma)\to(M,L)$ is a pseudoholomorphic map which represents the class β. Let $\mathcal{M}_{k+1}^{\rm main}(\beta)$ be the subset of $\mathcal{M}_{k+1}(\beta)$ consisting of elements $((\Sigma,\vec{z}),w)$ where the order of the marked points is cyclic. (See Definition 2.1). For given

$$(P_i,f_i)\in C^{g_i}(L;\mathbf{Q}),\quad i=1,\cdots,k,$$

we consider

$$\mathcal{M}_{k+1}^{\rm main}(\beta)_{(ev_1,\cdots,ev_k)}\times_{f_1\times\cdots\times f_k}(P_1\times\cdots\times P_k).$$

Proposition 4.10. *Suppose L is relatively spin. Then*

$$\mathcal{M}_{k+1}^{\rm main}(\beta)_{(ev_1,\cdots,ev_k)}\times_{f_1\times\cdots\times f_k}(P_1\times\cdots\times P_k)$$

has an oriented Kuranishi structure. Its dimension is $n-\sum(g_i-1)+\mu(\beta)-2$, where $n=\dim L$.

As in Definition 3.1, we define $\mathcal{M}_1^{\rm main}(\beta;P_1,\cdots,P_k)$ by the following.

Definition 4.11.

$$\mathcal{M}_1^{\rm main}(\beta;P_1,\cdots,P_k):=(-1)^{\epsilon}\mathcal{M}_{k+1}^{\rm main}(\beta)_{(ev_1,\cdots,ev_k)}\times_{f_1\times\cdots\times f_k}(P_1\times\cdots\times P_k),$$

where

$$\epsilon = (n+1)\sum_{j=1}^{k-1}\sum_{i=1}^{j}\deg P_i.$$

Now we define the maps $\mathfrak{m}_k$. We recall that we have the element $\beta_0 = 0 \in G_{++}(L)$ which satisfies $\mu(\beta_0) = 0$ and $\mathcal{A}(\beta_0) = 0$.

Definition 4.12. (1) For $(P, f) \in C^k(L, \mathbb{Q})$, we define

$$\mathfrak{m}_{0,\beta}(1) = \begin{cases} (\mathcal{M}_1(\beta), ev_0) & \text{for } \beta \neq \beta_0 \\ 0 & \text{for } \beta = \beta_0, \end{cases}$$

$$\mathfrak{m}_{1,\beta}(P, f) = \begin{cases} (\mathcal{M}_1^{\text{main}}(\beta; P), ev_0) & \text{for } \beta \neq \beta_0 \\ (-1)^n \partial P & \text{for } \beta = \beta_0, \end{cases}$$

and

$$\mathbf{e} = [L] \quad \text{(the fundamental cycle)}.$$

The notation ∂ in the definition of $\mathfrak{m}_{1,\beta_0}$ is the usual (classical) boundary operator.
(2) For each $k \geq 2$ and $(P_i, f_i) \in C^{g_i}(L, \mathbb{Q})$, we define $\mathfrak{m}_{k,\beta}$ by

$$\begin{aligned}\mathfrak{m}_{k,\beta}((P_1, f_1), \ldots, (P_k, f_k)) &= \mathfrak{m}_{k,\beta}((P_1, f_1) \otimes \cdots \otimes (P_k, f_k)) \\ &= (\mathcal{M}_1^{\text{main}}(\beta; P_1, \cdots, P_k), ev_0).\end{aligned}$$

(3) Then we define $\mathfrak{m}_k$ $(k \geq 0)$ by

$$\mathfrak{m}_k = \sum_{\beta \in \pi_2(M,L)} \mathfrak{m}_{k,\beta} \otimes [\beta] = \sum_{\beta \in \pi_2(M,L)} \mathfrak{m}_{k,\beta} \otimes T^{\omega(\beta)} e^{\frac{\mu(\beta)}{2}}.$$

Remark 4.13. (1) By definition, $\overline{\mathfrak{m}}_0 = 0$. (This is the case corresponding to $\beta = \beta_0$.) But $\mathfrak{m}_0 \neq 0$.

(2) In the definition of $\mathfrak{m}_{1,\beta_0}$ above, we see P as a *chain*. If we see P as a *cochain* (or a differential form), then we have

$$\mathfrak{m}_{1,\beta_0}(P) = (-1)^{n+\deg P+1} dP, \tag{4.13.2}$$

where $\deg P$ is the degree of P as a cochain. This is because we can see the following general formula (under certain our conventions [FOOO] about orientations of boundary and of normal bundle). For an s-dim chain S in L, we have

$$P.D.(\partial S) = (-1)^{\deg S+1} d(P.D.(S)).$$

Here $\deg S = n - s$ and $P.D.$ denotes the Poincaré duality. Of course, this sign depends on a convention about the Poincaré duality. Actually, we use the following convention. For a chain S in L, the Poincaré dual $P.D.(S)$ satisfies

$$\int_S \alpha|_S = \int_L P.D.(S) \wedge \alpha$$

for any $\alpha \in \Omega^{\dim S}(L)$. We also note that the universal constant $n+1$ in the power of the sign in (4.13.2) does not affect in the A_∞ relations in the case $\mathfrak{m}_0 = 0$. In this sense, this is consistent with Remark 4.9 (2).

By using the $\{\mathfrak{m}_k\}_{k\geq 0}$, we define

$$d_k : \bigoplus_n B_n(C[1](L;\Lambda_{0,nov})) \to \bigoplus_n B_{n-k+1}(C[1](L;\Lambda_{0,nov}))$$

as in (4.4). Then the following is our main theorem in this section.

Theorem 4.14. ([FOOO]) *Suppose L is a relatively spin Lagrangian submanifold. Then $(C(L;\Lambda_{0,nov}),\mathfrak{m})$ is a filtered A_∞ algebra (with unit $\mathbf{e}$). Furthermore, $(C(L;\Lambda_{0,nov}),\mathfrak{m})$ satisfies the gap condition (4.2.8).*

(Strictly speaking, $\mathbf{e}$ is not a unit, but a *homotopy unit.* (We can deform (unitarize) $\mathbf{e}$ to be unit.) But we omit the details, see [FOOO].) Moreover, by using moduli space of *metric ribbon trees*, we can construct an A_∞ algebra $(\mathbf{Q}\mathcal{X},\overline{\mathfrak{m}})$ over $\mathbf{Q}$ with $\overline{\mathfrak{m}}_0 = 0$, such that it describes the rational homotopy type of L and the cohomology of $\overline{\mathfrak{m}}_1$ is isomorphic to the cohomology of L [FOOO]. Then we can show the following.

Theorem 4.15. ([FOOO]) $(C(L;\Lambda_{0,nov}),\mathfrak{m})$ *is an A_∞ deformation of the A_∞ algebra $(\mathbf{Q}\mathcal{X},\overline{\mathfrak{m}})$.*

Sketch of the Proof of Theorem 4.14: To prove $\widehat{d}\circ\widehat{d} = 0$, we analyze the boundary of $\mathcal{M}_1^{\text{main}}(\beta;P_1,\cdots,P_k)$. We find that its boundary is the sum of

$$\sum_i \mathcal{M}_1^{\text{main}}(\beta;P_1,\cdots,\partial P_i,\cdots,P_k)$$

and the terms described by the bubbling off holomorphic discs. On the other hand, in order to prove $\widehat{d}\circ\widehat{d} = 0$, we note that it is enough to show that

$$(4.16.1) \quad \sum_{\beta_1+\beta_2=\beta}\ \sum_{k_1+k_2=k+1}\ \sum_i (-1)^{\deg P_1+\cdots+\deg P_{i-1}+i-1} \mathfrak{m}_{k_1,\beta_1}(P_1,\cdots,\mathfrak{m}_{k_2,\beta_2}(P_i,\cdots,P_{i+k_2-1}),\cdots,P_k) = 0.$$

(See Remark 4.9.) We divide the left hand side into 3 terms, according as $\beta_1 = 0(=\beta_0)$ and $k_1 = 1$, $\beta_2 = 0(=\beta_0)$ and $k_2 = 1$, and the other cases. Then we can rewrite the left hand side in (4.16.1) as follows:

$$\begin{aligned}
&\mathfrak{m}_{1,0}\mathfrak{m}_{k,\beta}(P_1,\cdots,P_k)\\
(4.16.2)\qquad &+\sum_i(-1)^{\deg P_1+\cdots+\deg P_{i-1}+i-1}\mathfrak{m}_{k,\beta}(P_1,\cdots,\mathfrak{m}_{1,0}(P_i),\cdots,P_k)\\
&+\sum_{\substack{\beta_1+\beta_2=\beta,\ k_1+k_2=k+1;\\ \beta_1\neq 0 \text{ or } k_1\neq 1,\\ \beta_2\neq 0 \text{ or } k_2\neq 1}}\sum_i(-1)^{\deg P_1+\cdots+\deg P_{i-1}+i-1}\\
&\qquad\qquad \mathfrak{m}_{k_1,\beta_1}(P_1,\cdots,\mathfrak{m}_{k_2,\beta_2}(P_i,\cdots,P_{i+k_2-1}),\cdots,P_k).
\end{aligned}$$

By Definition 4.12, we have $\mathfrak{m}_{1,0} = (-1)^n\partial$, where ∂ is the classical boundary map. Hence the first term in (4.16.2) is nothing but

$$(4.16.2.1)\qquad (-1)^n\partial(\mathcal{M}_1^{\text{main}}(\beta : P_1,\cdots,P_k),ev_0),$$

and the second term in (4.16.2) is the sum of

$$(4.16.2.2)\quad (-1)^{\sum_{j=1}^{i-1}(\deg P_j+1)}(-1)^n(\mathcal{M}_1^{\text{main}}(\beta : P_1,\cdots,\partial P_i,\cdots,P_k),ev_0).$$

The third term in (4.16.2) geometrically corresponds to moduli spaces described by bubbling off holomorphic discs. This is the sum of

$$\begin{aligned}
(4.16.2.3)\qquad &(-1)^{\sum_{j=1}^{i-1}(\deg P_j+1)}(\mathcal{M}_1^{\text{main}}(\beta_1;P_1,\cdots\\
&\qquad\cdots,P_{i-1},\mathcal{M}_1^{\text{main}}(\beta_2;P_i,\cdots,P_{i+k_2-1}),\cdots,P_k),ev_0).
\end{aligned}$$

Moreover, as for the orientations of these spaces, we can show the following:

$$\begin{aligned}
(4.16.3.1)\qquad &(-1)^{\sum_{j=1}^{i-1}(\deg P_j+1)}(-1)^n\mathcal{M}_1^{\text{main}}(\beta : P_1,\cdots,\partial P_i,\cdots,P_k)\\
&\subseteq (-1)^{n+1}\partial\mathcal{M}_1^{\text{main}}(\beta;P_1,\cdots,P_k)
\end{aligned}$$

and

(4.16.3.2)

$$\begin{aligned}
&(-1)^{\sum_{j=1}^{i-1}(\deg P_j+1)}\mathcal{M}_1^{\text{main}}(\beta_1;P_1,\cdots,\mathcal{M}_1^{\text{main}}(\beta_2;P_i,\cdots,P_{i+k_2-1}),\cdots,P_k)\\
&\subseteq (-1)^{n+1}\partial\mathcal{M}_1^{\text{main}}(\beta;P_1,\cdots,P_k).
\end{aligned}$$

Therefore we find that (4.16.2.1) and the sum of (4.16.2.2) and (4.16.2.3) cancel each other. Namely (4.16.2) is zero. This implies $\widehat{d}\circ\widehat{d} = 0$. We recall that we have an element $\beta_1 \in \pi_2(M, L)$ such that it is represented by J holomorphic disc with the minimal (non zero) area. We take $\lambda'' > 0$ such that

$$\lambda'' < \omega[\beta_1].$$

Then we can find that $\{\mathfrak{m}_k\}$ satisfies the gap condition (4.2.8).

§5. Bounding cochains and deformation.

From now on, we are working on cohains (or cohomologies), not on chains (or homologies) via the Poincaré duality, because they are fitted with the framework of obstruction theory.

5.A) Bounding cochains and the master equation.

For a cochain $b \in C[1]^0(L, \Lambda_{0,nov})$ with the shifted degree 0, we put

$$e^b = 1 + b + b \otimes b + b \otimes b \otimes b + \cdots \in \widehat{B}(C[1](L, \Lambda_{0,nov}). \tag{5.1}$$

(We do not put the factorials here unlike definition of the exponential, because we use only the main component among $(k+1)!$ components of $\mathcal{M}_{k+1}$ to define the map $\mathfrak{m}_k$.)

Definition 5.2. We say that b is a *bounding cochain* if $\widehat{d}e^b = 0$. A filtered A_∞ algebra is said to be *unobstructed* if there exists a bounding cochain and *obstructed* otherwise. Similarly, we call a Lagrangian submanifold L *unobstructed* if the associated filtered A_∞ algebra $(C(L; \Lambda_{0,nov}), \mathfrak{m})$ constructed in Section 4.B) is unobstructed. We denote by $\widehat{\mathcal{M}}(L) = \widehat{\mathcal{M}}(L; J, \Xi)$ the set of all bounding cochains b. Here J stands for a compatible almost complex structure and Ξ for a parameter of the Kuranishi structure.

From the construction in §3, we put

$$b = \sum \mathcal{B}(\beta_i) \otimes [\beta_i] = \sum \mathcal{B}(\beta_i) \otimes T^{\omega(\beta_i)} e^{\frac{\mu(\beta_i)}{2}} \in C[1]^0(L; \Lambda_{0,nov}). \tag{5.3}$$

Then we can show that

Lemma 5.4. *The chains $\mathcal{B}(\beta_i)$ bound $o_i(L)$ inductively if and only if b in* (5.3) *is a bounding cochain, i.e.,* $\widehat{d}e^b = 0$.

Remark 5.5. The equation $\widehat{d}e^b = 0$ is equivalent to

$$\mathfrak{m}_0(1) + \mathfrak{m}_1(b) + \mathfrak{m}_2(b, b) + \mathfrak{m}_3(b, b, b) + \cdots = 0.$$

If $\mathfrak{m}_0 = 0$ and $\mathfrak{m}_k = 0$ for $k \geq 3$, by putting $\mathfrak{m}_1 = d$ and $\mathfrak{m}_2 = \wedge$, the equation is equivalent to

$$db + b \wedge b = 0,$$

which is nothing but the classical Maurer-Cartan equation for DGA. Our equation $\widehat{d}e^b = 0$ is an inhomogeneous A_∞-version of Maurer-Cartan or Batalin-Vilkovisky master equation [BV]. The relation of Batalin-Vilkovisky master equation to the deformation theory is discussed in [Sch], [ASKZ], [BK], [$\mathbb{K}$].

The deformation of the Floer coboundary operators in §3 can be interpreted as follows. Here for simplicity, we discuss the case for one Lagrangian submanifold L. (The case for two Lagrangian submanifolds L_0 and L_1 is similar, but needs more notations and argument, e.g., we have to use Λ_{nov}.) Suppose that the filtered A_∞ algebra $(C(L;\Lambda_{0,nov}),\mathfrak{m})$ we constructed in §4 is unobstructed in the sense of Definition 5.2. Then we have bounding cochains $b_1, b_2 \in C[1]^0(L,\Lambda_{0,nov})$. (They may coincide.) By using these cochains b_1, b_2, we define

$$\delta_{b_1,b_2} : C(L;\Lambda_{0,nov}) \to C(L;\Lambda_{0,nov})$$

by

$$\delta_{b_1,b_2}(x) = \sum_{k_1,k_2\geq 0} \mathfrak{m}_{k_1+k_2+1}(\underbrace{b_1,\cdots,b_1}_{k_1},x,\underbrace{b_2,\cdots,b_2}_{k_2}). \tag{5.6}$$

Then we can find that

$$\widehat{d}(e^{b_1}xe^{b_2}) = e^{b_1}\delta_{b_1,b_2}(x)e^{b_2} + \widehat{d}(e^{b_1})xe^{b_2} + (-1)^{\deg x+1}e^{b_1}x\widehat{d}(e^{b_2}).$$

The second and the third term vanishes if $\widehat{d}(e^{b_1}) = \widehat{d}(e^{b_2}) = 0$. Thus we have

Proposition 5.7. *If $\widehat{d}(e^{b_1}) = \widehat{d}(e^{b_2}) = 0$, then $\delta_{b_1,b_2}\circ\delta_{b_1,b_2} = 0$.*

5.B) Deformation of A_∞ algebra.

Now let $b \in C[1]^0(L,\Lambda_{0,nov})$ be a cochain, which is not necessary a bounding cochain. For the cochain b, we next deform our filtered A_∞ algebra as follows.

Definition 5.8. Using this cochain b, we put

$$\begin{aligned}\mathfrak{m}_k^b(x_1,\cdots,x_k) &= \sum_{\ell_0,\cdots,\ell_k} \mathfrak{m}_{k+\sum\ell_i}(\underbrace{b,\cdots,b}_{\ell_0},x_1,\underbrace{b,\cdots,b}_{\ell_1},\cdots,\underbrace{b,\cdots,b}_{\ell_{k-1}},x_k,\underbrace{b,\cdots,b}_{\ell_k})\\ &= \mathfrak{m}(e^bx_1e^bx_2\cdots x_{k-1}e^bx_ke^b)\end{aligned}$$

for $k = 0,1,2\cdots$.

We note that $\mathfrak{m}_0^b(1) = \mathfrak{m}(e^b)$. Since we can find that

$$\widehat{d}e^b = e^b\mathfrak{m}_0^b(1)e^b,$$

we have the following.

Proposition 5.9. *$(C, \mathfrak{m}^b)$ is also a filtered A_∞ algebra. In addition, $\widehat{d}(e^b) = 0$ is equivalent to $\mathfrak{m}_0^b = 0$.*

This implies that an unobstructed filtered A_∞ algebra can be deformed to a filtered A_∞ algebra with $\mathfrak{m}_0 = 0$.

5.C) Homotopy equivalence, dependence and independence.

Let $(C_i, \mathfrak{m}^i)$, $i = 1, 2$, be filtered A_∞ algebras over the ring $\Lambda_{0,nov}$. For $k = 0, 1, 2, \cdots$, let us consider the family of maps

$$\mathfrak{f}_k : B_k(C_1[1]) \to C_2[1]$$

of degree 0 such that

$$\mathfrak{f}_k(F^\lambda B_k(C_1[1])) \subseteq F^\lambda C_2[1] \tag{5.10.1}$$

and

$$\mathfrak{f}_0(1) \in F^{\lambda'} C_2[1] \quad \text{for some } \lambda' > 0. \tag{5.10.2}$$

Note that $\mathfrak{f}_0 : \Lambda_{0,nov} \to C_2[1]$. These maps induce

$$\varphi_k : B_k(C_1[1]) \to \widehat{B}(C_2[1]),$$

by

$$\begin{aligned}\varphi_k(x_1 \otimes \cdots \otimes x_k) = \sum_{0 \le k_1 \le \cdots \le k_n \le k} & \mathfrak{f}_{k_1}(x_1, \cdots, x_{k_1}) \otimes \cdots \\ & \cdots \otimes \mathfrak{f}_{k_{i+1}-k_i}(x_{k_i+1}, \cdots, x_{k_{i+1}}) \otimes \cdots \\ & \cdots \otimes \mathfrak{f}_{k-k_n}(x_{k_n+1}, \cdots, x_k),\end{aligned} \tag{5.11}$$

and (5.10) implies that $\sum \varphi_k = \widehat{\varphi}$ converges. We note that when $\mathfrak{f}_0$ appears in the right hand side of (5.11), we put $\mathfrak{f}_0(1)$ there. Thus, in particular, φ_0 is given by

$$\varphi_0(1) = 1 + \mathfrak{f}_0(1) + \mathfrak{f}_0(1) \otimes \mathfrak{f}_0(1) + \cdots = e^{\mathfrak{f}_0(1)}$$

in our notation. (See (5.1)). Then it is easy to see that $\widehat{\varphi} : \widehat{B}(C_1[1]) \to \widehat{B}(C_2[1])$ is a coalgebra homomorphism.

Definition 5.12. We call $\mathfrak{f} = \{\mathfrak{f}_k\}_{k \ge 0}$ a *filtered A_∞ homomorphism from C_1 to C_2* if $\widehat{\varphi} \circ \widehat{d^1} = \widehat{d^2} \circ \widehat{\varphi}$.

Let $\mathbf{e}_{i_j}$ be a basis of C_1 as in (4.2.5). We say that $\mathfrak{f}$ satisfies the *gap condition* if

$$(5.12.1) \qquad \mathfrak{f}_k(\mathbf{e}_{i_1}, \ldots, \mathbf{e}_{i_k}) - \overline{\mathfrak{f}}_k(\mathbf{e}_{i_1}, \ldots, \mathbf{e}_{i_k}) \in F^{\lambda''} C_2[1]$$

where $\lambda'' > 0$ is independent of i_j and k. (Here $\overline{\mathfrak{f}}$ denotes the induced map on the (not filtered) A_∞ algebra over $\Lambda_{0,nov}/\Lambda^+_{0,nov} \cong \mathbf{Q}$, (see §4.))

Let $\mathfrak{f}^i_k : B_k(C_i[1]) \to C_{i+1}[1]$ $(i = 1, 2)$ define a filtered A_∞ homomorphism. Then the composition $\mathfrak{f}^2 \circ \mathfrak{f}^1 = \{(\mathfrak{f}^2 \circ \mathfrak{f}^1)_k\}$ of $\mathfrak{f}^1$ and $\mathfrak{f}^2$ is

$$\begin{aligned}&(\mathfrak{f}^2 \circ \mathfrak{f}^1)_k(x_1, \cdots, x_k)\\ &= \sum_m \sum_{k_1+\cdots+k_m=k} \mathfrak{f}^2_m(\mathfrak{f}^1_{k_1}(x_1, \cdots, x_{k_1}), \cdots, \mathfrak{f}^1_{k_m}(x_{k-k_m+1}, \cdots, x_k)).\end{aligned}$$

which defines a filtered A_∞ homomorphism from C_1 to C_3.

Let $(C_i, \mathfrak{m}^i)$ $(i = 1, 2)$ be filtered A_∞ algebras over $\Lambda_{0,nov}$ and $\mathfrak{f} : (C_1, \mathfrak{m}^1) \to (C_2, \mathfrak{m}^2)$ a filtered A_∞ homomorphism. Then $\mathfrak{f}$ naturally induces an A_∞ homomorphism $\overline{\mathfrak{f}} : (\overline{C}_1, \overline{\mathfrak{m}}^1) \to (\overline{C}_2, \overline{\mathfrak{m}}^2)$, where $(\overline{C}_i, \overline{\mathfrak{m}}^i)$ are the A_∞ algebras over $\mathbf{Q} \cong \Lambda_{0,nov}/\Lambda^+_{0,nov}$. If $\overline{\mathfrak{m}}_0 = 0$, then we note that $\overline{\mathfrak{m}}_1 \overline{\mathfrak{m}}_1 = 0$, (see Remark 4.9).

Definition 5.13. Let $(C_i, \mathfrak{m}^i)$ $(i = 1, 2)$ be filtered A_∞ algebras over $\Lambda_{0,nov}$ such that $\overline{\mathfrak{m}}^i_0 = 0$. For these filtered A_∞ algebras, we say that a filtered A_∞ homomorphism $\mathfrak{f} : (C_1, \mathfrak{m}^1) \to (C_2, \mathfrak{m}^2)$ is a ***weak homotopy equivalence***, if the induced A_∞ homomorphism $\overline{\mathfrak{f}} : (\overline{C}_1, \overline{\mathfrak{m}}^1) \to (\overline{C}_2, \overline{\mathfrak{m}}^2)$ induces an isomorphism $\overline{\mathfrak{f}}_1 : H^*(\overline{C}_1, \overline{\mathfrak{m}}^1_1) \to H^*(\overline{C}_2, \overline{\mathfrak{m}}^2_1)$.

We recall that the condition $\overline{\mathfrak{m}}_0 = 0$ is satisfied in our filtered A_∞ algebra $(C(L, \Lambda_{0,nov}), \mathfrak{m})$, see Definition 4.12.

Hereafter we assume that the filtered A_∞ algebras $(C_i, \mathfrak{m}^i)$ are *unobstructed* and *weakly finite*. Here, a filtered A_∞ algebra $(C, \mathfrak{m})$ is called *unobstructed* if there exists a bounding cochain $b \in C[1]^0$ such that $\widehat{d}(e^b) = 0$, and *weakly finite* if there exists a finite $\Lambda_{0,nov}$ module cochain complex (C', δ') such that there is a filtered A_∞ homomorphism $\mathfrak{f}' : (C', \delta') \to (C, \mathfrak{m})$ with satisfying the gap condition (5.12.1) which induces an isomorphism between $H^*(C', \delta')$ and $H^*(C[1], \mathfrak{m}^b_1)$, (see Definition 5.8 and Proposition 5.9 for $\mathfrak{m}^b$. We also note that our unobstructed filtered A_∞ algebra $(C(L, \Lambda_{0,nov}), \mathfrak{m})$ associated to Lagrangian submanifold L is weakly finite, see [FOOO] Theorem A4.28 in §A4). Under these assumptions, we can obtain the following lemma. (Kontsevich shows a similar lemma in the case of L_∞ algebra [K]).

Lemma 5.14. *Let $(C_i, \mathfrak{m}^i)$ be unobstructed and weakly finite filtered A_∞ algebras ($i = 1, 2$). If a filtered A_∞ homomorphism $\mathfrak{f}^1$ is a weak homotopy equivalence and if it satisfies the gap condition (5.12.1), then there exists a filtered A_∞ homomorphism $\mathfrak{f}^2$ such that both of the compositions $(\mathfrak{f}^1 \circ \mathfrak{f}^2)_1$ and $(\mathfrak{f}^2 \circ \mathfrak{f}^1)_1$ induce the identities on the cohomologies $H^*(\overline{C}_i[1], \overline{\mathfrak{m}}_1^i)$.*

Definition 5.15. Let $(C_i, \mathfrak{m}^i)$ be filtered A_∞ algebras over $\Lambda_{0,nov}$ with $\overline{\mathfrak{m}}_0^i = 0$ ($i = 1, 2$). We assume that $(C_i, \mathfrak{m}^i)$ are unobstructed and weakly finite. Then $(C_1, \mathfrak{m}^1)$ and $(C_2, \mathfrak{m}^2)$ are said to be *weakly homotopy equivalent* if there exist filtered A_∞ homomorphisms $\mathfrak{f}^1$ and $\mathfrak{f}^2$ from C_1 to C_2 and C_2 to C_1 respectively such that the compositions $\mathfrak{f}^1 \circ \mathfrak{f}^2$, $\mathfrak{f}^2 \circ \mathfrak{f}^1$ are weakly homotopy equivalences.

Now we recall from Proposition 5.7 that two bounding cochains b_1, b_2 on filtered A_∞ algebra $(C, \mathfrak{m})$ induce a coboundary map

$$\delta_{b_1, b_2} : C \to C$$

as in (5.6). We next prove that a weak homotopy equivalence induces a natural isomorphism between the cohomology of (C, δ_{b_1, b_2}). We first note the following lemma.

Lemma 5.16. *For a non-zero element $\mathbf{x}$ of $\widehat{B}(C[1])$, $\mathbf{x} = e^b$ for some $b \in B_1(C[1]) = C[1]$ if and only if $\Delta \mathbf{x} = \mathbf{x} \otimes \mathbf{x}$, where Δ is the coproduct as in (4.3).*

Since $\widehat{\varphi} : \widehat{B}(C[1]) \to \widehat{B}(C[1])$ is a coalgebra homomorphism, if $\mathbf{x}$ satisfies $\Delta \mathbf{x} = \mathbf{x} \otimes \mathbf{x}$, so does $\widehat{\varphi}(\mathbf{x})$ and so we have an element $\varphi(b_0)$ such that $\widehat{\varphi}(e^{b_0}) = e^{\varphi(b_0)}$ by Lemma 5.16. More explicitly, we have

$$\varphi(b_0) = \mathfrak{f}_0(1) + \mathfrak{f}_1(b_0) + \mathfrak{f}_2(b_0, b_0) + \cdots . \tag{5.17}$$

Proposition 5.18. *Let $(C_0, \mathfrak{m}^0)$ and $(C_1, \mathfrak{m}^1)$ be the filtered A_∞ algebras such that $\overline{\mathfrak{m}}_0^i = 0$ ($i = 0, 1$). We assume that $(C_i, \mathfrak{m}^i)$ are unobstructed and weakly finite. Let $\mathfrak{f} = \{\mathfrak{f}_k\}_{k \geq 0}$ define a weak homotopy equivalence between them. Let $\widehat{d}^0$, $\widehat{d}^1$ be obtained from $\mathfrak{m}^0$, $\mathfrak{m}^1$, and φ obtained from $\mathfrak{f}_k$. Let $b_0 \in (C_0[1])^0$. Then $\widehat{d}^0 e^{b_0} = 0$ if and only if $\widehat{d}^1 e^{\varphi(b_0)} = 0$.*

If we moreover assume the gap condition for $(C_0, \mathfrak{m}^0)$, $(C_1, \mathfrak{m}^1)$ and $\mathfrak{f}$, then the cohomology of $\delta^0_{b_1, b_2}$ is isomorphic to that of $\delta^1_{\varphi(b_1), \varphi(b_2)}$, $\widehat{d}^0 e^{b_0} = 0$.

For the proof of the last assertion, we need a spectral sequence argument. To construct the spectral sequence, we need the gap condition. See [FOOO] for more details.

Next, we can define an equivalence relation $\sim$ in the set of bounding cochains $\widehat{\mathcal{M}}(L)$, (see Definition 5.2). This can be regarded as a sort of "gauge equivalence" relation. A similar notion is introduced in [K] for L_∞ algebra. We do not explain it here. See [FOOO]. Anyway, we can show that the following.

Theorem 5.19. *Let (L, L') be a pair of relative spin Lagrangian submanifolds. Let $b_0, b_1 \in \widehat{\mathcal{M}}(L)$ and $b_0', b_1' \in \widehat{\mathcal{M}}(L')$. Assume that $b_0 \sim b_1$ and $b_0' \sim b_1'$. Then the deformed Floer cohomology $HF((L, b_0), (L', b_0'))$ is canonically isomorphic to $HF((L, b_1), (L', b_1'))$*

More generally, we can show the followings. We set $\mathcal{M}(L) = \widehat{\mathcal{M}}(L)/\sim$.

Theorem 5.20. *Let (L, L') be a pair of relative spin Lagrangian submanifolds of M. Then we have the following:*

(5.20.1) *$\mathcal{M}(L; J, \Xi)$ is independent of the choice of J, Ξ. Namely there exists a canonical isomorphism $\mathcal{M}(L; J, \Xi) \cong \mathcal{M}(L; J', \Xi')$. (Hereafter we omit J, Ξ and write $\mathcal{M}(L)$ in case no confusion can occur.)*

(5.20.2) *Floer cohomology is also independent of J, Ξ. More precisely we have the following : Let $b_0 \in \mathcal{M}(L; J_0, \Xi_0)$, $b_0' \in \mathcal{M}(L'; J_0, \Xi_0')$. Let $b_1 \in \mathcal{M}(L; J_1, \Xi_1)$, and $b_1' \in \mathcal{M}(L'; J_1, \Xi_1')$ corresponds to them by the isomorphism in (5.20.1). Then there exists a canonical isomorphism*

$$HF((L, b_0), (L', b_0'); J_0, \Xi_0, \Xi_0') \cong HF((L, b_1), (L', b_1'); J_1, \Xi_1, \Xi_1').$$

Hereafter we write $HF((L, b_0), (L', b_0'))$ in place of

$$HF((L, b_0), (L', b_0'); J_0, \Xi_0, \Xi_0')$$

when no confusion can occur.

(5.20.3) *Any Hamiltonian diffeomorphism ψ induces a map*

$$\psi_* : \mathcal{M}(L) \simeq \mathcal{M}(\psi(L)),$$

which depends only on the homotopy class of the Hamiltonian diffeomorphism $\psi : L \to \psi(L)$. Namely if ψ^s be a family of Hamiltonian diffeomorphisms such that $\psi^s(L)$ is independent of s then $\psi_*^0 = \psi_*^1$.

(5.20.4) *Let $\Psi = \{\psi^\tau\}_{0 \le \tau \le 1}$ and $\Psi' = \{\psi'^\tau\}_{0 \le \tau \le 1}$ be Hamiltonian isotopies with $\psi^0 = \psi'^0 = id$. Let $b_0 \in \mathcal{M}(L)$, and $b_0' \in \mathcal{M}(L')$. We put $b_1 = \psi_*^1(b_0) \in \mathcal{M}(\psi^1(L), J_1)$, $\psi_*'^1(b_0') = b_1' \in \mathcal{M}(\psi'^1(L'))$. Then Ψ, Ψ' induces an isomorphism*

$$(\Psi, \Psi')_* : HF((L, b_0), (L', b_0')) \cong HF((\psi^1(L), b_1), (\psi'^1(L'), b_1')),$$

which depends only on homotopy types of Hamiltonian isotopies between id and ψ^1 and between id and ψ'^1

Theorem 5.20 says, up to ambiguity of the choice of $\mathcal{B}$, the obstruction class and Floer cohomology are independent of the Hamiltonian isotopy and of the almost complex structure.

Moreover, we can show that $\mathcal{M}(L)$ can be described as some quotient space of the zero set of certain formal map (Kuranishi map). So it describes the deformation space. See [FOOO].

§6. Some applications.

In this last section, we give some applications of our theory to some concrete problems in symplectic geometry. For the proofs, see [FOOO].

The first one is the Arnold conjecture for Lagrangian intersections.

Theorem 6.1. *Assume that L is relatively spin closed Lagrangian submanifold of (M,ω) and that the natural map $H_*(L;\mathbb{Q}) \to H_*(M;\mathbb{Q})$ is injective. Then for any Hamiltonian diffeomorphism $\psi : M \to M$ such that L and $\psi(L)$ intersect transversally, we have*

$$\sharp(L \cap \psi L) \geq \sum_k \mathrm{rank} H_k(L;\mathbb{Q}).$$

The assumption that the natural map $H_*(L;\mathbb{Q}) \to H_*(M;\mathbb{Q})$ is injective implies that all our obstruction classes vanish. (See Theorem 2.6). We remark that this theorem implies the Arnold conjecture for the fixed point sets of Hamiltonian diffeomorphisms (over $\mathbb{Q}$-coefficients) which is proved by [FO], [LT] etc. Namely, let us consider $L = \Delta$ (the diagonal set) in $(M \times M, \omega \oplus -\omega)$. Then the intersection points are nothing but the fixed points of ψ. The relative spinness for Δ and the assumption above are automatically satisfied by the Kunneth formula.

More generally, by using our spectral sequence, we can get the following.

Theorem 6.2. *Let L be relatively spin and assume that the associated A_∞ algebra is unobstructed. Denote $A = \sum \mathrm{rank}\ H(L;\mathbb{Q})$ and*

$$B = \sum \mathrm{rank}\ker(H(L;\mathbb{Q}) \to H(M;\mathbb{Q})).$$

Then we have

$$\sharp(L \cap \psi(L)) \geq A - 2B$$

for any Hamiltonian diffeomorphism $\psi : M \to M$ such that L and $\psi(L)$ intersect transversally.

Next application is so called Arnold-Givental conjecture, which is a variant of Arnold conjecture. In general, the most naive statement such as

$$\sharp(L \cap \phi(L)) \geq \sum \mathrm{rank} H_*(L; \mathbb{Z}/2\mathbb{Z}) \tag{6.3}$$

is *not* true for general L and general Hamiltonian diffeomorphism ϕ. In this respect, Givental made a conjecture that (6.3) is true at least if L is the fixed point set of an anti-symplectic involution. However a careful analysis on the orientation of the moduli space shows that this cancellation does *not* happen over $\mathbb{Q}$ (or over $\mathbb{Z}$) but works only over $\mathbb{Z}/2\mathbb{Z}$-coefficient in general. Now we can prove the following :

Theorem 6.4. *Let $L = \mathrm{Fix}\,\tau$ be the fixed point set of an anti-symplectic involution $\tau : (M, \omega) \to (M, \omega)$ and L be semi-positive. Then the inequality* (6.3) *holds.*

Here the n dimensional Lagrangian submanifold L in (M, ω) is called semi-positive, if $\omega(\beta) \leq 0$ for any β with

$$3 - n \leq \mu_L(\beta) < 0.$$

Note that if $n \leq 3$, the semi-positivity automatically holds. This condition plays a role similar to the case of absolute case. The reason why we need to assume the semi-positivity is to handle the negative multiple cover problem. We recall that we should use $\mathbb{Z}/2\mathbb{Z}$-coefficient to have the cancellation of quantum effects in general which forces us to use integral cycles rather than rational cycles. We would like to emphasize that since we use $\mathbb{Z}/2\mathbb{Z}$-coefficients, we do not have to assume our Lagrangian submanifold is relatively spin.

The third application is so called the Maslov class conjecture. The general folklore conjecture says that the Maslov class $\mu_L \in H^1(L; \mathbb{Z})$ of Lagrangian embedding $L \subset \mathbb{C}^n$ is non-trivial for any compact Lagrangian embedding in $\mathbb{C}^n$. (We note that if the ambient symplectic manifold (M, ω) satisfies $c_1(TM) = 0$, then μ_L can be regarded as an element of $H^1(L, \mathbb{Z})$.) We can give a new partial answer.

Theorem 6.5. *Let L be a compact embedded Lagrangian submanifold of $\mathbb{C}^n$ that satisfies $H^2(L; \mathbb{Z}/2\mathbb{Z}) = 0$. Then its Maslov class $\mu_L \in H^1(L; \mathbb{Z})$ is nonzero.*

Moreover we can show the following estimate.

Theorem 6.6. *Let L be a compact embedded Lagrangian submanifold of $\mathbb{C}^n$. Suppose that it is unobstructed in the sense of Definition 5.2. Then we have the following inequality;*

$$1 \leq \Sigma_L \leq n + 1.$$

Here Σ_L *is a non-negative integer defined by* Image $(\mu_L) = \Sigma_L \mathbb{Z}$, *where* μ_L *is the Maslov index homomorphism in* (2.0).

References

[ASKZ] M. Alexandrov, A. Schwarz, M. Kontsevich and O. Zaboronsky, The geometry of the master equation and topological quantum field theory, *Intern. J. Modern Phys.* A, **12** (1997) 1405 - 1429.

[BV] I. Batalin and G. Vilkovsky, Quantization of gauge theories with linearly dependent generators, *Phys. Lett.* B, **311** (1993) 123 - 129.

[B] K. Behrend, Gromov-Witten invariants in algebraic geometry, *Invent. Math.*, **127** (1997) 604 - 617.

[FOOO] K. Fukaya, Y-G Oh, H. Ohta and K. Ono, Lagrangian intersection Floer theory – Anomaly and Obstruction –, to appear.

[FO] K. Fukaya and K. Ono, Arnold conjecture and Gromov-Witten invariants, *Topology* **38** (1999) 933 - 1048.

[GJ] E. Getzler and J. Jones, A_∞ algebra and cyclic bar complex, *Illinois J. Math.* **34** (1990) 256 - 283.

[K] M. Kontsevich, Deformation quantization of Poisson manifolds, q-alg/9709040

[LT] G. Liu and G. Tian, Floer homology and Arnold conjecture, *J. Diff. Geom.* **49** (1998) 1 - 74.

[Oh] Y.-G. Oh, Floer cohomology of Lagrangian intersections and pseudo-holomorphic disks I, II, *Comm. Pure and Appl. Math.* **46** (1993) 949 - 994 and 995 - 1012. *Addenda, ibid,* **48** (1995), 1299 - 1302.

[Sch] V. Schechtman, Remarks on formal deformations and Batalin-Vilkovsky algebras, math/9802006

[St] J. Stasheff, Homotopy Associativity of H-Spaces I, II, *Trans. Amer. Math. Soc.* (1966) 275 - 312.

TOPOLOGICAL OPEN P-BRANES

JAE-SUK PARK
Department of Physics, Columbia University
New York, N.Y. 10027, U.S.A.
E-mail: jspark@phys.columbia.edu

By exploiting the BV quantization of topological bosonic open membrane, we argue that flat 3-form C-field leads to deformations of the algebras of multi-vectors on the Dirichlet-brane world-volume as 2-algebras. This would shed some new light on geometry of M-theory 5-brane and associated decoupled theories. We show that, in general, topological open p-brane has a structure of $(p+1)$-algebra in the bulk, while a structure of p-algebra in the boundary. The bulk/boundary correspondences are exactly as of the generalized Deligne conjecture (a theorem of Kontsevich) in the algebraic world of p-algebras. It also imply that the algebras of quantum observables of $(p-1)$-brane are "close to" the algebras of its classical observables as p-algebras. We interpret above as deformation quantization of $(p-1)$-brane, generalizing the $p = 1$ case. We argue that there is such quantization based on the direct relation between BV master equation and Ward identity of the bulk topological theory. The path integral of the theory will lead to the explicit formula. We also discuss some applications to topological strings and conjecture that the homological mirror symmetry has further generalizations to the categories of p-algebras.

1 Introduction

The discovery of D-branes – extended objects carrying RR charge, has greatly enhanced our understanding of string theory [65]. D-branes can be realized as certain Dirichlet boundary condition of the fundamental open string. One can also say that the fundamental open strings describe excitations of the D-brane. The low energy dynamics of a D-brane is described by the maximally supersymmetric Yang-Mills (SYM) theory on the D-brane worldvolume. The open string can naturally be coupled to flat NS 2-form B-fields. It is by now well-known that there are suitable decoupling limits of the bulk degrees of freedom, and the dynamics of the D-brane worldvolume is described eitherby non-commutative SYM or by the non-commutative open string [73,72,33]. The basic picture is that the B-field induces a non-commutative deformation of the algebra $\mathcal{O}(X)$ of functions on the D-brane world-volume X; it makes the world-volume non-commutative [20,66,73].

For the non-commutativity the supersymmetry and the metric are secondary. We can consider open bosonic strings in an arbitrary number of Euclidean dimensions coupled only with the B-field (we may allow a Poisson bi-vector in general). Then the problem becomes equivalent to the quantization of the boundary particle theory, which has the "D-brane" world-volume as its classical phase space [17]. This consideration leads to the path integral derivation of the celebrated solution of deformation quantization by Kontsevich [50]. A crucial physical insight in the above approach is that the physical consistency of the bulk open string theory implies associativity of the non-commutative deformation of the algebra of functions on the D-brane world-volume [17]. A simple generalization of the above leads to a physical proof of the formality theorem of Kontsevich. This implies a deep connection between open strings and the world of associative algebras.

Strominger [77] and Townsend [82] showed that the M theory 5-brane can be interpreted as a D-brane of the open super-membrane and all D-branes of Type IIA string can be obtained by S^1-compactification. They also argued that the boundary dynamics of the above system is controlled by a six-dimensional self-dual string [96]. The open membrane can naturally be coupled with a flat 3-form C-field. The presence of the M5-brane requires self-duality for the parallel C-field. Recently the authors of [3,73,11,43,34] showed that there is suitable decoupling limit such that the bulk theory becomes topological and only the modes in the brane are left. The resulting theory is now called OM theory, which is related to other decoupled theories by a web of dualities [34]. An open question is the algebraic or geometrical meaning of turning on such a background, the resulting boundary dynamics, etc.

The main purpose of this paper is to uncover the basic picture on the role of C-field in more mundane situations. For this it is suffices to study the bosonic open membrane coupled with the C-field only. We call the resulting theory the topological open membrane theory. The topological open membrane makes sense in arbitrary dimensions, as does the topological open string. Actually we will start from one further step back by considering the open membrane without background. Then we interpret the topological open membrane theory as a certain deformation of the theory without background. The theory without background will tell us that the underlying algebraic structure of the boundary string theory is the Gerstenhaber algebra (G-algebra in short [30]) of polyvectors on X. More appropriately it is the algebra $\mathcal{O}(\Pi T^*X)$ of functions on the superspace ΠT^*X, which is the total space of the cotangent bundle over X after a parity change of the fiber. Then quantum consistency of the theory requires that the C-field must be flat, which corresponds to the infinitesimal deformation of the above G-algebra as a strongly homotopy G-algebra (a G_∞-algebra in short [79,81] or 2-algebra [50]). We call the resulting algebra the 2-algebra of X. Then one may specialize to the 6-dimensional case and consider the deformation by a self-dual C-field only. We may call this the self-dual 2-algebra of six dimensions X. An interesting point of OM theory is that it requires a non-vanishing constant self-dual C-field [34]. Thus the theory from the beginning should involve the deformed 2-algebra or self-dual 2-algebra of X. We will leave the detailed study of path integrals (deformed algebra) and applications to physics for a future publication [37].

Our approach also has a natural generalization to higher dimensional topological open p-branes. We shall see that open p-branes have a deep connection with the world of p-algebras. The bulk and the boundary correspondence of open p-brane theory follows exactly the generalized Deligne conjecture involving $(p+1)$ and p-algebra [51]. The crucial tools for our approach are the Feynman path integrals á la BV quantization [8,89,67,68,91,2,69]. We will also discuss some applications to the homological mirror conjecture [49]. We should mention that the 2-algebra is not an alien to string theory. It already appeared in the closed and open-closed string field theories of Zwiebach [102,103], VOA, TCFT and $D = 2$ string theory [100,85,101,58,14,64,32] (see also [76] for a nice review).

Now we begin a rather detailed introduction or sketch of our program, treating all topological open p-branes uniformly and emphasizing the more mathematical side of our story.

A sketch of our program

The basic idea of deformation quantization is that the algebra of observables in quantum mechanics is close, as an associative algebra, to the commutative algebra of functions on the classical phase space [9,50]. Thus the program reduces to finding formal deformations of the commutative algebra along non-commutative directions as an associative algebra. This is realized as deformations of the usual products of functions to star products, whose associativity automatically implies that an infinitesimal should be a Poisson bi-vector. A surprise of Kontsevich's result is that the quantization of the particle somehow requires open string theory [50]. Catteneo and Felder showed that Kontsevich's formula is the perturbative expansion of the path integral of bosonic topological open string theory [17].

A novelty of this approach is that the problem of deformation quantization of particles (thus the deformation of an associative algebra) is equated to quantum consistency of the bosonic open string theory. This maps the set of equivalence classes of Poisson structures on the target space X to the set of equivalence classes of deformations of the open string theory satisfying the BV master equation. Then the BV master equation automatically implies, via the Ward identity, that a suitable path integral of the theory on the disk defines a bijection from the above equivalence class to the set of isomorphism classes of associative star products. In general this approach leads to a string theoretic derivation of the formality theorem of Kontsevich. The unifying mathematical notion behind the above correspondences is the operads of little intervals and associated Swiss-Cheese operads, which relate associative algebras with 2-algebras (Deligne conjecture) [79,80,51]. On the other hand the Swiss-Cheese operad [86] is closely related to the tree level open-closed string field theory of Zwiebach [103]. Actually the path integral approach shows that the Deligne conjecture is just the bulk/boundary correspondence. Recently Hofman and Ma discussed those interrelations in a more general class of topological open-closed string theory [38] (see also [56,63] on some recent development on topological open-closed string theory).

We shall see that the topological open p-brane for any $p > 0$ is closely related with the world of $(p+1)$ and p-algebras. This generalizes the relation in the $p = 1$ case, where 1-algebra is another name for associative algebra. We may use topological open p-brane theory to define deformation quantization of the boundary closed bosonic $(p-1)$-brane. We shall see that the problem is equivalent to the problem of deformation of p-algebra as a p-algebra. Our basic tool is the method of BV quantization for Feynman path integrals. Here we sketch the general principle of our program.

Bosonic open p-brane theory is a theory of maps

$$\phi : N_p \to X,$$

where N_p is a $(p+1)$-dimensional manifold with boundary, regarded as the open p-brane world volume, and X is the target space. The bosonic topological open p-brane coupled with a $(p+1)$-form c in X (open p-brane in a closed NS p-brane background) is described by the action functional

$$I' = \int_{N_p} \phi^*(c) + \int_{\partial N_p} \mathcal{V} \tag{1.1}$$

where $\mathcal{V}$ denotes a possible boundary interaction. We will regard the above theory as describing deformations of bosonic open p-brane theory ***without background*** defined by certain action functional I_o being first order in derivatives and invariant under affine transformations of X. After BV quantization one may obtain the BV master action functional $\boldsymbol{S}_o$ of I_o. Then we examine consistent deformations, modulo equivalence, of $\boldsymbol{S}_o$ using the BV master equation. The resulting theory $\boldsymbol{S}'$ will be identified with the BV quantization of the theory I' if the deformation preserves the ghost number symmetry.

In general the undeformed theory $\boldsymbol{S}_o$ tells us which mathematical structure (associated with the boundary degrees) we want to deform. It can be determined by correlation functions of observables inserted on the boundary

∂N_p. Now the bulk deformation term in the action functional tells us how to deform the mathematical structure by specifying only the infinitesimal deformation. And the perturbative expansion with the bulk term determines the deformation of the mathematical structure to all orders. Note that the consistent bulk deformation term is determined by the BV master equation, while the correlation functions should satisfy the BV Ward identity of the theory, which is a direct consequence of the BV master equation.

It turns out that the BV quantized topological open p-brane theory is a theory of maps

$$\phi : \Pi T N_p \to M_p(X) \tag{1.2}$$

between two superspaces $\Pi T N_p$ and $M_p(X)$ associated with N_p and X, respectively.

The superspace $\Pi T N_p$ is the total space of the tangent bundle of N_p after parity change of the fiber. We denote a set of local coordinates on $\Pi T N_p$ by $(\{x^\mu\}|\{\theta^\mu\})$, $\mu = 1, \dots, p+1$, where θ^μ are odd constants. We introduce the ghost number or degree $U \in \mathbb{Z}$. We assign $U = 1$ to θ^μ.

The target superspace $M_p(X)$ of the p-brane for $p \geq 1$ can most easily be described recursively. M_p for $p \geq 1$ is the total space of the twisted by $[p]$ cotangent bundle $T^*[p]M_{p-1}$ over M_{p-1} and $M_0 = X$ is the target space X. For example

$$\begin{aligned} M_1 &= T^*[1]X \equiv \Pi T^* X, \\ M_2 &= T^*[2]M_1 = T^*[2]\left(\Pi T^* X\right), \\ M_3 &= T^*[3]M_2 = T^*[3]\left(T^*[2]\left(\Pi T^* X\right)\right), \end{aligned} \tag{1.3}$$

etc. Note that the base space M_{p-1} of the target superspace M_p of the p-brane is the target superspace of the $(p-1)$-brane. Physically for open the p-brane, M_{p-1} corresponds to the target superspace of boundary the $(p-1)$-brane. The iterative nature of target superspace is due to the degeneracy of the first order formalism, which requires to introduce ghosts for ghosts etc.

Now we explain the notation $T^*[p]M_{p-1}$. We introduce the following set of local coordinates $(base|fiber)$ on $M_p \to M_{p-1}$;

$$(\{q^\alpha\}|\{p_\alpha\}), \tag{1.4}$$

where $\alpha = 1, \dots, 2^{p-1} \times \dim(X)$. We assign ghost number U or degree of such coordinates by the formula

$$U(q^\alpha) + U(p_\alpha) = p, \qquad U(q^\alpha) \geq 0, \quad U(p_\alpha) \geq 1. \tag{1.5}$$

Thus the twisted by $[p]$ cotangent bundle $T^*[p]M_{p-1}$ over M_{p-1} is the cotangent bundle over M_{p-1} with the above assignment of ghost number. A coordinate is commuting or even if the ghost number is even. A coordinate is anti-commuting or odd if the ghost number is odd. An index α can be either a tangent or a cotangent index of X and indices for q^α and p_β should be (up, down) or (down,up) for $\alpha = \beta$. The ghost number of various coordinates can be determined recursively by assigning $U = 0$ to local coordinates on $M_0 = X$. For instance the ghost numbers of, say, base coordinates q^α of M_p can be different in general for each α. We also note that $M_p(X)$ for $p \geq 2$ can be identified with the total space of $(p-1)$th iterated supertangent bundle over ΠT^*X, i.e., $M_p(X) \simeq \Pi T(\Pi T(\ldots(\Pi T(\Pi T^*X))\ldots))$.

We describe a map $\boldsymbol{\phi} : \Pi TN_p \to M_p$ locally by local coordinates on M_p

$$(\boldsymbol{q}^\alpha, \boldsymbol{p}_\alpha) := (q^\alpha(x^\mu, \theta^\mu), p_a(x^\mu, \theta^\mu)) \tag{1.6}$$

which are functions on ΠTN_p. The superfields $(\boldsymbol{q}^\alpha, \boldsymbol{p}_\alpha)$ combine all the "fields" and "anti-fields" of the theory. The assignment of ghost numbers (1.5) are a consequence of BV quantization.

We note that the target superspace $M_p(X)$, for $p \geq 1$, always has the following non-degenerate canonical symplectic form ω_p for any manifold X as the total space of the (twisted by $[p]$) cotangent bundle over M_{p-1};

$$\omega_p = dp_\alpha \wedge dq^\alpha. \tag{1.7}$$

The symplectic form ω_p carries degree $U = p$. The parity of ω_p is the same as the parity of p. Now the degree $U = p$ symplectic structure ω_p on M_p defines a degree $U = -p$ (odd or even) graded Poisson bracket $[.,.]_{p+1}$ on functions on M_p. The BV bracket $(.,.)_{BV}$ of the p-brane theory is an odd Poisson bracket with degree $U = 1$ among local functions on the space $\mathcal{A}$ of all maps $\boldsymbol{\phi} : \Pi TN_p \to M_p$. The corresponding odd symplectic form $\boldsymbol{\omega}$ with degree $U = -1$ on $\mathcal{A}$ originates from a degree $U = p$ (odd or even as the same parity of p) symplectic form ω_p on the (super)-target space M_p by the formula

$$\boldsymbol{\omega} := \int_{N_p} d^{p+1}\theta\, \boldsymbol{\phi}^*(\omega_p). \tag{1.8}$$

The super-integral shifts the degree U by $(-p-1)$. The above considerations lead us to the following crucial relation

$$\left(\int_{N_p} d^{p+1}\theta\, \boldsymbol{\phi}^*(\gamma), \int_{N_p} d^{p+1}\theta\, \boldsymbol{\phi}^*(\gamma)\right)_{BV} = \int_{N_p} d^{p+1}\theta\, \boldsymbol{\phi}^*([\gamma,\gamma]_{p+1}), \tag{1.9}$$

where γ is a local function on M_p.

The BV action functional $\mathcal{S}_o$ of the topological open p-brane without background is defined as follows

$$\mathcal{S}_o = \int_{N_p} d^{p+1}\theta \left(\boldsymbol{p}_\alpha D\boldsymbol{q}^\alpha + \phi^* \left(h(q^\alpha, p_a) \right) \right), \tag{1.10}$$

where $D = \theta^\mu \partial_\mu$ and $h(q^\alpha, p_a)$ is a degree $U = p+1$ function on M_p. The function $h(q^\alpha, p_a)$ is invariant under affine transformations on X and constant on X. It satisfies

$$[h, h]_{p+1} = 0, \qquad h(q^\alpha, p_\alpha)|_{M_{p-1}} = 0. \tag{1.11}$$

Thus h generates a differential Q_o with $U = 1$ and $Q_o^2 = 0$ via the bracket,

$$[h, \ldots]_{p+1} = Q_o. \tag{1.12}$$

Futhermore $Q_o|_{M_{p-1}} = 0$. It turns out that $h = 0$ if $p = 1$ and h for general p can be determined recursively. Note also that

$$\left(\int_{N_d} d^{d+1}\theta \ (\boldsymbol{p}_\alpha D\boldsymbol{q}^\alpha), \ldots \right)_{BV} = D. \tag{1.13}$$

The BV BRST charge $\boldsymbol{Q}_o$ carrying $U = 1$ corresponds to an odd Hamiltonian vector of $\mathcal{S}_o$ on the space $\mathcal{A}$ of all fields, i.e., $\boldsymbol{Q}_o = (\mathcal{S}_o, \ldots)_{BV}$. Then we obtain another crucial relation;

$$\boldsymbol{Q}_o = D + \phi^*(Q_o). \tag{1.14}$$

Combining (1.9), (1.9), and (1.14), we see that the action functional satisfies the quantum master equation if the boundary conditions are such that $\boldsymbol{p}_\alpha(x) = 0$ "in directions tangent" to ∂N_p for $x \in \partial N_p$.[a]

Now we describe the possible bulk deformations. We consider any function (or sum of functions in general) $\gamma(q^\alpha, p_\alpha)$ of M_p, whose degree $U = |\gamma|$ has the same parity as $p+1$. The action functional $\mathcal{S}_\gamma$ deformed by γ is given by

$$\mathcal{S}_\gamma = \mathcal{S}_o + \int_{N_p} d^{p+1}\theta \ \phi^* \left(\gamma \left(p^\alpha, q_\alpha \right) \right). \tag{1.15}$$

The above corresponds to an even function on $\mathcal{A}_p$. Combining (1.9) and (1.13), we see that the deformed action functional satisfies the master equation

[a]It means that the worldvolume (N_p) scalar components $p_\alpha(x^\mu)$ of $\boldsymbol{p}_\alpha$ vanish at the boundary, the vector components in directions tangent to ∂N_p vanish at the boundary, etc.

$(\boldsymbol{S}_\gamma, \boldsymbol{S}_\gamma)_{BV} = 0$[b] if and only if

$$\int_{\partial N_p} d^p\theta\ \boldsymbol{\phi}^*(\gamma(q^\alpha, p_\alpha) = 0,$$
$$[h+\gamma, h+\gamma]_{p+1} = 0. \tag{1.16}$$

Note that the first condition gives a reason to call $\int_{N_p} d^{p+1}\theta\ \boldsymbol{\phi}^*\ (\gamma)$ the bulk term. Now the boundary conditions and the definition (1.12) imply that the above conditions are equivalent to

$$\gamma|_{M_{p-1}} = 0, \tag{1.17}$$

Thus the set of equivalence classes of bulk deformations of the topological p-brane theory, satisfying the BV master equation, is isomorphic to the set of equivalence classes of solutions of the Maurer-Cartan (MC) equation for functions γ on M_{p+1} satisfying the condition $\gamma|_{M_{p-1}} = 0$. We denote the Hamiltonian vector of $h+\gamma$ by Q_γ;

$$Q_\gamma := [h+\gamma, \ldots]_{p+1}, \tag{1.18}$$

which satisfies $Q_\gamma^2 = 0$ due to the second equation in (1.16). We note that the restriction $Q_\gamma|_{M_{p-1}}$ of Q_γ to the base space M_{p-1} corresponds to a first order differential acting on functions on the base space M_{p-1}.

The BRST charge $\boldsymbol{Q}_\gamma \equiv (\boldsymbol{S}_\gamma, ...)$ or the odd Hamiltonian vector of $\boldsymbol{S}_\gamma$ is given by

$$\boldsymbol{Q}_\gamma = D + \boldsymbol{\phi}^*(Q_\gamma), \tag{1.19}$$

which satisfies $\boldsymbol{Q}_\gamma^2 = 0$ due to the master equation. This implies that, for any function κ on M_p satisfying $Q_\gamma \kappa = 0$, we have

$$\boldsymbol{Q}_\gamma \boldsymbol{\phi}^*(\kappa) = D\boldsymbol{\phi}^*(\kappa). \tag{1.20}$$

This gives us so called the descent equations. Related to the above we note that the action functional $\boldsymbol{S}_\gamma$ has another fermionic symmetry generated by an odd vector $\mathcal{K}_\mu = -\frac{\partial}{\partial\theta^\mu}$ with $U = -1$ since it is written as a superspace integral. Then (1.14) implies that

$$\{\boldsymbol{Q}_\gamma, \boldsymbol{Q}_\gamma\} = 0, \qquad \{\boldsymbol{Q}_\gamma, \mathcal{K}_\mu\} = -\partial_\mu, \qquad \{\mathcal{K}_\mu, \mathcal{K}_\nu\} = 0. \tag{1.21}$$

Now we turn to the boundary interactions and boundary observables. The boundary interaction due to the boundary conditions is given by a certain local

[b]In the present model the classical master equation also implies the quantum master equation.

functional $\mathcal{V}(\boldsymbol{q}^\alpha, \boldsymbol{D}\boldsymbol{q}^\alpha)$ of $\boldsymbol{q}^\alpha$ and $\boldsymbol{D}\boldsymbol{q}^\alpha$;

$$\boldsymbol{S}'_o = \boldsymbol{S}_o + \int_{\partial N_p} d^d\theta\, \mathcal{V}(\boldsymbol{q}^\alpha, \boldsymbol{D}\boldsymbol{q}^\alpha). \tag{1.22}$$

The above action functional satisfies the quantum master equation for any such $\mathcal{V}$ satisfying $(\phi^*(Q_\gamma))\mathcal{V} = 0$. Clearly $\mathcal{V}$ should have $U = p$ to preserve the ghost number symmetry. We consider N_p as a $(p+1)$-dimensional disk with boundary $\partial N_p = S^p$. On the boundary we have $n+1$ punctures $x_0, \dots x_n$. We also consider S_i^{p-1} surrounding a puncture. We consider a local function $f(q^\alpha)$ of the base of $M_p \to M_{p-1}$ satisfying $Q_\gamma|_{M_{p-1}} f(q_\alpha) = 0$. Now we let $\boldsymbol{f} := f(\boldsymbol{q}_\alpha) := f(\boldsymbol{q}_\alpha(x^\mu, \theta^\mu))$ be the corresponding function of superfields $\boldsymbol{q}^\alpha$. The descent equation (1.20) implies that $\boldsymbol{Q}_\gamma \boldsymbol{f} = \boldsymbol{D}\boldsymbol{f}$ and we obtain the following non-trivial BV observables

$$\begin{aligned} \mathcal{O}^0_f(x_i) &= \boldsymbol{f}(x_i), \\ \mathcal{O}^{(p-1)}_f(x_i) &= \oint_{S_i^{p-1}} d^{p-1}\theta\, \boldsymbol{f}, \\ \mathcal{O}^{(p)}_f &= \int_{\partial N_p} d^p\theta\, \boldsymbol{f}. \end{aligned} \tag{1.23}$$

The last one above may be regarded as part of the boundary interaction.

Now we turn to the role of the path integral. For simplicity we ignore boundary interactions and consider the action functional $\boldsymbol{S}_\gamma$

$$\boldsymbol{S}_\gamma = \int_{N_d} d^{d+1}\theta\, (\boldsymbol{p}_\alpha \boldsymbol{D}\boldsymbol{q}^\alpha) + \int_{N_d} d^{d+1}\theta \left(h(\boldsymbol{q}^\alpha, \boldsymbol{p}_a) + \gamma(\boldsymbol{q}^\alpha, \boldsymbol{p}_a) \right). \tag{1.24}$$

The first term is the kinetic term and the remaining terms may be regarded as "interaction" terms. Note that the interaction terms are coming from the function $(h + \gamma)$ on M_p with $(h + \gamma)|_{M_{p-1}} = 0$, which we called bulk terms. After a suitable gauge fixing one may evaluate correlation functions of observables supported on the boundary ∂N_p using perturbation theory. Note that such observables originate from functions on the base space $M_{p-1}(X)$ of $M_p \to M_{p-1}(X)$.

Now we introduce the following definition

> We call the algebra $\mathcal{O}(M_p(X))$ of functions on $M_p(X)$ with the bracket $[.,.]_{p+1}$, and ordinary (super-commutative and associative) product the classical $(p+1)$-algebra $Cl_{p+1}(X)$ of X. Thus, by definition, the classical p algebra $Cl_p(X)$ of X is the algebra $\mathcal{O}(M_{p-1}(X))$ of functions on the base space of $M_p \to M_{p-1}$.

Note that the classical 1-algebra $Cl_1(X)$ ($p = 0$) is the algebra $\mathcal{O}(X)$ of functions on X without a bracket (since we do not have ω_0). The classical 2-algebra $Cl_1(X)$ ($p = 1$) is the algebra of polyvectors on X with the wedge product and Schouten-Nijenhuis bracket.

Combining altogether, we showed that

> The BV structure of the topological bosonic open p-brane theory originates from the classical $(p+1)$-algebra $Cl_{p+1}(X) = \mathcal{O}(M_p(X)$. A BV master action functional of the theory is determined by a bulk term associated with $(h+\gamma)$, which is a function on M_p with $(h+\gamma)|_{M_{p-1}} = 0$ solving the MC equation of $Cl_{p+1}(X)$. The classical algebra of observables in the boundary is the classical p-algebra $Cl_p(X)$, which is the algebra $\mathcal{O}(M_{p-1}(X))$ of functions on the base space M_{p-1} of $M_p \to M_{p-1}$.

The perturbative expansion of the theory can be viewed as a certain morphism $Cl_{p+1}(X) \to Qh_{p+1}(X)$ satisfying suitable Ward identities controlled by the bulk theory. We call $Qh_{p+1}(X)$ the quantum $(p+1)$ algebra. Note that the Ward identity is a direct consequence of the BV master equation. Thus the set of equivalence classes of dolutions to the BV master equation is isomorphic to the set of equivalence classes of of solutions to the Ward identity. This implies that the morphism is a quasi-isomorphism. Now the quantum algebra of observables (defined by the correlation functions) should be a deformation of the classical p-algebra $Cl_p(X)$. We call the resulting algebra of quantum observables the quantum p-algebra $Qh_p(X)$. This is again controlled by the bulk Ward identity.

It turns out that the classical p-algebra $Cl_p(X)$ is an example of the so called cohomological p-algebra $H^*(A_p)$ [31,50]. Kontsevich defined a p-algebra as an algebra over the operad $\mathtt{Chains}(\mathtt{C}_p)$, where $\mathtt{C}_p$ is the p-dimensional little disk operad. According to Kontsevich an algebra over the cohomology $H_*(\mathtt{C}_p)$ (cohomological p-algebra in short) is a twisted Gerstenhaber algebra with Lie bracket with degree $1-p$ for $p = 2k$, where k is a positive integer, and a twisted Poisson algebra with Lie bracket with degree $1-p$ for $p = 2k+1$, both with the commutative associative product of degree 0 and zero differential [51]. This agrees with our definition of a classical p-algebra $Cl_p(X)$.[c]

The generalized Deligne conjecture says that for every p-algebra there exists an universal $p+1$ algebra acting on it. For example there is a structure of $(p+1)$ algebra on the generalized Hochschild complex $\mathtt{Hoch}(A_p)$ of a p-algebra A_p [51,80]. The picture is that there is a $(p+1)$-algebra controlling deformations

---

[c]Note, however, that we generally have a differential.

of a p-algebra as a p-algebra. Kontsevich, generalizing Tarmarkin [79], proved the above conjecture as well as the formality of p-algebra, i.e., $\mathtt{Chains}(\mathtt{C}_p) \otimes \mathbb{R}$ is quasi-isomorphic to its cohomology $H_*(\mathtt{C}_p) \otimes \mathbb{R}$ endowed with zero differential. In particular the two sets of equivalence classes of solutions to the Maurer-Cartan (MC) equations on $\mathtt{Chains}(\mathtt{C}_p) \otimes \mathbb{R}$ and $H_*(\mathtt{C}_p) \otimes \mathbb{R}$ are isomorphic. Kontsevich suggested that the generalized Deligne conjecture seems to be related to quantum field theories with boundaries. He also conjectured the existence of a structure of p-algebra on p-dimensional conformal field theories [51].

It is amusing to see that all the above is beautifully realized in the topological open p-brane theory.[d] Recall that the bulk theory is determined by a cohomological $(p+1)$-algebra as the algebra of functions on $M_p(X)$, while the boundary observables are associated with a cohomological p-algebra as the algebra of functions on the base space M_{p-1} of $M_p \to M_{p-1}$. Note that the algebra of polynomials $\mathbb{R}[\{q^\alpha\}]$ – the polynomials in the coordinates of the base space $M_{p-1}(X)$ – can be viewed as a cohomological p-algebra without differential. It is easy to see (following sect.3.4. of [50]) that the Hochschild cohomology of the above cohomological p-algebra is the algebra $\mathbb{R}[\{q^\alpha\}, \{p^\alpha\}]$ – the polynomials in the coordinates of the total space M_p, without bracket. Note also that our assignment of ghost numbers is consistent with [50]. One may endow $\mathbb{R}[\{q^\alpha\}]$ with a bracket. Then the the Hochschild cohomology should have a differential. It turns out that the differential is exactly the differential Q_o associated with the term h that appeared in the definition of the open p-brane without background. In fact the term $\int_{N_p} d^{p+1}\theta\ \phi^*(h)$ in the action functional is responsible for the appearance of the bracket of the p-algebra by correlation functions.

It is also not difficult to see the appearance of $\mathtt{Hoch}(A_p)$ and the structure of $(p+1)$-algebra. Applying a theorem in [51] it is easy to see that the Hochschild cohomology $H^*(\mathtt{Hoch}(Cl_p(X)))$ of the classical algebra $Cl_p(X)$ is the classical $(p+1)$-algebra $Cl_{p+1}(X)$. Thus a bulk "interaction" term of the p-brane theory is an element of $H^*(\mathtt{Hoch}(Cl_p(X)))$.[e] The bulk "interaction" term (in the path integral) generates elements of the Hochschild complex $\mathtt{Hoch}(Cl_p(X))$ of the classical p-algebra $Cl_p(X)$ by perturbative expansions.

Thus all together we may conclude that the path integral of the topological open p-brane serves as a morphism between the two $(p+1)$-algebras associated

---

[d]We note that such a possibility was considered before [38,21], though without actual realizations.

[e]It seems to be more natural to modify Kontsevich's definition (Sect. 3.4 in [50]). In our case $H^*(\mathtt{Hoch}(Cl_p(X)))$ is an algebra of functions on $M_{p+1}(X)$ but with an additional condition that such a function should vanish after restriction to $M_{p-1}(X)$.

with the target space X; $H^*(\mathtt{Hoch}(Cl_p(X))$ and $\mathtt{Hoch}(Cl_p(X))$. As we argued before the morphism must be a quasi-isomorphism due to the BV master equation of the topological open p-brane theory. One should check this by working out the BV Ward identity of the theory by carefully working out the compactification of the moduli space, related with the higher dimensional Swiss-Cheese operads. Furthermore the Ward identity will tell us that the quantum algebra $Qh_p(X)$ of observables has the structure of a p-algebra. This motivates us to define the deformation quantization of the $(p-1)$-brane as the deformation of the algebra $\mathcal{O}(M_{p-1}(X)) \equiv Cl_p(X)$ of functions on $M_{p-1}(X)$ as a p-algebra.

We note the path integral derivation of the formality theorem for the $p = 1$ case goes along the same lines as the WDDV equation [95,23], and as that of 2D strings [100,85] involving compactification of the moduli space. Similarly the topological open p-brane is determined essentially by the bulk BV master action functional involving $Cl_{p+1}(X)$, boundary observables involving $Cl_p(X)$ and the moduli space of disks with boundary punctures. The rest is determined by the Feynman path integral. It is highly unlikely that the quantum algebra $Cl_p(X)$ is not a p-algebra defined by Kontsevich based on the same data [51]. Now there are several mathematical proofs of the generalized Deligne conjecture of formality of $p+1$ algebra [79,50,81,52]. However none of those proofs seem to give an explicit quasi-isomorphism except for the $p = 1$ case [49]. The topological open p-brane theory will lead to such an explicit formula. The details will appear elsewhere [37]. It will be also interesting to see if the path integrals of topological open p-brane "confirm" another conjecture of Kontsevich on the action of motivic Galois group [51].

We like to mention that our approach is quite similar to that of Witten in his attempt to formulate background independent open string field theory, suggesting a generalization for the closed string case [94]. It seems to be also closely related to the use of Chern-Simons theory in 3 dimensions on rational conformal field theory and vise versa [98]. The model for $p = 2$ is related with the so called extended or BV Chern-Simons theory [48,5,2,69] and the BV quantized higher dimensional BF theory [42,18]. The interplay between bulk and boundary obviously reminds of the Maldacena conjecture [60,36,99], which is perhaps a purely algebraic (or geometrical) remnant. There is long history on deep relations between open string theory and 1-algebras (A_∞-algebras [74]). A central example is the open string field theory based on a non-commutative and strictly associative algebra [88] and its generalization up to homotopy [29]. The associativity up to homotopy allows one to construct an action functional of open string field theory satisfying the Batalin-Vilkovisky (BV in short) master equation, therefore admitting a consistent quantization.

Similar structures also appear as Fukaya's A_∞-category [25] in the open string version of the so called A model of the topological sigma model [93]. The open-closed string field theory [103] is related to the algebra over Swiss-Cheese operads, while the closed string field theory is based on the L^∞ part of the homotopy 2-algebra [102,45,46]. It is also natural to expect that the homological mirror conjecture [49] can eventually be the physical equivalence of open-closed string field theory. It seems to be also reasonable to suspect that one may define open-closed p-brane field theory using higher dimensional Swiss-Cheese operads with additional decoration as in the $p = 1$ case. Actually the above is true only for the string tree level. The string field theory is based on a certain loop generalization of the above. For higher dimensional branes it is practically impossible to consider any "loops" as summing over all topologies and geometries. We may content with the theory defined on tree level.

The structure of this paper

This paper is organized as follows. In Sect. 2 we review the BV quantization method, emphasizing the underlying super-geometrical structure and relations with deformation theory. Then we reformulate the Catteno-Felder model and the A model in a language suitable for our purpose. This section will set up notations and our general strategy. In Sect. 3 we study BV quantization of the topological open membrane. This is a detailed example of the general structure discussed in the introduction for $p = 2$. We also present the leading deformations of $Cl_2(X)$. We also discuss different choices of boundary conditions and their implications to possible generalizations of homological mirror symmetry to the category of homotopy 2-algebras (open membrane). In Sect. 4 we return to the $p = 1$ case (string). We introduce an unified topological sigma model, which has the A, B and Catteneo-Felder models as special limits. We construct an extended B model parametrized by the extended moduli space of complex structures and show how the non-commutativity appears for the open string case. We briefly discuss applications of the extended B model to the homological mirror conjecture. We also conjecture that the homological mirror conjecture can be generalized to the category of any p-algebra of X or the physical equivalence of any topological open p-brane theory.

2 Preliminary

In this section we begin by reviewing the method of BV quantization. For details we refer to [8,91,67,68,94,2]. We compare it with the modern deformation theory and argue, as a general statement, that deformation problems of certain mathematical structures may be represented as a BV quantization problem of a suitable quantum field theory and vice versa. Then we reconsider the formulation of Kontsevich-Catteneo-Felder as a deformation problem of bosonic string theory. We also discuss BV quantization of the A model, generalizing [2], for a later purpose. We do not assume any originality in this section, but perhaps some new interpretations.

2.1 BV quantization and deformation theory

A path integral is a formal integral of certain observables over the space of all fields of a classical theory weighted by the exponential of the classical action functional I. An observable of the theory is a function on the space of all fields invariant under the symmetries of the classical action functional. Consequently one needs to mod out volume of the orbit of symmetry group. Thus we need to construct a "well-defined" quotient measure for the path integral. The BRST-BV quantization is a systematic and versatile way for doing this. The BV-BRST quantization can be done by the following steps.

The first step is to fermionize the symmetry by introducing anticommuting ghost fields for the infinitesimal parameters of the symmetry. We call the corresponding charge of the fermionic symmetry the BRST charge $\boldsymbol{Q}$. One introduces an additive quantum number U called ghost number or degree and assign $U = 1$ to $\boldsymbol{Q}$. We call the set of original fields and ghosts the set of "fields". One introduces a set of "anti-fields" for the set of "fields" such that in the space $\mathcal{A}$ of all fields (thus "fields" and "anti-fields") one has a natural odd symplectic structure $\boldsymbol{\omega}$. Then we have a corresponding odd Poisson bracket $(.,.)_{BV}$ called the BV bracket among the functions on $\mathcal{A}$. The idea is that one regards "fields" as coordinates, while "anti-fields" are regarded as corresponding conjugate momenta but with the opposite parities (commuting≡even and anti-commuting≡odd) in a certain infinite dimensional phase space. More precisely, one assigns integral ghost number U, or degree, to each field such that $U(\overline{\phi}) = -1 - U(\phi)$, where $\overline{\phi}$ is the "anti-field" of a "field" ϕ. The parity of a field is the same as the parity of its ghost number U. It follows that the odd symplectic form $\boldsymbol{\omega}$ carries $U = -1$, while the BV bracket carries $U = 1$. A BV bracket has the following properties

$$
\begin{aligned}
(A,B)_{BV} &= -(-1)^{(|A|+1)(|B|+1)}(B,A)_{BV},\\
(A,(B,C)_{BV})_{BV} &= ((A,B)_{BV},C)_{BV} + (-1)^{(|A|+1)(|B|+1)}(B,(A,C)_{BV})_{BV},
\end{aligned}
\tag{2.1}
$$

where A, B, C are (even or odd) local functions on $\mathcal{A}$ and $|A| = U(A)$, etc. We also have the Leibniz law, stating that the BV bracket behaves as a derivation on the ordinary product of functions on $\mathcal{A}$;

$$(A, BC)_{BV} = (A, B)_{BV}C + (-1)^{(|A|+1)|B|}B(A, C)_{BV}. \tag{2.2}$$

Such a product is (super)-commutative and associative and has degree 0.

One requires, in addition to the odd symplectic structure ω, that $\mathcal{A}$ has a volume element specified by a density ρ compatible with ω. Then one has a BV Laplacian Δ_ρ defined by $\Delta_\rho A = \frac{1}{2}\text{div}_\rho V_A$, where A is a function on $\mathcal{A}$, V_A is the Hamiltonian vector and div_ρ is the divergence calculated with respect to ρ. The compatibility is the condition $\Delta_\rho^2 = 0$. Note that Δ_ρ carries ghost number $U = 1$. From now one we shall not specify the density ρ explicitly. In local coordinates $(\{\phi^\alpha\}, \{\overline{\phi}_\alpha\})$ on $\mathcal{A}$ we have $\Delta = \partial^2/\partial\phi^\alpha\partial\overline{\phi}_\alpha$. Now the BV bracket can be defined by the failure of Δ being a derivation of the product of functions on $\mathcal{A}$ by the formula

$$(A, B)_{BV} = (-1)^{|A|}\Delta(AB) - (-1)^{|A|}\Delta(A)B + A\Delta(B). \tag{2.3}$$

The algebra of functions on $\mathcal{A}$ endowed, by the relations (2.1), (2.2) and (2.3), with the bracket $(.,.)_{BV}$ with $U = 1$ generated by Δ as well as with the (super)-commutative and associative product with $U = 0$, is called a BV algebra.

The BV action functional $\boldsymbol{S}$ is an even function on $\mathcal{A}$ with vanishing ghost number[f], such that (i) its restriction to the subspace of "fields" is the original classical action functional I, (ii) it generates the BRST symmetry via the BV bracket i.e.,

$$(\boldsymbol{S}, \ldots)_{BV} = \boldsymbol{Q}. \tag{2.4}$$

Equivalently $\boldsymbol{Q}$ is the odd Hamiltonian vector of $\boldsymbol{S}$. (iii) it satisfies the quantum master equation

$$(\boldsymbol{S}, \boldsymbol{S})_{BV} - 2\hbar\Delta\boldsymbol{S} = 0 \equiv \Delta e^{-\frac{\boldsymbol{S}}{\hbar}} = 0, \tag{2.5}$$

where $\hbar$ denotes Planck constant. A BRST-BV observable $\mathcal{O}$ is a function on $\mathcal{A}$ annihilated by $\boldsymbol{Q} - \hbar\Delta$. The BV master equation is the condition that the expectation value

$$\langle \mathcal{O} \rangle = \int_{\mathcal{L}} d\mu\, \mathcal{O}e^{-\boldsymbol{S}/\hbar} \tag{2.6}$$

of a BV observable $\mathcal{O}$ is invariant under continuous deformations of the Lagrangian subspace $\mathcal{L}$ with respect to ω in $\mathcal{A}$. It also implies that the following

[f] One may allow the action functional to be any even function. Our action functional will always have vanishing ghost number unless specified otherwise.

path integral identically vanishes

$$\langle(-\hbar\Delta + \mathcal{Q})A\rangle := \int_{\mathcal{L}} d\mu\,(-\hbar\Delta A + \mathcal{Q}A)e^{-\frac{1}{\hbar}\mathcal{S}} = 0, \tag{2.7}$$

for any product of functions A on $\mathcal{A}$. Picking a homology class of a Lagrangian subspace $\mathcal{L}$ of $\mathcal{A}$ is called a gauge fixing.

There are cases in which the classical master equation $(\mathcal{S}, \mathcal{S}) = 0$ implies the quantum master equation (2.5), i.e., $\Delta\mathcal{S} = 0$. Then the master equation implies that $\mathcal{S}$ has a fermionic symmetry generated by a nilpotent, $\mathcal{Q}^2 = 0$, BRST charge $\mathcal{Q}$, which acts on functions on $\mathcal{A}$ as an odd derivation. That is, from (2.1), (2.4), and the classical master equation,

$$\begin{aligned} &\mathcal{Q}^2 = 0, \\ &\mathcal{Q}(A, B)_{BV} = (\mathcal{Q}A, B)_{BV} + (-1)^{|A|+1}(A, \mathcal{Q}B)_{BV}. \end{aligned} \tag{2.8}$$

The above structure induces the on the BV algebra a structure of differential BV (dBV) algebra. Remark that $\mathcal{Q}$ can be identified with an odd nilpotent vector on $\mathcal{A}$.[g] Now we may further specialize to the case that there are classes of BV observables A_i satisfying $\Delta A_i = 0$. Such observables satisfy $\mathcal{Q}A_i = 0$. Note that the BRST charge $\mathcal{Q}$ transforms as a scalar under the rotation group of the manifold on which the field theory is defined. In other words the fermionic symmetry is global. Then the BV quantized theory, under the above assumption, is a cohomological field theory, first introduced by Witten. The fixed point theorem of Witten [92] implies that the path integral of $\mathcal{Q}$-invariant observables is further localized to an integral over the $\mathcal{Q}$-fixed point locus $\mathcal{M}$ in $\mathcal{L}$.

Now we turn to the BV Ward identity. We consider expectation values $\langle A_1 \dots A_n\rangle$ of products of functions A_i on $\mathcal{A}$. Note that the space $\mathcal{A}$ of all fields is a graded (by the ghost number) superspace. Thus observables also carry ghost numbers. In general any correlation function has vanishing ghost number. Usually the path integral measure carries a ghost number anomaly although the BV action functional has vanishing ghost number. Consequently the net ghost number of $A_1 \dots A_n$ should cancel the ghost number anomaly in order to have non-vanishing correlation function. Now the identity (2.7) implies that

$$\hbar\,\langle\Delta(A_1 \dots A_n)\rangle = \langle\mathcal{Q}(A_1 \dots A_n)\rangle\,. \tag{2.9}$$

The above identity is called the BV Ward identity, which is non-empty if the net ghost number of $A_1 \dots A_n$ plus 1 is the total ghost number anomaly. Now

[g] We will always denote, as is the convention, a derivation with the corresponding vector field by the same symbol.

consider the case that $\triangle A_i = 0$. From (2.1) and (2.3), we have

$$\hbar \sum_{1\le j<k\le n} \sigma_{jk} \left\langle (A_j, A_k)_{BV} \prod_{i\neq j,k} A_i \right\rangle = \langle Q(A_1 \dots A_n)\rangle \,, \tag{2.10}$$

where σ_{jk} is a sign factor. Now we assume that the classical master equation implies the quantum master equation. Then Q behaves like the exterior derivative on field space. Thus the right hand side of above receives contributions only from the boundary in field space. For this one should introduce an appropriate compactification.[h]

2.1.1. Deformations of BV quantized theory

Now we consider deformations of the BV quantized field theory, defined by the triple $(\mathcal{A}, \omega, S)$. Note that such a field theory is typically defined on a manifold N and $\mathcal{A}$ is the space of all fields under consideration (the space of all sections of a certain bundle over N), and ω is an odd symplectic structure on $\mathcal{A}$ with $U = -1$. The action functional S is a function with $U = 0$ defined as an integral of a top-form on N over N, satisfying the BV master equation. For simplicity we assume that $(S, S) = \triangle S = 0$.

Two BV quantized theories $(\mathcal{A}_1, \omega_1, S_1)$ and $(\mathcal{A}_2, \omega_2, S_2)$ are physically equivalent if there is a diffeomorphism $F : \mathcal{A}_1 \to \mathcal{A}_2$ such that $F^*\omega_2 = \omega_1$ and $F^* S_2 = S_1$. It also follows that the odd Hamiltonian vectors Q_1 and Q_2 of S_1 and S_2, respectively, are related as $Q_2 = F_* Q_1$ [29]. Now we consider physically inequivalent deformations of a given theory $(\mathcal{A}, \omega, S)$.

We consider a basis $\{\Gamma_a\}$ of Q-cohomology among functions on $\mathcal{A}$, which are integrals of top-forms on N over N. Then we can consider a family of action functionals given by $S + t^a\Gamma_a$, where $\{t^\alpha\}$ are formal parameters on a dual basis of Q-cohomology. Since $Q\Gamma_a \equiv (S, \Gamma_a) = 0$ the family of action functionals again satisfies the BV master equation up to first order in t^a. To go beyond the first order we should imagine a certain function $\Gamma(t)$, satisfying $\triangle\Gamma(t) = 0$, which has the following formal expansions

$$\Gamma(t) = t^a\Gamma_a + \sum_{n>1} t^{a_1} \dots t^{a_n} \Gamma_{a_1 \dots a_n}. \tag{2.11}$$

Then we consider the family of action functionals $S(t)$ defined by

$$S(t) = S + \Gamma(t). \tag{2.12}$$

[h]For example, all the WDDV type equations (the associativity of quantum cohomology rings, the path integral derivation of Kontsevich's L^∞-quasi-isomorphism, and the A^∞-structure of Fukaya category) are based on elaborations of the above idea.

The above deformation is said to be well-defined if $S(t)$ satisfies the BV master equation

$$(S(t), S(t)) = 0 \equiv Q\Gamma(t) + \frac{1}{2}(\Gamma(t), \Gamma(t)) = 0. \tag{2.13}$$

Obviously $\Gamma_{a_1\ldots a_n}$ for $n > 1$ can not be an element of Q-cohomology. We call the deformation (2.12) unobstructed if the solution of $Q\Gamma + \frac{1}{2}(\Gamma, \Gamma) = 0$ has the expansion (2.11). We call the deformation (2.12) classical (or unextended) if $t^a \neq 0$ only for Γ_a with ghost number $U = 0$, while otherwise $t^a = 0$. We define the extended moduli space $\mathfrak{M}$ of (the well-defined deformations of) the theory by the solution space of (2.13) modulo equivalences. We note that the deformed theory (2.12) has a new BRST charge $Q(t)$ defined by $(S(t), \ldots) \equiv Q(t)$. Then the equation (2.13) is equivalent to $Q(t)^2 = 0$ and

$$Q(t)\Gamma(t) = 0. \tag{2.14}$$

We denote call thesubspace $\mathfrak{M}_{cl} \subset \mathfrak{M}$ consisting of solutions with $U = 0$ the classical moduli space. Obviously the tangent space of $\mathfrak{M}$ at a classical point is isomorphic to the Q cohomology group, provided that the deformations are unobstructed.

The above discussion on BV quantization is closely related to modern deformation theory. Details and precise definitions of deformation theory can be found in [50,7]. Relations with BV quantization are also discussed in [75]. Modern deformation theory associates a certain differential graded Lie algebra (dgLa in short) $\mathfrak{g} = \oplus \mathfrak{g}^k$, modulo quasi-isomorphism, with a mathematical structure being deformed. This gives rise to a formal supermoduli space $\mathfrak{M}_\mathfrak{g}$ defined by the solution space of the Maurer-Cartan (MC in short) equation modulo equivalences,

$$d\gamma + \frac{1}{2}[\gamma, \gamma] = 0/\sim \tag{2.15}$$

where $\gamma \in \mathfrak{g}$ and $(d, [.,.])$ denote the differential and the bracket of $\mathfrak{g}$ with degree 1 and 0, respectively;

$$d : \mathfrak{g}^k \to \mathfrak{g}^{k+1}, \qquad [.,.] : \mathfrak{g}^k \otimes \mathfrak{g}^\ell \to \mathfrak{g}^{k+\ell}. \tag{2.16}$$

The differential and the bracket have the following properties

$$\begin{aligned} d^2 &= 0, \\ d[\gamma_1, \gamma_2] &= [d\gamma_1, \gamma_2] + (-1)^{|\gamma_1|}[\gamma_1, d\gamma_2], \\ [\gamma_1, \gamma_2] &= -(-1)^{|\gamma_1||\gamma_2|}[\gamma_2, \gamma_1], \\ [\gamma_1, [\gamma_2, \gamma_3]] &= [[\gamma_1, \gamma_2], \gamma_3] + (-1)^{|\gamma_1||\gamma_2|}[\gamma_2, [\gamma_1, \gamma_3]]. \end{aligned} \tag{2.17}$$

A crucial idea is that one imagines an underlying formal supermanifold $M_{\mathfrak{g}}$ with a nilpotent odd vector $Q_{\mathfrak{g}}$ with degree 1 such that d and $[.,.]$ are the first and second coefficient in the Taylor expansion of $Q_{\mathfrak{g}}$. Then the condition that $Q_{\mathfrak{g}}^2 = 0$ is equivalent to those in (2.17) [2]. Thus

$$Q_{\mathfrak{g}}\gamma = d\gamma + \frac{1}{2}[\gamma,\gamma]. \tag{2.18}$$

Then the moduli space $\mathfrak{M}_{\mathfrak{g}}$ can be identified with the space of $Q_{\mathfrak{g}}$-cohomology $\mathcal{M}_{\mathfrak{g}} = \text{Ker } Q_{\mathfrak{g}}/\text{Im } Q_{\mathfrak{g}}$, where the quotient by Im $Q_{\mathfrak{g}}$ plays the role of dividing out the equivalence in (2.15). In general, $Q_{\mathfrak{g}}$ may have Taylor components beyond the quadratic terms and those define an L_∞-structure. We may have

$$Q_{\mathfrak{g}}\gamma = d\gamma + \frac{1}{2}[\gamma,\gamma] + \frac{1}{3!}[\gamma,\gamma,\gamma] + \ldots \tag{2.19}$$

A quasi-isomorphism between two dgLas $\mathfrak{g}_1$ and $\mathfrak{g}_2$ is a Q-equivariant map between the two associated formal supermanifolds with base points such that the first Taylor coefficients of the map induces an isomorphism between the cohomologies of $(\mathfrak{g}_1, d_1)$ and $(\mathfrak{g}_2, d_2)$. It follows that the $Q_{\mathfrak{g}_1}$ cohomology is isomorphic to that of $Q_{\mathfrak{g}_1}$. Thus the two formal supermoduli spaces $\mathfrak{M}_{\mathfrak{g}_1}$ and $\mathfrak{M}_{\mathfrak{g}_2}$ are isomorphic.

We note that after shifting the degree by -1, the relations in (2.17) are identical to those in (2.1) and (2.8). This implies that one may "represent" a deformation theory by a suitable BV quantized field theory. That is, for a given mathematical structure one wants to deform and the associated dgLa $\mathfrak{g}$, one may consider a BV quantized field theory whose moduli space is isomorphic to the formal super-moduli space $\mathfrak{M}_{\mathfrak{g}}$.[i] Then the quantum field theory gives ***additional*** information about the deformation theory via correlation functions. What would be the additional information? We do not know the general answer. It may, for example, be possible that deformation problems of different mathematical structures lead to quantum field theories which are physically equivalent, etc. etc. Such phenomena is usually called "dualities" or "mirror symmetry".

2.2 Bosonic string theory as a deformation theory

String theory in the sigma model approach is a theory of maps $u : \Sigma \to X$ from a Riemann surface Σ to a target space X. Picking local coordinates $\{u^I\}$ on X such a map is described by a set of functions $\{u^I(x^\mu)\}$ on Σ, where x^μ,

[i] We also note that the BV quantization can be generalized to deal with deformation problems involving the full L_∞-algebra [10].

$\mu = 1, 2$, denotes a set of local coordinates on Σ. The bosonic string action in a NS B-field background is given by

$$I' = \frac{1}{2}\int_\Sigma \left(b_{IJ} du^I \wedge du^J\right) \tag{2.20}$$

where b_{IJ} denote the components of the NS 2-form B-field, i.e. $B = \frac{1}{2}b_{IJ}dx^I \wedge dx^J$. Classically a B-field gives a certain structure to the target space X. For instance a non-degenerate B-field corresponds to an almost symplectic structure on X, while a non-degenerate closed (flat $dB = 0$) B-field corresponds to a symplectic structure on X. For our purpose it is convenient to introduce a first order formalism starting from a bivector field $\pi = \frac{1}{2}b^{IJ}\partial_I \wedge \partial_J \in \Gamma(\wedge^2 TX)$, which is not necessarily non-degenerate. We have

$$I = \int_\Sigma \left(H_I \wedge du^I + \frac{1}{2}b^{IJ} H_I \wedge H_J\right), \tag{2.21}$$

where H_I is an "auxiliary" 1-form field in Σ taking values in $u^*(T^*X)$. The boundary condition for $\partial\Sigma \neq 0$ is that H_I vanishes along the tangential direction in $\partial\Sigma$. For a non-degenerate bi-vector, we can integrate out the auxiliary field H_I and we obtain the original action functional (2.20). We note that the action functional I was originally studied for closed Rieman surface in [41,71]. The authors of [4] also found relation wiht geometric quantization of the symplectic leaves in the target. The path integral approach of Catteo-Felder is based on BV quantization of the above action functional [17]. Here we find it more convenient to make a detour.

The first order formalism above allows us to consider a notion of string theory without an NS background, defined by the following action functional

$$I_o = \int_\Sigma H_I \wedge du^I. \tag{2.22}$$

The resulting theory is obviously topological in two dimensions. It is also invariant under affine transformations of X and does not depends on any other structure in the target space. We may view the string theory in the NS B-field background as a certain deformation of the theory I_o along the "direction" of the bi-vector. We have a criterium for well-defined deformations by requiring that a deformed theory should be a consistent quantum theory. For this we first quantize the theory with action functional I_0 using the BRST-BV formalism. Then we deform the theory along a certain direction and examine if the deformed theory satisfies the quantum master equation.

2.2.1. BV quantization

The quantization of I_0 is rather simple. We note that the action functional I_0 has the following symmetry

$$\delta u^I = 0, \qquad \delta H_I = -d\chi_I, \tag{2.23}$$

where χ_I is an infinitesimal gauge parameter taking values in $u^*(T^*X)$, which vanishes on $\partial\Sigma$ if the boundary of Σ is non-empty. In the BRST quantization one promotes the symmetry to a fermionic one by taking χ_I anticommuting with ghost number $U = 1$. This is equivalent to regarding χ_I as taking value in $u^*(\Pi T^*X)$, where ΠT^*X denotes the parity change of the fiber of T^*X. Now we have the following fermionic symmetry with charge $\boldsymbol{Q}_o$ carrying $U = 1$

$$\boldsymbol{Q}_o u^I = 0, \qquad \boldsymbol{Q}_o H_I = -d\chi_I, \qquad \boldsymbol{Q}_o \chi_I = 0, \tag{2.24}$$

and satisfying $\boldsymbol{Q}_o^2 = 0$. In the present case we introduce a set of "anti-fields" (η_I, ρ^I, v^I) with the ghost numbers $U = (-1,-1,-2)$ for the set of "fields" (u^I, H_I, χ_I). Let $\mathcal{A}$ denote the space of all fields. The "fields" and the "anti-fields" should have a pairing defining a two-form in Σ such that one can integrate over Σ to get a two-form in $\mathcal{A}$. Thus the "anti-fields" (η_I, ρ^I, v^I) are $(2,1,2)$-forms on Σ. We can easily find the BV action functional $\boldsymbol{S}_o$ satisfying all the requirements stated in Sect. 2.1 as follows

$$\boldsymbol{S}_o = \int_\Sigma \left(H_I \wedge du^I - \rho^I \wedge d\chi_I\right). \tag{2.25}$$

Thus the BRST transformation laws are

$$\begin{aligned} \boldsymbol{Q}_o u^I &= 0, & \boldsymbol{Q}_o \rho^I &= -du^I, & \boldsymbol{Q}_o v^I &= -d\rho^I, \\ \boldsymbol{Q}_o \chi_I &= 0, & \boldsymbol{Q}_o H_I &= -d\chi_I, & \boldsymbol{Q}_o \eta_I &= -dH_I, \end{aligned} \tag{2.26}$$

satisfying $\boldsymbol{Q}_o^2 = 0$. It is trivial to check the BV master equation

$$(\boldsymbol{S}_o, \boldsymbol{S}_o)_{BV} = 0, \qquad \Delta \boldsymbol{S}_o = 0, \tag{2.27}$$

provided that we have the right boundary conditions mentioned earlier.

A more conceptual and compact formulation can be obtained by combing "fields" and "anti-fields" into two superfields $(\boldsymbol{u}^I, \boldsymbol{\chi}_I)$ carrying $U = (0,1)$;

$$\begin{aligned} \boldsymbol{u}^I &:= \boldsymbol{u}^I(x^\mu, \theta^\mu) = u^I(x^\mu) + \rho^I_\nu(x^\mu)\theta^\nu - \frac{1}{2}v^I_{\nu_1\nu_2}(x^\mu)\theta^{\nu_1}\theta^{\nu_2}, \\ \boldsymbol{\chi}_I &:= \boldsymbol{\chi}_I(x^\mu, \theta^\mu) = \chi_I(x^\mu) + H_{I\nu}(x^\mu)\theta^\nu + \frac{1}{2}\eta_{I\nu_1\nu_2}(x^\mu)\theta^{\nu_1}\theta^{\nu_2}, \end{aligned} \tag{2.28}$$

where θ^μ is an anti-commuting vector with $U = 1$. In terms of the superfields we have

$$\boldsymbol{S}_o = \int d^2\theta\, \boldsymbol{\chi}_I D\boldsymbol{u}^I, \tag{2.29}$$

and

$$Q_o u^I = Du^I, \qquad Q_o \chi_I = D\chi_I, \tag{2.30}$$

where $D = \theta^\mu \partial_\mu$. The natural odd symplectic form $\boldsymbol{\omega}$ in $\mathcal{A}$ is defined by

$$\boldsymbol{\omega} = \int_\Sigma d^2\theta \left(\delta u^I \delta \chi_I\right), \tag{2.31}$$

where δ denotes the exterior differential on $\mathcal{A}$. The odd symplectic form $\boldsymbol{\omega}$ has degree $U = -1$ since $d^2\theta$ shift the degree by $U = -2$. The BV bracket is the odd Poisson bracket among functions in $\mathcal{A}$ with respect to $\boldsymbol{\omega}$. The BRST charge $\boldsymbol{Q}_o$ can be identified with the odd Hamiltonian vector of $\boldsymbol{S}_o$;

$$i_{\boldsymbol{Q}_o}\boldsymbol{\omega} = \delta \boldsymbol{S}_o, \text{ equivalently } (\boldsymbol{S}_o, \ldots)_{BV} = \boldsymbol{Q}_o. \tag{2.32}$$

Now we observe that the superfields (u^I, χ_I) parametrize maps

$$\phi : \Pi T\Sigma \to \Pi T^* X \tag{2.33}$$

between the two superspaces. Here $\Pi T\Sigma$ denotes the total space of the tangent bundle of Σ after parity change of the fiber. We regard $(\{x^\mu\}, \{\theta^\mu\})$ as a set of local coordinates on $\Pi T\Sigma$. With the target superspace ΠT^*X we mean the total space of the cotangent bundle of X after parity change of the fiber. We denote a set of local coordinates on ΠT^*X by $(\{u^I\}, \{\chi_I\})$ carrying the ghost number $U = (\{0\}, \{1\})$.[j] Thus the space $\mathcal{A}$ of all fields is the space of all maps above. The odd symplectic form $\boldsymbol{\omega}$ (2.31) on $\mathcal{A}$ is the unique extension of the odd symplectic form $\omega = du^I d\chi_I$ with degree $U = 1$ on ΠT^*X to $\mathcal{A}$. More precisely

$$\boldsymbol{\omega} \equiv \int_\Sigma d^2\theta \; \phi^*(\omega), \tag{2.34}$$

where ϕ^* is the pull-back of a map $\phi : \Pi T\Sigma \to \Pi T^* X$. The odd Poisson bracket $[.,.]_S$ with degree $U = -1$ among functions γ_i on ΠT^*X is called the Schouten-Nijenhuis bracket;

$$[\gamma_1, \gamma_2]_S := \frac{\partial \gamma_1}{\partial \chi_I}\frac{\partial \gamma_2}{\partial u^I} - (-1)^{(|\gamma_1|-1)(|\gamma_2|-1)}\frac{\partial \gamma_2}{\partial \chi_I}\frac{\partial \gamma_1}{\partial u^I} \tag{2.35}$$

where $|\gamma|$ denotes the degree of γ. We remark that a function on ΠT^*X with $U = p$ is a p-vector (an element of $\Gamma(X, \wedge^p TX)$) after parity change. We also remark that functions on ΠT^*X after shifting the degree by 1 together with

[j] By abuse of notation, u^I may denote both a coordinate on X or a function $u^I(x^\mu)$ on Σ etc. This should not cause any confusion in the present context.

the Schouten-Nijenhuis bracket form a dgLa with zero differential. The product of functions on ΠT^*X corresponds to wedge products of multivectors on X. We also note that a function γ on ΠT^*X induces a function $\int_\Sigma d^2\theta\ \boldsymbol{\phi}^*(\gamma)$ on $\mathcal{A}$ with $U = |\gamma| - 2$. It follows that

$$\left(\int_\Sigma d^2\theta\ \boldsymbol{\phi}^*(\gamma), \int_\Sigma d^2\theta\ \boldsymbol{\phi}^*(\gamma)\right)_{BV} = \int_\Sigma d^2\theta\ \boldsymbol{\phi}^*([\gamma,\gamma]_S). \tag{2.36}$$

2.2.2. Deformations

Now we examine deformations of the theory. A 0-dimensional observable with ghost number $U = p$ may be any function on $\mathcal{A}$ of the form

$$\gamma^{(0)} = \frac{1}{p!}\gamma^{I_1\ldots I_p}(u^I(x^\mu))\chi_{I_1}(x^\mu)\ldots\chi_{I_p}(x^\mu). \tag{2.37}$$

Any such $\gamma^{(0)}$ satisfies $\boldsymbol{Q}_o\gamma^{(0)} = 0$ since $\boldsymbol{Q}_o u^I = \boldsymbol{Q}_o\chi_I = 0$. It is obvious that no such $\gamma^{(0)}$ can be $\boldsymbol{Q}_o$-exact. Thus an arbitrary functional $\gamma(u^I(x^\mu), \chi_J(x^\mu))$ belongs to the $\boldsymbol{Q}_o$ cohomology group. It is also obvious that $\Delta\gamma^{(0)} = 0$. The coefficients $\gamma^{I_1\ldots I_p}$ can be identified with coefficients of a multivector $\Gamma(X, \wedge^p TX)$. Thus the space of 0-dimensional observables is isomorphic to the space of all multivectors on X. Now we denote by $\boldsymbol{\gamma} := \gamma^{(0)}(\boldsymbol{u}^I, \boldsymbol{\chi}_I)$ the corresponding function of superfields $(\boldsymbol{u}^I, \boldsymbol{\chi}_I)$. Equivalently $\boldsymbol{\gamma} := \boldsymbol{\phi}^*(\gamma)$. Then we have the following expansions

$$\boldsymbol{\gamma} = \gamma^{(0)} + \left(\mathcal{K}_\mu\gamma^{(0)}\right)\theta^\mu + \frac{1}{2}\left(\mathcal{K}_\mu\mathcal{K}_\nu\gamma^{(0)}\right)\theta^\mu\theta^\nu, \tag{2.38}$$

where $\mathcal{K}_\mu := -\frac{\partial}{\partial\theta^\mu}$. We note that $\mathcal{K}_\mu$ is anticommuting and carries $U = -1$. Together with $\boldsymbol{Q}_o$ we have the following anticommutation relations

$$\{\boldsymbol{Q}_o, \boldsymbol{Q}_o\} = 0, \qquad \{\boldsymbol{Q}_o, \mathcal{K}_\mu\} = -\partial_\mu, \qquad \{\mathcal{K}_\mu, \mathcal{K}_\nu\} = 0. \tag{2.39}$$

The above relations and (2.38) imply that $\gamma^{(n)} := \frac{1}{n!}\left(\mathcal{K}_{\mu_1}\ldots\mathcal{K}_{\mu_n}\boldsymbol{\gamma}\right)dx^{\mu_1}\wedge\ldots\wedge dx^{\mu_n}$, where $n = 0, 1, 2$, satisfy the descent equations

$$\begin{aligned}
\boldsymbol{Q}_o\gamma^{(0)} &= 0,\\
\boldsymbol{Q}_o\gamma^{(1)} &= -d\gamma^{(0)},\\
\boldsymbol{Q}_o\gamma^{(2)} &= -d\gamma^{(1)}.
\end{aligned} \tag{2.40}$$

Consequently $\int_\Sigma \gamma^{(2)} \equiv \int_\Sigma d^2\theta\ \boldsymbol{\gamma}$ is invariant under $\boldsymbol{Q}_o$, if $\gamma^{(1)} = 0$ at the boundary $\partial\Sigma$ of Σ, i.e., $\boldsymbol{Q}_o\int_\Sigma\gamma^{(2)} = -\int_{\partial\Sigma}\gamma^{(1)}$. It is not difficult to check $\Delta\int_\Sigma d^2\theta\ \boldsymbol{\gamma} = 0$, with suitable regularization [17].

Provided that $\gamma^{(1)} = 0$ at the boundary $\partial\Sigma$, $\int_\Sigma d^2\theta\ \gamma$ is a 2-dimensional observable, which can be used to deform the theory;

$$\mathcal{S}_\gamma = \mathcal{S}_o + \int_\Sigma d^2\theta\ \gamma. \tag{2.41}$$

The above action functional satisfies $\mathcal{Q}_o\mathcal{S}_\gamma = \triangle\mathcal{S}_\gamma = \mathcal{K}_\mu\mathcal{S}_\gamma = 0$. It follows, from the relation (2.36), that the deformed action functional satisfies the BV master equation if $[\gamma,\gamma]_S = 0$. Note that the ghost number of $\int_\Sigma d^2\theta\ \gamma$ is shifted by 2 from the ghost number of γ. We can view γ as the pull-back of a function on ΠT^*X by a map $\Pi T\Sigma \to \Pi T^*X$. Thus an even (odd) function on ΠT^*X leads to an even (odd) function on $\mathcal{A}$.

Now the boundary condition introduced earlier implies that $\gamma^{(1)} = 0$ at the boundary iff $\gamma(\boldsymbol{u}^I, \chi_I)|_{\chi_I=0} = 0$ for all I. Equivalently $\gamma|_X = 0$. Then we may use $\int_\Sigma d^2\theta\ \gamma$ to deform the theory. On the other hand such a function γ on ΠT^*X does not give non-trivial observables supported on the boundary. Non-trivial boundary observables must come from functions on the base space X of ΠT^*X, which can not be used to deform the bulk theory.

The above considerations lead us to consider the following action functional with $U = 0$

$$\mathcal{S} = \mathcal{S}_o + \int_\Sigma d^2\theta\ \boldsymbol{\pi}, \tag{2.42}$$

where $\boldsymbol{\pi} = \frac{1}{2}b^{IJ}(\boldsymbol{u}^K)\chi_I\chi_J$ satisfying

$$\left(\int_\Sigma d^2\theta\ \boldsymbol{\pi}, \int_\Sigma d^2\theta\ \boldsymbol{\pi}\right)_{BV} = 0. \tag{2.43}$$

The above condition implies that the action functional $\mathcal{S}$ satisfies the master equation $(\mathcal{S},\mathcal{S})_{BV} = 0$. A parity changed bivector $\pi = \frac{1}{2}b^{IJ}(u^K)\chi_I\chi_J$ on X satisfying $[\pi,\pi]_S = 0$ is called a Poisson bi-vector. Since $\mathcal{S}$ is an even function on the space of all fields (the space of all maps $\Pi T\Sigma \to \Pi T^*X$) $\mathcal{A}$ we have an odd Hamiltonian vector $\mathcal{Q}$ defined by $\boldsymbol{i}_{\mathcal{Q}}\omega = \boldsymbol{\delta}\mathcal{S}$ or, equivalently $(\mathcal{S},\ldots)_{BV} = \mathcal{Q}$, regarding the Hamiltonian vector $\mathcal{Q}$ as an odd derivation. Thus $\mathcal{Q}$ is the BV-BRST charge of the action functional $\mathcal{S}$. Explicitly

$$\mathcal{Q} = \left(\frac{\partial\boldsymbol{\pi}}{\partial\chi_I} + D\boldsymbol{u}^I\right)\frac{\partial}{\partial\boldsymbol{u}^I} + \left(\frac{\partial\boldsymbol{\pi}}{\partial\boldsymbol{u}^I} + D\chi_I\right)\frac{\partial}{\partial\chi_I}. \tag{2.44}$$

The BV master equation is equivalent to the condition that $\mathcal{Q}$ is nilpotent. Finally we note that

$$\{\mathcal{Q},\mathcal{Q}\} = 0, \qquad \{\mathcal{Q},\mathcal{K}_\mu\} = -\partial_\mu, \qquad \{\mathcal{K}_\mu,\mathcal{K}_\nu\} = 0. \tag{2.45}$$

The bosonic part of the action functional S is given by the classical action functional I in (2.21). By construction, compare with [17], S is obtained by BV quantization of I. We also mention that the anti-commutation relation (2.45) should be compared with the relations $L_{-1} = \{Q, b_{-1}\}$ and $\bar{L}_{-1} = \{Q, \bar{b}_{-1}\}$ in 2-dimensional conformal field theory (see in particular [100] on closely related issues with this paper) or with a twisted $N_{ws} = (2,2)$ world-sheet supersymmetry [89].

2.3 Open string and formality

Now we recall the solution of deformation quantization based on the path integral approach.

We first recall some basic properties of associative algebras (see [50,83]). On any associative algebra A we have the Hochschild co-chain complex $(\delta, \oplus_n C^n(A,A))$ where $C^n(A,A)$ is the space of linear maps $A^{\otimes n} \to A$ and $\delta : C^\bullet(A,A) \to C^{\bullet+1}(A,A)$ with $\delta^2 = 0$ is the Hochschild differential. For $A = C^\infty(X)$ the Hochschild co-chains are identified with the space of multidifferential operators. It is known that the Hochschild cohomology $HH^\circ(A,A)$ is isomorphic to the space of multivectors $\Gamma(X, \wedge^\bullet TX)$.[k] We recall that the space $\Gamma(X, \wedge^{\bullet+1}TX) \equiv HH^{\bullet+1}(A,A)$ together with the Schouten-Nijenhuis bracket $[.,]_S$ and zero-differential form forms the dgLa called $T^\bullet_{poly}(X)$. This dgLa originates from another dgLa called $D_{poly}(X)$ on the space $C^{\circ+1}(A,A)$ with the Gerstenhaber (G in short) bracket $[.,.]_G$ and differential δ and associative cup product. The G bracket

$$[.,.]_G : C^n(A,A) \otimes C^m(A,A) \to C^{m+n-1}(A,A) \tag{2.46}$$

is defined as follows. For $\Phi_1 \in C^n(A,A)$ and $\Phi_2 \in C^m(A,A)$

$$[\Phi_1, \Phi_2]_G = \Phi_1 \circ \Phi_2 - (-1)^{(n-1)(m-1)}\Phi_2 \circ \Phi_1, \tag{2.47}$$

where

$$\begin{aligned}
&\Phi_1 \circ \Phi_2(a_1, \ldots, a_{n+m-1}) \\
&\quad = \sum_{j=0}^{n-1} (-1)^{(m-1)j}\Phi_1(a_1, \ldots, a_j, \Phi_2(a_{j+1}, \ldots, a_{j+m}), \ldots).
\end{aligned} \tag{2.48}$$

The differential $\delta : C^\bullet(A,A) \to C^{\bullet+1}(A,A)$ is defined by $\delta = [m, \ldots]_G$ where $m(a,b) = ab$ for $a, b \in A$. Then $[m,m]_G = 0$ is equivalent to m being associative, which implies that $\delta^2 = 0$.

---

[k]Thus the Q_o-cohomology is isomorphic to the total Hochschild cohomology $\oplus_n HH^n(A,A)$.

Now for a given manifold X a star product $f \star g$ among two functions f and g is defined by

$$f \star g = fg + \Pi(f,g), \tag{2.49}$$

where $\Pi \in D^1_{poly}(X)$ has a formal expansion

$$\Pi(f,g) = \Pi_1(f,g)\hbar + \Pi_2(f,g)\hbar^2 + \dots. \tag{2.50}$$

The associativity of the star product can be written in terms of the MC equation for $D^1_{poly}(X)$ as

$$\delta\Pi + \frac{1}{2}[\Pi,\Pi]_G = 0 \equiv [m+\Pi, m+\Pi]_G = 0. \tag{2.51}$$

To first approximation the associativity implies that $\Pi(f,g)_1 = \langle \pi, df \otimes dg\rangle \equiv \frac{1}{2}\{f,g\}$ and the bracket $\{.,.\}$ satisfies the Jacobi identity. Namely the bivector π is Poisson and $\{.,.\}$ is the Poisson bracket. Equivalently

$$[\pi,\pi]_S = 0, \tag{2.52}$$

which is the MC equation for $T^1_{poly}(X)$.

Consequently the problem of deformation quantization for an arbitrary Poisson manifold X is equivalent to proving isomorphism between two moduli spaces defined by the two MC equations (2.51) and (2.52). Kontsevich proved the more general theorem (L_∞ formality) that the dgLa of multi-vectors on X with vanishing differential and Schouten-Nijenhuis bracket is quasi-isomorphic to the dgla of multi-differential operators on X. He also gave an explicit expression for his quasi-isomorphism in the case of $X = \mathbb{R}^n$, essentially summing over contributions from certain graphs resembling Feynman path integral.

Catteneo-Felder studied the path integral of the theory defined by the action functional S on the disk $\Sigma = D$ with boundary punctures, where observables constructed from functions on X are inserted. They obtained Kontsevich's explicit formula by perturbative expansions around constant maps [17]. This formulation essentially maps the space of polyvectors on the target space X with vanishing Schouten-Nijenhuis bracket modulo equivalence to the space of solution of the BV master equation of the theory deformed along the direction of polyvectors modulo equivalences. The BV master equation then implies according to (2.10) that the perturbative expansion of the theory defines a quasi-isomorphism. That is, the MC equation (2.52) for $T^1_{poly}(X)$ is the BV master equation, while the MC equation (2.51) is the Ward identity.

We remark that an associative algebra A is a 1-algebra and its Hochschild cochain complex $(C^\bullet(A,A),\delta)$ has the structure of a 2-algebra by the cup product and the G-bracket $[.,.]_G$. We note that the cup product is (super)-commutative only up to homotopy. The cohomology of the Hochschild complex is isomorphic to the space of polyvectors on X and also has a structure of a 2-algebra with the Schouten-Nijenhuis bracket (induced from G-bracket) and the wedge product (induced from the cup product), which is (super)-commutative. We call the latter the cohomological 2-algebra. The formality theorem of Kontsevich means that the 2-algebra and its cohomology are equivalent up to homotopy as a L_∞ algebra, thus forgetting the product structure, for $A = C(X)$. Tarmakin generalized the formality to the category of G_∞ algebras for $X = \mathbb{R}^n$ [79]. Note that the Schouten-Nijenhuis bracket can be generated by $\Delta = \frac{\partial^2}{\partial u^i \partial \chi_i}$ for $X = \mathbb{R}^n$. Thus the cohomological 2-algebra is a BV algebra. The BV_∞-formality is conjectured in [55]. We do not want to review here all those purely algebraic approaches (see however [51]). We should also mention that those are closely related with the open-closed string field theory of Zwiebach. For us it should suffice to meditate OS.

2.4 A model

In this subsection we reexamine the A model of the topological sigma model. The A model was originally defined for an arbitrary almost complex manifold [89]. Our presentation for the A model will be similar to that of the Catteneo-Felder model in the previous section. Here, however, we will take the opposite direction by starting from a dGBV algebra associated with a symplectic manifold X. Then we will construct the corresponding two-dimensional sigma model, which leads to the A model after gauge fixing. A similar approach to the A model in the Kähler case is discussed in [2]. We recall that the A model for the Kähler case is a twisted version of a $N_{ws} = (2,2)$ world-sheet supersymmetric sigma model.

2.4.1. Covariant Schouten-Nijenhuis bracket

Consider a manifold X with a Poisson bi-vector π, which corresponds to even function $\pi = \frac{1}{2}b^{IJ}\chi_I\chi_J$ on ΠT^*X with $[\pi,\pi]_S = 0$. The associated odd nilpotent Hamiltonian vector Q_π is given by

$$Q_\pi = b^{IJ}\chi_J \frac{\partial}{\partial u^I} + \frac{1}{2}\partial_K b^{IJ}\chi_I\chi_J\frac{\partial}{\partial \chi_K}. \tag{2.53}$$

Since $Q_\pi^2 = 0$, Q_π defines a cohomology on $\Pi\Gamma(\wedge^\bullet TX)$. The resulting cohomology is called the Poisson cohomology $H_\pi^\circ(X)$ of X [59]. For the symplectic case we are considering the Poisson cohomology is isomorphic to de Rham cohomology.

Now we consider the dual picture in ΠTX. We introduce natural local coordinates (u^I, ψ^I) on ΠTX. For any differentiable manifold X we have a distinguished odd vector Q in ΠTX

$$Q = \psi^I \frac{\partial}{\partial u^I} \tag{2.54}$$

of degree $U = 1$ with $Q^2 = 0$. It is obtained by the parity change of the exterior derivative d on X. Any differential form on X corresponds to a function on ΠTX. The Q-cohomology is isomorphic to the de Rham cohomology. We consider a Poisson bi-vector π on X and define the associated contraction operator i_π;

$$i_\pi = \frac{1}{2} b^{IJ} \frac{\partial^2}{\partial\psi^I \partial\psi^J}. \tag{2.55}$$

Then we define an odd second order differential operator Δ with degree $U = -1$ by

$$\Delta_\pi := [i_\pi, Q] = b^{IJ} \frac{\partial^2}{\partial u^I \partial\psi^J} + \frac{1}{2}\partial_I b^{JK} \psi^I \frac{\partial^2}{\partial\psi^I \partial\psi^J}. \tag{2.56}$$

It is not difficult to show that the condition $[\pi, \pi]_S = 0$ implies

$$\Delta_\pi^2 = [Q, \Delta_\pi] = 0. \tag{2.57}$$

The operator Δ_π acting on polynomial functions in ΠTX is the parity changed version of the Koszul-Brylinski boundary operator, which defines the Poisson homology $H^\pi_\bullet(X)$ [54,15]. If X is an unimodular Poisson manifold (including the symplectic case) with dimensions n we have the duality $H^\pi_p(X) = H^{n-p}_\pi(X)$. It follows that an element $\alpha \in H^\circ(X)$ satisfies $Q\alpha = \Delta_\pi \alpha = 0$ for a symplectic manifold X. For a Kähler manifold we have $\Delta_\pi = i\partial^* - i\bar{\partial}^*$ by the Kähler identity. From now on we omit the subscript π from Δ_π.

The operator Δ allows us to define an odd Poisson structure on ΠTX, which is the covariant version of the Schouten-Nijenhuis bracket [54]. Let a, b denote functions on ΠTX with degree $U = |a|, |b|$. The ordinary product $a \cdot b$ originates from the wedge product on $\Gamma(\wedge^\bullet T^* X)$. It follows that Q is a derivation while Δ fails to be a derivation. One defines the covariant Schouten-Nijenhuis bracket by the formula

$$[a \bullet b] = (-1)^{|a|}\Delta(a \cdot b) - (-1)^{|a|}\Delta a \cdot b - a \cdot \Delta b. \tag{2.58}$$

It is not difficult to check that $(Q, \Delta, \cdot, \mathcal{O}(\Pi TX)$ form a dGBV algebra [62].

From now on we assume that X is a symplectic manifold. A symplectic manifold is a Poisson manifold with a non-degenerated Poisson bi-vector π Let b_{IJ} be the inverse of b^{IJ}. Then the condition $[\pi, \pi]_S = 0$ is equivalent to

$d\omega = 0$, where $\omega = \frac{1}{2}b_{IJ}du^I \wedge du^J$. Let $B = \frac{1}{2}b_{IJ}\psi^I\psi^J$ be the parity changed symplectic form. It is not difficult to show that

$$\Delta B = 0, \qquad [B \bullet B] = 0, \qquad [B \bullet a] = Qa. \tag{2.59}$$

Thus Q is the Hamiltonian vector of B.

2.4.2. A BV sigma model

Now we consider a two dimensional sigma model which is a theory of maps $\Pi T\Sigma \to \Pi TX$. We denote local coordinates in $\Pi T\Sigma$ by (x^μ, θ^μ), $\mu = 1, 2$. We denote $(\boldsymbol{u}^I := u^I(x^\mu, \theta^\mu), \boldsymbol{\psi}^I := \psi^I(x^\mu, \theta^\mu)$ as the extension of (u^I, χ^I) to functions on $\Pi T\Sigma$. Let $\mathcal{A}$ denote the space of all maps $\boldsymbol{\phi} : \Pi T\Sigma \to \Pi TX$. The odd vector Q on ΠTX can naturally be extended to an odd vector $\boldsymbol{Q}$ on $\mathcal{A}$;

$$\boldsymbol{Q} = \boldsymbol{\psi}^I \frac{\partial}{\partial \boldsymbol{u}^I}. \tag{2.60}$$

It is obvious that $\boldsymbol{Q}^2 = 0$. We define the action functional $\boldsymbol{S}$ of the theory by $\int_\Sigma d^2\theta\, \boldsymbol{\phi}^*(B)$;

$$\boldsymbol{S} = \frac{1}{2}\int d^2\theta \left(b_{IJ}\boldsymbol{\psi}^I\boldsymbol{\psi}^J\right). \tag{2.61}$$

The odd symplectic structure on ΠTX induces an odd Poisson structure on $\mathcal{A}$. The BV operator Δ is defined by the formula $\Delta := [i_\pi, \boldsymbol{Q}]$. The associated BV bracket, denoted by $(.,.)_{BV}$, is the BV bracket of the theory. It follows from (2.59) that

$$\Delta\boldsymbol{S} = 0, \qquad (\boldsymbol{S}, \boldsymbol{S})_{BV} = 0, \qquad (\boldsymbol{S}, \ldots)_{BV} = \boldsymbol{Q}. \tag{2.62}$$

Thus the action functional satisfies the quantum master equation.

We may expand the two superfields $(\boldsymbol{u}^I, \boldsymbol{\psi}^I)$ carrying $U = (0,1)$; as follows

$$\begin{aligned} \boldsymbol{u}^I &= u^I + \rho^I_\mu\theta^\mu + \frac{1}{2}v^I_{\mu\nu}\theta^\mu\theta^\nu, \\ \boldsymbol{\psi}^I &= \psi^I + H^I_\mu\theta^\mu + \frac{1}{2}\eta^I{}_{\mu\nu}\theta^\mu\theta^\nu. \end{aligned} \tag{2.63}$$

As before the ghost number of θ^μ is assigned the ghost number $U = 1$. The $\boldsymbol{Q}$ transformation law is

$$\boldsymbol{Q}\boldsymbol{u}^I = \boldsymbol{\psi}^I, \qquad \boldsymbol{Q}\boldsymbol{\psi}^I = 0. \tag{2.64}$$

The explicit form of the action functional S is

$$S = \int_{\Sigma} d\theta^2 \Big(\frac{1}{2} b_{IJ} H^I \wedge H^J + 2 b_{IJ} \eta^I \psi^J + \partial_K b_{IJ} \rho^K \wedge H^I \psi^J \\ + \frac{1}{2} \partial_K \partial_L b_{IJ} \rho^K \wedge \rho^L \psi^I \psi^J + \partial_K b_{IJ} v^K \psi^I \psi^J \Big). \tag{2.65}$$

2.4.3. Gauge fixing

Recall that the quantum master equation is the condition that the path integral of the theory $\langle \mathcal{O} \rangle = \int_{\mathcal{L}} d\mu\, \mathcal{O} e^{-\frac{1}{\hbar} S}$ restricted to a Lagrangian submanifold $\mathcal{L} \subset \mathcal{A}$ is invariant under smooth deformations of $\mathcal{L}$, provided that $\mathcal{O}$ is a BV observable, $(Q - \hbar\Delta)\mathcal{O} = 0$. Picking a homology class of $\mathcal{L}$ is called BV gauge fixing. In the present case all the observables $\mathcal{O}_\alpha$ will satisfy $\Delta \mathcal{O}_\alpha = 0$. Thus $\mathcal{O}_\alpha$ should be invariant under Q. Note also that the action functional S is also invariant under Q. Then we can apply the fixed point theorem of Witten [92], since Q generates a global fermionic symmetry on Σ. According to this theorem the path integral is localized to an integral over the fixed point locus of Q. Consequently the path integral is localized to an integral over the fixed point locus $\mathcal{L}_0$ in the Lagrangian subspace $\mathcal{L}$. Note that the space of all fields $\mathcal{A}$ and its Lagrangian subspace $\mathcal{L}$ are both infinite dimensional superspaces. Thus the path integral is difficult to make sense of. However we can reduce the integral to an finite dimensional subspace $\mathcal{L}_0$ due to the fixed point theorem. Now we look for an appropriate gauge fixing.

On any symplectic manifold (X, ω) there is an almost complex structure $J \in \Gamma(End(TX))$ compatible with B. The almost complex structure J^I_J obeying

$$J^I{}_K J^K{}_J = -\delta^I{}_K \tag{2.66}$$

is compatible with b_{IJ} in the sense that

$$b_{IJ} = J^K{}_I J^L{}_J b_{KL}. \tag{2.67}$$

Thus b^{IJ} is of type $(1,1)$. The above relation is equivalent to the following condition

$$g_{IJ} = g_{JI}, \tag{2.68}$$

where $g_{IJ} = b_{IK} J^K{}_J$ is a Riemannian metric with torsion free connection. We want to gauge fix the theory such that the path integral is localized to the moduli space of pseudo-holomorphic maps

$$du^I + J^I{}_K * du^K = 0, \tag{2.69}$$

where d and $*$, $*^2 = -1$ denote the exterior derivative and the Hodge star in Σ. Note that the transformation laws of the anti-ghost multiplet (ρ^I, H^I) are

$$Q\rho^I = -H^I, \qquad QH^I = 0. \tag{2.70}$$

Thus $H^I = 0$ at the fixed point locus. The appropriate gauge fixing should be $H^I = du^I + J^I{}_K * du^K = 0$. For this purpose we should impose the "self-duality" condition of Witten [89]

$$\begin{aligned} \rho^I &= *J^I{}_K\rho^K, \\ H^I &= *J^I{}_K H^K. \end{aligned} \tag{2.71}$$

Then, as explained by Witten, the Q transformation law on the "self-dual" part ρ^{+I} of ρ^I should be

$$Q\rho^{+I} = -H^{+I} - \frac{1}{2}\left(\mathcal{D}_K J^I{}_J\right)\psi^K * \rho^{+J} + \Gamma^I{}_{JK}\psi^J\rho^{+K}, \tag{2.72}$$

where $\mathcal{D}_K$ denotes the covariant derivative. In the above the term proportional to $\mathcal{D}^K J^I{}_L$ is needed for the compatibility of Q with the "self-duality" (2.71).

$$Q\rho^{+I} = *\left(QJ^I{}_K\right)\rho^{+K} + J^I{}_K Q\rho^{+K}. \tag{2.73}$$

The last term is needed for covariance with respect to reparameterizations of u^I. For this we regard (H_I^-, ρ^{+I}) as fields, while (ρ^{-I}, H_I^+) are the corresponding anti-fields. Now we pick the following gauge fermion Ψ with $U = -1$

$$\Psi = \int_\Sigma b_{IJ}\rho^{+I} \wedge \left(du^J + \frac{1}{2}\left(\mathcal{D}_K J^J{}_L\right)\psi^K * \rho^{+L} - \Gamma^J{}_{KL}\psi^K\rho^{+L}\right). \tag{2.74}$$

Now the Lagrangian submanifold $\mathcal{L}$ is determined by the following equations;

$$b_{IJ}v^J = \frac{\delta\Psi}{\delta\psi^I}, \quad b_{IJ}\rho^{-I} := \frac{\delta\Psi}{\delta H^{-I}}, \quad b_{IJ}H^{+I} := \frac{\delta\Psi}{\delta\rho^{+I}}, \quad b_{IJ}\eta^J := \frac{\delta\Psi}{\delta u^I}. \tag{2.75}$$

Thus we have, for instance,

$$\begin{aligned} &\rho^{-I} = 0, \\ &H^I = \left(du^I + *J^I{}_K du^K\right) + \frac{1}{2}\left(\mathcal{D}_K J^I{}_L\right)\psi^K * \rho^{+L} - \Gamma^I{}_{KL}\psi^K\rho^{+L}, \end{aligned} \tag{2.76}$$

as desired. Then the gauge fixed action functional (or the action restricted as a function on $\mathcal{L}$) is exactly the action functional of Witten's topological sigma model [89]. The transformation laws of ρ^I after the gauge fixing (restricted to $\mathcal{L}$) is then

$$Q\rho^I = -\left(du^I + *J^I{}_K du^K\right) - \frac{1}{2}\left(\mathcal{D}_K J^I{}_J\right)\psi^K * \rho^J + \Gamma^I{}_{JK}\psi^J\rho^K. \tag{2.77}$$

Since $\psi^I = 0$ at the fixed point the fixed point locus $\mathcal{L}_0$ in $\mathcal{L}$ is the space of pseudo holomorphic maps.

3 Topological Open Membrane

We want to study the theory of membranes in the background of a C-field but no metric. The theory is defined by maps from a 3-dimensional world-volume N with boundary to the target space X. We assume that the canonical line bundle $\det(T^*N)$ is trivial so that we have a well-defined volume element. We describe a map $u : N \to X$ locally by functions $u^I(x^\mu)$ where $\{x^\mu\}$, $\mu = 1,2,3$ are local coordinates on N. We consider a 3-form $C = \frac{1}{3!}c_{IJK}(u^L)du^I \wedge du^J \wedge du^K$ in X, where $\{u^I\}$ denote local coordinates on X. Now the action functional of the membrane coupled to a C-field is, for $\partial N = 0$,

$$I_1' = \int_N u^*(C) = \frac{1}{3!}\int_N c_{IJK}(u^L)du^I \wedge du^J \wedge du^K. \tag{3.1}$$

To adopt the same strategy as for the string theory, we rewrite the action in the following first order formalism:

$$I_1 = \int_N \left(F_I^{(2)} \wedge (A^I - du^I) - A^I \wedge dH_I \; + \frac{1}{3!}c_{IJK}(u^L)A^I \wedge A^J \wedge A^K \right), \tag{3.2}$$

where A^I are 1-forms in N taking values in $u^*(TX)$, while $F_I^{(2)}$ and H_I are 2 and 1-forms, respectively, taking values in $u^*(T^*X)$. The algebraic equation of motion of $F_I^{(2)}$ imposes the constraint $A^I - du^I = 0$ and the H_I equation of motion $dA^I = 0$ gives the integrability condition. Thus we have a well-defined first order formalism. On shell we recover the action functional I_1' in (3.1). We might try to do BV quantization of the first order action functional I_1. It is however more appropriate to consider the membrane theory without background and study consistent deformations.

We may define the bosonic membrane theory without background by the following action functional

$$I_o = \int_N \left(F_I^{(2)} \wedge (A^I - du^I) - A^I \wedge dH_I \right), \tag{3.3}$$

where all fields above carry ghost number $U = 0$. From now on we remove the restriction $\partial N = 0$. The boundary conditions are such that $A^I(x)$ and $*F_I^{(2)}$ vanish along the directions tangent to ∂N while $H_I(x)$ vanishes along the direction normal to ∂N for $x \in \partial N$. On shell we have the boundary string without background,

$$I_o|_{\text{on shell}} = \int_{\partial N} H_I \wedge du^I, \tag{3.4}$$

as a bonus of the integrability of the first order formalism. We note that I_o is invariant, up to a total derivative, under the following BRST symmetry (the bosonic symmetry after fermionization)

$$\begin{aligned} Q_o u^I &= \psi^I, & Q_o F_I^{(2)} &= -d\eta_I, \\ Q_o A^I &= -d\psi^I, & Q_o H_I &= -d\chi_I + \eta_I, \end{aligned} \tag{3.5}$$

where ψ^I and χ_I are 0-forms taking values in $u^*(\Pi TX)$ and $u^*(\Pi T^*X)$, respectively, with $U = 1$ while η_I are 1-forms taking value in $u^*(\Pi T^*X)$ with $U = 1$. We have $Q_o I_o = \int_{\partial N}(\eta_I \wedge du^I + \psi^I dH_I)$. The boundary conditions are such that $\psi^I(x) = 0$ and $\eta_I(x)$ vanish along the directions tangent to ∂N, while $H_I(x) = 0$ along the direction normal to ∂N for $x \in \partial N$. The above BRST transformation laws should be completed by demanding $Q_o^2 = 0$;

$$Q_o\psi^I = 0, \qquad Q_o\eta_I = dF_I, \qquad Q_o\chi_I = F_I, \qquad Q_o F_I = 0. \tag{3.6}$$

Note that we introduce a new scalar bosonic field F_I taking values in $u^*(T^*X)$ with $U = 2$.[l] The boundary condition is that $F_I(x) = 0$ for $x \in \partial\Sigma$.

Now all those fields appearing above are regarded as "fields". One then introduces "anti-fields" as follows,

$$\begin{array}{ccccccccc} \text{Fields} & u^I & F_I & \eta_I & F_I^{(2)} & \chi_I & H_I & \psi^I & A^I \\ \text{Anti-Fields} & \eta_I^{(3)} & \rho_{(3)}^I & u_{(2)}^I & \rho^I & A_{(3)}^I & \psi_{(2)}^I & H_I^{(3)} & \chi_I^{(2)}. \end{array} \tag{3.7}$$

Here we used the convention that Latin letters denote bosonic (or even) fields while Greek letters denote fermionic (or odd) fields. If a "field" is a n-form on N its "anti-field" is a $(3-n)$-form. The ghost numbers U of a "field" ϕ and its "anti-field" $\bar{\phi}$ are relates as $U(\bar{\phi}) = -1 - U(\phi)$. Now we follows the usual steps to find a BV master action functional. The resulting theory is described in the following subsection.

3.1 BV quantized membrane without background

We start from the total space of ΠT^*X, with local coordinates $(\{u^I\}, \{\chi_I\})$. Next, we consider the total space of $\Pi T(\Pi T^*X)$, with local coordinates

[l]From $Q_o^2 F_I^{(2)} = -d(Q_o\eta_I) = 0$, we see that the general solution for $Q_o\eta_I$ is an exact 1-form. The moral is that wherever there is an ambiguity, which is actually a gauge symmetry, there should be a new field (or ghost for ghost).

$(\{\psi^I\},\{F_I\})$ on the the fiber. We assign degrees or ghost numbers $U = (0,1,1,2)$ to (u^I,χ_I,ψ^I,F_I). Now we define the following *even* symplectic structure on the space $\Pi T(\Pi T^*X)$ with degree $U=2$:

$$\omega = dF_I du^I + d\psi^I d\chi_I. \tag{3.8}$$

We note that the total space of $\Pi T(\Pi T^*X)$ can be identified with the total space of $T^*[2](\Pi T^*X)$. Then the degree $U=2$ symplectic structure ω can be identified with the canonical symplectic form on $T^*[2](\Pi T^*X)$ as a "cotangent bundle".

Based on the even symplectic structure ω we define an even Poisson bracket among functions in $(\{\phi^I\},\{\chi_I\}|\{\psi^I\},\{F_I\})$, i.e. functions on $\Pi T(\Pi T^*X)$, by the formula

$$\{\gamma_1,\gamma_2\}_P = \frac{\partial\gamma_1}{\partial F_I}\frac{\partial\gamma_2}{\partial u^I} - (-1)^{|\gamma_1||\gamma_2|}\frac{\partial\gamma_2}{\partial F_I}\frac{\partial\gamma_1}{\partial u^I} + \frac{\partial\gamma_1}{\partial\psi^I}\frac{\partial\gamma_2}{\partial\chi_I} - (-1)^{|\gamma_1||\gamma_2|}\frac{\partial\gamma_2}{\partial\psi^I}\frac{\partial\gamma_1}{\partial\chi_I}, \tag{3.9}$$

where $|\gamma|$ denotse the degree U of γ. We note that the graded Poisson bracket has degree $U=-2$, i.e., $|\{\gamma_1,\gamma_2\}_P| = |\gamma_1|+|\gamma_2|-2$. It is not difficult to check the following properties:

$$\begin{aligned}
\{\gamma_1,\gamma_2\}_P &= -(-1)^{|\gamma_1||\gamma_2|}\{\gamma_2,\gamma_1\}_P,\\
\{\gamma_1,\gamma_2\gamma_3\}_P &= \{\gamma_1,\gamma_2\}_P\gamma_3 + (-1)^{|\gamma_1||\gamma_2+1|}\gamma_2\{\gamma_1,\gamma_3\}_P,\\
\{\gamma_1,\{\gamma_2,\gamma_3\}_P\}_P &= \{\{\gamma_1,\gamma_2\}_P,\gamma_3\}_P + (-1)^{|\gamma_1||\gamma_2|}\{\gamma_2,\{\gamma_1,\gamma_3\}_P\}_P.
\end{aligned} \tag{3.10}$$

The second relation above (the Leibniz law) implies that the bracket behaves as a derivation of the ordinary product of functions. Such a product is (super)-commutative and associative and carries degree 0. Thus functions on $\Pi T(\Pi T^*X)$ form an algebra with a degree -2 Poisson bracket, a (super)-commutative associative product with degree 0 and a vanishing differential (no differential operator). We call such an algebra a cohomological 3-algebra.

On the space $\Pi T(\Pi T^*X)$ we have a canonical nilpotent odd vector Q_o with $U=1$

$$Q_o = \psi^I\frac{\partial}{\partial u^I} + F_I\frac{\partial}{\partial\chi_I}, \tag{3.11}$$

owhich riginates from the exterior derivative on ΠT^*X. We find that Q_o is the Hamiltonian vector of the following odd function

$$h = F_I\psi^I \tag{3.12}$$

carrying degree $U=3$. Note that an odd (even) function on $\Pi T(\Pi T^*X)$ has odd (even) Hamiltonian vector since our symplectic form is even. We have

$$\{h,h\}_P = 0, \qquad \{h,\ldots\}_P = Q_o, \qquad Q_o^2 = 0. \tag{3.13}$$

The BV quantized topological open membrane is a theory of maps

$$\phi : \Pi TN \to \Pi T(\Pi T^*X), \tag{3.14}$$

where ΠTN is the parity change of the total space of tangent bundle TN of N. We denote local coordinates on ΠTN by $(\{x^\mu\}, \{\theta^\mu\})$, $\mu = 1, 2, 3$. We parameterize a map by functions

$$(\boldsymbol{u}^I, \boldsymbol{\chi}_I, \boldsymbol{\psi}^I, \boldsymbol{F}_I) := (u^I, \chi_I, \psi^I, F_I)(x^\mu, \theta^\mu). \tag{3.15}$$

Now we consider the space $\mathcal{A}$ of all maps (3.14). For any function γ on $\Pi T(\Pi T^*X)$ we denote the corresponding function of $(\{x^\mu\}, \{\theta^\mu\})$ by $\boldsymbol{\gamma}$, i.e., $\boldsymbol{\gamma} = \boldsymbol{\phi}^*(\gamma) = \gamma(x^\mu, \theta^\mu)$. We have an expansion

$$\boldsymbol{\gamma} = \gamma(x^\sigma) + \gamma_\mu(x^\sigma)\theta^\mu + \frac{1}{2}\gamma_{\mu\nu}(x^\sigma)\theta^\mu\theta^\nu + \frac{1}{3!}\gamma_{\mu\nu\rho}(x^\sigma)\theta^\mu\theta^\nu\theta^\rho. \tag{3.16}$$

We denote $\gamma^{(n)} = \frac{1}{n!}\gamma_{m_1\ldots\mu_n} dx^{\mu_1} \wedge \ldots \wedge dx^{\mu_n}$, where $n = 0, \ldots, 3$. We obtain functions on $\Pi T\mathcal{A}$ by $\int_{c_n} \gamma^{(n)} \equiv \int_{c_n} d^n\theta\, \boldsymbol{\gamma}$, where c_n is a n-dimensional cycle in N. We see that $\int_M d^3\theta\, \boldsymbol{\gamma}$ is an even (odd) function on $\mathcal{A}$ if γ is an odd (even) function on $\Pi T(\Pi T^*X)$, with the degree shifted by -3.

On the space $\mathcal{A}$ of all fields we have the following odd symplectic structure $\boldsymbol{\omega}$

$$\boldsymbol{\omega} = \int_N d^3\theta \left(\delta \mathbb{F}_I \delta \boldsymbol{u}^I + \delta \boldsymbol{\psi}^I \delta \boldsymbol{\chi}_I\right). \tag{3.17}$$

We note that a n-form field is paired with a $(3-n)$-form field. For instance the 0-form part of $\boldsymbol{u}^I$, which is even, is paired with the 3-form part of $\mathbb{F}_I$, which is odd. We also remark that θ^μ carry degree $U = -1$ and the degree of $\boldsymbol{\omega}$ is $U = -1$. Then we can define the BV bracket $(.,.)_{BV}$ as the odd Poisson bracket with respect to $\boldsymbol{\omega}$ among functions on $\mathcal{A}$. The degree of the BV bracket is $U = 1$. We observe that the *odd* symplectic form $\boldsymbol{\omega}$ (of degree $U = -1$) on $\mathcal{A}$ is induced from the *even* symplectic structure ω (of degree $U = 2$) on $\Pi T(\Pi T^*X)$. Similarly the BV bracket (of degree $U = 1$) among functions on $\mathcal{A}$ is induced from the even Poisson bracket $\{.,.\}_{PB}$ (of degree $U = -2$) among functions on $\Pi T(\Pi T^*X)$. We also note that the relations in (3.10) after shifting the degree by -3 become the usual relations for the BV bracket. It follows that

$$\{\gamma, \gamma\}_P = 0 \text{ if and only if } \left(\textstyle\int_N d^3\theta\, \boldsymbol{\gamma}, \int_N d^3\theta\, \boldsymbol{\gamma}\right)_{BV} = 0.$$

We note that the superfields in (3.15) contain all the "fields" and "anti-fields" (3.7) in the BV quantization of membrane. For the explicit identifications we expand the superfields as follows:

$$\begin{aligned}
\boldsymbol{u}^I &= u^I + \rho^I_\mu\theta^\mu + \frac{1}{2}u^I_{\mu\nu}\theta^\mu\theta^\nu + \frac{1}{3!}\rho^I_{\mu\nu\rho}\theta^\mu\theta^\nu\theta^\sigma, \\
\boldsymbol{F}_I &= F_I + \eta_{I\mu}\theta^\mu + \frac{1}{2}F_{I\mu\nu}\theta^\mu\theta^\nu - \frac{1}{3!}\eta_{I\mu\nu\sigma}\theta^\mu\theta^\nu\theta^\sigma, \\
\boldsymbol{\psi}^I &= \psi^I + A^I_\mu\theta^\mu - \frac{1}{2}\psi^I_{\mu\nu}\theta^\mu\theta^\nu - \frac{1}{3!}A^I_{\mu\nu\rho}\theta^\mu\theta^\nu\theta^\sigma, \\
\boldsymbol{\chi}_I &= \chi_I + H_{I\mu}\theta^\mu + \frac{1}{2!}\chi_{I\mu\nu}\theta^\mu\theta^\nu + \frac{1}{3!}H_{I\mu\nu\sigma}\theta^\mu\theta^\nu\theta^\sigma.
\end{aligned} \tag{3.18}$$

The ghost number of the superfields $(\boldsymbol{u}^I, \boldsymbol{\chi}_I, \boldsymbol{\psi}^I, \boldsymbol{F}_I)$ are $U = (0,1,1,2)$, the same as the ghost numbers of $(u^I, \chi_I, \psi^I, F_I)$. Note that the ghost number of θ^μ is $U = 1$. As a differential form we write, for example, $\rho^I = \rho^I_\mu dx^\mu$, $u^I_{(2)} = \frac{1}{2}u^I_{\mu\nu}dx^\mu \wedge dx^\nu$ and $\rho^I_{(3)} = \frac{1}{3!}\rho^I_{\mu\nu\sigma}dx^\mu dx^\nu dx^\sigma$, etc. Thus the assignments of "fields" and "anti-fields" in (3.7) and the ghost numbers are consistent with our definition of the odd symplectic structure ω (3.17) and the decompositions of the superfields. One can also check that for any function γ on $\Pi T(\Pi T^* X)$, we have

$$\begin{aligned}
\triangle \int_N d^3\theta\, \boldsymbol{\phi}^*(\gamma) &= (-1+3-3+1)C\int_N dv \left(\frac{\delta^2}{\delta F_I \delta u^I} + \frac{\delta^2}{\delta\psi^I\delta\chi_I}\right)\gamma^{(0)} \\
&= 0,
\end{aligned} \tag{3.19}$$

where C is an infinite constant and dv is the volume form on N.

The BV quantized version $\boldsymbol{S}_o$ of the action functional I_o in (3.3) is given by

$$\boldsymbol{S}_o = \int_M d^3\theta \left(\boldsymbol{\psi}^I D\boldsymbol{\chi}_I + \boldsymbol{F}_I D\boldsymbol{u}^I + \boldsymbol{F}_I\boldsymbol{\psi}^I\right). \tag{3.20}$$

The BV BRST charge $\boldsymbol{Q}_o$ corresponds to the odd Hamiltonian vector $\boldsymbol{Q}_o = (\boldsymbol{S}_o, \ldots)_{BV}$ of $\boldsymbol{S}_o$. We have

$$\boldsymbol{Q}_o = D + \boldsymbol{\phi}^*(Q_o), \tag{3.21}$$

where $D = \theta^\mu\partial_\mu$. Explicitly,

$$\boldsymbol{Q}_0 = \left(D\boldsymbol{u}^I + \boldsymbol{\psi}^I\right)\frac{\partial}{\partial \boldsymbol{u}^I} + D\boldsymbol{\psi}^I\frac{\partial}{\partial\boldsymbol{\psi}^I} + (D\boldsymbol{\chi}_I + \boldsymbol{F}_I)\frac{\partial}{\partial\boldsymbol{\chi}_I} + D\boldsymbol{F}_I\frac{\partial}{\partial \boldsymbol{F}_I}. \tag{3.22}$$

In components we see that the above BRST charge leads to the transformation laws (3.5) and (3.6) for the "fields' as well as for the "anti-fields".

It is trivial to check that $\boldsymbol{S}_o$ satisfies the quantum master equation,

$$(\boldsymbol{S}_0, \boldsymbol{S}_0)_{BV} = 0, \qquad \triangle\boldsymbol{S}_0 = 0. \tag{3.23}$$

Now we consider the case that the boundary of N is non-empty. Then we should impose suitable boundary conditions such that $D\boldsymbol{S}_o = 0$.

$$\begin{aligned} \int_{\partial N} d^2\theta \, \boldsymbol{F}_I \boldsymbol{\psi}^I &= 0, \\ \int_{\partial N} d^2\theta \, \left(\boldsymbol{\psi}^I D\boldsymbol{\chi}_I + \boldsymbol{F}_I D\boldsymbol{u}^I \right) &= 0. \end{aligned} \tag{3.24}$$

The above equations are satisfied by the boundary conditions we introduced earlier. We note that the $\boldsymbol{F}_I$ enter into the action functional linearly. Thus the integration over $\boldsymbol{F}_I$ leads to a delta function like constraint

$$D\boldsymbol{u}^I + \boldsymbol{\psi}^I = 0. \tag{3.25}$$

Then the on-shell action functional reduces to the BV quantized boundary closed string theory without background,

$$\boldsymbol{S}_o|_{\text{on shell}} = \int_{\partial N} d^2\theta \left(\boldsymbol{\chi}_I D\boldsymbol{u}^I \right). \tag{3.26}$$

3.2 Bulk deformations and boundary observables

Now we consider a deformation the BV action functional $\boldsymbol{S}_o$ preserving the ghost number symmetry. A deformed action functional is of the following form

$$\boldsymbol{S}_\gamma = \boldsymbol{S}_o + \int_N d^3\theta \, \boldsymbol{\phi}^*(\gamma), \tag{3.27}$$

where γ is a function on $\Pi T(\Pi T^* X)$. The idea and procedure behind the above deformation are exactly the same as those of the string case discussed extensively in Sect. 2.2.2. The Brezin integral $\int d^3\theta$ will decrease the ghost number by $U = 3$. Thus γ should have degree $U = 3$ to preserve the ghost number symmetry. Here we will impose such a condition.

The above deformation is well-defined or admissible if $\boldsymbol{S}_\gamma$ satisfies the quantum master equation,

$$-\hbar \Delta \boldsymbol{S}_\gamma + \frac{1}{2} \left(\boldsymbol{S}_\gamma, \boldsymbol{S}_\gamma \right)_{BV} = 0. \tag{3.28}$$

Since $\Delta \boldsymbol{S}_\gamma = \Delta \int_N d^3\theta \, \boldsymbol{\gamma} = 0$, the quantum master equation (3.28) reduces to

$$\left(\boldsymbol{S}_\gamma, \boldsymbol{S}_\gamma \right)_{BV} = 0. \tag{3.29}$$

The above is equivalent to the following two conditions:

$$\int_N d^3\theta\, D\phi^*(\gamma) = 0,$$
$$\int_N d^3\theta\, \phi^*(Q_o\gamma + \frac{1}{2}\{\gamma,\gamma\}_P) = 0. \tag{3.30}$$

Note that the boundary conditions are such that $\psi^I(x), F_I(x) = 0$ in directions tangent to ∂N_p for $x \in \partial N_p$. Thus a consistent deformation of the theory is determined by a degree $U = 3$ function γ on $\Pi T(\Pi T^* X)$ satisfying

$$\gamma|_{\Pi T^* X} = 0,$$
$$Q_o\gamma + \frac{1}{2}\{\gamma,\gamma\}_P = 0. \tag{3.31}$$

The first condition means that the restriction of $\gamma(u^I, \chi_I, F^I, \psi_I)$ to the base space $\Pi T^* X$ of $\Pi T(\Pi T^* X)$ vanishes. We denote the set of equivalence classes of all solution of (3.31) by $\mathfrak{M}^{cl}$. Thus the space $\mathfrak{M}^{cl}$ is isomorphic to the set of equivalence classes of all consistent bulk deformations (or backgrounds) of the membrane theory $\boldsymbol{S}_o$ preserving the ghost number symmetry U.

From now on we assume that γ is a solution of (3.31). We denote by Q_γ the degree $U = 1$ odd nilpotent Hamiltonian vector of $h + \gamma$;

$$\{h + \gamma, \ldots\}_P = Q_\gamma. \tag{3.32}$$

Explicitly

$$Q_\gamma = \left(\psi^I + \frac{\partial\gamma}{\partial F_I}\right)\frac{\partial}{\partial u^I} + \left(F_I + \frac{\partial\gamma}{\partial\psi^I}\right)\frac{\partial}{\partial\chi_I}$$
$$+ \frac{\partial\gamma}{\partial u^I}\frac{\partial}{\partial F_I} + \frac{\partial\gamma}{\partial\chi_I}\frac{\partial}{\partial\psi^I}. \tag{3.33}$$

Then the action functional $\boldsymbol{S}_\gamma$ in (3.27) has a BRST symmetry generated by the odd nilpotent vector $\boldsymbol{Q}_\gamma$ with $U = -1$ defined by $(\boldsymbol{S}_\gamma, \ldots)_{BV} = \boldsymbol{Q}_\gamma$. Equivalently[m]

$$\boldsymbol{Q}_\gamma = D + \phi^*(Q_\gamma). \tag{3.34}$$

[m]We should emphasize that the BV master equation is equivalent to the condition that $\boldsymbol{Q}_\gamma^2 = 0$, which implies that $Q_\gamma^2 = 0$ and the first equation of (3.30). Thus the moduli space of the theory is defined *not* by the non-linear cohomology of Q_γ but by the non-linear cohomology of the BV BRST charge $\boldsymbol{Q}_\gamma$. The difference is precisely encoded in the boundary degrees due to the relation

$$\boldsymbol{Q}_\gamma \int_N d^3\theta\phi^*(\alpha) - \int_N d^3\theta\phi^*(Q_\gamma\alpha) = \int_{\partial N} d^2\theta\phi^*(\alpha)|_{\partial N}.$$

Denoting $\gamma = \phi^*(g(u^I, \chi_I, F_I, \psi^I) = \gamma(\boldsymbol{u}^I, \boldsymbol{\chi}_I, \boldsymbol{F}_I, \boldsymbol{\psi}^I)$ we have

$$\begin{aligned} \boldsymbol{Q}_\gamma = &\left(D\boldsymbol{u}^I + \boldsymbol{\psi}^I + \frac{\partial \gamma}{\partial \boldsymbol{F}_I}\right)\frac{\partial}{\partial \boldsymbol{u}^I} + \left(D\boldsymbol{\chi}_I + \boldsymbol{F}_I + \frac{\partial \gamma}{\partial \boldsymbol{\psi}^I}\right)\frac{\partial}{\partial \boldsymbol{\chi}_I} \\ &+ \left(D\boldsymbol{F}_I + \frac{\partial \gamma}{\partial \boldsymbol{u}^I}\right)\frac{\partial}{\partial \boldsymbol{F}_I} + \left(D\boldsymbol{\psi}^I + \frac{\partial \gamma}{\partial \boldsymbol{\chi}_I}\right)\frac{\partial}{\partial \boldsymbol{\psi}^I}. \end{aligned} \tag{3.35}$$

The action functional $\boldsymbol{S}_\gamma$ also has a fermionic symmetry generated by $\mathcal{K}_\mu = -\frac{\partial}{\partial \theta^\mu}$, acting on superfields, with $U = -1$. Together with $\boldsymbol{Q}_\gamma$ they satisfy the following anti-commutation relations,

$$\{\boldsymbol{Q}_\gamma, \boldsymbol{Q}_\gamma\} = 0, \qquad \{\boldsymbol{Q}_\gamma, \mathcal{K}_\mu\} = -\partial_\mu, \qquad \{\mathcal{K}_\mu, \mathcal{K}_\nu\} = 0. \tag{3.36}$$

Given a $\boldsymbol{Q}_\gamma$ invariant function $A^{(0)}$ on $\mathcal{A}$, which is a 0-form on N, the n-form $A^{(n)} := \frac{1}{n!}\left(\mathcal{K}_{\mu_1}\ldots\mathcal{K}_{\mu_n}A^{(0)}\right)dx^{\mu_1}\wedge\ldots\wedge dx^{\mu_n}$ for $n = 1, 2, 3$ give the canonical set of solutions of the descent equations

$$\begin{aligned} \boldsymbol{Q}_\gamma A^{(0)} &= 0, \\ \boldsymbol{Q}_\gamma A^{(1)} + dA^{(0)} &= 0, \\ \boldsymbol{Q}_\gamma A^{(2)} + dA^{(1)} &= 0, \\ \boldsymbol{Q}_\gamma A^{(3)} + dA^{(2)} &= 0, \\ dA^{(3)} &= 0. \end{aligned} \tag{3.37}$$

The above is a direct consequence of (3.34).

A BV observable in general is a function $\mathcal{O}$ on $\mathcal{A}$ satisfying $(-\hbar\Delta + \boldsymbol{Q}_\gamma)\mathcal{O} = 0$. A zero-dimensional observable $\mathcal{O}^{(0)}(x)$ can be inserted at a point in the interior or at the boundary of N. Since $\mathcal{O}^{(0)}$ is a scalar on N it is a function of the scalar components of the superfields only. It follows that $\Delta\mathcal{O}^{(0)} = 0$ and $\boldsymbol{Q}_\gamma\mathcal{O}^{(0)} = 0$. The latter, together with (3.34), implies that $\mathcal{O}^{(0)}(x)$, $x \notin \partial\Sigma$, must derive from a Q_γ cohomology class among functions on $\Pi T\Pi(T^*X)$. Recall that the boundary condition for $\psi^I(x^\mu)$ and $F_I(x^\mu)$ is that they vanish identically at the boundary. Thus a zero observable inserted at a boundary point originates from a function $f(u^I, \chi_I)$ of the base $\Pi(T^*X)$ of $\Pi T(\Pi(T^*X))$. Note that no such function can be used to deform the bulk action functional due to the first condition in (3.31). We may take for N a three-dimensional disk with boundary $\partial N = S^2$, having a number of marked points x_i where zero-dimensional observables are inserted. We can also consider a small circle $C_i \subset \partial\Sigma$ surrounding a marked point x_i. Then $\oint_{C_i} d^\theta f(\boldsymbol{u}^I, \boldsymbol{\chi}_I)$ defines an observable, whose expectation value does not depend on the contour. We have other observables $\int_{\partial N} d^2\theta\, f(\boldsymbol{u}^I, \boldsymbol{\chi}_I)$, which may be viewed as boundary interactions.

Now our goal is to determine the general γ with $U = 3$ satisfying (3.31). The explicit form of the above equation is easy to write down, though complicated, by the general form of a degree $U = 3$ function γ on on $\Pi T(\Pi T^* X)$.

3.2.1. Turning on a C-field

Now we consider a bulk deformation leading to the BV quantized version of the action functional of the membrane with a flat 3-form C-field (3.2). For a 3-form $C = \frac{1}{3!}c_{IJK}(u^L)du^I \wedge du^J \wedge du^K$ on X, we obtain a degree $U = 3$ function c on $\Pi T(\Pi T^*(X))$

$$c = \frac{1}{3!}c_{IJK}(u^L)\psi^I\psi^J\psi^K, \tag{3.38}$$

satisfying $c|_{\Pi T^* X} = 0$. The condition $Q_o c + \frac{1}{2}\{c, c\}_P = 0$ is equivalent to $dC = 0$. Thus we obtain the following action functional satisfying the BV master equation

$$\boldsymbol{S}_c = \int_N d^3\theta \left(D\boldsymbol{\chi}_I\boldsymbol{\psi}^I + \boldsymbol{F}_I D\boldsymbol{u}^I + \boldsymbol{F}_I\boldsymbol{\psi}^I\right) + \frac{1}{3!}\int_N d^3\theta \left(\boldsymbol{c}_{IJK}\boldsymbol{\psi}^I\boldsymbol{\psi}^J\boldsymbol{\psi}^K\right). \tag{3.39}$$

This is the BV quantized action functional of (3.2). The action functional $\mathbb{S}_c$ is the Hamiltonian function, $(\boldsymbol{S}_c, \ldots)_{BV} = \mathbb{Q}_c$, of the odd vector $\mathbb{Q}$ given by (in terms of BRST transformation laws)

$$\begin{aligned}
\mathbb{Q}_c\boldsymbol{u}^I &= D\boldsymbol{u}^I + \boldsymbol{\psi}^I, \\
\mathbb{Q}_c\boldsymbol{\chi}_I &= D\boldsymbol{\chi}_I + \boldsymbol{F}_I + \frac{1}{2}\boldsymbol{c}_{IJK}\boldsymbol{\psi}^J\boldsymbol{\psi}^K, \\
\mathbb{Q}_c\boldsymbol{\psi}^I &= D\boldsymbol{\psi}^I \\
\mathbb{Q}_c\boldsymbol{F}_I &= D\boldsymbol{F}_I - \frac{1}{3!}\partial_I\boldsymbol{c}_{JKL}\boldsymbol{\psi}^J\boldsymbol{\psi}^K\boldsymbol{\psi}^L.
\end{aligned} \tag{3.40}$$

On-shell we can eliminate $\boldsymbol{F}_I$ and $\boldsymbol{\psi}^I$ using the $\boldsymbol{F}_I$ equation of motion;

$$\mathbb{S}_c|_{\text{on shell}} = -\frac{1}{3!}\int_N d^3\theta \left(\boldsymbol{c}_{IJK}D\boldsymbol{u}^I D\boldsymbol{u}^J D\boldsymbol{u}^K\right) + \int_{\partial N} d^2\theta\, \boldsymbol{\chi}_I D\boldsymbol{u}^I. \tag{3.41}$$

Now consider an arbitrary function $f(u^I, \chi_I)$ on the base $\Pi T^* X$ of $\Pi T(\Pi T^* X)$. The corresponding function $\boldsymbol{f}$ of the superfields has an expansion

$$\boldsymbol{f} = f^{(0)}(x^\mu) + \ldots. \tag{3.42}$$

Inserting $f^{(0)}(x_i)$ at a boundary puncture x_i we find $\mathbb{Q}_c f^{(0)} = 0$. It also follows that $\oint_{C_i} d\theta \boldsymbol{f}$ and $\int_{\partial N} d\theta \boldsymbol{f}$ are BV observables.

Now we discuss more general deformations. We consider the following non-linear transformations of the fiber coordinates of $\Pi T(\Pi T^* X) \to \Pi T^* X$,

$$
\begin{aligned}
F_I \to F'_I &:= F_I - \frac{\partial \pi}{\partial u^I}, \\
\psi^I \to \psi'^I &:= \psi^I - \frac{\partial \pi}{\partial \chi_I},
\end{aligned} \tag{3.43}
$$

with u^I and χ unchanged. Here π is an arbitrary degree $U = 2$ function on the base space $\Pi T^* X$ of $\Pi T(\Pi T^* X)$, i.e., $\pi = \frac{1}{2} b^{IJ}(u^L)\chi_I \chi_J$. Under the above transformation we have

$$
h + c \to h + \gamma,
$$

where

$$
\begin{aligned}
\gamma = &- F_I \frac{\partial \pi}{\partial \chi_I} - \frac{\partial \pi}{\partial u^I}\psi^I + \frac{1}{2}[\pi, \pi]_S \\
&+ \frac{1}{3!} c_{IJK} \left(\psi^I - \frac{\partial \pi}{\partial \chi_I}\right)\left(\psi^J - \frac{\partial \pi}{\partial \chi_J}\right)\left(\psi^K - \frac{\partial \pi}{\partial \chi_K}\right)
\end{aligned} \tag{3.44}
$$

It can be easily shown that $\{h + \gamma, h + \gamma\}_P = 0$ if and only if $dC = 0$. The above class of solutions of (3.31) is determined by an element of $H^3(X)$ and a bi-vector on X. Now the condition $\gamma|_{\Pi T^* X} = 0$ implies that

$$
\frac{1}{2}[\pi, \pi]_S - \frac{1}{3!} c_{IJK} \frac{\partial \pi}{\partial \chi_I} \frac{\partial \pi}{\partial \chi_I} \frac{\partial \pi}{\partial \chi_I} = 0. \tag{3.45}
$$

More explicitly

$$
h^{LMN} - c_{IJK} b^{IL} b^{JM} b^{KN} = 0, \tag{3.46}
$$

where $h^{IJK} = b^{LI}\partial_L b^{JK} + cyclic$. Such a pair (c, π) leads to the following degree $U = 3$ function on $T^*\Pi(T^* X)$ satisfying (3.31),

$$
\begin{aligned}
\gamma = &- b^{IJ} F_I \chi_J - \frac{1}{2}\left(\partial_I b^{JK} + c_{IMN} b^{MJ} b^{NK}\right)\chi_J \chi_K \psi^I \\
&- \frac{1}{2} c_{IJK} b^{IL} \chi_L \psi^J \psi^K + \frac{1}{3!} c_{IJK} \psi^I \psi^J \psi^K.
\end{aligned} \tag{3.47}
$$

It follows that the action functional

$$
\begin{aligned}
S_\gamma = \int_N d^3\theta \Big(& D\chi_I \psi^I + F_I D u^I + F_I \psi^I + \frac{1}{3!} c_{IJK} \psi^I \psi^J \psi^K - b^{IJ} F_I \chi_J \\
& - \frac{1}{2} c_{IJK} b^{IL} \chi_L \psi^J \psi^K - \frac{1}{2}\left(\partial_I b^{JK} + c_{IMN} b^{MJ} b^{NK}\right)\chi_J \chi_K \psi^I \Big)
\end{aligned} \tag{3.48}
$$

satisfies the quantum BV master equation provided that $dC = 0$.

The BV BRST transformation laws, with the condition (3.46), are given by

$$
\begin{aligned}
Q u^I =& D u^I + \psi^I - b^{IJ}\chi_J,\\
Q\chi_I =& D\chi_I + F_I - \frac{1}{2}\partial_I b^{JK}\chi_J\chi_K + \frac{1}{2}c_{IJK}\left(\psi^J - b^{JM}\chi_M\right)\left(\psi^K - b^{KN}\chi_N\right),\\
Q\psi^I =& D\psi^I - b^{IJ}F_J - \partial_K b^{IJ}\psi^K\chi_J - \frac{1}{2}h^{IJK}\chi_J\chi_K\\
&- \frac{1}{2}c_{LJK}b^{LI}\left(\psi^J - b^{JM}\chi_M\right)\left(\psi^K - b^{KN}\chi_N\right),\\
QF_I =& DF_I + \partial_I b^{JK}F_J\chi_K + \frac{1}{2}\partial_I\partial_K b^{LJ}\psi^K\chi_L\chi_J + \frac{1}{3!}\partial_I h^{JKL}\chi_J\chi_K\chi_L\\
&- \frac{1}{3!}\partial_I c_{PJK}\left(\psi^P - b^{PL}\chi_L\right)\left(\psi^J - b^{JM}\chi_M\right)\left(\psi^K - b^{KN}\chi_N\right)\\
&+ \frac{1}{2}c_{PJK}\partial_I b^{PL}\chi_L\left(\psi^J - b^{JM}\chi_M\right)\left(\psi^K - b^{KN}\chi_N\right).
\end{aligned}
\tag{3.49}
$$

We have $Q_\gamma^2 = 0$, which follows from $(S_\gamma, S_\gamma)_{BV} = 0$.

Now we examine the use of the above deformation. From the F_I equation of motion

$$
Du^I - b^{IJ}\chi_J + \psi^I = 0 \tag{3.50}
$$

we can eliminate ψ^I from the action functional, leading to the following on-shell action functional

$$
\begin{aligned}
S_\gamma|_{\text{on shell}} =& -\frac{1}{3!}\int_N d^3\theta\left(c_{IJK}Du^I Du^J Du^K\right)\\
&+ \int_{\partial N} d^2\theta\left(\chi_I Du^I + \frac{1}{2}b^{IJ}\chi_I\chi_J\right).
\end{aligned}
\tag{3.51}
$$

We obtain a boundary string theory in an arbitrary bivector background, while a closed 3-form C-field is coupled to the membrane in the bulk. Note that the boundary action functional may be viewed as the closed string version of the Catteneo-Felder model. Recall that such an action functional satisfies the quantum master equation (but with a different BV bracket) if and only if π is Poisson. We may identify the open membrane theory defined by the action functional S_γ (3.48) together with the condition (3.46) as the off-shell closed string theory coupled to an arbitrary B-field.

Now we assume that the bi-vector is non-degenerate, and therefore has an inverse. We then have a corresponding 2-form or anti-symmetric tensor field on X, $B = \frac{1}{2}b_{IJ}du^I \wedge du^J$. Then the condition (3.46) implies that $C = dB$. Hence the 3-form C is the field strength of the B-field.

3.3 The first approximation

In this subsection we discuss the first approximation of the path integral for a manifold X with $c_1(X) = 0$. Our presentation will be indirect and the actual path integral calculations will appear elsewhere [37].

The first order problem can be viewed as a "quantization" of $\Pi T(\Pi T^*X)$ viewed as a classical phase space with respect to the even symplectic structure ω in (3.8). Here we regard (u^I, χ_I) as the canonical coordinates and (F_I, ψ^I) as the conjugate momenta. Now recall that the bulk term is determined by the function $h+\gamma$ on $\Pi T\Pi(T^*X)$

$$
\begin{aligned}
h+\gamma = &- b^{IJ}F_I\chi_J - \frac{1}{2}\left(\partial_I b^{JK} + c_{IMN}b^{MJ}b^{NK}\right)\chi_J\chi_K\psi^I \\
&+ F_I\psi^I - \frac{1}{2}c_{IJK}b^{IL}\chi_L\psi^J\psi^K + \frac{1}{3!}c_{IJK}\psi^I\psi^J\psi^K
\end{aligned}
\tag{3.52}
$$

satisfying the condition (3.45). By "quantization" of $\Pi T(\Pi T^*X)$ we mean the following replacements

$$
F_I \to -\hbar\frac{\partial}{\partial u^I}, \qquad \psi^I \to -\hbar\frac{\partial}{\partial \chi_I}, \tag{3.53}
$$

where $\hbar$ is regarded as a formal parameter with $U = 2$. From $(h+\gamma)/\hbar$ we obtain the following differential operators acting on functions on ΠT^*X;

$$
\mathcal{D} = \mathcal{D}_1 + \hbar\mathcal{D}_2 + \hbar^2\mathcal{D}_3, \tag{3.54}
$$

where

$$
\begin{aligned}
\mathcal{D}_1 &= b^{IJ}\chi_J\frac{\partial}{\partial u^I} + \frac{1}{2}\left(\partial_I b^{JK} + c_{IMN}b^{MJ}b^{NK}\right)\chi_J\chi_K\frac{\partial}{\partial \chi_I}, \\
\mathcal{D}_2 &= \frac{\partial^2}{\partial u^I\partial\chi_I} - \frac{1}{2}c_{IJK}b^{IL}\chi_L\frac{\partial^2}{\partial\chi_J\partial\chi_K}, \\
\mathcal{D}_3 &= \frac{1}{3!}c_{IJK}\frac{\partial^3}{\partial\chi_I\partial\chi_J\partial\chi_K}.
\end{aligned}
\tag{3.55}
$$

Note that $\mathcal{D}_i$ is an order i differential operator and has degree $U = 3 - 2i$. Now the conditions to satisfy the BV master equation, $dC = 0$ and (3.45), imply that $\mathcal{D}^2 = 0$;

$$\begin{aligned} \mathcal{D}_1^2 &= 0, \\ \mathcal{D}_1\mathcal{D}_2 + \mathcal{D}_2\mathcal{D}_1 &= 0, \\ \mathcal{D}_1\mathcal{D}_3 + \mathcal{D}_3\mathcal{D}_1 + \mathcal{D}_2^2 &= 0, \\ \mathcal{D}_2\mathcal{D}_3 + \mathcal{D}_3\mathcal{D}_2 &= 0, \\ \mathcal{D}_3^2 &= 0. \end{aligned} \tag{3.56}$$

The differential operators above defines various structures on the algebra $\mathcal{O}(\Pi T^*X)$ of functions on ΠT^*X (multivectors on X). For $c = \pi = 0$, therefore $\gamma = 0$, we have $\mathcal{D}_1 = \mathcal{D}_3 = 0$, and $\mathcal{D}_2 = \Delta$ generates the Schouten-Nijenhuis bracket $[,.,]_S$ on functions on ΠT^*X. Together with the ordinary product we get the cohomological 2-algebra or GBV-algebra of X. For $c = 0$ the condition (3.45) implies that $[\pi,\pi]_s = 0$. Now we have $\mathcal{D}_1 = Q_\pi \equiv [\pi,\ldots]_S$, $\mathcal{D}_2 = \Delta$ and $\mathcal{D}_3 = 0$. Then $\mathcal{D}_1$ induces a differential on the cohomological 2-algebra. Forgetting the product we have a structure of dgLa on $\mathcal{O}(\Pi T^*X)$, or in general a structure of dGBV algebra.

Akman [1] (see also [10]), motivated by VOSA and generalizing Koszul, introduced the concept of higher order differential operators on a general superalgebra. Using such a differential operator, say D, he considered the following recursive definition of higher brackets,

$$\begin{aligned} \Phi_1(a) =& D(a), \\ \Phi_2(a,b) =& D(a\cdot b) - (Da)\cdot b - (-1)^{|a||D|}a\cdot \mathcal{D}(b), \\ \Phi_3(a,b,c) =& \Phi_2(a,b\cdot c) - \Phi_2(a,b)\cdot c - (-1)^{|b|(|a|+|D|}b\cdot\Phi_2(a,c), \\ &\vdots \\ \Phi_{r+1}(a_1,\ldots,a_{r+1}) =& \Phi_r(a_1,\ldots,a_r\cdot a_{r+1}) - \Phi_r(a_1,\ldots,a_r)\cdot a_{r+1} \\ &- (-1)^{|a_r|(|a_1|+\ldots+|a_{r-1}+|D|}a_r\cdot\Phi_r(a_1,\ldots,a_{r-1},a_{r+1}), \end{aligned} \tag{3.57}$$

such that for D of order $\leq r$ Φ_{r+1} vanishes identically. He examined the general properties of those higher brakets.

Using the above we define the following 2- and 3-brackets associated with $\mathcal{D}$ among functions on ΠT^*X

$$\begin{aligned} [a,b] &= (-1)^{|a|}\Phi_2(a,b), \\ [a,b,c] &= (-1)^{|b|}\Phi_3(a,b,c), \end{aligned} \tag{3.58}$$

while the higher brackets all vanish. Explicitly

$$
\begin{aligned}
[a,b] =& \hbar\frac{\partial a}{\partial \chi_I}\frac{\partial b}{\partial u^I} - \frac{\hbar}{2}c_{IJK}b^{IL}\chi_L\frac{\partial a}{\partial \chi_J}\frac{\partial b}{\partial \chi_K} \\
&- (-1)^{(|a|+1)(|b|+1)} \times (a \leftrightarrow b), \qquad (3.59)\\
[a,b,c] =& \hbar^2 c_{IJK}\frac{\partial a}{\partial \chi_I}\frac{\partial b}{\partial \chi_J}\frac{\partial c}{\partial \chi_K}.
\end{aligned}
$$

These brackets have the following properties:

(i) super-commutativity,

$$[a,b] = -(-1)^{(|a|+1)(|b|+1)}[b,a] \tag{3.60}$$

(ii) deformed Leibniz law,

$$[a,bc] - [a,b]c - (-)^{(|a|+1)|b|}b[a,c] = (-1)^{|a|+|b|}[a,b,c]. \tag{3.61}$$

(ii) derivation,

$$\mathcal{D}[a,b] = [\mathcal{D}a,b] + (-1)^{|a|+1}[a,\mathcal{D}b]. \tag{3.62}$$

(iii) Jacobi identity up to homotopy,

$$
\begin{aligned}
[[a,b],c]+&(-1)^{(|a|+1)(|b|+1)}[b,[a,c]] - [a,[b,c]] \\
=&\mathcal{D}[a,b,c] + (-1)^{|b|}[\mathcal{D}a,b,c] \qquad (3.63)\\
&+ (-1)^{|a|+|b|}[a,\mathcal{D}b,c] + (-1)^{|a|}[a,b,\mathcal{D}c].
\end{aligned}
$$

Now we return to some special cases. For $c = \pi = 0$ the 2-bracket $[.,.]$ becomes the Schouten-Nijenhuis bracket. Together with the product we have

$$
\begin{aligned}
[a,b] - (-1)^{|a|}\Delta(ab) + (-1)^{|a|}(\Delta a)b + a(\Delta b) &= 0,\\
[a,b] + (-1)^{(|a|+1)(|b|+1)}[b,a] &= 0,\\
[a,bc] - [a,b]c - (-)^{(|a|+1)|b|}b[a,c] &= 0, \qquad (3.64)\\
[[a,b],c] + (-1)^{(|a|+1)(|b|+1)}[b,[a,c]] - [a,[b,c]] &= 0,\\
\Delta[a,b] - [\Delta a,b] - (-1)^{|a|+1}[a,\Delta b] &= 0.
\end{aligned}
$$

For $c = 0$ with Poisson bi-vector π we have the Schouten-Nijenhuis bracket above and the differential Q_π satisfying;

$$
\begin{aligned}
Q_\pi^2 &= 0,\\
Q_\pi[a,b] - [Q_\pi a,b] - (-1)^{|a|+1}[a,Q_\pi b] &= 0.
\end{aligned} \tag{3.65}
$$

Thus in the $c = 0$ case we have the usual structue of a dGVB algebra for a Poisson manifold X with $c_1(X) = 0$ on $\mathcal{O}(\Pi T^*X)$.

Hence the actual deformation comes from non-zero c. We remark that Kravchenko defined a BV_∞-algebra by a sum of differentials of degree $3-2i$ and order i, whose square is zero. Hence the deformed algebra $\mathcal{O}(\Pi T^*X)$ of functions on ΠT^*X induced by the differential operator $\mathcal{D}$ in (3.54) has a structure of BV_∞-algebra. It has a L_∞ structure with brackets and a first order differential.

3.4 Generalization

We can relax the requirement that the deformation terms should have ghost number $U=0$, and allow them to have any even ghost number. Now we let $C_k = \frac{1}{(2k+1)!} c_{I_1 \ldots I_{2k+1}} du^{I_1} \wedge \ldots \wedge du^{I_{2k+1}}$ be a closed $(2k+1)$-form, $k=0,1,\ldots$ on X;

$$dC_k = 0. \tag{3.66}$$

Then we have the corresponding degree $U=2k+1$ function c_k on $\Pi(T\Pi T^*X)$;

$$c_k = \frac{1}{(2k+1)!} c_{I_1 \ldots I_{2k+1}} \psi^{I_1} \ldots \psi^{I_{2k+1}}, \tag{3.67}$$

satisfying

$$\begin{aligned} \sum_k c_k|_{\Pi T^*X} &= 0, \\ Q_o \sum_k c_k + \frac{1}{2}\left\{\sum_k c_k, \sum_k c_k\right\}_P &= 0, \end{aligned} \tag{3.68}$$

Now we let π_ℓ be an arbitrary degree $U=2\ell$ function on the base of $\Pi T(\Pi T^*X) \to \Pi T^*X$, i.e., $\pi_\ell = \frac{1}{(2\ell)!} b^{I_1 \ldots I_{2\ell}}(u^L)\chi_{I_1} \cdots \chi_{I_{2\ell}}$. Then we consider the following non-linear transformations of the fiber coordinates of $\Pi T(\Pi T^*X) \to \Pi T^*X$,

$$\begin{aligned} F_I \to F'_I &:= F_I - \sum_\ell \frac{\partial \pi_\ell}{\partial u^I}, \\ \psi^I \to \psi'^I &:= \psi^I - \sum_\ell \frac{\partial \pi_\ell}{\partial \chi_I}, \end{aligned} \tag{3.69}$$

with u^I and χ unchanged. Under the above transformation we have

$$h + \sum c_k \to h + \Gamma, \tag{3.70}$$

where

$$\Gamma = -\sum_k \left(F_I \frac{\partial \pi_k}{\partial \chi_I} + \frac{\partial \pi_k}{\partial u^I}\psi^I \right) + \frac{1}{2}\left[\sum_k \pi_k, \sum_k \pi_k \right]_S$$
$$+ \sum_k \frac{1}{(2k+1)!} c_{I_1 \ldots I_{2k+1}} \left(\psi^{I_1} - \sum_\ell \frac{\partial \pi_\ell}{\partial \chi_{I_1}} \right) \cdots \left(\psi^{I_{2k+1}} - \sum_\ell \frac{\partial \pi_\ell}{\partial \chi_{I_{2k+1}}} \right) \tag{3.71}$$

It can be easily shown that $Q_o\Gamma + \frac{1}{2}\{\Gamma, \Gamma\}_P = 0$ if and only if $dC_k = 0$ for all k. Now the condition $\Gamma|_{\Pi T^* X} = 0$ implies that

$$\frac{1}{2}\left[\sum_\ell \pi_\ell, \sum_\ell \pi_\ell \right]_S - \sum_k \frac{1}{(2k+1)!} c_{I_1 \ldots I_{2k+1}} \left(\frac{\sum_\ell \partial \pi_\ell}{\partial \chi_{I_1}} \right) \cdots \left(\frac{\sum_\ell \partial \pi_\ell}{\partial \chi_{I_{2k+1}}} \right) = 0. \tag{3.72}$$

Assuming the above conditions we get the following action functional

$$\boldsymbol{S}_\Gamma = \boldsymbol{S}_o + \int_N d^3\theta\, \boldsymbol{\phi}^*(\Gamma), \tag{3.73}$$

satisfying the BV master equation. Using the $\boldsymbol{F}_I$ equation of motion

$$Du^I - \sum_\ell \frac{\partial \pi_\ell}{\partial \chi_I} + \psi^I = 0, \tag{3.74}$$

we have the following on-shell action functional

$$\boldsymbol{S}_\Gamma|_{\text{on shell}} = -\sum_k \frac{1}{(2k+1)!} \int_N d^3\theta \left(c_{I_1 \ldots I_{2k+1}} Du^{I_1} \ldots Du^{I_{2k+1}} \right)$$
$$+ \int_{\partial N} d^2\theta \left(\chi_I Du^I + \sum_\ell \frac{1}{(2\ell)!} b^{I_1 \ldots I_{2\ell}} \chi_{I_1} \cdots \chi_{I_{2\ell}} \right). \tag{3.75}$$

We may also allow for arbitrary ghost numbers and the most general form for the solutions. Such a case may be used to determine explicit quasi-isomorphism of the 3-algebra.

3.5 Other boundary conditions

In this subsection we consider some variants of the topological open membrane by changing the boundary conditions. There can be more general boundary conditions than mentioned here. We pick two of them, which are relevant to mirror symmetry.

3.5.1. A boundary conditions

Now we consider the following general action functional

$$S_\Gamma = \int_N d^3\theta \Big(\psi^I D\chi_I + F_I Du^I + \mathbb{F}_I \psi^I + \phi^*(\Gamma) \Big), \tag{3.76}$$

where Γ is defined by (3.71) with the condition (3.66). Then the above action functional satisfies the BV master equation if and only if

$$\int_N d^3\theta \; D\Big(\psi^I D\chi_I + \mathbb{F}_I Du^I + \mathbb{F}_I \psi^I + \phi^*(\Gamma) \Big) = 0. \tag{3.77}$$

Thus we may exchange the boundary conditions of χ_I and ψ^I, such that $\chi_I(x) = 0$, and $H_I(x) = 0$ along the direction tangent to $\partial\Sigma$, while $A^I(x) = 0$ along the direction normal to $\partial\Sigma$ for $x \in \partial\Sigma$. Then we must set $c_k = 0$ for all k instead of (3.72) to satisfy the master equation. Hence the following action functional satisfies the BV master equation

$$\begin{aligned} S = &\int_N d^3\theta \Big(\psi^I D\chi_I + F_I Du^I + F_I \psi^I \Big) \\ &- \sum_k \int_N d^3\theta \Big(F_I \frac{\partial \pi_k}{\partial \chi_I} + \frac{\partial \pi_k}{\partial u^I} \psi^I \Big) \\ &+ \frac{1}{2} \int_N d^3\theta \Big(\phi^* \Big(\Big[\sum_k \pi_k, \sum_k \pi_k \Big]_S \Big) \Big). \end{aligned} \tag{3.78}$$

From the point of view of the target superspace, we are replacing $\Pi T(\Pi T^* X)$ with $\Pi T^*(\Pi T X)$. In general $T^*X \to TX$ is an isomorphism only when X is a symplectic manifold. In our case $\Pi T(\Pi T^* X) \to \Pi T^*(\Pi T X)$ is an isomorphism since our even symplectic structure ω does not see the difference. Now the boundary observables of the theory are derived from functions on ΠTX, which are differential forms on X. We denote by $\mathcal{O}(\Pi TX)$ the algebra of functions on ΠTX, which has the ordinary product (the wedge product).

Now we consider the role of the bulk deformations to first approximation. For simplicity we only consider the deformations preserving the ghost number symmetry;

$$\begin{aligned} S = &\int_N d^3\theta \Big(\psi^I D\chi_I + F_I Du^I \Big) \\ &+ \int_N d^3\theta \Big(F_I \psi^I - b^{IJ} F_I \chi_J + \frac{1}{2} \partial_I b^{JK} \psi^I \chi_J \chi_K + \frac{1}{3!} h^{IJK} \chi_I \chi_J \chi_K \Big). \end{aligned} \tag{3.79}$$

Then we obtain the following differential operators

$$\widetilde{\mathcal{D}} = \widetilde{\mathcal{D}}_1 + \hbar \widetilde{\mathcal{D}}_2 + \hbar^2 \widetilde{\mathcal{D}}_3 \tag{3.80}$$

where

$$
\begin{aligned}
\widetilde{\mathcal{D}}_1 &= \psi^I \frac{\partial}{\partial u^I}, \\
\widetilde{\mathcal{D}}_2 &= b^{IJ} \frac{\partial^2}{\partial u^I \partial \psi^J} + \frac{1}{2} \partial_I b^{JK} \psi^I \frac{\partial^2}{\partial \psi^J \partial \psi^K}, \\
\widetilde{\mathcal{D}}_3 &= \frac{1}{3!} h^{IJK} \frac{\partial^3}{\partial \psi^I \partial \psi^J \partial \psi^K}.
\end{aligned}
\tag{3.81}
$$

It is not difficult to show that $\widetilde{\mathcal{D}}^2 = 0$ for any bi-vector;

$$
\begin{aligned}
\widetilde{\mathcal{D}}_1^2 &= 0, \\
\widetilde{\mathcal{D}}_1 \widetilde{\mathcal{D}}_2 + \widetilde{\mathcal{D}}_2 \widetilde{\mathcal{D}}_1 &= 0, \\
\widetilde{\mathcal{D}}_1 \widetilde{\mathcal{D}}_3 + \widetilde{\mathcal{D}}_3 \widetilde{\mathcal{D}}_1 + \widetilde{\mathcal{D}}_2^2 &= 0, \\
\widetilde{\mathcal{D}}_2 \widetilde{\mathcal{D}}_3 + \widetilde{\mathcal{D}}_3 \widetilde{\mathcal{D}}_2 &= 0, \\
\widetilde{\mathcal{D}}_3^2 &= 0.
\end{aligned}
\tag{3.82}
$$

Thus we obtain another BV_∞ structure.

We encountered the operator $\widetilde{\mathcal{D}}_1$ and $\widetilde{\mathcal{D}}_2$ in Sect. 2.4 where we used the notations Q and Δ_π. The latter generated, for $[\pi, \pi]_S = 0$, the covariant Schouten-Nijenhuis bracket on the differential forms. Those operators induced a dGBV structure on the algebra $\mathcal{O}(\Pi TX)$ of functions on ΠTX. We remark that Manin associated a Frobenius structure to a special kind of dGBV algebra [61]. It is shown that the above dGBV algebra defines a Frobenius structure when X is symplectic manifold satisfying the strong Lefschetz condition [62]. Any Kähler manifold is such a manifold, and the above construction reproduces the so called AKS theory [13]. AKS theory is the A model version of Kodaria-Spencer gravity [12].

In general $[\pi, \pi]_S \neq 0$ and we find a homotopy version of a dGBV algebra. It will be interesting to examine the corresponding deformation of the Frobenius structure. We note that the relation between this subsection and Sect. 3.3 can be seen as the relation of cochains versus chains. We remark that there is also a non-commutative version of the above differential operators at least for $h^{IJK} = 0$ [83,81]. In such a case the cohomology of $\widetilde{\mathcal{D}}$ is closely related to the periodic cyclic homology of the Hochschild chain complex of the deformed algebra of functions on X with the star product. It will be interesting to see if $\widetilde{\mathcal{D}}$ in general has a non-commutative version, which is not necessarily associative. We also note that the periodic cyclic homology is the non-commutative version of the de Rham cohomology. Can there be an A model on non-commutative space? If so the extended moduli space of the theory may be identified with the periodic cyclic homology.

3.5.2. B boundary conditions

Now we assume that the target space X is a complex Calabi-Yau space. We introduce a complex structure and consider the open membrane theory without background,

$$\boldsymbol{S}_o = \int_N d^3\theta \left(\boldsymbol{\psi}^i D\boldsymbol{\chi}_i + \boldsymbol{\psi}^{\bar{i}} D\boldsymbol{\chi}_{\bar{i}} + \boldsymbol{F}_i D\boldsymbol{u}^i + \boldsymbol{F}_{\bar{i}} D\boldsymbol{u}^{\bar{i}} + \boldsymbol{F}_i \boldsymbol{\psi}^i + \boldsymbol{F}_{\bar{i}} \boldsymbol{\psi}^{\bar{i}} \right). \quad (3.83)$$

Now consider the original boundary conditions. We can exchange the boundary conditions of $\boldsymbol{\chi}_{\bar{i}}$ and $\boldsymbol{\psi}^{\bar{i}}$, while maintaining the original boundary conditions for $\boldsymbol{\chi}_i$ and $\boldsymbol{\psi}^i$. Then $\chi_{\bar{i}}(x) = \psi^i(x) = 0$ and the 1-forms $H_{\bar{i}}(x)$ and $A^i(x)$ vanish along the direction tangent to $\partial\Sigma$ for $x \in \partial\Sigma$. Then the action functional (3.83) satisfies the master equation. Now the boundary observables are derived from functions on $\Pi\mathcal{T}^*(\Pi\overline{\mathcal{T}}X)$, namely the elements of $\oplus\Omega^{0,\bullet}(X, \wedge^\bullet \mathcal{T}X)$. The general form of the ghost number $U = 2$ function on $\Pi\mathcal{T}^*(\Pi\overline{\mathcal{T}}X)$ is

$$\beta := a + \kappa + b, \qquad \begin{cases} a := a_{\bar{i}}{}^j \chi_j \psi^{\bar{i}}, \\ \kappa := \dfrac{1}{2}\kappa_{\bar{i}\bar{j}} \psi^{\bar{i}} \psi^{\bar{j}}, \\ b := \dfrac{1}{2} b^{ij} \chi_i \chi_j. \end{cases} \quad (3.84)$$

We note that the BRST transformation laws restricted to boundary are

$$\begin{aligned} \boldsymbol{Q}_o u^i &= 0, & \boldsymbol{Q}_o \chi_i &= 0, \\ \boldsymbol{Q}_o u^{\bar{i}} &= \psi^{\bar{i}}, & \boldsymbol{Q}_o \psi^{\bar{i}} &= 0. \end{aligned} \quad (3.85)$$

Hence the BRST charge acts like the $\bar{\partial}$-operators.

The undeformed theory induces the following differential operators acting on functions on $\Pi\mathcal{T}^*(\Pi\overline{\mathcal{T}}X)$ – which can be identified with the elements of $\oplus\Omega^{0,\bullet}(X, \wedge^\bullet \mathcal{T}X)$ –

$$\mathcal{D}_o = \bar{\partial} + \hbar\Delta_T, \qquad \begin{cases} \bar{\partial} = \psi^{\bar{i}} \dfrac{\partial}{\partial u^{\bar{i}}}, \\ \Delta_T = \dfrac{\partial^2}{\partial u^i \partial \chi_i}. \end{cases} \quad (3.86)$$

We have $\mathcal{D}_o^2 = 0$ and

$$\bar{\partial}^2 = \bar{\partial}\Delta_T + \Delta_T \bar{\partial} = \Delta_T^2 = 0. \quad (3.87)$$

The operators together with the wedge product endow $\oplus\Omega^{0,p}(X, \wedge^q \mathcal{T}X)$ with the structure of dGBV algebra [7,61,6]. The operator Δ_T generates a bracket (Tian-Todorov lemma) $[.,.]_T$, whose holomorphic Schouten-Nijenhuis bracket together with the wedge product on forms define a dGBV algebra. This forms the classical algebra of observables. We note that the solution space of the MC equation of this classical algebra modulo equivalences defines the so-called extended moduli space of complex structures on X [7,6]. This also induces a Frobenius structure [7,61], generalizing [12], which is relevant to mirror symmetry.

Now we consider a degree $U = 3$ function γ on $\Pi T(\Pi T^* X)$ satisfying

$$\begin{aligned} \gamma|_{\psi^i=\chi_{\bar{i}}=F_I=0} &= 0, \\ Q_o\gamma + \frac{1}{2}\{\gamma,\gamma\}_P &= 0, \end{aligned} \tag{3.88}$$

compare with (3.31). Then the action functional

$$\boldsymbol{S}_\gamma = \boldsymbol{S}_o + \int_N d^3\theta\ \boldsymbol{\phi}^*(\gamma) \tag{3.89}$$

satisfies the BV master equation. The deformation term will deform the dGBV algebra (the classical algebra of observables) as a 2-algebra. The deformation is controlled by the MC equation (3.88) of the 3-algebra. It will be interesting to examine the corresponding deformation of the Frobenius structure.

Here we determine a class of solutions of (3.88). The desired γ is given by

$$\begin{aligned} \gamma =& F_i\frac{\partial\beta}{\partial\chi_i} + \psi^i\frac{\partial\beta}{\partial u^i} + \psi^{\bar{i}}\frac{\partial\beta}{\partial u^{\bar{i}}} + \frac{1}{2}[\beta,\beta]_T \\ &+ \frac{1}{3!}c_{ijk}\left(\psi^i + \frac{\partial\beta}{\partial\chi_i}\right)\left(\psi^j + \frac{\partial\beta}{\partial\chi_j}\right)\left(\psi^k + \frac{\partial\beta}{\partial\chi_k}\right), \end{aligned} \tag{3.90}$$

satisfying

$$\begin{aligned} \bar{\partial}\beta + \frac{1}{2}[\beta,\beta]_T + \frac{1}{3!}[\beta,\beta,\beta]_T &= 0, \\ \bar{\partial}c + [\beta, c]_T &= 0. \end{aligned} \tag{3.91}$$

Here we used the following definitions

$$\begin{aligned} c &:= \frac{1}{3!}c_{ijk}\psi^i\psi^j\psi^k, \\ [\beta,\beta,\beta]_T &:= c_{ijk}\frac{\partial\beta}{\partial\chi_i}\frac{\partial\beta}{\partial\chi_j}\frac{\partial\beta}{\partial\chi_k}. \end{aligned} \tag{3.92}$$

The first condition in (3.91) comes from the first condition in (3.88), while the second condition in (3.91) comes from the second condition in (3.88).

We note that the first equation in (3.91) is the flatness condition $\overline{\partial}_\beta^2 = 0$ of the "covariant" derivative $\overline{\partial}_\beta$;

$$\overline{\partial}_\beta := \overline{\partial} + [\beta, .]_T + \frac{1}{2}[\beta, \beta, .]_T. \tag{3.93}$$

Thus the equations (3.91) become

$$\overline{\partial}_\beta^2 = 0, \qquad \overline{\partial}_\beta c = 0, \tag{3.94}$$

where we used $[\beta, \beta, c]_T = 0$. The equations in (3.91) can be written as follows;

$$\begin{aligned} \overline{\partial} a + \frac{1}{2}[a, a]_T = 0, &\qquad \overline{\partial}\kappa + [a, \kappa]_T + \frac{1}{3!}[a, a, a]_T = 0, \\ [b, \kappa]_T + \frac{1}{2}[b, a, a]_T = 0, &\qquad \overline{\partial} b + [a, b]_T + \frac{1}{2}[a, b, b]_T = 0, \\ \frac{1}{2}[b, b]_T + \frac{1}{3!}[b, b, b]_T = 0, &\qquad \overline{\partial} c + [a, c]_T = 0. \end{aligned} \tag{3.95}$$

Now the deformed theory induces the following differential operators acting on functions on $\Pi\mathcal{T}^*(\Pi\overline{\mathcal{T}}X)$, identified with the elements of $\oplus\Omega^{0,\bullet}(X, \wedge^\bullet \mathcal{T}X)$;

$$\mathcal{D}_\gamma = \mathcal{D}_1 + \hbar\mathcal{D}_2 + \hbar\mathcal{D}_3, \tag{3.96}$$

where

$$\begin{aligned} \mathcal{D}_1 &= \psi^{\bar{i}}\frac{\partial}{\partial u^{\bar{i}}} + \frac{\partial\beta}{\partial\chi_i}\frac{\partial}{\partial u^i} + \frac{\partial\beta}{\partial u^i}\frac{\partial}{\partial\chi_i} + \frac{1}{2}c_{ijk}\frac{\partial\beta}{\partial\chi_i}\frac{\partial\beta}{\partial\chi_j}\frac{\partial}{\partial\chi_k}, \\ \mathcal{D}_2 &= \frac{\partial^2}{\partial u^i\partial\chi_i} + \frac{1}{2}c_{ijk}\frac{\partial\beta}{\partial\chi_i}\frac{\partial^2}{\partial\chi_j\partial\chi_k}, \\ \mathcal{D}_3 &= c_{ijk}\frac{\partial^3}{\partial\chi_i\partial\chi_j\partial\chi_k}. \end{aligned} \tag{3.97}$$

Once again we have $\mathcal{D}_\gamma^2 = 0$ from (3.91);

$$\begin{aligned} \mathcal{D}_1^2 &= 0, \\ \mathcal{D}_1\mathcal{D}_2 + \mathcal{D}_2\mathcal{D}_1 &= 0, \\ \mathcal{D}_1\mathcal{D}_3 + \mathcal{D}_3\mathcal{D}_1 + \mathcal{D}_2^2 &= 0, \\ \mathcal{D}_2\mathcal{D}_3 + \mathcal{D}_3\mathcal{D}_2 &= 0, \\ \mathcal{D}_3^2 &= 0. \end{aligned} \tag{3.98}$$

The operators together with the wedge product endow $\oplus\Omega^{0,p}(X, \wedge^q \mathcal{T}X)$ with a structure of BV_∞ algebra.

We may consider more general boundary observables of the theory than β in (3.84). For this we consider an arbitrary function $\alpha(u^i, u^{\bar{i}}, \chi_i, \psi^{\bar{i}})$ of $(u^I, \chi_i, \psi^{\bar{i}})$ on $\Pi\mathcal{T}^*\Pi(\overline{\mathcal{T}}X)$, which is an element of $\oplus\Omega^{0,\bullet}(X, \wedge^\bullet \mathcal{T}X)$. Now we can replace β with α in the equations (3.90) and (3.91) to obtain a more general solution of the BV master equation (3.88). We can also replace β with α in (3.97). Then the BV master equation is given by

$$\begin{aligned} \bar{\partial}\alpha + \frac{1}{2}[\alpha, \alpha]_T + \frac{1}{3!}[\alpha, \alpha, \alpha]_T = 0, \\ \bar{\partial}c + [\alpha, c]_T = 0, \end{aligned} \tag{3.99}$$

while the differentials acting on functions on $\Pi\mathcal{T}^*(\Pi\overline{\mathcal{T}}X)$ are

$$\begin{aligned} \mathcal{D}_1 &= \psi^{\bar{i}} \frac{\partial}{\partial u^{\bar{i}}} + \frac{\partial \alpha}{\partial \chi_i} \frac{\partial}{\partial u^i} + \frac{\partial \alpha}{\partial u^i} \frac{\partial}{\partial \chi_i} + \frac{1}{2} c_{ijk} \frac{\partial \alpha}{\partial \chi_i} \frac{\partial \alpha}{\partial \chi_j} \frac{\partial}{\partial \chi_k}, \\ \mathcal{D}_2 &= \frac{\partial^2}{\partial u^i \partial \chi_i} + \frac{1}{2} c_{ijk} \frac{\partial \alpha}{\partial \chi_i} \frac{\partial^2}{\partial \chi_j \partial \chi_k}, \\ \mathcal{D}_3 &= c_{ijk} \frac{\partial^3}{\partial \chi_i \partial \chi_j \partial \chi_k}. \end{aligned} \tag{3.100}$$

Now we consider the special case that $c = 0$. Then we are left with only the boundary degrees of freedom and the master equation (3.88) or equivalently (3.99) reduces to

$$\bar{\partial}\alpha + \frac{1}{2}[\alpha, \alpha]_T = 0. \tag{3.101}$$

The differential operators $\mathcal{D}$ acting on functions on $\Pi\mathcal{T}^*(\Pi\overline{\mathcal{T}}X)$ become

$$\begin{aligned} \mathcal{D}_1 &= \psi^{\bar{i}} \frac{\partial}{\partial u^{\bar{i}}} + \frac{\partial \alpha}{\partial \chi_i} \frac{\partial}{\partial u^i} + \frac{\partial \alpha}{\partial u^i} \frac{\partial}{\partial \chi_i}, \\ \mathcal{D}_2 &= \Delta_T = \frac{\partial^2}{\partial u^i \partial \chi_i}, \\ \mathcal{D}_3 &= 0. \end{aligned} \tag{3.102}$$

We see that the boundary deformations correspond to deformations of the first order differential operator. We shall see that those correspond to deformations of the BRST operator of the B model of the topological closed string theory. We shall also see that the solution space of (3.101) modulo equivalences defines the extended moduli space of the B model of the topological string theory or the extended moduli space of complex structures. Now turning on the c-field gives rise to bulk deformations beyond the extended moduli space of the boundary theory.

4 Back to the strings

In this section we discuss some applications to homological mirror symmetry. In the previous section we already discussed about related issues, though on a level one step up (the open membrane).

4.1 Relation with $N_{ws} = (2,2)$ supersymmetric sigma models

We consider the following two functions carrying $U = 3$

$$\begin{aligned} h &= F_I \psi^I, \\ \widetilde{h} &= b^{IJ} F_I \chi_J + \frac{1}{2} \partial_K b^{IJ} \psi^k \chi_I \chi_J, \end{aligned} \tag{4.1}$$

where $\pi = \frac{1}{2} b^{IJ} \chi_I \chi_J$ is a Poisson bivector, i.e. $[\pi, \pi]_S = 0$. Then it is easy to check that

$$\{h, h\}_P = 0, \qquad \{h, \widetilde{h}\}_P = 0, \qquad \{\widetilde{h}, \widetilde{h}\}_P = 0. \tag{4.2}$$

Thus the Hamiltonian vectors $\{h.\ldots\}_P = Q_o$ and $\{\widetilde{h}, \ldots\}_P = \widetilde{Q}_o$ satisfy the following relations

$$\{Q, Q\} = 0, \qquad \{Q, \widetilde{Q}\} = 0, \qquad \{\widetilde{Q}, \widetilde{Q}\} = 0. \tag{4.3}$$

Explicitly

$$\begin{aligned} Q_o &= \psi^I \frac{\partial}{\partial u^I} + F_I \frac{\partial}{\partial \chi_I}, \\ \widetilde{Q}_o &= b^{IJ} \chi_J \frac{\partial}{\partial u^I} + \frac{1}{2} \partial_I b^{JK} \chi_J \chi_K \frac{\partial}{\partial \chi_I} - \left(b^{IJ} F_J + \partial_K b^{IJ} \psi^k \chi_J \right) \frac{\partial}{\partial \psi^I} \\ &- \left(\partial_I b^{JK} F_J \chi_K + \frac{1}{2} \partial_I \partial_J b^{LM} \psi^J \chi_L \chi_M \right) \frac{\partial}{\partial F^I}. \end{aligned} \tag{4.4}$$

We note that the two odd vectors Q_o and $\widetilde{Q}_o$ are the supercharges of a two dimensional $N_{ws} = (1,1)$ supersymmetric non-linear sigma model after dimensional reductions to zero dimension. Then the commuting coordinate field F_I correspond to the auxiliary fields. Now we assume that X is a Kähler manifold and the bivector is the inverse of the Kähler form. We introduce a complex structure and decompose the coordinates according to $u^I = u^i + u^{\bar{i}}$. We decompose the tangent vectors, $\psi^I = \psi^i + \psi^{\bar{i}}$, and cotangent vectors, $\chi_I =$

$\chi_{\bar{i}} + \chi_i$ and $F_I = F_{\bar{i}} + F_i$, accordingly. Then the odd vectors Q_o and $\widetilde{Q}_o$ are decomposed into holomorphic and anti-holomorphic vectors; $Q_o = Q'_o + Q''_o$, $\widetilde{Q}_o = \widetilde{Q}'_o + \widetilde{Q}''_o$ and all four odd vectors are mutually nilpotent. Again the four odd vectors correspond to the supercharges of a two dimensional $N_{ws} = (2,2)$ supersymmetric non-linear sigma model after dimensional reduction to zero dimensions. Now we imagine undoing the dimensional reduction. Then our four odd vectors transform as the left or right moving spinors under the two-dimensional rotation group of a Riemann surface Σ. They also carry so called $U(1)_{\mathcal{R}}$ charges.

As was originally proposed [21,57], the mirror symmetry is very natural from a physical viewpoint. On the other hand the symmetry becomes quickly mystifying once we translate it into mathematical language in terms of Gromov-Witten invariants and the variation of Hodge structures. An application of mirror symmetry in adopting the latter viewpoint marked the first grand success of the mirror conjecture in algebra-arithmetic geometry [16], after the first construction of a mirror pair [35]. Later Witten proposed a unified and effective viewpoint based on twisted versions (A and B models) of $\mathcal{N} = (2,2)$ supersymmetric sigma models [92]. Additionally, Witten also argued that both the original A and B models should be extended and the mirror symmetry would be more obvious in the extended models. In this language the mirror symmetry is the physical equivalence between the extended A model $A^e(X)$ on X and the extended B model $B^e(Y)$ on the mirror Y.

The twisting procedure is explained in detail in [92,12]. There are two different twistings, leading to the A and the B model, respectively. There is also a half-twisted model which we will not consider here. Note that the B model makes sense iff X is a Calabi-Yau space with Ricci-flat metric. After twisting two supercharges transform as scalars on Σ. The other two supercharges transform as vectors. The auxiliary fields F_i and $F_{\bar{i}}$ transform as a vector for the A model, while decomposing into a scalar and a two-form in the B model. For the A model, the two anti-commuting scalar supercharges (BRST charges) are given by

$$\begin{aligned} Q'_o &= \psi^i \frac{\partial}{\partial u^i} + \ldots, \\ Q''_o &= \psi^{\bar{i}} \frac{\partial}{\partial u^{\bar{i}}} + \ldots, \end{aligned} \tag{4.5}$$

where the omitted the part involving non-scalar fields. For the B model the two BRST charges are given by

$$\begin{aligned} Q''_o &= \psi^{\bar{i}}\frac{\partial}{\partial u^{\bar{i}}} + F_i\frac{\partial}{\partial \chi_i} + \dots, \\ \widetilde{Q}''_o &= b^{\bar{i}j}\chi_j\frac{\partial}{\partial u^{\bar{i}}} - \left(b^{\bar{i}j}F_j + \partial_{\bar{k}}b^{\bar{i}j}\psi^k\chi_j\right)\frac{\partial}{\partial \psi^{\bar{i}}} + \dots, \end{aligned} \tag{4.6}$$

where we again omitted the part involving non-scalar fields.

At this level we already see the sharp distinction between the A and the B model. As a general principle observables of the theory are constructed by BRST cohomology classes. The zero-dimensional observables, transforming as scalars, are constructed from scalar fields. For the A model the answer is obvious. The two BRST charges originated from the ∂ and $\overline{\partial}$ differentials of the target space X. Thus the zero-dimensional observables are the pull-backs of de Rham cohomology classes of X after a parity change.

On the other hand the Q''_o charge can better be interpreted as $\overline{\partial}$-operator on the total superspace ΠT^*X. Furthermore, the ∂ part of BRST charge does not exist. The other BRST charge $\widetilde{Q}''_o$ can be interpreted as the holomorphic part of the Poisson differential on the superspace ΠT^*X. Consequently the zero-dimensional observables of the B model should come from the Dolbeault cohomology of the superspace ΠT^*X. However it is not clear how it can be defined. The usual approach is based on an on-shell formalism[n] and taking the diagonal. The zero-dimensional observables are elements of the sheaf cohomology $H^{0,\bullet}(X, \wedge^{\bullet}T^{1,0}X)$, modulo equations of motion. Then one solves the descent equations to construct two dimensional observables. However some problems appear in constructing higher dimensional observables by solving the descent equations. Two dimensional observables are very important since one can add them by integrating over Σ to the action functional to define a family of the theory. Witten gave a recipe for the B model with some first order analysis and shows that the classical moduli space $\mathcal{M}_{cl}$ of complex structures should be extended [92]. He argued that the tangent space of the extended moduli space $\mathcal{M}$ at a classical point is $\oplus H^{0,\circ}(X, \wedge^{\circ}T^{1,0}X)$. A noble point of the above analysis is that the BRST transformation laws should be changed recursively. This is a perfect lead to the method of BV quantization.

The question on the extended moduli space $\mathcal{M}$ was the starting point of the homological mirror conjecture of Kontsevich [49]. Roughly there are three questions. (i) define (or find equations for) the moduli space $\mathcal{M}$. (ii) define generalized periods. (iii) what is the meaning of $\mathcal{M}$ or which kind of object does it parameterize?. The first and second questions are answered by

---

[n]On-shell the term $(b^{\bar{i}j}F_j + \partial_{\bar{k}}b^{\bar{i}j}\psi^k\chi_j)$ vanishes due to an equation of motion.

Barannikov and Kontsevich [7,6] using purely mathematical techniques based on modern deformation theory. They also constructed explicit formulas for the tree level potential, which is the generalization of the Kodaira-Spensor theory of Bershadsky et. al [12] (see also [61]). However the corresponding generalization for string loops is not yet achieved. The third question remains still mysterious. Kontsevich conjectured that $\mathcal{M}$ parameterizes the A_∞ deformation of $D^b(Coh(X))$. This conjecture was answered affirmatively in [6], based on the formality theorem of Kontsevich.

The homological mirror conjecture seems to be closely related to the SYZ conjecture [78,84,53]. In many respects the SYZ construction makes mirror symmetry obvious as a physical equivalence [78]. Recently Hori et. al. [39,40] developed another physically natural approach which has a highly constructive power.

4.1.1. Cohomological ABC model

It is possible to describe the A, B and Catteneo-Felder (C in short) models in a unified fashion. For this we consider a theory based on the following map

$$\varphi : \Pi T\Sigma \rightarrow \Pi T^*(\Pi(TX)). \tag{4.7}$$

We let $\mathcal{A}_\Sigma$ denotes the space of all maps. We denote a system of local coordinates on ΠTX by $(\{u\}^I, \{\psi^I\})$, while by $(\{\chi_I\}, \{F_I\})$ for the fiber of $\Pi T^*(TX)$. We assign the ghost numbers of the coordinates by $U(u^I, \psi_I, \chi_I, F_I) = (0,1,1,0)$. We may use the same component notations as for the membrane case (3.18) by simply setting the 3-form parts to zero. We describe a map (4.7) by the superfields

$$(\boldsymbol{u}^I, \boldsymbol{\psi}^I, \boldsymbol{\chi}_I, \boldsymbol{F}_I) := (u^I, \chi_I, \psi^I, F_I)(x^\mu, \theta^\mu), \tag{4.8}$$

where $(\{x^\mu\}, \{\theta^\mu\})$, $\mu = 1,2$ denote a set of local coordinates on $\Pi T\Sigma$. On $\Pi T^*\Pi(TX)$ we have the following odd symplectic form with $U = 1$,

$$\omega_\Sigma = du^I d\chi_I + d\psi^I dF_I. \tag{4.9}$$

The corresponding odd Poisson bracket is an extension of the Schouten-Nijenhuis bracket. On $\mathcal{A}_\Sigma$ we have an induced odd symplectic form $\boldsymbol{\omega}_\Sigma$ with $U = -1$

$$\boldsymbol{\omega}_\Sigma = \int_\Sigma d^2\theta \left(\delta \boldsymbol{u}^I \delta \boldsymbol{\chi}_I + \delta \boldsymbol{\psi}^I \delta \boldsymbol{F}_I\right). \tag{4.10}$$

Thus the BV bracket of the theory originates form the extended Schouten-Nijenhuis bracket. Now we may follow the usual steps, as in the previous sections.

We start from the following action functional,

$$\boldsymbol{S}_o = \int_\Sigma d^2\theta \Big(\chi_I D\boldsymbol{u}^I + \boldsymbol{F}_I D\boldsymbol{\psi}^I + \chi_I \boldsymbol{\psi}^I \Big). \tag{4.11}$$

The above action functional satisfies the BV master equation. The BRST transformation laws can be obtained by the odd Hamiltonian vector $\boldsymbol{Q}_o$ of $\boldsymbol{S}_o$;

$$\boldsymbol{Q}_o = \Big(D\boldsymbol{u}^I + \boldsymbol{\psi}^I \Big) \frac{\partial}{\partial \boldsymbol{u}^I} + D\boldsymbol{\psi}^I \frac{\partial}{\partial \boldsymbol{u}^I} + (D\boldsymbol{F}_I + \chi_I) \frac{\partial}{\partial \boldsymbol{F}_I} + D\chi_I \frac{\partial}{\partial \chi_I}. \tag{4.12}$$

Now we consider the following general deformation preserving the ghost number

$$\begin{aligned} \boldsymbol{S}_\Gamma &= \boldsymbol{S}_o + \int_\Sigma d^2\theta \varphi^*(\Gamma) \\ &= \boldsymbol{S}_o + \int_\Sigma d^2\theta \Big(a_I{}^J(\boldsymbol{u}^L)\chi_J \boldsymbol{\psi}^I + \frac{1}{2}\kappa_{IJ}(\boldsymbol{u}^L)\boldsymbol{\psi}^I\boldsymbol{\psi}^J + \frac{1}{2} b^{IJ}(\boldsymbol{u}^L)\chi_I\chi_J \Big), \end{aligned} \tag{4.13}$$

where $\Gamma = A + K + \pi \in \Omega^2(X) \oplus \Omega^1(X, TX) \oplus \Omega^0(X, \wedge^2 TX)$ with suitable parity changes;

$$\begin{aligned} A &= a_I{}^J(\boldsymbol{u}^L)\chi_I \boldsymbol{\psi}^J, \\ K &= \frac{1}{2}\kappa_{IJ}(\boldsymbol{u}^L)\boldsymbol{\psi}^I \boldsymbol{\psi}^J, \\ \pi &= \frac{1}{2} b^{IJ}(\boldsymbol{u}^L)\chi_I \chi_J. \end{aligned} \tag{4.14}$$

The above deformation terms do not have any dependence on $\boldsymbol{F}_I$, which simplifies our analysis a lots.[o] The above action functional satisfies the BV master equation if and only if

$$\begin{aligned} dK + [A, K]_S &= 0, \\ dA + \frac{1}{2}[A, A]_S + [\pi, K]_S &= 0, \\ d\pi + [A, \pi]_S &= 0, \\ [\pi, \pi]_S &= 0, \end{aligned} \tag{4.15}$$

where $d := \boldsymbol{\psi}^I \frac{\partial}{\partial \boldsymbol{u}^I}$ is the (parity changed) exterior derivative on X and the bracket above denotes the Schouten-Nijenhuis bracket together with the wedge

[o] Hence we are using $\boldsymbol{F}_I$ as auxiliary devices.

product on forms. The above equation can be summarized into a single equation,

$$d\Gamma + \frac{1}{2}[\Gamma, \Gamma]_S = 0. \tag{4.16}$$

It seems reasonable to relate the moduli space defined by the above equations to the moduli space of $N = (2,2)$ superconformal field theory in two-dimensions.

Now we consider the special cases of $\boldsymbol{S}_\gamma$. The C model[p] can be obtained by setting $\boldsymbol{\psi}^I = \boldsymbol{F}_I = 0$ for all I

$$\boldsymbol{S}^C = \int_\Sigma d^2\theta \left(\boldsymbol{\chi}_I D\boldsymbol{u}^I + \frac{1}{2} b^{IJ} \boldsymbol{\chi}_I \boldsymbol{\chi}_J\right). \tag{4.17}$$

The A model in Sect. 2.4. can be obtained by setting $\boldsymbol{\chi}_I = \boldsymbol{F}_I = 0$ for all I;

$$\boldsymbol{S}^A = \frac{1}{2}\int_\Sigma d^2\theta \left(\kappa_{IJ} \boldsymbol{\psi}^I \boldsymbol{\psi}^J\right). \tag{4.18}$$

We may also consider the open string version of the A model [93,27,25,28]. It is worth mentioning that the BV master equation does not tell anything on the possible boundary conditions, since the master action functional does not have a kinetic term before gauge fixing.

Next we turn to the B model.

4.2 Extended B model

The B model is more complicated. First of all X must be a Calabi-Yau manifold. We pick a complex structure on X and $\boldsymbol{\psi}^i = \boldsymbol{F}_{\bar{i}} = 0$ for all i. Then $\boldsymbol{S}_\gamma$ in (4.16) becomes

$$\boldsymbol{S}^B_\beta = \boldsymbol{S}^B_o + \int_\Sigma \varphi^*(\beta), \tag{4.19}$$

where

$$\boldsymbol{S}^B_o = \int_\Sigma d^2\theta \Big(\boldsymbol{\chi}_i D\boldsymbol{u}^i + \boldsymbol{\chi}_{\bar{i}} D\boldsymbol{u}^{\bar{i}} + \boldsymbol{\chi}_{\bar{i}} \boldsymbol{\psi}^{\bar{i}}\Big), \tag{4.20}$$

and

$$\int_\Sigma \varphi^*(\beta) = \int_\Sigma d^2\theta \Big(a_{\bar{i}}{}^j \boldsymbol{\chi}_j \boldsymbol{\psi}^{\bar{i}} + \frac{1}{2}\kappa_{\bar{i}\bar{j}} \boldsymbol{\psi}^{\bar{i}} \boldsymbol{\psi}^{\bar{j}} + \frac{1}{2} b^{ij} \boldsymbol{\chi}_i \boldsymbol{\chi}_j\Big). \tag{4.21}$$

---

[p] We note that the relation between the C model and the ABC model is analogous to the relation between cohomological field theory and balanced cohomological field theory [22].

The authors of [2] showed that the standard B model action functional can be obtained by a suitable gauge fixing of the master action functional $\boldsymbol{S}_o^B$ given by (4.20).[q] For us it suffices to mention that H_i, which is the 1-form component of χ_i, is replaced by

$$H_i = g_{i\bar{i}} * du^{\bar{i}} + \dots . \tag{4.22}$$

We will use this later. We also take the only non-vanishing components of the superfields $\chi_{\bar{i}}$ to be the two-form parts. Then the kinetic term $\int_\Sigma d^2\theta\, \chi_{\bar{i}} D u^{\bar{i}}$ in (4.20) vanishes. Thus the undeformed master action functional for the B model is actually given by

$$\boldsymbol{S}_o^B = \int_\Sigma d^2\theta \Big(\chi_i D u^i + \chi_{\bar{i}} \psi^{\bar{i}} \Big). \tag{4.23}$$

We note that the conditions $\psi^i = \boldsymbol{F}_{\bar{i}} = \chi_{\bar{i}} = H_{\bar{i}} = 0$ are equivalent to the B boundary conditions for the open membrane.

Now we regard the terms in (4.21) as deformations of the theory preserving the ghost number symmetry. The BV master equation reduces to

$$\bar{\partial}\beta + \frac{1}{2}[\beta, \beta]_T = 0, \tag{4.24}$$

where $\beta = \kappa + a + b \in \Omega^{0,2}(X) \oplus \Omega^{0,1}(X, \mathcal{T}X) \oplus \Omega^{0,0}(X, \wedge^2 \mathcal{T}X)$. Note that β is defined as (3.84). The bracket denotes a holomorphic version of the Schouten-Nijenhuis bracket and we have the wedge product on forms. By the Tian-Todorov lemma the bracket can be obtained from the holomorphic top-form on X, etc. [12,7].

We may consider a more general deformation of the theory (4.20). A zero-dimensional observable is a function of the scalar components of the superfields (4.8). For this we consider an arbitrary function $\alpha(u^i, u^{\bar{i}}, \chi_i, \psi^{\bar{i}})$ of $(u^I, \chi_i, \psi^{\bar{i}})$ on $\Pi\mathcal{T}^*\Pi(\overline{\mathcal{T}}X)$, which is an element of $\oplus\Omega^{0,\circ}(X, \wedge^\circ \mathcal{T}X)$. Then $\alpha(x^\mu)$ is a BV BRST observable if and only if $\boldsymbol{Q}_o\alpha = \psi^{\bar{i}}\partial_{\bar{i}}\alpha = 0$. Thus $\boldsymbol{Q}_o$ acts on α as the $\bar{\partial}$ operators on X; α is an element of $\oplus H^{0,\circ}(X, \wedge^\circ \mathcal{T}X)$. We consider a basis $\{\alpha_a\}$ of $\oplus H^{0,\bullet}(X, \wedge^\circ \mathcal{T}X)$. Then we have the following family of B models

$$\boldsymbol{S}^B(\{t^a\}) = \boldsymbol{S}_o^B + \sum t^a \int_\Sigma d^2\theta\, \varphi^*(\alpha_a). \tag{4.25}$$

The above action functional satisfies the BV master equation if and only if

$$\bar{\partial}(t^a\alpha_a) + \frac{1}{2}\left[\sum t^a\alpha_a, \sum t^b\alpha_b\right]_T = 0. \tag{4.26}$$

---

[q]They actually start from the action functional $\boldsymbol{S}_o$ in (4.11). They regard the conditions $\psi^i = \boldsymbol{F}_{\bar{i}} = 0$ as part of the gauge fixing.

It follows that the BV master equation holds up to first order in t^a. To go beyond the first order we consider a certain function $\alpha(t)$, which is an element of $\oplus\Omega^{0,\bullet}(X,\wedge^\bullet \mathcal{T}X)$, having the the following formal expansion:

$$\alpha(t) = t^a\alpha_a + \sum_{n>1} t^{a_1}\ldots t^{a_n}\alpha_{a_1\ldots a_n}. \tag{4.27}$$

Then we consider the family of action functionals $\boldsymbol{S}(t)$ defined by

$$\boldsymbol{S}^B(t) = \boldsymbol{S}^B_o + \int_\Sigma d^2\theta\, \boldsymbol{\varphi}^*(\alpha(t)). \tag{4.28}$$

The above deformation is said to be well-defined if $\boldsymbol{S}(t)$ satisfies the BV master equation

$$\left(\boldsymbol{S}^B(t), \boldsymbol{S}^B(t)\right)_{BV} = 0. \tag{4.29}$$

Equivalently,

$$\boldsymbol{Q}_o\int_\Sigma d^2\theta\, \boldsymbol{\varphi}^*(\alpha(t)) + \frac{1}{2}\left(\int_\Sigma d^2\theta\, \boldsymbol{\varphi}^*(\alpha(t)), \int_\Sigma d^2\theta\, \boldsymbol{\varphi}^*(\alpha(t))\right)_{BV} = 0. \tag{4.30}$$

The above condition is equivalent to the following MC equation,

$$\bar\partial\alpha(t) + \frac{1}{2}\left[\alpha(t),\alpha(t)\right]_T = 0. \tag{4.31}$$

Thus the success of BV quantization is equivalent to having solutions to the above MC equation in the form (4.27). We note that the BV BRST transformation laws for the deformed action functional (4.28) are

$$\begin{aligned}
\boldsymbol{Q}(t)\boldsymbol{u}^i &= D\boldsymbol{u}^i + \frac{\partial\boldsymbol{\alpha}(t)}{\partial\boldsymbol{\chi}_i},\\
\boldsymbol{Q}(t)\boldsymbol{u}^{\bar i} &= \boldsymbol{\psi}^{\bar i},\\
\boldsymbol{Q}(t)\boldsymbol{\psi}^{\bar i} &= 0,\\
\boldsymbol{Q}(t)\boldsymbol{\chi}_i &= D\boldsymbol{\chi}_i + \frac{\partial\boldsymbol{\alpha}(t)}{\partial\boldsymbol{u}^i},\\
\boldsymbol{Q}(t)\boldsymbol{\chi}_{\bar i} &= \frac{\partial\boldsymbol{\alpha}(t)}{\partial\boldsymbol{u}^{\bar i}}.
\end{aligned} \tag{4.32}$$

The condition that $\boldsymbol{Q}(t)^2 = 0$ is equivalent to the master equation (4.31). The MC equation or master equation (4.31) is precisely the equation for the extended moduli space of complex structures on X, defined by [7]. Here we derived the result as the consistency condition for quantization, completing the original analysis of Witten in [92].

We also obtained a family of B models, which is actually parametrized by the extended moduli space. We call the resulting theory the extended B model. We note that the deformations of the B model leading to the extended B model are deformations of the differential or the BRST charge of the theory. On the other hand the topological open membrane theory with B boundary conditions in Sect. 3.5.2 deforms the bracket as well by the bulk deformations. We also showed that the topological open membrane theory with B boundary conditions contains all the information of the extended B model of topological closed string theory.

Now we turn to the open string version of the extended B model.

4.2.1. Towards physical descriptions of A_∞-deformations of $D^bCoh(X)$

According to a conjecture of Kontsevich the extended moduli space $\mathcal{M}$ of complex structures parameterizes the A_∞ deformations of $D^b(Coh(X))$ [49,6]. In this section we sketch a program constructing $D^b(Coh(X))$ and its A_∞-deformations by path integral methods. For the A model side of the story we refer to [28] for a state of art construction. For this we consider the theory on a disc.

We begin by comparing the B model with the C model in Sect. 2. The undeformed C model is given by

$$S_o^C = \int_\Sigma d^2\theta \left(\chi_I D u^I\right) \tag{4.33}$$

while the undeformed B model is given by

$$S_o^B = \int_\Sigma d^2\theta \Big(\chi_i D u^i + \chi_{\bar{i}} \psi^{\bar{i}}\Big). \tag{4.34}$$

In both cases the boundary condition, in order to satisfy the BV master equation, should be that $\chi_I(x)$ vanish along the direction tangent to $\partial\Sigma$ for $x \in \partial\Sigma$. We note that the B model is just the holomorphic version of the C-model but with one additional term compared to the B model. Recall that the boundary observables of C model are given by arbitrary functions on X. The algebra $\mathcal{O}(X)$ of functions (cohomological 1-algebra of X) on X is determined only by the commutative and associative product. As we explained earlier, the undeformed action functional has only the kinetic term, accordingly.

For the B model above a boundary observable is given by a function on $\Pi\overline{T}X$, i.e., $\oplus\Omega^{0,\bullet}(X)$. The algebra $\mathcal{O}(\Pi\overline{T}X)$ of functions on $\Pi\overline{T}X$ has a supercommutative, an associative product, and a differential. The undeformed B model action functional tells us that the differential is the $\bar{\partial}$ operator. This can be seen by the BRST transformation laws restricted to the boundary;

$$\begin{aligned} \boldsymbol{Q}_o \boldsymbol{u}^i &= D\boldsymbol{u}^i, \\ \boldsymbol{Q}_o \boldsymbol{u}^{\bar{i}} &= \psi^{\bar{i}}, \\ \boldsymbol{Q}_o \psi^{\bar{i}} &= 0. \end{aligned} \tag{4.35}$$

Then $\boldsymbol{Q}_o u^i = 0$, $\boldsymbol{Q}_o u^{\bar{i}} = \psi^{\bar{i}}$ and $\boldsymbol{Q}_o \psi^{\bar{i}} = 0$, which defines the $\bar{\partial}$ operator. Thus the algebra of classical observables is $\mathcal{O}(\Pi\overline{\mathcal{T}}X)$, with a supercommutative and associative product (wedge product) and a differential $\bar{\partial}$, which equip $\mathcal{O}(\Pi\overline{\mathcal{T}}X)$ with a structure of dgLa without bracket.

Now we turn on a deformation preserving the ghost number symmetry. For the C model we have

$$\delta \boldsymbol{S}^C = \int_\Sigma d^2\theta\, \varphi^*(\pi), \tag{4.36}$$

where $\pi = \frac{1}{2} b^{IJ} \chi_I \chi_J$ denotes a bi-vector on X. The bivector is an element of the 2nd Hochschild cohomology of $\mathcal{O}(X)$, i.e. $\pi \in H^2(\mathtt{Hoch}(\mathcal{O}(X)))$. Now the bulk deformation above deforms the product in a non-commutative direction. The master equation

$$[\pi, \pi]_S = 0, \tag{4.37}$$

implies that the deformed product is associative.

For the B model we have

$$\delta \boldsymbol{S}^B = \int_\Sigma d^2\theta\, \varphi^*(a + b), \tag{4.38}$$

where

$$\begin{aligned} a &= a_{\bar{i}}{}^{j} \chi_j \psi^{\bar{i}}, \\ b &= \frac{1}{2} b^{ij} \chi_i \chi_j. \end{aligned} \tag{4.39}$$

Note that the deformation by $\frac{1}{2}\kappa_{\bar{i}\bar{j}} \psi^{\bar{i}} \psi^{\bar{j}}$ in (4.21) is not allowed due to the master equation. As before it is not a bulk term. The two objects above correspond to elements of the 2nd Hochschild cohomology of $\mathcal{O}(\Pi\overline{\mathcal{T}}X)$, i.e. $a, b \in H^2(\mathtt{Hoch}(\mathcal{O}(\overline{\mathcal{T}}X))$. According to the mathematical definition $\frac{1}{2}\kappa_{\bar{i}\bar{j}} \psi^{\bar{i}} \psi^{\bar{j}} \in \Omega^{0,2}(X)$ also belongs to $H^2(\mathtt{Hoch}(\mathcal{O}(\Pi\mathcal{T}X)))$ [49]. However it is physically unnatural. In general we identify $H^*(\mathtt{Hoch}(\mathcal{O}(\Pi\overline{\mathcal{T}}X)))$ with $\oplus_{p\geq 1, q} \Omega^{0,q}(X, \wedge^p \overline{\mathcal{T}}X)$. The meaning of a deformation by a is obvious; it is a deformation of the complex structure. Comparing with the C model we see that the b term in the bulk generates a non-commutative deformation of the product. The master equation is

$$\begin{aligned}\bar{\partial}a + \frac{1}{2}[a,a]_T &= 0,\\ \bar{\partial}b + [a,b]_T &= 0,\\ [b,b]_T &= 0.\end{aligned} \tag{4.40}$$

The first equation implies that the deformation of the complex structure is integrable. The last equation implies that the non-commutative deformation is associative. The middle equation may be viewed as a compatibility of the two deformations. The mirror picture for Abelian varieties is discussed in [26] (see also [6] for a non-commutative version of the variation of Hodge structures more for closed string case).

Note that the deformation by a changes the BRST transformation laws restricted to the boundary as follows

$$\begin{aligned}\boldsymbol{Q}\boldsymbol{u}^i &= D\boldsymbol{u}^i + \boldsymbol{a}_i{}^{\bar{j}}\boldsymbol{\psi}_{\bar{j}},\\ \boldsymbol{Q}\boldsymbol{u}^{\bar{i}} &= \boldsymbol{\psi}^{\bar{i}},\\ \boldsymbol{Q}\boldsymbol{\psi}^{\bar{i}} &= 0.\end{aligned} \tag{4.41}$$

Thus the boundary observables should be changed accordingly. In the above case it is just the change of complex structure. We may consider the general deformation

$$\delta \mathbb{S}^B = \int_\Sigma d^2\theta\varphi^*(\gamma(u^i, u^{\bar{i}}, \chi_i, \psi^{\bar{i}}) \tag{4.42}$$

where $\gamma \in \oplus_{p\geq 1,q}\Omega^{0,q}(X, \wedge^p\mathcal{T}X) \equiv H^*(\mathtt{Hoch}(\mathcal{O}(\Pi\bar{\mathcal{T}}X))$. Then the boundary BRST transformation laws are changed as follows

$$\begin{aligned}\boldsymbol{Q}\boldsymbol{u}^i &= D\boldsymbol{u}^i + \frac{\partial\gamma}{\partial\chi_i}|_{\chi_i=0},\\ \boldsymbol{Q}\boldsymbol{u}^{\bar{i}} &= \boldsymbol{\psi}^{\bar{i}},\\ \boldsymbol{Q}\boldsymbol{\psi}^{\bar{i}} &= 0.\end{aligned} \tag{4.43}$$

Thus only the element in $\oplus_p\Omega^{0,p}(X, \mathcal{T}X)$ affects the boundary BRST transformation laws. Such elements correspond to the classical deformation of complex structures. Thus, without loss of generality, we say that the boundary observables are determined by the $\bar{\partial}$ cohomology among functions on $\Pi\bar{\mathcal{T}}X$.

Kontsevich's formality theorem implies that the dgLa on $\oplus_{p,q}\Omega^{0,q}(X, \wedge^p\mathcal{T}X)$ or, equivalently, on $H^*(\mathtt{Hoch}(\mathcal{O}(\Pi\bar{\mathcal{T}}X)))$ is quasi-isomorphic to the Lie algebra of local Hochschild cochains $\mathtt{Hoch}(\mathcal{O}(\Pi\bar{\mathcal{T}}X))$.

According to [6] this also implies that the extended moduli space parameterizes A_∞-deformations of $D^b(Coh(X))$ [6]. The solutions of the MC equation

$$\overline{\partial}\gamma + \frac{1}{2}[\gamma,\gamma]_T = 0, \tag{4.44}$$

also parameterizes the consistent deformations of the B model, through the solutions of BV master equation. The path integral of the theory can be viewed as a morphism from $H^*(\mathtt{Hoch}(\mathcal{O}(\Pi\overline{\mathcal{T}}X)))$ to $\mathtt{Hoch}(\mathcal{O}(\Pi\overline{\mathcal{T}}X))$. A path integral proof that it is a quasi-isomorphism can be done as for the C model. An explicit computation of the path integral will be useful even for very simple manifolds. We expect that both the product and differential of $\mathcal{O}(\Pi\overline{\mathcal{T}}X)$ deform.

Now we make some remarks on adding gauge fields. We regard the target space X as the D-brane world-volume, which has gauge fields on it. The background gauge field can be coupled to the theory by adding the following term to the action

$$\oint_{\partial\Sigma} du^I A_I(u^L). \tag{4.45}$$

For the C model an arbitrary gauge field A can be added without destroying the master equation since $\boldsymbol{Q}\boldsymbol{u}^I = \boldsymbol{D}\boldsymbol{u}^I$ on the boundary. Seiberg-Witten showed that the effective theory is governed by a non-commutative gauge theory, at least for constant B field [73]. For the B model the situation is different, since $\boldsymbol{Q}\boldsymbol{u}^{\bar{i}} = \boldsymbol{D}\boldsymbol{u}^{\bar{i}} + \boldsymbol{\psi}^{\bar{i}}$ on the boundary. Following Witten we consider

$$\oint_{\partial\Sigma} \left(du^I A_I(u^L) - \psi^{\bar{i}} F_{\bar{i}j}(A)\rho^i\right) \tag{4.46}$$

where ρ^i is the 1-form component of $\boldsymbol{u}^i$. Using $\boldsymbol{Q}\psi^{\bar{i}} = 0$ and $\boldsymbol{Q}\rho^i = -du^i$ we see that the above is $\boldsymbol{Q}$-invariant if the $(0,2)$ part of the curvature of the connection 1-form A vanish. In general we can couple connection of $U(N)$ (Hermitian) holomorphic bundles E on X. Similar to the C model case we expect that the D-brane world-volume effective theory is governed by a non-commutative gauge theory due to the bulk term

$$\frac{1}{2}\int_\Sigma d^2\theta \boldsymbol{b}^{ij}\boldsymbol{\chi}_i\boldsymbol{\chi}_j = \frac{1}{2}\int_\Sigma \left(b^{ij}H_i \wedge H_j + \ldots\right). \tag{4.47}$$

The above bulk term, after the gauge fixing (4.22), contains

$$\int_\Sigma \frac{1}{2} b_{\bar{i}\bar{j}} du^{\bar{i}} \wedge du^{\bar{j}} := \int_\Sigma u^*(B^{0,2}) \tag{4.48}$$

where $b_{\bar{i}\bar{j}} := b^{ij} g_{i\bar{i}} g_{j\bar{j}}$ and $g_{i\bar{j}}$ is a Ricci-flat Kähler metric. The presence of the above term implies, due to the gauge invariance, that the $(0,2)$ part of the curvature 2-form $F^{0,2}$ should be replaced by $F^{0,2} + B^{0,2}$ everywhere in the effective theory.

For X a Calabi-Yau 3-fold, without the bulk deformation by b, Witten determined the effective theory to be the holomorphic Chern-Simons theory,

$$I_{HCS} = \int_X \omega^{3,0} \wedge \mathrm{Tr}\left(A \wedge \bar{\partial} A + \frac{2}{3} A \wedge A \wedge A \right). \tag{4.49}$$

Now by turning on the bulk deformation by b we know that the wedge product among elements of $\Omega^{0,\bullet}(X)$ must be deformed to a suitable non-supercommutative product, say $\star$. We also expect that the differential $\bar{\partial}$ will be deformed in general, say to $\overline{Q}$. Thus the effective theory would look like

$$I = \int_X \omega^{3,0} \wedge \mathrm{Tr}\left(A \star (\overline{Q} A + B^{0,2}\mathbb{I}) + \frac{2}{3} A \star A \star A \right), \tag{4.50}$$

which is more similar to the open string field theory [88].

We may consider arbitrary deformations and more general boundary interactions by including first descendents (1-dimensional observables) of observables. Being inserted in cyclic order in the boundary with punctures, the general correlation functions are no-longer expected to show strict associativity, but will do so only up to homotopy (See a general discussion in [38]). Recall that all the boundary observables originate from $\oplus H^{0,q}(X)$. We also let those 1-dimensional observables take values in $(E_i^\circ)^* \otimes E_{i+1}^\bullet$, i.e. $Ext(E_i^\circ, E_{i+1}^\circ)$. Those may be viewed as the A_∞-category structure on $D^b(Coh(X))$, whose compositions are given by correlation functions.[r]. Being an A_∞-category (as the collection of all A_∞ modules over the A_∞-algebra $(\mathcal{O}(\Pi\overline{T}X), \bar{\partial})$) it satisfies a certain MC equation. Now turning on all bulk deformations leads to certain deformations of $D^b(Coh(X))$. The BV master equation, via the general structure of the BV Ward identity, may imply that the deformed category is again an A_∞ category. Since the extended moduli space $\mathcal{M}$ parameterizes all consistent bulk deformations of the extended B-model, we see that $\mathcal{M}$ indeed parameterizes the A_∞ deformations of $D^b(Coh(X))$.

We expect that the above discussions can be made more precise and explicit by carefully studying the extended B-model. We also expect to have explicit results on correlation functions at least for simple cases. The final effective theory will be much more complicated than (4.50), resembling the A_∞-version of open string field theory [29]. In principle the path integral of

[r] It is rather correlation maps than correlation functions, though we will stick to the original terminology

the extended B-model can lead to explicit answers for all higher compositions (string products). Then it will be the first explicit construction of open string field theory introduced in [29].

4.3 Topological open p-branes and generalized homological mirror conjectures

We showed that there is a unified description of A, B and C models based on essentially the same geometrical structure of the topological open membrane. Those models are special limits (or projections or gauge fixing) of the same (ABC) model. More fundamentally we showed that topological open membrane theory can describes and control A, B and C models of topological closed string as different boundary theories. The relation between A and B models are much like different twistings of the same superconformal field theory. Furthermore the topological open membrane theory leads to further (bulk) deformations of the mathematical or physical structure of the boundary string theory. Then it is natural to conjecture that the homological mirror symmetry can be generalized to the category of homotopy 2-algebra.

We may further generalize the topological membrane theory the way we generalized topological strings. The natural candidate is to use essentially the same geometrical structure as the topological open 3-brane. The resulting theory will have a quite large internal symmetry group, and many topological membrane theories in different limits. Then we may use open membrane versions of the resulting theories to state a homotopy 2-algebraic mirror conjecture. Note that the mirror symmetry between the A and B models is a remnant of the duality between type IIA and IIB strings. It would be interesting to examine if the internal symmetry of the generalized topological membrane theory is related to duality in M theory. In general we may consider topological open p-branes on X, based on the p-th iterated superbundle

$$M_p \to (M_{p-1} \to (M_{p-1} \to \ldots \to (M_1 \to X)\ldots).$$

We recall that the total space M_p has a degree $U = p$ symplectic structure ω_p and the base space M_{p-1} is a Lagrangian submanifold with respect to ω_p. The space M_p has various discrete symmetries preserving ω_p. We may take different Lagrangian subspaces and corresponding boundary conditions. Then we may obtain different theories, which can be physically equivalent in a suitable sense. We conjecture that the homological mirror symmetry can be generalized accordingly, at least for Calabi-Yau spaces X. More specifically, we conjecture that for a given homological mirror symmetry we have a corresponding mirror symmetry in the category of p-algebras. Then the mirror symmetry is generalized to physical equivalences of all topological open p-brane theories.

Acknowledgement

I am grateful to Robbert Dijkgraaf, Kenji Fukaya, Brian Greene, Christiaan Hofman, Seungjoon Hyun, Daniel Kabat, Calin Lazaroiu, Sangmin Lee, Yong-Geun Oh, Jongwon Park, Richard Thomas, David Tong and Herman Verlinde for discussions or communications in the various stage of this paper. I am also grateful to Christiaan Hofman, Calin Lazaroiu, Sangmin Lee and Jongwon Park for collaborations in related subjects. I am grateful to Christiaan Hofman for a proof reading of this manuscript as well as for useful suggestions. Some results in this paper have been announced in the conference "Symplectic geometry and mirror symmetry, 2000" held in KIAS. I am grateful for the organizers and KIAS for financial support and hospitality. This research is supported by DOE Grant # DE-FG02-92ER40699.

References

1. F. Akman, *On some generalizations of Batalin-Vilkovisky algebras*, q-alg/950627.
2. M. Alexandrov, M. Kontsevich, A. Schwarz and O. Zaboronsky, *The geometry of the master equation and topological quantum field theory*, Int. J. Mod. Phys. A12 (1997) 1405-1430, hep-th/9502010
3. M. Alishahiha, Y. Oz and M.M. Sheikh-Jabbari, *Supergravity and the large N noncommutative field theories*, hep-th/9909215
4. A.Yu. Alekseev, P. Schaller and T. Strobl, *The topological G/G WZW Model in the generalized momentum representation* , Phys. Rev. D52 (1995) 7146-7160, hep-th/9505012.
5. S. Axelord and I.M. Singer, Chern-Simons perturbation theory I, hep-th/9110056; Chern-Simons perturbation theory II, J. Diff. Geom. 39 (1994) 173-213, hep-th/9304087
6. S. Barannikov , *Generalized periods and mirror symmetry in dimensions* $n > 3$, math.AG/9903124 ;
Quantum periods - I. Semi-infinite variations of Hodge structures, math.AG/0006193
7. S. Barannikov and M. Kontsevich, *Frobenius Manifolds and Formality of Lie Algebras of Polyvector Fields*, alg-geom/9710032
8. I. Batalin and G. Vilkovisky, *Gauge algebra and quantization*, Phys. Lett. 102 B (1981), 27;

Quantization of gauge theories with linearly dependent generators, Phys. Rev. D29 (1983), 2567

9. F. Bayen, M. Flato, C. Fronsdal, A. Lichnerowicz and D. Sternheimer, *Deformation theory and quantization I, II*, Ann. Phys. 111 (1978)61–110, 111–151
10. K. Bering, P.H. Damgaard and J. Alfaro, *Algebra of Higher Antibrackets*, Nucl. Phys. B478 (1996) 459–504 `hep-th/9604027`
11. E. Bergshoeff, D.S. Berman, J.P. van der Schaar, and P. Sundell *A noncommutative M-theory five-brane*, `hep-th/0005026`; *Critical fields on the M5-brane and noncommutative open strings*, `hep-th/0006112`.
12. M. Bershadsky, S. Cecotti, H. Ooguri and C. Vafa, *Kodaira-Spencer theory of gravity*, Commun. Math. Phys. 165 (1994) 311
13. M. Bershadsky and V. Sadov, *Theory of Kähler gravity*, Int. J. Mod. Phys. A11 (1996) 4689-4730, tt hep-th/9410011
14. P. Bouwknegt, J. McCarty and J. Pilch, *BRST analysis of physical states for 2D gravity coupled to $c \leq 1$ matter*, Commu. Math. Phys. 145 (1992) 541
15. J.-L. Brylinski, *A differential complex for Poisson manifolds*, J. Diff. Geom. 28 (1988) 93
16. P. Candelas, X.C. De La Ossa, P.S. Green and L. Parkes, *A pair of Calabi-Yau manifolds as an exactly soluble superconformal theory*, Nucl. Phys. **B359**, 21 (1991).
17. A.S. Cattaneo and G. Felder, *A path integral approach to the Kontsevich quantization formula*, `math.QA/9902090`
18. A.S. Cattaneo and C.A. Rossi, *Higher-dimensional BF theories in the Batalin-Vilkovisky formalism: The BV action and generalized Wilson loops*, `math.QA/0010172`
19. A. Connes, *Noncommutative geometry*, Academic Press (1994).
20. A. Connes, M.R. Douglas and A. Schwarz, *Noncommutative geometry and matrix theory: compactification on tori*, JHEP **9802** (1998) 003, `hep-th/9711162`.
21. R. Dijgraaf, private communication.
22. R. Dijkgraaf and G. Moore, *Balanced topological field theories*, Commun. Math. Phys. 185 (1997) 411, `hep-th/9608169`
23. R. Dijkgraaf, H. Verlinde and E. Verlinde, *Topological strings in $d < 1$*, Nucl. Phys. B 352 (1991), 59–86
24. R.P. Feynman, *Space-time approach to non-relativistic quantum mechanics*, Rev. Mod. Phys. **20** (1948) 367-387.
25. K. Fukaya, *Morse Homotopy, A^{∞}-category and Floer Homologies*, MSRI

preprint No. 020-94 (1993)
26. K. Fukaya, *Floer homology of Lagrangian foliation and noncommutative mirror symmetry I*, ;
Mirror symmetry of Abelian variety and multi theta function, preprints
27. K. Fukaya and Y.-G. Oh, *Zero-Loop open string in the cotangent bundle and Morse homotopy*, Asian J. of Math. 1 (1998)96 - 180
28. K. Fukaya, Y.-G. Oh, H. Ohta and K. Ono, *Lagrangian intersection Floer theory - anomaly and obstruction*, preprint
29. M.R. Gaberdiel and B. Zwiebach, *Tensor Constructions of Open String Theories I: Foundations* Nucl. Phys. B505 (1997) 569-624 hep-th/9705038
30. M. Gerstenhaber, *The cohomology structure of an associative rings*, Ann. Math. 78 (1963) 59
31. E. Getzler and J.D.S. Jones, *Operads, homotopy algebra and iterated integrals for double loop spaces*, hep-th/9403055
32. E. Getzler, *Batalin-Vilkovisky algebras and two dimensional topological field theories*, hep-th/9212043
33. R. Gopakumar, J. Maldacenda, S. Minwalla and A. Strominger, *S-duality and noncommutative gauge theory*, hep-th/0005048
34. R. Gopakumar, S. Minwalla, N. Seiberg and A. Strominger, *OM theory in diverse dimensions*, hep-th/0006062.
35. B.R. Greene and M.R. Plesser, *Duality in Calabi-Yau moduli space*, Nucl. Phys. B338 (1990) 15
36. S.S. Gubser, I.R. Klebanov and A.M. Polyakov *Gauge theory correlators from non-critical string theory*, Nucl. Phys. B534 (1998) 202-222, hep-th/9802109
37. C. Hofman, S. Lee and J.-S. Park, *to appear*
38. C. Hofman and W.K. Ma, *Deformation of topological open strings*, hep-th/0006120
39. K. Hori and C. Vafa, *Mirror symmetry*, hep-th/0002222
40. K. Hori, A. Iqbal and C. Vafa, *D-Branes And Mirror Symmetry*, hep-th/0005247
41. N. Ikeda, *Two-dimensional gravity and nonlinear gauge theory*, Ann. Phys. 235 (1994) 435-464, hep-th/9312059
42. N. Ikeda *A Deformation of Three Dimensional BF Theory*, JHEP 0011 (2000) 0091, hep-th/0010096
43. S. Kawamoto and N. Sasakura, *Open membranes in a constant C-field background and noncommutative boundary strings*, hep-th/0005123.
44. J. Khoury and H. Verlinde, *On open/closed string duality*, hep-th/0001056

45. T. Kimura, J. Stasheff and A.A. Voronov, *On operad structures of moduli spaces and string theory*, hep-th/9307114
46. T. Kimura, J. Stasheff and A.A. Voronov *Homology of moduli spaces of curves and commutative homotopy algebras*, alg-geom/9502006.
47. T. Kimura, A.A. Voronov, G.J. Zuckerman, *Homotopy Gerstenhaber algebras and topological field theory* q-alg/9602009 ;
A.A. Voronov, *Homotopy Gerstenhaber algebras*, math.QA/9908040
48. M. Kontsevich, *Graph Cohomology*
49. M. Kontsevich, *Homological Algebra of Mirror Symmetry*, talk at ICM, Zurich 1994, alg-geom/9411018
50. M. Kontsevich, *Deformation quantization of Poisson manifolds, I*, q-alg/9709040
51. M. Kontsevich, *Operads and Motives in Deformation Quantization*, math.QA/9904055
52. M. Kontsevich and Y. Soibelman, *Deformations of algebras over operads and Deligne's conjecture*, math.QA/0001151
53. M. Kontsevich and Y. Soibelman, *Homological mirror symmetry and torus fibrations*, math.SG/0011041.
54. J.L. Koszul, *Crochet de Schouten-Nijenhuis et cohomologie*, Astérisque, hors série, (1985) 257
55. O. Kravchenko, *Deformations of Batalin-Vilkovisky Algebra*, math.QA/9903191
56. C.I. Lazaroiu, *On the structure of open-closed topological field theory in two dimensions*, hep-th/0010269.
57. W. Lerche, C. Vafa and N.P. Warner, *Chiral rings in $N = 2$ superconformal theories*, Nucl. Phys. B 324 (1989) 427
58. B.H. Lian, G.J. Zuckerman, *New Perspectives on the BRST-algebraic Structure of String Theory*, Commun. Math. Phys. 154 (1993) 613-646, hep-th/9211072
59. A. Lichnerowicz, *Les variétés de Poisso et leurs algébres de Lie associées*, J. Diff. Geom. 12 (1977) 253
60. J. Maldacena, *The large N limit of superconformal field theories and supergravity*, Adv. Theor. Math. Phys. 2 (1998) 231-252, hep-th/9711200
61. Y.I. Manin, *Three constructions of Frobenius manifolds: a comparative study*, math.QA/9801006
62. S.A. Merkulov, *Formality of canonical symplectic complexes and Frobenius manifolds*, math.SG/9805072
63. G. Moore, *Some comments on branes, G-flux, and K-theory*, hep-th/0012007.
64. M. Penkava and A. Schwarz, *On some algebraic structures arising in*

string theory, hep-th/9212072
65. J. Polchinski, *Dirichlet-branes and Ramond-Ramond charges*, Phys. Rev. Lett. 75 (1995) 4724, hep-th/9510017
66. V. Schomerus *D-branes and deformation quantization*, JHEP 9906 (1999) 030, hep-th/9903205.
67. A. Schwarz, *Geometry of Batalin-Vilkovisky quantization*, Commun. Math. Phys. 155 (1993) 249-260, hep-th/9205088
68. A. Schwarz, *Semiclassical approximation in Batalin-Vilkovsky formalism*, Commun. Math. Phys. 158 (1993) 373-396, hep-th/9210115
69. A. Schwarz, *Quantum observables, Lie algebra homology and TQFT*, Lett. Math. Phys. 49 (1999) 115-122, hep-th/9904168
70. B. Shoikhet, *A proof of the Tsygan formality conjecture for chains*, math.QA/0010321.
71. P. Schaller, T. Strobl, *Poisson structure induced (topological) field theories*, Mod. Phys. Lett. A9 (1994) 3129-3136, hep-th/9405110; *Poisson-sigma-models: A generalization of 2-D gravity Yang-Mills-systems*, hep-th/9411163.
72. N. Seiberg, L. Susskind and N. Toumbas, *Strings in background electric field, space/time noncommutativity and a new noncritical string theory*, JHEP 0006 (2000) 036, hep-th/0005040
73. N. Seiberg and E. Witten, *String theory and noncommutative geometry*, JHEP 9909 (1999) 032, hep-th/9908142.
74. J. Stasheff, *On the homotopy associativity of H-spaces I, II*, Trans. AMS 108 (1963) 275
75. J. Stasheff, *Deformation theory and the Batalin-Vilkovisky master equation*, q-alg/9702012 ;
The (secret?) homological algebra of the Batalin-Vilkovisky approach, Secondary calculus and cohomological physics (Moscow, 1997), 195-21-, Contemp. Math., 219, AMS, Providence, RI, 1998
76. J. Stasheff, *Closed string field theory, strong homotopy Lie algebras and the operad actions of moduli space*, hep-th/9304061
77. A. Strominger, *Open p-branes*, Phys. Lett. B383 (1996) 44, hep-th/9512059
78. A. Strominger, S.-T. Yau and E. Zaslow, *Mirror symmetry is a T-duality*, Nucl. Phys. B 479 (1996) 243, hep-th/9612121
79. D.E. Tamarkin, *Another proof of M. Kontsevich formality theorem*, math.QA/9803025;
Formality of Chain Operad of Small Squares, math.QA/9809164
80. D.E. Tamarkin, *Deformation complex of a d-algebra is a (d+1)-algebra*, math.QA/0010072.

81. D.E. Tamarkin and B. Tsygan, *Noncommutative differential calculus, homotopy BV algebras and formality conjectures*, math.KT/0002116
82. P.K. Townsend, *D-branes from M-branes*, Phys. Lett. B373 (1996) 68, hep-th/9512062
83. B. Tsygan, *Formality conjectures for chains*, math.QA/9904132
84. C. Vafa, *Extending mirror conjecture to Calabi-Yau with bundles* hep-th/9804131
85. E. Verlinde, *The Master Equation of 2D String Theory*, Nucl. Phys. B381 (1992) 141-157 hep-th/9202021
86. A.A. Voronov, *The Swiss-Cheese Operad*, math.QA/9807037
87. A.A. Voronov, *Homotopy Gerstenhaber algebras*, math.QA/9908040
88. E. Witten, *Non-commutative geometry and string field theory*, Nucl. Phys. B268 (1986) 253
89. E. Witten, *Topological Sigma Models*, Commun. Math. Phys. 118 (1988) 411
90. E. Witten, *Ground ring of the two dimensional string theory*, Nucl. Phys. B373 (1992) 187
91. E. Witten, *A note on the antibracket formalism*, Modern Phys. Lett. A 5 (1990), no. 7, 487–494
92. E. Witten, *Mirror Manifolds And Topological Field Theory*, hep-th/9112056
93. E. Witten, *Chern-Simons Gauge Theory As A String Theory*, hep-th/9207094
94. E. Witten, *On Background Independent Open-String Field Theory*, Phys. Rev. D46 (1992) 5467-5473 hep-th/9208027
95. E. Witten, *On the structure of the topological phase of two-dimensional gravity*, Nucl. Phys. B 340 (1990) 281–332
96. E. Witten, *Some comments on string dynamics*, hep-th/9507012
97. E. Witten, *Duality relations among topological effects in string theory*, JHEP 0005 (2000) 031, hep-th/9912086
98. E. Witten, *Quantum field theory and the Jones polynomials*, Commun. Math. Phys. 121 (1989) 351
99. E. Witten, *Anti de Sitter space and holography*, Adv. Theor. Math. Phys. 2 (1998) 253-291, hep-th/9803002
100. E. Witten and B. Zwiebach, *Algebraic structures and differential geometry in 2D string theory*, Nucl. Phys. B377 (1992) 55-112, hep-th/9201056
101. Y. Wu and C. Zhu, *The complete structure of the cohomology ring and associated symmetries in $d = 2$ string theory*, Nucl. Phys. B404 (1993) 245, hep-th/9209011.
102. B. Zwiebach, *Closed string field theory: quantum action and the BV mas-*

ter equation, Nucl. Phys. B390 (1993) 33 hep-th/9206084
103. B. Zwiebach, *Oriented Open-Closed String Theory Revisited*, Ann. Phys. 267 (1998) 193-248, hep-th/9705241

LAGRANGIAN TORUS FIBRATIONS AND MIRROR SYMMETRY OF CALABI-YAU MANIFOLDS

WEI-DONG RUAN
Department of mathematics, University of Illinois, Chicago, IL 60607
Email: ruan@math.uic.edu

1 Introduction

A very beautiful part of Kodaira's complex surface theory is the theory of elliptic surfaces. Elliptic surfaces lie somewhere between "positive" (rational) surfaces and "negative" (general type) surfaces, where many wonderful things happen. Kodaira's elliptic surface theory, especially his classification of singular fibres, have lots of deep relations with other parts of mathematics, including the theory of elliptic curves and singularity theory. In recent years, a rather surprising higher dimensional generalization of such classical and fundamental nature to Calabi-Yau manifolds was discovered in the context of mirror symmetry!

Mirror Symmetry conjecture originated from physicists' work in conformal field theory and string theory. It proposes that for a Calabi-Yau 3-fold X there exists a Calabi-Yau 3-fold Y as its mirror. The quantum geometry of X and Y are closely related. In particular one can compute the number of rational curves in X by solving the Picard-Fuchs equation coming from variation of Hodge structure of Y [8]. Before mirror symmetry, Calabi-Yau manifolds were rather abstract and mysterious in mathematics. The study of mirror symmetry, especially the exciting development of counting of rational curves gave us a lot of insight of quantum geometry of Calabi-Yau manifolds or more precisely the geometry of the moduli spaces of Calabi-Yau manifolds. However the understanding of classical geometry and topology of general Calabi-Yau manifolds and how mirror symmetry is reflected in the classical level was still lacking. Strominger-Yau-Zaslow's mirror conjecture, which is very much in the spirit of Kodaira elliptic surface theory, changed all these.

In 1996 Strominger, Yau and Zaslow ([38]) proposed a geometric construction of mirror manifold via special Lagrangian torus fibration. According to their program (we will call it SYZ construction), a Calabi-Yau 3-fold should

[0]Partially supported by NSF Grant DMS-9703870 and AMS Centennial Fellowship.

admit a special Lagrangian torus fibration. The mirror manifold can be obtained by dualizing the fibres. Or equivalently, the mirror manifold of X is the moduli space of special Lagrangian 3-torus in X with a flat $U(1)$ connection.

Notice that despite its physical root, the statement of SYZ conjecture is purely mathematical and rather classical in nature. It makes the structure of Calabi-Yau manifolds rather explicit and mirror symmetry a rather concrete duality. In the K3 surface case, K3 with SYZ fibration is equivalent to elliptic K3 under the hyperKähler twist.

However, in spite of the great promise of SYZ conjecture, our understanding on special Lagrangian submanifolds and therefore the SYZ fibration is very limited. The known examples are mostly explicit local examples or examples coming from 2-dimensional case. There are very few examples of special Lagrangian submanifold or special Lagrangian fibration for dimension higher than two. M. Gross, P.M.H. Wilson and N. Hitchin ([17][18][21][23]) did some important work in this area in recent years. These works mainly concern local geometric structure of the special Lagrangian fibration and cases that can be reduced to 2-dimensional situation. On the other extreme, in [40], Zharkov constructed some non-Lagrangian torus fibration of Calabi-Yau hypersurface in toric variety. Despite all these efforts, SYZ construction still remains a beautiful dream to us.

Given the general lack of knowledge for special Lagrangian, our approach to SYZ conjecture is to relax the special Lagrangian condition and consider Lagrangian fibration, which we feel like to be a good compromise and is interesting in its own. Special Lagrangians are very rigid, while Lagrangian submanifolds are more flexible and can be modified locally by Hamiltonian deformation. However for many application to mirror symmetry, especially those concerning (symplectic) topological structure of fibration, Lagrangian fibration will provide quite sufficient information. In our work [29, 30, 31, 32], we mainly concern Lagrangian torus fibrations of Calabi-Yau hypersurfaces in toric variety, namely the symplectic topological aspect of SYZ mirror construction.

In light of the recent important explicit local examples of generic special Lagrangian fibrations constructed by Dominic Joyce [24], the SYZ fibration map of Calabi-Yau manifolds are likely to be only piecewise C^∞ (Lipschitz) instead of C^∞, and the singular locus instead of being a codimension two graph in S^3 is likely to be of codimension one as a fattening of a graph. This is very much in line with the codimension one singular locus pictures that naturally come out of our gradient flow construction in [29]. As pointed out by Joyce, the actual special Lagrangian fibration probably can be constructed by perturbing our Lagrangian fibration with codimension one singular locus. SYZ duality of the fibres are likely to be precise duality only at the large complex (radius) limit. In my point of view, this does not necessarily destroy the beauty of SYZ construction, rather makes it much richer and more intriguing. For the sake of computation, understanding the geometric and topological structure of Calabi-Yau manifolds, especially, constructing the mirror Calabi-Yau symplectic topologically, it is still rather convenient to have a Lagrangian fibration with codimension two graph singular locus.

In [32], we were able to construct Lagrangian torus fibrations with codimension two graph singular locus of generic Calabi-Yau hypersurfaces in toric varieties correponding to reflexive polyhedra in complete generality. With these detailed understanding of Lagrangian torus fibrations of generic Calabi-Yau hypersurfaces in toric varieties, we were able to prove the symplectic topological version of SYZ mirror conjecture for Calabi-Yau hypersurfaces in toric varieties, although the origional SYZ conjecture for special Lagrangian fibrations probably need some modification. More precisely we have:

Theorem 1.1 *For generic Calabi-Yau hypersurface X in the toric variety corresponding to a reflexive polyhedron Δ and its mirror Calabi-Yau hypersurface Y in the toric variety corresponding to the dual reflexive polyhedron $\Delta^\vee$ near their corresponding large complex limit and large radius limit, there exist corresponding Lagrangian torus fibrations*

$$\begin{array}{ccccccc} X_{\phi(b)} & \hookrightarrow & X & \qquad & Y_b & \hookrightarrow & Y \\ & & \downarrow & & & & \downarrow \\ & & \partial\Delta_v & & & & \partial\Delta_w \end{array}$$

with singular locus $\Gamma \subset \partial\Delta_v$ and $\Gamma' \subset \partial\Delta_w$, where $\phi : \partial\Delta_w \to \partial\Delta_v$ is a natural homeomorphism that satisfies $\phi(\Gamma') = \Gamma$. For $b \in \partial\Delta_w \backslash \Gamma'$, the corresponding fibres $X_{\phi(b)}$ and Y_b are naturally dual to each other.

Our work essentially indicates that the Batyrev-Borisov mirror construction [3, 7], which was proposed purely from toric geometry stand point, can also be understood and justified by the SYZ mirror construction. This should give us greater confidence on SYZ mirror conjecture for general Calabi-Yau manifolds.

In an earlier physics work of Leung and Vafa [25], they discussed a heuristic derivation of Batyrev's mirror construction via T-duality. Although we do not fully understand the physics argument presented in [25], we later find that some ideas in [25] seem to be closely related to our approach toward SYZ mirror symmetry in the Batyrev case and may be helpful to better understand the relation of our mathematical construction and its underlying physics.

Around the same time as our paper [31], in which Lagrangian torus fibrations were constructed for generic quintic Calabi-Yau and corresponding symplectic SYZ was proved, M. Gross independently (using completely different method) constructed certain ***non-Lagrangian*** torus fibrations for quintic Calabi-Yau in [19], which exhibit similar topological structure as our Lagrangian torus fibration in [31]. His idea, very much in line with part of our previous work [29], was to use the monodromy information to guess the structure of the fibration topologically. (Here he got the expected monodromy information from the resolution of orbifold singularities in the mirror.) He then constructed the topological manifold accordingly and used a theorem of C.T.C. Wall to prove the topological manifold constructed is homeomorphic to quintic Calabi-Yau, therefore constructing a topological torus fibration on the quintic Calabi-Yau. On the other hand, our completely different approach in [31] constructed the Lagrangian torus fibration on quintic Calabi-Yau directly (without going to the mirror) using gradient flow based on the graphlike behavior of ameoba/string diagram discussed in [33, 27]. We note that our approach (in [31]) also provides an alternative way (from [19]) to construct ***topological*** torus fibrations on quintic Calabi-Yau directly. The technical results required for such alternative topological construction will be much easier than the technical results required for the construction of Lagrangian torus fibration discussed in [30, 33].

The purpose of this paper is to give a coherent overview of our work on Lagrangian torus fibration and symplectic SYZ mirror correspondence in various cases, starting from the most famous case of quintic Calabi-Yau and its mirror. Through our discussion, we will pay special attention to exploring the relation between various changes of singular locus graphs of the fibrations and various geometric changes of the corresponding Calabi-Yau manifolds.

In the elliptic surface case, the singular locus is just a finite set of isolated points. (For a generic elliptic K3, there are 24 points.) Generic type of singular fibre is the nodel $\mathbf{CP^1}$. Several singular locus points can come together in non-generic elliptic surfaces to form more complex singular fibres. Analogously for the case of Lagrangian torus fibrations of Calabi-Yau 3-folds over S^3, the singular locus is genericly a graph Γ with only 3-valent vertex points Γ^0, which are seperated into two sets — the positive (negative) vertex points Γ^0_+ (Γ^0_-)— depending on whether Euler number of the singular fibre over it is +1 or −1 as described in [32]. The singular locus graph and singular fibres change from region to region in the moduli space by passing through some walls of non-generic Calabi-Yau's. The most generic of non-generic graphs in 3-space is a graph with only 3-valent vertex points except one 4-valent vertex. From the point of view of deforming graphs in 3-space, there are the following three natural ways to degenerate generic graphs into such non-generic graph as indicated in the following picture.

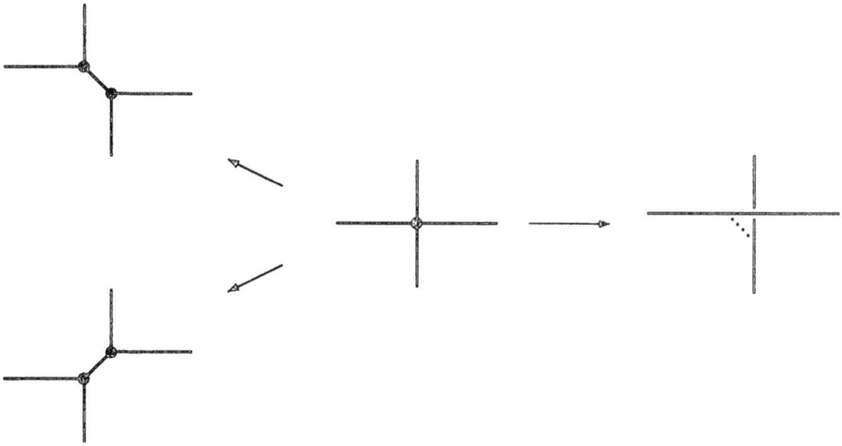

Figure 1: Generic degeneration of graph

Notice that the graphs in the above picture are all located in parallel planes. (In our case planes parallel to the paper.) We will see the reason for this constrain later. There are two ways to collide two 3-valent points into one 4-valent point (as indicated on the left) and there is one way for two smooth lines to meet and form the 4-valent vertex. It turns out these graph operations canonically correspond to some natural geometric or topological change of Calabi-Yau manifolds. In the ground breaking work of Aspinwall, Greene and Morrison [2] on topological change in mirror symmetry, Kähler moduli of different birational equivalent Calabi-Yau as well as other non-Calabi-Yau regions are unified to form the Kähler moduli that can be identified with the complex moduli of the mirror Calabi-Yau via the monomial-divisor mirror map ([1]). Different regions in the Kähler moduli corresponding to different birational Calabi-Yau models are related by flops. The transition between the two graphs on the left of Figure 1 in the Kähler side exactly corresponds to such a flop. Another way to extend the moduli space of Calabi-Yau is through a conifold transition (or black hole condensation in physics literature) that is related to the so-called Reid's fantasy [28], which conjectures that all the moduli spaces of Calabi-Yau are connected through such transition. As suggested in [9, 16, 26, 11], the conifold transition can be used to find mirrors beyond toric cases. The transitions between graphs on the left and the graph on the right of Figure 1 exactly corresponds to a conifold transition. (Gross also observed such transition in [20].) These correspondences further indicate the crucial importance of Lagrnagian torus fibration in understanding mirror symmetry. In Section 5 and 6, we will indicate these rather intriging relations through examples. We will also describe the corresponding changes of singular fibres.

2 Quintic case

Our idea of construction is a very natural one. We try to use gradient flow to get Lagrangian torus fibration from a known Lagrangian torus fibration at the "Large Complex Limit". This method will in principle be able to produce Lagrangian torus fibration in general Calabi-Yau hypersurfaces or complete intersections in toric variety.

To illustrate our idea, it is helpful to explore the historically most famous case of quintic Calabi-Yau threefolds in $\mathbb{CP}^4$ in detail. Most of the essential features of the general case already show up here. Let

$$z^m = \prod_{k=1}^{5} z_k^{m_k}, \; |m| = \sum_{k=1}^{5} m_k, \; \text{for } m = (m_1, m_2, m_3, m_4, m_5) \in \mathbb{Z}^5_{\geq 0}.$$

Then a general quintic can be denoted as

$$p(z) = \sum_{|m|=5} a_m z^m.$$

Let $m_0 = (1, 1, 1, 1, 1)$, and denote $a_{m_0} = \psi$. Consider the quintic Calabi-Yau familly $\{X_\psi\}$ in $\mathbf{P}^4$ defined as

$$p_\psi(z) = p_a(z) + \psi \prod_{k=1}^{5} z_k = \sum_{m \neq m_0, |m|=5} a_m z^m + \psi \prod_{k=1}^{5} z_k = 0.$$

When ψ approaches ∞, the familly approach its "Large Complex Limit" X_∞ defined by

$$p_\infty = \prod_{k=1}^{5} z_k = 0.$$

X_∞ is a union of five $\mathbf{CP}^3$'s. There is a natural degenerate T^3 fibration structure for X_∞ given by the natural moment map $F : \mathbf{CP}^4 \longrightarrow \Delta$. $\Delta = \text{Image}(F)$ is a 4-simplex. X_∞ is naturally fibered over $\partial\Delta$ via this map F with general fibre being T^3. This is in a sense precisely the SYZ special Lagrangian T^3 fibration for X_∞.

Consider the meromorphic function

$$s = \frac{p_\infty(z)}{p_a(z)} = f + ih$$

defined on $\mathbf{CP}^4$. Let ∇f denote the gradient vector field of real function $f = Re(s)$ with respect to the Kähler metric g.

Notice that

$$\nabla f = H_h,$$

where H_h is the Hamiltonian vector field generated by $h = Im(s)$. This implies the following

Lemma 2.1 *The gradient flow of f leaves the set $\{Im(s) = 0\}$ invariant and deforms Lagrangian submanifolds in X_∞ to Lagrangian submanifolds in X_ψ.*

With this lemma in mind, the construction of Lagrangian torus fibration of X_ψ for ψ large is immediate. Deforming the canonical Lagrangian torus fibration of X_∞ over $\partial\Delta$ along the gradient flow of f will naturally induce a

Lagrangian torus fibration of X_ψ over $\partial\Delta$ for ψ large and real.

The key advantage of the gradient flow method is that once it is run, the Lagrangian fibrations are automatically produced almost effortlessly. There are no adhoc manual construction involved to this point. Of course, for such Lagrangian fibration to be of any use, it is necessary to understand the detailed structure such as the singular fibres, singular set and the singular locus, etc. For this purpose, it is necessary to understand the details of dynamics of the gradient flow and how the fibrations evolve under the flow. There are a lot of rather delicate technical issues to be addressed here. First of all, the gradient flow in our situation is rather non-conventional. The critical points of the function are usually highly degenerate and often non-isolated. Worst of all the function is not even smooth (it has infinities along some subvarieties). In [30] we discussed the local behavior of our gradient flow near critical points and infinities of f, and also the dependence on the metric to make sure that they behave the way we want.

Secondly, our gradient flow method naturally produces piecewise smooth (Lipschitz) Lagrangian fibrations. The singular locus is of codimension one. This is quite to the contrary of the conventional wisdom, where people expect the special Lagragian fibration to be smooth and the singular locus to be of codimension two. Nerverthless, in light of the recent examples of Joyce [24], Lagrangian fibrations with codimension one singular locus might reflect the structure of the actual special Lagrangian fibration after all. This further indicates that it might be a good idea to consider Lagrangian fibrations, which for most purposes would be as good as, or even more convenient to use than the actural special Lagrangian fibrations. In [30], we squeeze the codimension one singular locus by symplectic geometry technique to get Lagrangian torus fibration with codimension two graph singular locus. For computation purposes and to construct mirror manifolds symplectic topologically, such models clearly are rather desirable. In particular, the SYZ duality can be made more precise in the symplectic category using such models. For details, please refer to [30].

With all the technical points taken care of, it is not hard to observe that the singular set of our fibration $F_\psi : X_\psi \to \partial\Delta$ is the curve

$$C = X_\psi \cap \mathrm{Sing}(X_\infty).$$

The singular locus

$$\tilde{\Gamma} = F_{\psi}(C)$$

is located in the 2-skeleton of Δ, which is a union of 2-simplices. More precisely, since C is reducible, and each irreducible component is in a $\mathbb{CP}^2$, according to [30], the study of singular locus can be isolated to each $\mathbb{CP}^2$ and reduced to the following problem.

Problem: Let $F : \mathbf{CP}^2 \to \Delta$ be the standard moment map with respect to the Fubini-Study metric. For what kind of quintic curve C in $\mathbb{CP}^2$, $\tilde{\Gamma} = F(C)$ is a fattening of some graph Γ?

This problem was discussed in [33] for curves C of arbitrary degrees in $\mathbf{CP}^2$, more generally for curves in arbitrary 2-dimensional toric varieties. There is a notion of **near the large complex limit**, which corresponds to the coefficients of the polynomial defining the curve satisfying some convexity condition with respect to the Newton polygon of the polynomial. For curves in $\mathbf{CP}^2$, we have

Theorem 2.1 *When a degree d curve C_{p_d} defined by polynomial $p_d(z)$ is near the large complex limit, $F(C_{p_d})$ will have exactly $g = \dfrac{(d-1)(d-2)}{2}$ holes and d external points in each edge of Δ.*

Different convexity condition Z of the coefficients of $p_d(z)$ corresponds to different large complex limit chambers, which determine different graphs $\Gamma_Z \subset \Delta$, of which $\tilde{\Gamma} = F(C_{p_d})$ is a fattening. More precisely, we have

Theorem 2.2 *For $t \in \mathbb{R}_+$ small, $F_{t^w}(C_{p_d})$ is a fattening of Γ_Z.*

Here t^w is the coefficient set of $p_d(z)$. w satisfies the convexity condition Z. The theorem is roughly saying that when $p_d(z)$ is approaching the large complex limit corresponding to Z, $\tilde{\Gamma} = F_{t^w}(C_{p_d})$ will resemble a fattening of the corresponding graph Γ_Z. The following is the picture of Γ_Z for the standard Z.

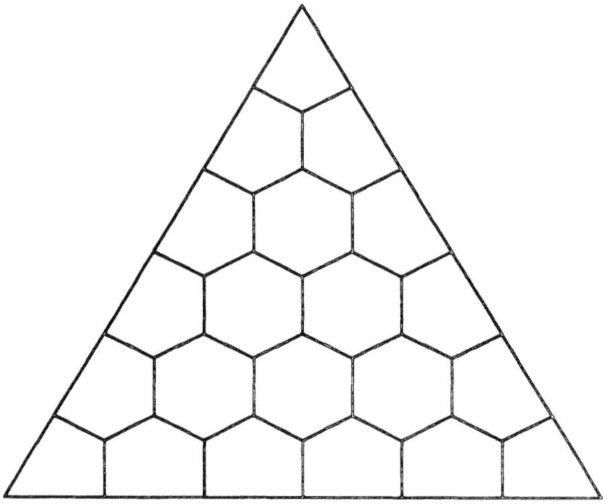

Figure 2: standard Γ_Z when degree $d = 5$

With different convex condition Z, the corresponding Γ_Z could change to

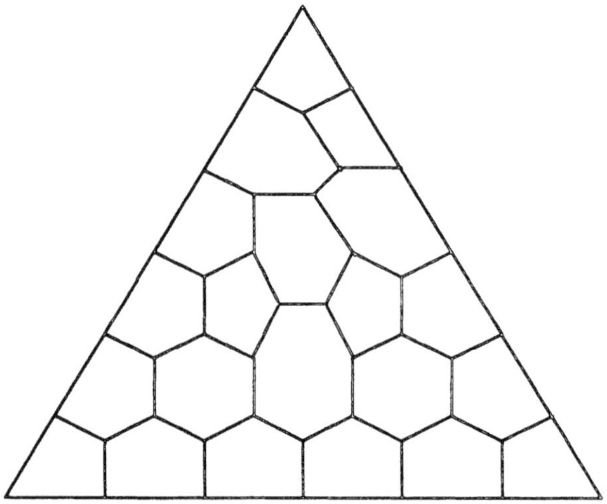

Figure 3: a different Γ_Z

For details of notations and results, please refer to [33]. In [33], these results are actually proved in the more general context of curves in arbitrary

2-dimensional toric varieties. After our work was finished, Prof. Oh pointed out to me (during the KIAS conference) the references [27, 12] originated from the fundamental work of Gelfand, Kapranov, Zelevinsky [13], where the authors aimed at very different applications arrived at similar results as in [33] (except our symplectic deformation to graph image). The images of curves under moment map are called amoebas in their work. For more detail on relation to their work, please refer to [33].

The detailed structure of the resulting Lagrangian torus fibration of X_ψ is described in the following theorem. For detail of the proof, please refer to [30, 31].

Theorem 2.3 *The gradient flow will produce a Lagrangian fibration $\tilde{F} : X_\psi \to \partial\Delta$ with singular locus $\tilde{\Gamma} = \tilde{\Gamma}^0 \cup \tilde{\Gamma}^1 \cup \tilde{\Gamma}^2$. There are 4 types of fibres.*

(i). For $p \in \partial\Delta\backslash\tilde{\Gamma}$, $\tilde{F}^{-1}(p)$ is a smooth Lagrangian 3-torus.
(ii). For $p \in \tilde{\Gamma}^2$, $\tilde{F}^{-1}(p)$ is a Lagrangian 3-torus with 2 circles collapsed to 2 singular points.
(iii). For $p \in \tilde{\Gamma}^1$, $\tilde{F}^{-1}(p)$ is a Lagrangian 3-torus with 1 circles collapsed to 1 singular points.
(iv). For $p \in \tilde{\Gamma}^0$, $\tilde{F}^{-1}(p)$ is a Lagrangian 3-torus with 1 2-torus collapsed to 1 singular points.

In order to get codimension two singular locus, it is necessary to perturb the moment map, which is a Lagrangian fibration, such that the image of the curve under the perturbed moment map is exactly the graph Γ_Z. This construction is also carried out in [33]. We have

Theorem 2.4 *There exists a perturbed Lagrangian fibration $\hat{F}$ of the moment map F satisfying $\hat{F}(C_{s_t}) = \Gamma_Z$.*

With these facts from [33] in hand, general methods developed in [29], [30] will enable us to construct Lagrangian torus fibration with one-dimensional singular locus for generic quintic Calabi-Yau near the large complex limit.

More precisely, when the generic quintic is near the large complex limit, with the help of theorem 2.4, we can produce a Lagrangian fibration $\hat{F} : X_\infty \to \partial\Delta$ such that $\hat{F}(C) = \Gamma$. Γ is a graph in $\partial\Delta \cong S^3$. Its part in each 2-simplex is the kind of graphs described in theorem 2.4. Let $\Gamma = \Gamma^1 \cup \Gamma^2 \cup \Gamma^3$, where Γ^1 is the smooth part of Γ, Γ^2 is the singular part of Γ in the interior of the 2-skeleton of Δ, Γ^3 is the singular part of Γ in the 1-skeleton of Δ. Then we have

Theorem 2.5 *When X_ψ is near the large complex limit, start with Lagrangian fibration $\hat{F}$ the flow of V will produce a Lagrangian fibration $\hat{F}_\psi : X_\psi \to \partial\Delta$. There are 4 types of fibres.*
(i). For $p \in \partial\Delta\backslash\Gamma$, $\hat{F}_\psi^{-1}(p)$ is a smooth Lagrangian 3-torus.
(ii). For $p \in \Gamma^1$, $\hat{F}_\psi^{-1}(p)$ is a type I singular fibre.
(iii). For $p \in \Gamma^2$, $\hat{F}_\psi^{-1}(p)$ is a type II singular fibre.
(iv). For $p \in \Gamma^3$, $\hat{F}_\psi^{-1}(p)$ is a type III singular fibre.

Type I singular fibre in the theorem refers to a two-dimensional singular fibre (in this case a nodal $\mathbf{CP}^1$) times S^1. This type of singular fibres have Euler number zero. Other types of singular fibres are illustrated in the following pictures. Notice that type II and III singular fibres have Euler numbers equal to -1 and 1 respectively.

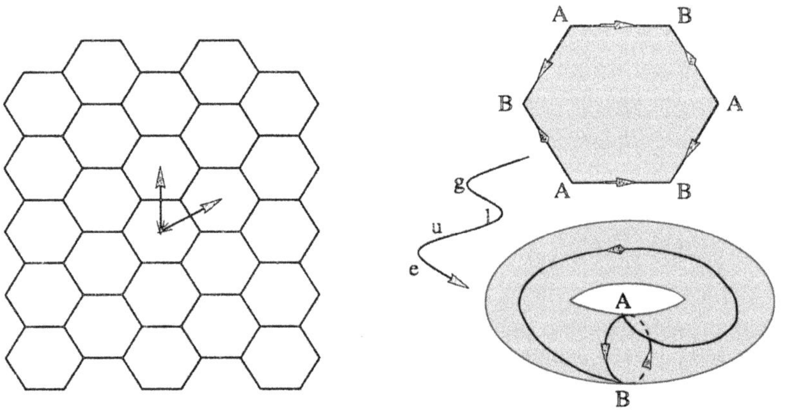

Figure 4: Type III_5 and type III fibres

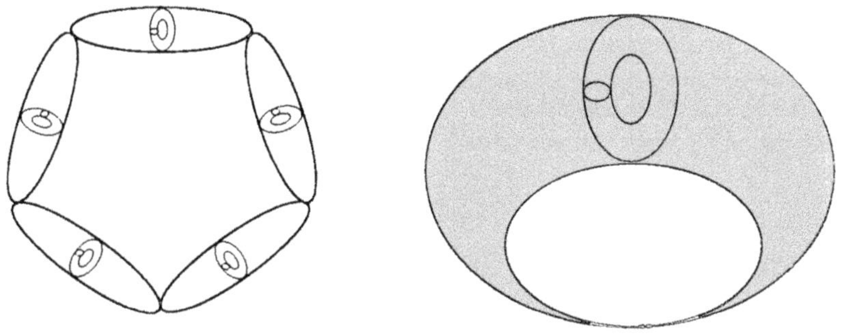

Figure 5: Type $II_{5\times 5}$ and type II fibres

3 The mirror of quintic

In the mirror side, the nontrivial part is in the 101 dimensional Kähler moduli near the large radius limit. As we know the anti-canonical model of the mirror of quintic is the quotient of Fermat type Calabi-Yau

$$Y_\psi = X_\psi/(\mathbf{Z}_5)^3 \subset P_{\Sigma_{\Delta^\vee}} \cong \mathbf{CP}^4/(\mathbf{Z}_5)^3.$$

Here $\Delta^\vee$ is the dual polyhedron of the polyhedron Δ corresponding to $\mathbb{CP}^4$. $\Sigma_{\Delta^\vee}$ is the fan corresponding to $\Delta^\vee$ via normal cone construction. The corresponding toric variety $P_{\Sigma_{\Delta^\vee}}$ is naturally equivalent to $\mathbf{CP}^4/(\mathbf{Z}_5)^3$.

The mirror of quintic Calabi-Yau are the crepant resolutions of Y_ψ. Different crepant resolutions correspond to different chambers of the Kähler moduli connected by flops. These crepant resolutions of Y_ψ are naturally induced from crepant resolutions of $P_{\Sigma_{\Delta^\vee}} \cong \mathbb{CP}^4/(\mathbb{Z}_5)^3$, which can be described via crepant subdivision of fan $\Sigma_{\Delta^\vee}$ into different simplical fans. Let w denote the Kähler class of one such crepant resolution, w determines a simplical fan Σ^w. $P_{\Sigma^w} \to P_{\Sigma_{\Delta^\vee}}$ is a crepant resolution. The pullback of Y_ψ (we still use the same notation) is the mirror Calabi-Yau with Kähler class corresponding to w.

The gradient flow of the Fermat type quintic Calabi-Yau family $\{X_\psi\}$ is invariant under the action of $(\mathbf{Z}_5)^3$. The quotient gives us the corresponding gradient flow on $P_{\Sigma_{\Delta^\vee}} \cong \mathbf{CP}^4/(\mathbf{Z}_5)^3$ of the family $\{Y_\psi\}$. This flow pulled back to P_{Σ^w} will flow Y_∞ to Y_ψ and induce Lagrangian torus fibration structure on

$Y_\psi \subset P_{\Sigma^w}$. Let $F_\psi : Y_\psi \to \partial\Delta_w$ denote the Lagrangian fibration of Y_ψ. Then

Proposition 3.1 *The singular set $C \subset Y_\psi$ of the fibration F_ψ is exactly the intersection of Y_ψ with the complex 2-skeleton of P_{Σ^w}. In another word,*

$$C = Y_\psi \cap \mathrm{Sing}(Y_\infty).$$

The singular locus $\tilde{\Gamma} = F_\psi(C) = F_\infty(C)$ is a fattening of some graph Γ. We will again use techniques in [30, 33] to modify $F_\infty(C)$ to $\hat{F}_\infty(C)$ so that $\hat{F}_\infty(C)$ exactly equal to the one-dimensional graph Γ. To describe this graph Γ, let's recall from the last section of [29] that the singular locus of the Lagrangian fibreation of $Y_\psi \subset P_{\Sigma_{\Delta^\vee}}$ is a fattening of a graph $\hat{\Gamma} \subset \partial\Delta^\vee$, where

$$\hat{\Gamma} = \bigcup_{\{ijklm\}=\{12345\}} \overline{P_{ij}P_{klm}}.$$

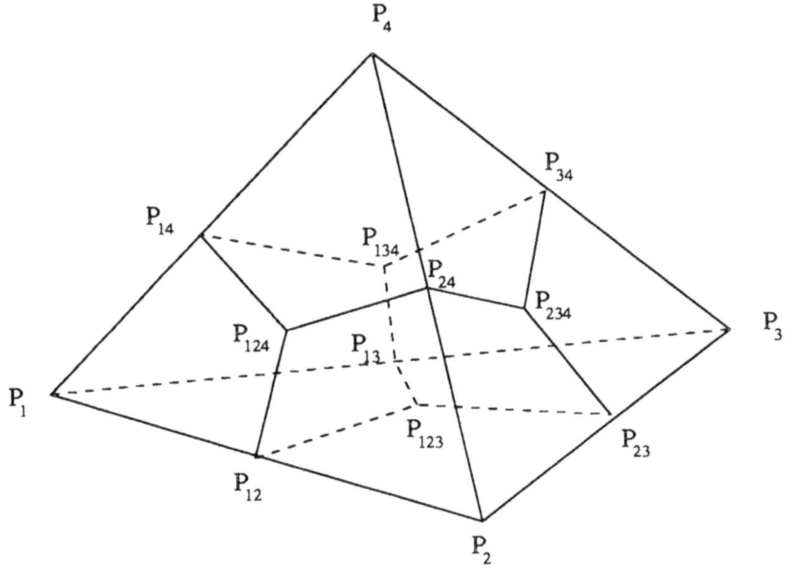

Figure 6: $\hat{\Gamma} \subset \partial\Delta^\vee$.

It is interesting to observe that

$$\mathrm{Sing}(P_{\Sigma_{\Delta^\vee}}) = \mathrm{Sing}(Y_\infty).$$

Hence

$$\mathrm{Sing}(Y_\psi) = C = Y_\psi \cap \mathrm{Sing}(Y_\infty).$$

Let $\tilde{P}_{ij}$ be the point in C that maps to P_{ij} in $\Delta^\vee$. Notice that $\mathrm{Sing}(C) = \{\tilde{P}_{ij}\}$. Along smooth part of C, Y_ψ has A_5-singularity. Under the unique crepant resolution, C is turned into 5 copies of C. Around $\tilde{P}_{ij} \in \mathrm{Sing}(C)$, singularity of Y_ψ is much more complicated and crepant resolution is not unique (depending on the Kähler moduli w). The following is a picture (from [29]) of fan of such singularity and the subdivision fan of the standard crepant resolution.

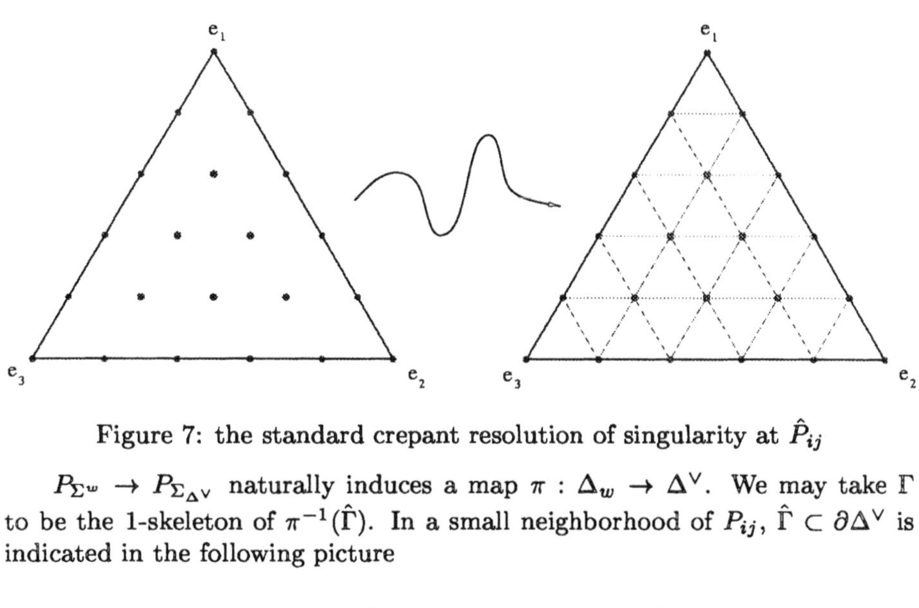

Figure 7: the standard crepant resolution of singularity at $\hat{P}_{ij}$

$P_{\Sigma^w} \to P_{\Sigma_{\Delta^\vee}}$ naturally induces a map $\pi : \Delta_w \to \Delta^\vee$. We may take Γ to be the 1-skeleton of $\pi^{-1}(\hat{\Gamma})$. In a small neighborhood of P_{ij}, $\hat{\Gamma} \subset \partial\Delta^\vee$ is indicated in the following picture

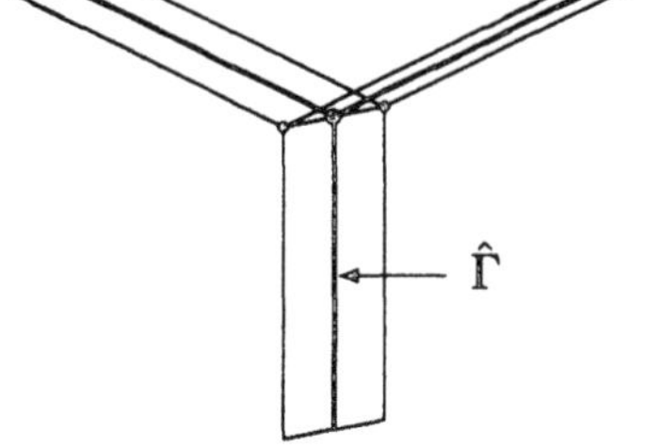

Figure 8: $\hat{\Gamma} \subset \partial\Delta^\vee$ near P_{ij}

Under the standard crepant resolution, we get $\Gamma \subset \partial\Delta_w$ as indicated in the following picture

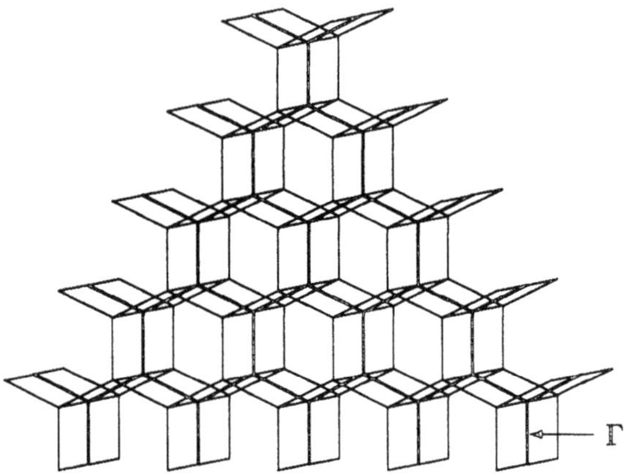

Figure 9: Γ for the standard crepant resolution

For a different crepant resolution, we can get alternative picture for $\Gamma \subset \partial\Delta_w$.

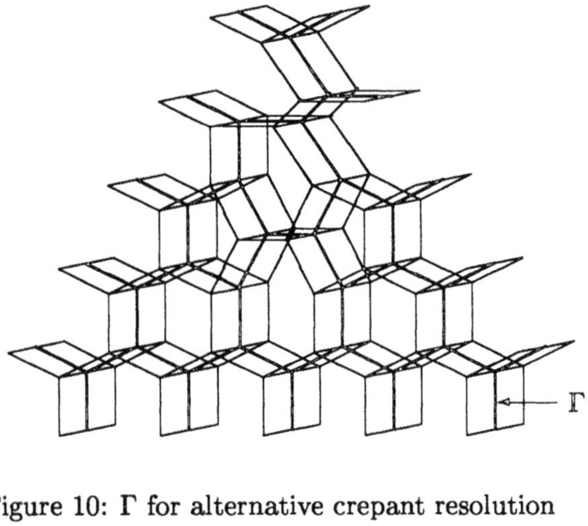

Figure 10: Γ for alternative crepant resolution

Notice that such graph changes are caused by flops of the corresponding Calabi-Yau manifolds and locally are of the type

Figure 11: Graph degeneration corresponding to flop

These singular locus graphs clearly resemble singular locus graphs in figure 2 and 3 obtained via string diagram construction, although the two constructions are quite different.

Let $\Gamma = \Gamma^1 \cup \Gamma^2 \cup \Gamma^3$, where Γ^1 is the smooth part of Γ, Γ^3 is the singular part of Γ in the 1-skeleton of Δ_w, Γ^2 is the rest of singular part of Γ. With help of some other result in [33], we can produce a Lagrangian fibration $\hat{F} : Y_\infty \to \partial\Delta_w$ such that $\hat{F}(C) = \Gamma$. Then we have

Theorem 3.1 *For $Y_\psi \subset P_{\Sigma^w}$, when $w = (w_m)_{m\in\Delta^0} \in \tau$ is generic and near the large radius limit of τ, start with Lagrangian fibration $\hat{F}$ the flow of V will produce a Lagrangian fibration $\hat{F}_\psi : Y_\psi \to \partial\Delta_w$. There are 4 types of fibres.*
(i). For $p \in \partial\Delta_w \backslash \Gamma$, $\hat{F}_\psi^{-1}(p)$ is a smooth Lagrangian 3-torus.
(ii). For $p \in \Gamma^1$, $\hat{F}_\psi^{-1}(p)$ is a type I singular fibre.
(iii). For $p \in \Gamma^2$, $\hat{F}_\psi^{-1}(p)$ is a type II singular fibre.
(iv). For $p \in \Gamma^3$, $\hat{F}_\psi^{-1}(p)$ is a type III singular fibre.

With these constructions of Lagrangian torus fibrations for generic quintics and their mirrors, in [31], we were able to prove the symplectic topological version of SYZ conjecture for quintic Calabi-Yau.

Theorem 3.2 *For generic quintic Calabi-Yau X near the large complex limit, and its mirror Calabi-Yau Y near the large radius limit, there exist corresponding Lagrangian torus fibrations*

$$\begin{array}{ccccccc} X_{s(b)} & \hookrightarrow & X & & Y_b & \hookrightarrow & Y \\ & & \downarrow & & & & \downarrow \\ & & \partial\Delta & & & & \partial\Delta_w \end{array}$$

with singular locus $\Gamma \subset \partial\Delta$ and $\Gamma' \subset \partial\Delta_w$, where $s : \partial\Delta_w \to \partial\Delta$ is a natural homeomorphism with $s(\Gamma') = \Gamma$. For $b \in \partial\Delta_w \backslash \Gamma'$, the corresponding fibres $X_{s(b)}$ and Y_b are naturally dual to each other.

4 Calabi-Yau hypersurfaces and complete intersections in toric variety

In [32], we generalize our work in [31] to the case of general Calabi-Yau hypersurfaces in toric varieties with respect to reflexive polyhedra, which is exactly the situation of the Batyrev dual polyhedron mirror construction. In a forthcoming paper [34], we will further generalize our construction to the case of general Calabi-Yau complete intersections in toric varieties, where mirror construction was proposed by Borisov. Compared to the quintic case, in general toric hypersurface case usually both the Kähler moduli and the complex moduli of a Calabi-Yau hypersurface are non-trivial. The construction of the Lagrangian torus fibration has to depend on both the Kähler form and the complex structure of the Calabi-Yau hypersurface. We also need the most general monomial-divisor map to carry out the discussion of symplectic topological SYZ mirror construction for general Calabi-Yau hypersurfaces or complete intersections in toric variety. In the case of complete intersection, The singular locus graph actually exhibit certain knotting phenomenon similar to the flag manifold case we will discuss in the last section. We will only state the result for hypersurfaces here.

Theorem 4.1 *For generic Calabi-Yau hypersurface X in the toric variety corresponding to a reflexive polyhedron Δ and its mirror Calabi-Yau hypersurface Y in the toric variety corresponding to the dual reflexive polyhedron $\Delta^\vee$ near their corresponding large complex limit and large radius limit, there exist corresponding Lagrangian torus fibrations*

$$\begin{array}{ccccccc} X_{\phi(b)} & \hookrightarrow & X & & Y_b & \hookrightarrow & Y \\ & & \downarrow & & & & \downarrow \\ & & \partial\Delta_v & & & & \partial\Delta_w \end{array}$$

with singular locus $\Gamma \subset \partial\Delta_v$ *and* $\Gamma' \subset \partial\Delta_w$*, where* $\phi : \partial\Delta_w \to \partial\Delta_v$ *is a natural homeomorphism,* $\phi(\Gamma') = \Gamma$*. For* $b \in \partial\Delta_w \backslash \Gamma'$*, the corresponding fibres* $X_{\phi(b)}$ *and* Y_b *are naturally dual to each other. For* $b \in \Gamma'$*, if* Y_b *is a singular fibre of type* I*,* II*,* III*, then* $X_{\phi(b)}$ *is a singular fibre of type* I*,* III*,* II*.*

The original SYZ mirror conjecture was rather sketchy in nature, with no mention of singular locus, singular fibres and duality of singular fibres, which is essential if one wants to use SYZ to construct mirror manifold. Our discussions on the construction of Lagrangian torus fibrations and symplectic topological SYZ of generic Calabi-Yau hypersurfaces and complete intersections in toric varieties explicitly produce the three types of generic singular fibres (type I, II, III as described in [31]) and exhibit how they are dual to each other under the mirror symmetry. It gives clear indications what should happen in general. In particular, it suggests that type II singular fibre with Euler number -1 should be dual to type III singular fibre with Euler number 1. This together with the knowledge of singular locus from our construction will enable us to give a more precise formulation of SYZ mirror conjecture. This precise formulation naturally suggests a way to construct mirror manifold from a generic Lagrangian torus fibration of a Calabi-Yau manifold in general. We will state the symplectic version of SYZ conjecture, where duality relation is more precise. In light of examples from Joyce [24], for the origional special Lagrangian version, singular locus should be a fattening of our graph, singular fibres should also change accordingly, and the identification of singular locus and duality of smooth fibres will be subject to certain quantum effects centered around the large complex (radius) limit, which is not yet fully understood.

Precise symplectic SYZ mirror conjecture

For any Calabi-Yau 3-fold X, with Calabi-Yau metric ω_g and holomorphic volume form Ω, there exists a Lagrangian fibration of X over S^3

$$\begin{array}{ccc} T^3 & \hookrightarrow & X \\ & & \downarrow \\ & & S^3 \end{array}$$

with a Lagrangian section and codimension two singular locus $\Gamma \subset S^3$, such that general fibres (over $S^3 \backslash \Gamma$) are 3-torus. For generic such fibration, Γ is a graph with only 3-valent vertices. Let $\Gamma = \Gamma^1 \cup \Gamma^2 \cup \Gamma^3$, where Γ^1 is the smooth part of Γ, $\Gamma^2 \cup \Gamma^3$ is the set of the vertices of Γ. For any leg $\gamma \subset \Gamma^1$, the monodromy of $H_1(X_b)$ of fibre under suitable basis is

$$T_\gamma = \begin{pmatrix} 1 & 1 & 0 \\ 0 & 1 & 0 \\ 0 & 0 & 1 \end{pmatrix}$$

Singular fibre along γ is of type I.

Consider a vertex $P \in \Gamma^2 \cup \Gamma^3$ with legs γ_1, γ_2, γ_3. Correspondingly, we have monodromy operators T_1, T_2, T_3.

For $P \in \Gamma^2$, under suitable basis we have

$$T_1 = \begin{pmatrix} 1 & 1 & 0 \\ 0 & 1 & 0 \\ 0 & 0 & 1 \end{pmatrix} \quad T_2 = \begin{pmatrix} 1 & 0 & -1 \\ 0 & 1 & 0 \\ 0 & 0 & 1 \end{pmatrix} \quad T_3 = \begin{pmatrix} 1 & -1 & 1 \\ 0 & 1 & 0 \\ 0 & 0 & 1 \end{pmatrix}$$

Singular fibre over P is of type II.

For $P \in \Gamma^3$, under suitable basis we have

$$T_1 = \begin{pmatrix} 1 & 0 & 0 \\ 1 & 1 & 0 \\ 0 & 0 & 1 \end{pmatrix} \quad T_2 = \begin{pmatrix} 1 & 0 & 0 \\ 0 & 1 & 0 \\ -1 & 0 & 1 \end{pmatrix} \quad T_3 = \begin{pmatrix} 1 & 0 & 0 \\ -1 & 1 & 0 \\ 1 & 0 & 1 \end{pmatrix}$$

Singular fibre over P is of type III.

The Lagrangian fibration for the mirror Calabi-Yau manifold Y has the same base S^3 and singular locus $\Gamma \subset S^3$. For $b \in S^3 \backslash \Gamma$, Y_b is the dual torus of $X_b \cong T^3$. In another word, the T^3-fibrations

$$\begin{array}{ccc} T^3 & \hookrightarrow & X \\ & & \downarrow \\ & & S^3 \backslash \Gamma \end{array} \qquad \begin{array}{ccc} T^3 & \hookrightarrow & Y \\ & & \downarrow \\ & & S^3 \backslash \Gamma \end{array}$$

are dual to each other. In particular the monodromy operator will be dual to each other.

For the fibration of Y, singular fibres over Γ^1 should be type I, singular fibres over Γ^2 should be type III, singular fibres over Γ^2 should be type II. Namely, dual singular fibre of a type I singular fibre is still type I. Type II and III singular fibres are dual to each other.

Our work in [29, 30, 31, 32] verified such symplectic SYZ duality for Calabi-Yau hypersurfaces in toric varieties. Our work in progress [34] will treat the case of complete intersections in toric varieties. [36] will further push it to the case of Calabi-Yau complete intersections in flag manifolds.

5 Conifold transition

In this section we will discuss the behavior of Lagrangian torus fibrations under conifold transitions of Calabi-Yau mainifolds. Let X be the Calabi-Yau manifold that is a smoothing of a Calabi-Yau conifold X_0 with p nodes. Let α be the number of relations among the vanishing cycles in X coming from deformation of the nodes in X_0. The conifold transition Y of X is a Calabi-Yau manifold that is a small resolution of X_0. The corresponding birational map $\pi : Y \to X_0$ contracts p $\mathbb{CP}^1$'s in Y to p nodes in X_0. The topology of X and Y are related as follows [10].

Proposition 5.1

$$h^{1,1}(X) = h^{1,1}(Y) - \alpha.$$

$$h^{2,1}(X) = h^{2,1}(Y) + p - \alpha.$$

We will first describe some local models to indicate the behavior of the Lagrangian fibration near the conifold transition point. Since all these local models have certain T^2 or S^1 symmetries, it is natural to use symplectic reduction technique to construct fibrations whose fibres are T^2 or S^1 invariant. Symplectic reduction techniques have been widely used in many fields of mathematics. Recently, Goldstein ([14, 15]) applied such techniques to construct certain local special Lagrangian fibrations. Gross in a preprint discussed similar construction. Recently Gross expanded his preprint into [20], where he also discussed further local examples of generalized special Lagrangian fibrations. For example, among other fibration examples, he briefly discussed the special Lagrangian $T^2 \times \mathbb{R}$ fibration of conifold transition (not as detailed as below) and a diagram similar to Figure 1 of this paper describing corresponding singular locus graph transition. Further discussion of the literature of symplectic reduction techniques can also be found in [20]. (Our gradient flow method also has such flavor.) We are using similar ideas here. Symplectic reduction technique is very powerful when the Calabi-Yau 3-fold has T^2 symmetry, while its usefulness will be much limited when the Calabi-Yau 3-fold has merely S^1 symmetry. This is probably related to the fact pointed out by Joyce [24] (and also indicated by our gradient flow construction [30]) that the special Lagrangian fibrations for such manifolds are not necessarily C^∞.

Example: ($T^2 \times \mathbf{R}$ fibrations) Let

$$X_\epsilon = \{z = (z_1, z_2, z_3, z_4) \in \mathbf{C}^4 | p(z) = z_1 z_2 - z_3 z_4 = \epsilon\}.$$

Then

$$Y = \{(z, [t_1, t_2]) \in \mathbf{C}^4 \times \mathbf{CP}^1 | t_1 z_1 = t_2 z_3, t_2 z_2 = t_1 z_4\}$$

is a small resolution of X_0. Let $\pi : Y \to X_0$. For $\epsilon > 0$

$$C_1^\epsilon = \{z = (z_1, z_2, z_3, z_4) \in X_\epsilon | z_2 = \bar{z}_1,\ z_4 = -\bar{z}_3\} \cong S^3$$

is the vanishing cycle.

$$C_2^\epsilon = \{z = (z_1, z_2, z_3, z_4) \in X_\epsilon | \mathrm{Re}(z_1) > 0,\ \mathrm{Im}(z_1) = 0,\ z_4 = \bar{z}_3\} \cong \mathbf{R}^3$$

is the transversal cycle such that $(C_1^\epsilon, C_2^\epsilon) = 1$. Let $t = t_2/t_1$, then $\pi^{-1}(C_2^0) \subset Y$ satisfies $x_1 \geq 0$ and

$$x_1 = t z_3,\ t x_2 = \bar{z}_3 = z_4.$$

For fixed t, $(x_1, x_2, z_3, \bar{z}_3)$ is uniquely determined up to a positive multiple. Fix $x_1 = 1$, we have $(1, 1/(|t|^2), 1/t, 1/\bar{t})$. Therefore $\partial(\pi^{-1}(C_2^0)) = \pi^{-1}(0) \cong \mathbf{CP}^1$.

On the other hand, Let

$$S = \{(z, [1,0]) \in \mathbf{C}^4 \times \mathbf{CP}^1 | z_1 = z_4 = 0\} \cong \mathbf{C}^2$$

be a hypersurface in Y such that $(S, \pi^{-1}(0)) = 1$. Let $p(z) = f + ih$ and ∇f denote the gradient vector field for f. Consider

$$V = \frac{\nabla f}{|\nabla f|^2}.$$

The flow of V will determine a family of symplectic "blow up" $\phi_\epsilon : X_\epsilon \to X_0$. Under the flow, $\pi(S)$ is deformed to symplectic surface $S^\epsilon = \phi_\epsilon^{-1}(\pi(S)) \subset X_\epsilon$

with boundary $\partial(S^\epsilon) = C_1^\epsilon$.

Consider a T^2 action on Y. For $\xi = (\xi_1, \xi_2) \in T^2$, define the action

$$\xi \circ t = \xi_1 \xi_2^{-1} t;$$
$$\xi \circ z_1 = \xi_1 z_1, \quad \xi \circ z_3 = \xi_2 z_3;$$
$$\xi \circ z_2 = \xi_1^{-1} z_2, \quad \xi \circ z_4 = \xi_2^{-1} z_4.$$

The vector fields corresponding to the two generators are

$$v_1 = 2\mathrm{Im}\left(z_1 \frac{\partial}{\partial z_1} - z_2 \frac{\partial}{\partial z_2} + t\frac{\partial}{\partial t}\right),$$
$$v_2 = 2\mathrm{Im}\left(z_3 \frac{\partial}{\partial z_3} - z_4 \frac{\partial}{\partial z_4} - t\frac{\partial}{\partial t}\right).$$

The set where the stablizer of T^2 is non-trivial is a union of 5 irreducible curves

$$\Delta = \bigcup_{i=0}^{4} \Delta_i \subset Y,$$

where

$$\Delta_0 = \{(z,t) \in Y | z = 0\} \cong \mathbb{CP}^1,$$
$$\Delta_i = \{(z,t) \in Y | z_j = 0, \text{ for } j \neq i\} \cong \mathbb{C}, \quad \text{for } 1 \leq i \leq 4.$$

Clearly,

$$\Delta_2 \cup \Delta_3 = \{(z,t) \in Y | t = 0, z_2 z_3 = 0\} \subset \{(z,t) \in Y | t = 0\} \cong \mathbb{C}^2,$$
$$\Delta_1 \cup \Delta_4 = \{(z,t) \in Y | t = \infty, z_1 z_4 = 0\} \subset \{(z,t) \in Y | t = \infty\} \cong \mathbb{C}^2.$$

On Y, one can consider the metric

$$\omega_g = \frac{i}{2}\left(\sum_{i=1}^{4} dz_i \wedge d\bar{z}_i + \delta \frac{dt \wedge d\bar{t}}{(1+|t|^2)^2}\right).$$

We can compute

$$\imath(v_1)\omega_g = \frac{1}{2} d\left(|z_1|^2 - |z_2|^2 - \frac{\delta}{1+|t|^2}\right),$$

$$\iota(v_2)\omega_g = \frac{1}{2}d\left(|z_3|^2 - |z_4|^2 + \frac{\delta}{1+|t|^2}\right).$$

Y/T^2 can be naturally identified with $\mathbb{R}^4$ via

$$\left(|z_1|^2 - |z_2|^2 - \frac{\delta}{1+|t|^2}, |z_3|^2 - |z_4|^2 + \frac{\delta}{1+|t|^2}, z_1z_2 + z_3z_4\right).$$

$\rho : Y \to \mathbf{R}^3$ defined by

$$\rho(z) = \left(|z_1|^2 - |z_2|^2 - \frac{\delta}{1+|t|^2}, |z_3|^2 - |z_4|^2 + \frac{\delta}{1+|t|^2}, \mathrm{Re}(z_1z_2 + z_3z_4)\right)$$

is a $T^2 \times \mathbf{R}$ fibration. The singular set of the fibration is exactly Δ. The singular locus $\rho(\Delta) \subset \{\rho_3 = 0\} \cong \mathbf{R}^2$ is a graph as follows.

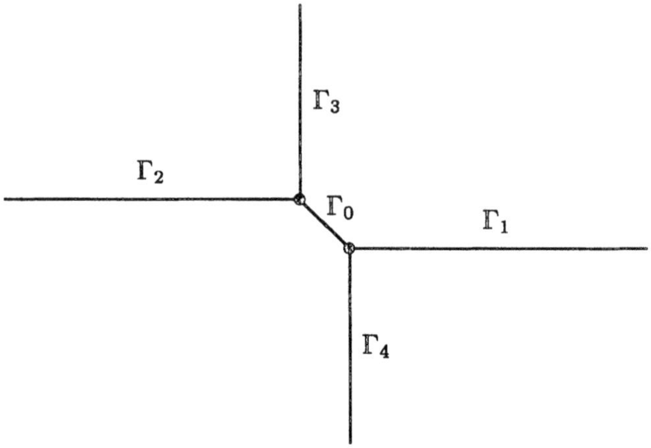

Figure 12: Singular locus of the $T^2 \times \mathbf{R}$ fibration of Y

When δ shrinks to 0, Y is blown down to X_0 and the singular locus of the fibration changes to

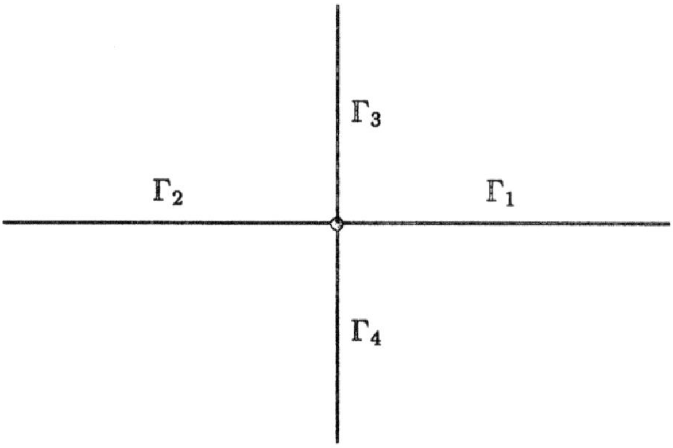

Figure 13: Singular locus of the $T^2 \times \mathbf{R}$ fibration of X_0

The T^2 action on Y can naturally be carried over to a T^2 action on X_ϵ, defined as

$$\xi \circ z_1 = \xi_1 z_1, \quad \xi \circ z_3 = \xi_2 z_3;$$
$$\xi \circ z_2 = \xi_1^{-1} z_2, \quad \xi \circ z_4 = \xi_2^{-1} z_4.$$

X_ϵ / T^2 can be naturally identified with $\mathbb{R}^4$ via $(|z_1|^2 - |z_2|^2, |z_3|^2 - |z_4|^2, z_1 z_2 + z_3 z_4)$.

$$\rho_\epsilon : X_\epsilon \to \mathbb{R}^3$$

defined by

$$\rho(z) = (|z_1|^2 - |z_2|^2, |z_3|^2 - |z_4|^2, \mathrm{Re}(z_1 z_2 + z_3 z_4))$$

is a $T^2 \times \mathbf{R}$ fibration. The singular set is now union of 2 curves

$$\Delta^\epsilon = \Delta_{12}^\epsilon \cup \Delta_{34}^\epsilon$$

where

$$\Delta_{12}^\epsilon = \{z \in X_\epsilon | z_3 = z_4 = 0, \ z_1 z_2 = \epsilon\} \cong \mathbb{C}^*,$$
$$\Delta_{34}^\epsilon = \{z \in X_\epsilon | z_1 = z_2 = 0, \ z_3 z_4 = -\epsilon\} \cong \mathbb{C}^*.$$

The singular locus $\rho(\Delta^\epsilon)$ consists of two lines not in a plane.

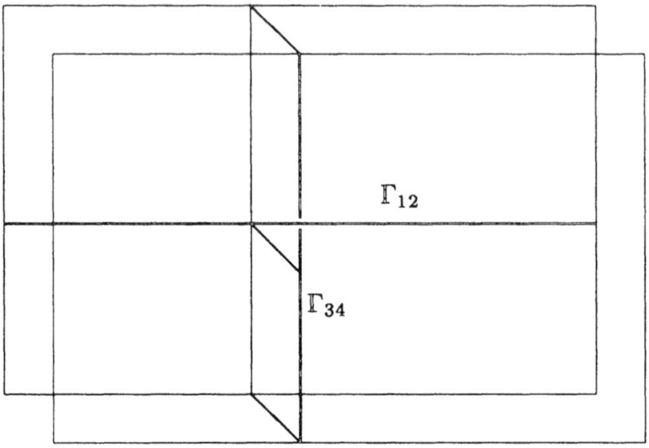

Figure 14: Singular locus of the $T^2 \times \mathbb{R}$ fibration of X_ϵ

□

It is not hard to verify that this fibration is actually the so-called generalized special Lagrangian fibration.

Example: ($S^1 \times \mathbb{R}^2$ fibrations)
Consider the same situation as in the previous example

$$X_\epsilon = \{z = (z_1, z_2, z_3, z_4) \in \mathbb{C}^4 | p(z) = z_1 z_2 - z_3 z_4 = \epsilon\}$$

$$Y = \{(z, [t_1, t_2]) \in \mathbb{C}^4 \times \mathbb{CP}^1 | t_1 z_1 = t_2 z_3, t_2 z_2 = t_1 z_4\}$$

is a small resolution of X_0.

Consider an S^1 action on Y. For $\xi \in S^1$, define the action

$$\xi \circ t = \xi^{-1} t;$$

$$\xi \circ z_1 = z_1, \quad \xi \circ z_3 = \xi z_3;$$

$$\xi \circ z_2 = z_2, \quad \xi \circ z_4 = \xi^{-1} z_4.$$

The vector field corresponding to the generator is

$$v = 2\mathbb{I}\mathrm{m}\left(z_3\frac{\partial}{\partial z_3} - z_4\frac{\partial}{\partial z_4} - t\frac{\partial}{\partial t}\right).$$

The set where the stablizer of S^1 is non-trivial is a union of 2 irreducible curves

$$\Delta = \Delta_1 \cup \Delta_2 \subset Y,$$

where

$$\Delta_i = \{(z,t) \in Y | z_j = 0, \text{ for } j \neq i\} \cong \mathbf{C}, \quad \text{for } 1 \leq i \leq 2.$$

On Y, one can consider the metric

$$\omega_g = \frac{i}{2}\left(\sum_{i=1}^{4} dz_i \wedge d\bar{z}_i + \delta\frac{dt \wedge d\bar{t}}{(1+|t|^2)^2}\right).$$

We have

$$\imath(v)\omega_g = \frac{1}{2}d\left(|z_3|^2 - |z_4|^2 + \frac{\delta}{1+|t|^2}\right).$$

Y/S^1 can be naturally identified with $\mathbf{R}^5$ via

$$\left(z_1, z_2, |z_3|^2 - |z_4|^2 + \frac{\delta}{1+|t|^2}\right).$$

$\rho : Y \to \mathbf{R}^3$ defined by

$$\rho(z) = \left(\mathrm{Re}(z_1), \mathrm{Re}(z_2), |z_3|^2 - |z_4|^2 + \frac{\delta}{1+|t|^2}\right)$$

is an $S^1 \times \mathbf{R}^2$ fibration. The singular set of the fibration is exactly Δ. The singular locus $\Gamma = \rho(\Delta) = \Gamma_1 \cup \Gamma_2 \subset \mathbf{R}^2$ consists of two lines not in a plane ($\Gamma_1 = \rho(\Delta_1) = \{\rho_2 = \rho_3 = 0\}$ and $\Gamma_2 = \rho(\Delta_2) = \{\rho_1 = \rho_3 = 0\}$).

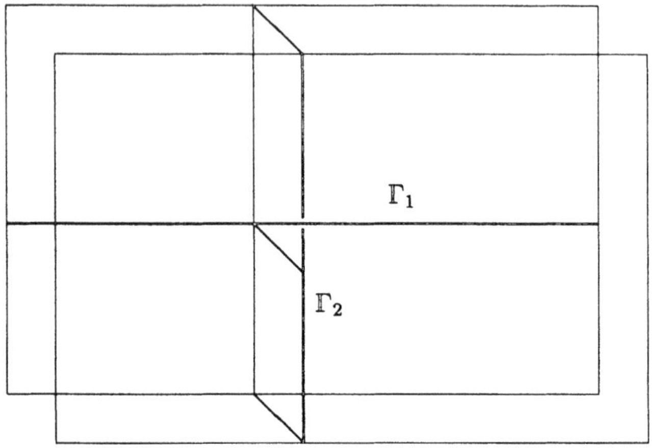

Figure 15: Singular locus of the $S^1 \times \mathbb{R}^2$ fibration of Y

When δ shrinks to 0, Y is blown down to X_0 and the singular locus of the fibration changes to

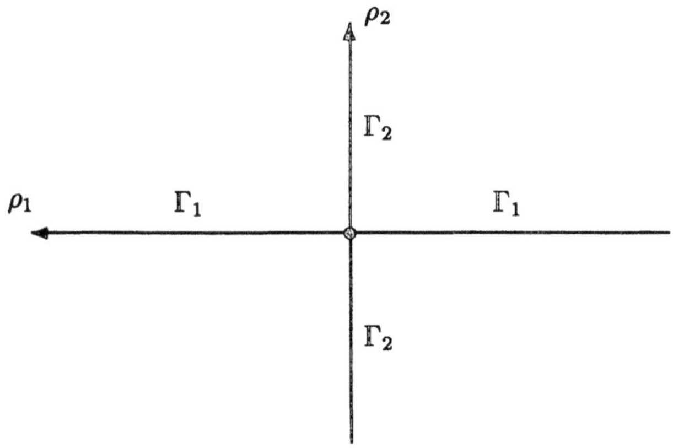

Figure 16: Singular locus of the $S^1 \times \mathbb{R}^2$ fibration of X_0

The S^1 action on Y can naturally be carried over to an S^1 action on X_ϵ, defined as

$$\xi \circ z_1 = z_1, \quad \xi \circ z_3 = \xi z_3;$$

$$\xi \circ z_2 = z_2, \quad \xi \circ z_4 = \xi^{-1} z_4.$$

X_ϵ / S^1 can be naturally identified with $\mathbf{R}^5$ via $(z_1, z_2, |z_3|^2 - |z_4|^2)$.

$$\rho_\epsilon : X_\epsilon \to \mathbf{R}^3$$

defined by

$$\rho(z) = (\mathrm{Re}(z_1), \mathrm{Re}(z_2), |z_3|^2 - |z_4|^2)$$

is an $S^1 \times \mathbf{R}^2$ fibration. The singular set is now

$$\Delta^\epsilon = \{z \in X_\epsilon | z_3 = z_4 = 0, \ z_1 z_2 = \epsilon\} \cong \mathbf{C}^*.$$

Assume that ϵ is positive, then the singular locus of ρ is

$$\tilde{\Gamma} = \rho(\Delta^\epsilon) = \{0 \leq \rho_1 \rho_2 \leq \epsilon, \rho_3 = 0, \} \subset \{\rho_3 = 0\}.$$

It contains two parts

$$\tilde{\Gamma}_+ = \{\rho_1 \geq 0, \rho_2 \geq 0, \rho_1 \rho_2 \leq \epsilon, \rho_3 = 0\},$$

$$\tilde{\Gamma}_- = \{\rho_1 \leq 0, \rho_2 \leq 0, \rho_1 \rho_2 \leq \epsilon, \rho_3 = 0\}.$$

The vanishing circle of Δ^ϵ is

$$S^1_\epsilon = \{(\sqrt{\epsilon} e^{i\theta}, \sqrt{\epsilon} e^{i\theta}, 0, 0)\}.$$

Under the fibration map

$$\Gamma_0 = \rho(S^1_\epsilon) = \{(t, t, 0) | -1 \leq t \leq 1\}.$$

The following is a picture of $\tilde{\Gamma}$.

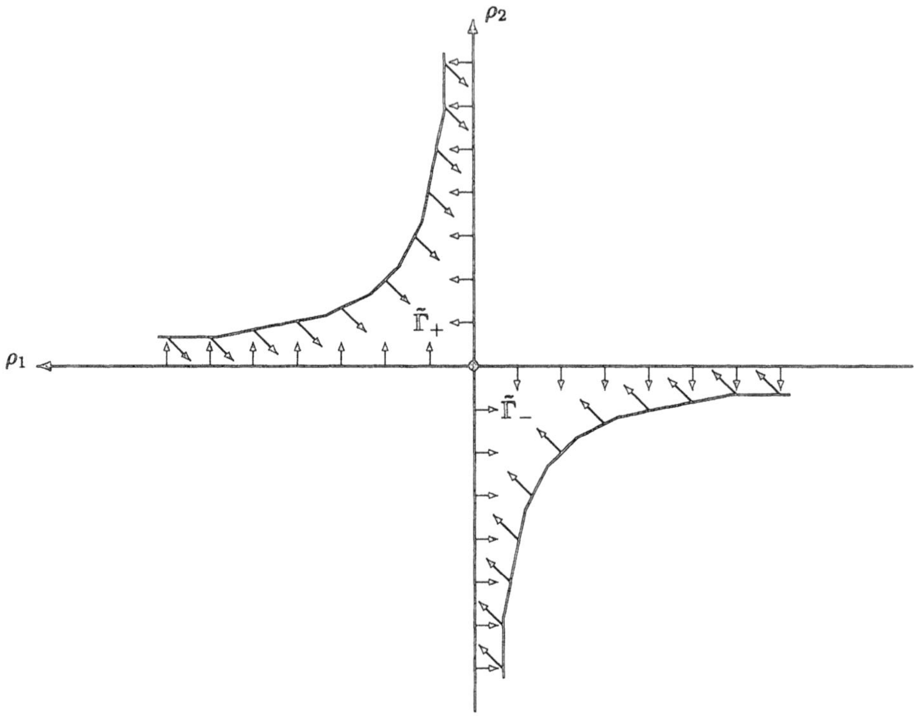

Figure 17: Singular locus of the $S^1 \times \mathbb{R}^2$ fibration of X_ϵ

Deforming the fibration along the arrows, we can get an $S^1 \times \mathbb{R}^2$ fibration with singular locus

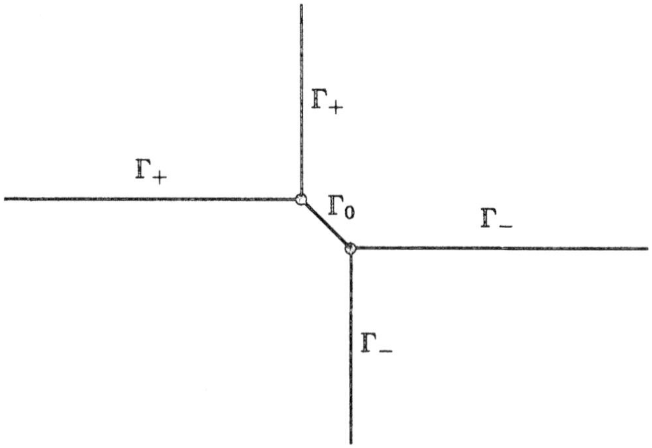

Figure 18: Modified singular locus of the $S^1 \times \mathbb{R}^2$ fibration of X_ϵ

The singular set of the fibration in a fibre over Γ_+ and Γ_- is a line. The singular set of the fibration in a fibre over Γ_0 is a union of two parallel lines. The singular set of the fibration in a fibre over the 3-valent points of Γ is a graph identical to the graph Γ.

Unfortunately, these $S^1 \times \mathbf{R}^2$ fibrations are not even Lagrangian fibrations. In [35], we will discuss this local conifold transition for more general metrics and modify these explicit non-Lagrangian fibrations into explicit Lagrangian ones with the same topological structure.

From these $S^1 \times \mathbf{R}^2$ fibrations, it is also easy to see why we required the components of the singular locus to be located in parallel planes as mentioned in the introduction. Over the singular locus, the invariant S^1 is vanishing, which implies that the components of the singular locus will be located in some level planes of $\rho_3 = |z_3|^2 - |z_4|^2$.

6 Calabi-Yau complete intersection in flag manifold

In [5, 6], Batyrev *et. al.* were able to generalize the mirror construction to the case of Calabi-Yau complete intersections in partial flag manifolds, which had much to do with conifolds transitions. It is interesting to see how we can generalize our construction of Lagrangian fibration to this case and what new phenomenon might ocurr.

With the above local models in mind, using the gradient flow approach together with some symplectic patching technique, we will be able to construct Lagrangian torus fibration for Calabi-Yau complete intersections in flag manifolds and their mirror, therefore proving the symplectic SYZ mirror symmetry for such classes of Calabi-Yau. In this section we will illustrate the key ideas through an example (the case of Grassmannian $\mathbb{Gr}(2,4)$). As one might notice, $\mathbf{Gr}(2,4)$ is actually a conic hypersurface in $\mathbb{CP}^5$. The Calabi-Yau hypersurface in $\mathbf{Gr}(2,4)$ is actually a complete intersection in toric variety. In the Lagrangian torus fibration point of view, this is a rare coincidence. As we will show in [34, 36], Lagrangian torus fibration structure for Calabi-Yau complete intersections in toric varieties and Calabi-Yau hypersurfaces in flag manifolds have very distinct characters. In fact, for this example, the large complex limits from the flag manifold point of view and from the toric point of view are entirely different in conventional sense. Consequently the Lagrangian torus fibrations constructed from the two points of view are also entirely different. In any case, our method here will not really rely on the fact that $\mathbb{Gr}(2,4)$ is a conic hypersurface in $\mathbf{CP}^5$, and key ideas for the general case already show up here. We will provide details of the general case in the forthcoming paper [36].

Example: Consider the case of $\mathbf{Gr}(2,4)$, $z \in \Lambda^2\mathbf{H_C}$ can be expressed as

$$z = \sum_{1\leq i\leq j\leq 4} z_{ij} e_i \wedge e_j.$$

$$z \wedge z = (z_{12}z_{34} - z_{13}z_{24} + z_{14}z_{23}) e_1 \wedge e_2 \wedge e_3 \wedge e_4.$$

$\mathbb{Gr}(2,4) \subset \mathbf{P}(\Lambda^2\mathbf{H_C}) \cong \mathbf{CP}^5$ is a quadric defined by

$$z_{12}z_{34} - z_{13}z_{24} + z_{14}z_{23} = 0.$$

Let $P(2,4) \subset \mathbf{CP}^5$ be the 4-dimensional Gorenstein toric Fano variety defined by the quadratic equation

$$z_{13}z_{24} - z_{14}z_{23} = 0.$$

Then $P(2,4)$ is a degeneration of $\mathrm{Gr}(2,4)$ in terms of the family

$$z_{13}z_{24} - z_{14}z_{23} - \epsilon z_{12}z_{34} = 0.$$

Consider the fan description of $\mathbb{CP}^5$ with homogeneous coordinates $[z_{jk}]_{1\leq j<k\leq 4}$. Then M is naturally

$$M = \left\{ z^I = \prod_{1\leq j<k\leq 4} z_{jk}^{i_{jk}} \;\middle|\; I = (i_{jk})_{1\leq j<k\leq 4} \in \mathbf{Z}^6,\ |I| = \sum_{1\leq j<k\leq 4} i_{jk} = 0 \right\}.$$

Let

$$\Delta = \{ z^I \in M \mid I + I_0 \geq 0 \}$$

where $I_0 = (1,1,\cdots,1)$. Let

$$s_{m_1} = z_{13}z_{24}, \quad s_{m_2} = z_{14}z_{23}.$$

Then $P(2,4) \cong \{s = s_{m_1} - s_{m_2} = 0\}$. $m_1 - m_2 = (i_{jk})_{1\leq j<k\leq 4} \in \mathbf{Z}^6$, where $i_{12} = i_{34} = 0$, $i_{13} = i_{24} = 1$ and $i_{14} = i_{23} = -1$. Naturally

$$N = \{ w = (w_{jk})_{1\leq j<k\leq 4} \in \mathbf{Z}^6 \} / \mathbf{Z}\cdot w_0,$$

where $w_0 = (1,1,\cdots,1)$.

$$\Delta^\vee = \{ [w] \in N \mid w \geq 0,\ |w| \leq 1 \}.$$
$$M_s = M/(m_1 - m_2).$$
$$N_s = (m_1 - m_2)^\perp = \{ [w] \in N \mid w_{13} + w_{24} = w_{14} + w_{23} \}.$$

1-cones in Σ_s come from 1-cones and 2-cones in Σ that intersect $(m_1 - m_2)^\perp$. Assume $w = (w_{12}, w_{34}, w_{13}, w_{24}, w_{14}, w_{23})$. The generating vectors are

$$w_1 = (1,0,0,0,0,0),\ w_3 = (0,0,1,0,1,0),\ w_5 = (0,0,0,1,1,0),$$

$$w_2 = (0,1,0,0,0,0),\ w_4 = (0,0,1,0,0,1),\ w_6 = (0,0,0,1,0,1).$$

These are the vertices of $\Delta_s^\vee$.

Up to symmetries, there are 3 distinguished 1-simplices $\overline{w_1w_2}$, $\overline{w_1w_3}$, $\overline{w_3w_4}$. There are no additional integral points in any one of them. Up to symmetries, there are 3 distinguished 2-faces $\overline{w_1w_2w_3}$, $\overline{w_1w_3w_4}$, $\overline{w_3w_4w_6w_5}$. There are no additional integral points in any one of them. Up to symmetries, there are 2 distinguished 3-faces $\overline{w_1w_2w_3w_4}$, $\overline{w_1w_3w_4w_6w_5}$. There are no additional integral points in any one of them.

The fan Σ_s corresponding to $\Delta_s \subset M_s$ consists of cones over subfaces of $\Delta_s^\vee$. Recall that

$$P(2,4) \cong P_{\Sigma_s} = \bigcup_{\sigma\in\Sigma_s} T_\sigma.$$

The singular set of P_{Σ_s} is a line $l = \{z_{11} = z_{13} = z_{14} = z_{23} = 0\} \cong \mathbf{CP}^1$ which is an union of T_σ corresponding to non-simplicial cones σ over $\overline{w_1w_3w_4w_6w_5}$, $\overline{w_3w_4w_6w_5}$, $\overline{w_2w_3w_4w_6w_5}$. Singularity along l is 3-dimensional conifold singularity characterized by the cone over $\overline{w_3w_4w_6w_5}$.

Assume $I = (i_{12}, i_{34}, i_{13}, i_{24}, i_{14}, i_{23})$. The vertices of Δ_s are

$$I_1 = (1,-3,1,1,0,0),\ I_3 = (1,1,1,-3,0,0),\ I_5 = (1,1,0,0,1,-3),$$
$$I_2 = (-3,1,1,1,0,0),\ I_4 = (1,1,0,0,-3,1),\ I_6 = (1,1,-3,1,0,0).$$

Under the moment map, l is mapped to $\overline{I_1I_2}$. Small resolution of l will result in a modified polyhedron $\hat{\Delta}_s$ of Δ_s, where edge $\overline{I_1I_2}$ is replaced by a 2-dimensional parallelgram.

A Calabi-Yau hypersurface X in $P(2,4)$ is the intersection of a quartic and $P(2,4)$ in $\mathbf{CP}^5$. Using the gradient flow approach developed in [29, 30, 31, 32], when X is near the large complex limit, we can construct a Lagrangian torus fibration of X over $\hat{\Delta}_s$ with codimension 2 singular locus Γ. We will describe Δ_s and Γ in detail.

Up to symmetries, there are 3 distinguished 1-simplices $\overline{I_1I_2}$, $\overline{I_1I_3}$, $\overline{I_3I_4}$. On $\overline{I_1I_2}$, there are 3 other integral points

$$(0,-2,1,1,0,0),\ (-1,-1,1,1,0,0),\ (-2,0,1,1,0,0).$$

On $\overline{I_1 I_3}$, there are 3 other integral points

$$(1,-2,1,0,0,0),\ (1,-1,1,-1,0,0),\ (1,0,1,-2,0,0).$$

On $\overline{I_3 I_4}$, there are 3 other integral points

$$(1,1,1,-2,-1,0),\ (1,1,1,-1,-2,0),\ (1,1,0,-1,-2,1).$$

Up to symmetries, there are 3 distinguished 2-faces $\overline{I_1 I_2 I_3}$, $\overline{I_1 I_3 I_4}$, $\overline{I_3 I_4 I_6 I_5}$. On $\overline{I_1 I_2 I_3}$ (or $\overline{I_1 I_3 I_4}$), there are 15 integral points. Graph Γ in $\overline{I_1 I_2 I_3}$ is determined by certain simplicial decomposition of $\overline{I_1 I_2 I_3}$. For the standard sympli-cial decomposition, we have the following graph Γ

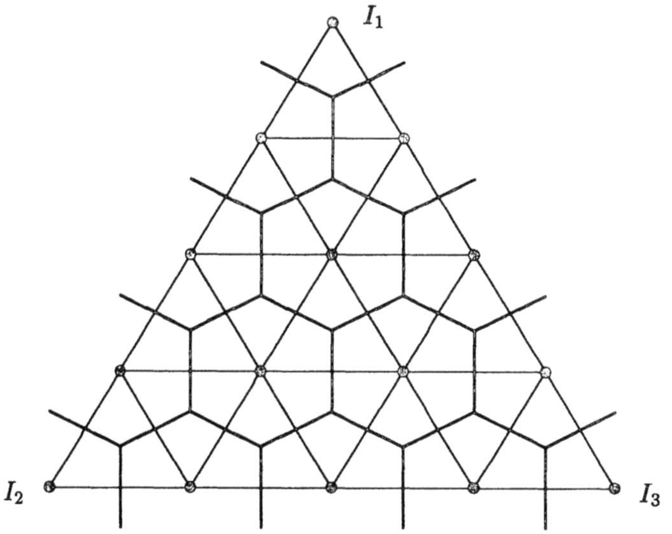

Figure 19: Possible singular locus in $\overline{I_1 I_2 I_3}$

If the simplicial decomposition is changed, Γ will change accordingly

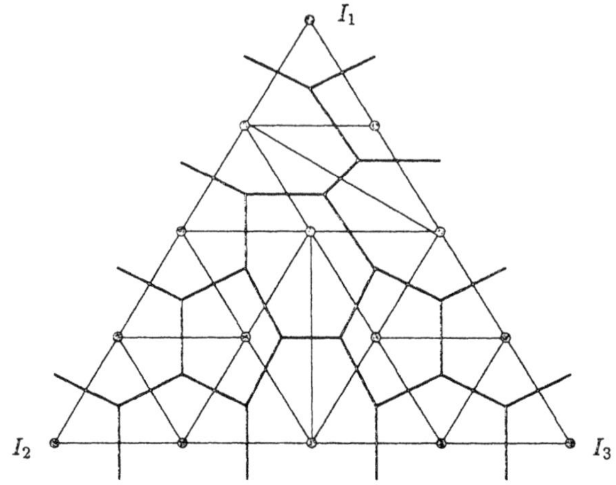

Figure 20: Possible singular locus in $\overline{I_1 I_2 I_3}$

On $\overline{I_3 I_4 I_6 I_5}$, there are 25 integral points. For the standard symplicial decomposition, we have the following graph Γ

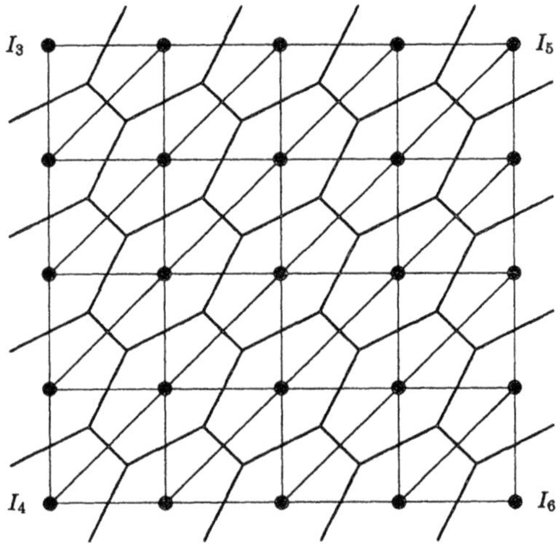

Figure 21: Possible singular locus in $\overline{I_3 I_4 I_6 I_5}$

Of course, suitable changes of simplicial decomposition will result in changes of corresponding Γ.

Singular set l of P_{Σ_s} is a $\mathbb{CP}^1$ corresponding to $\overline{I_1 I_2}$. Around $\overline{I_1 I_2}$, before the small resolution, Γ looks like

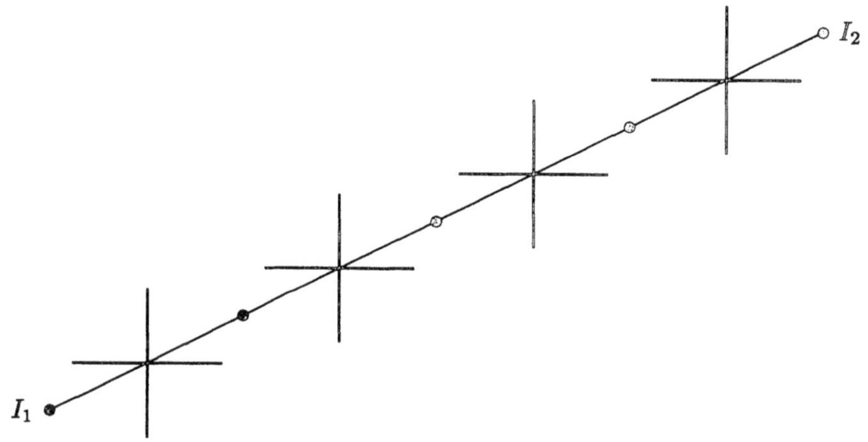

Figure 22: Singular locus around $\overline{I_1 I_2}$

Γ in $\overline{I_1 I_2 I_3}$, $\overline{I_1 I_2 I_4}$, $\overline{I_1 I_2 I_5}$, $\overline{I_1 I_2 I_6}$ intersect at 4 points in $\overline{I_1 I_2}$. When one resolve singular set l by small resolution, then Γ around $\overline{I_1 I_2}$ changes to

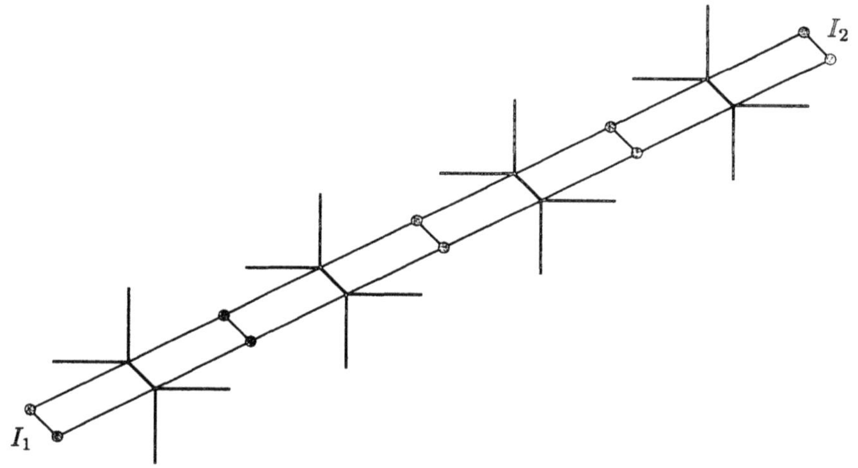

Figure 23: Singular locus around $\overline{I_1 I_2}$ after small resolution

Instead, if one smooths out the singular set l of $P_{\Sigma_s} \subset \mathbf{CP}^5$ by deformation in $\mathbf{CP}^5$, then Γ around $\overline{I_1 I_2}$ changes to

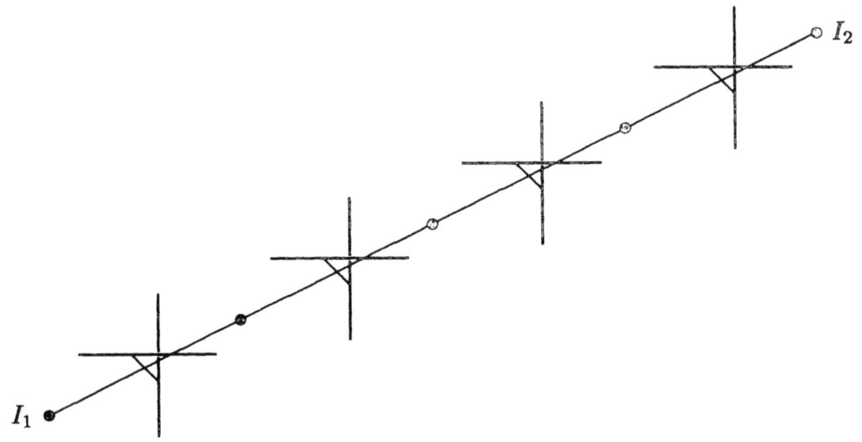

Figure 24: Singular locus around $\overline{I_1 I_2}$ after smoothing

Γ in $\overline{I_1 I_2 I_3} \cup \overline{I_1 I_2 I_4} \cup \overline{I_1 I_2 I_5} \cup \overline{I_1 I_2 I_6}$, which used to intersect at 4 points in $\overline{I_1 I_2}$, now form two pieces $\overline{I_1 I_6 I_2 I_3}$ and $\overline{I_1 I_5 I_2 I_4}$ (like in the following picture) that avoid each other and intertwine near $\overline{I_1 I_2}$ as in the above picture.

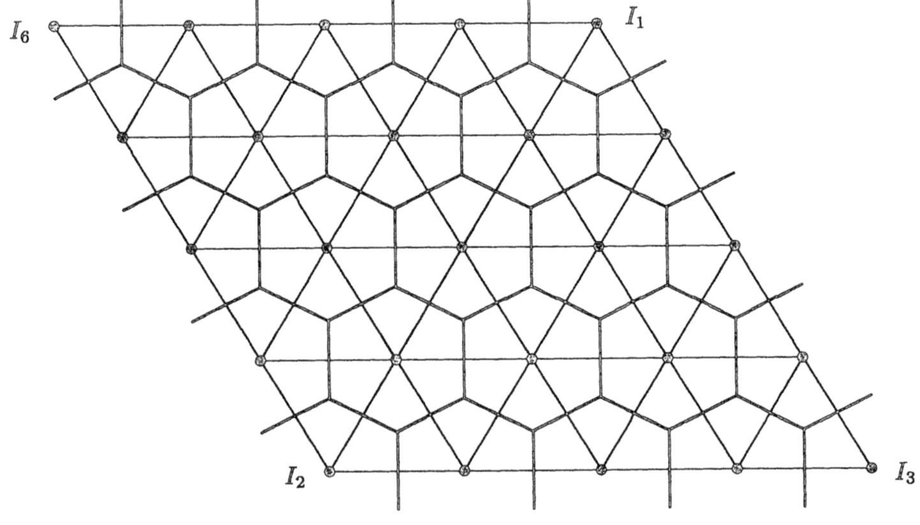

Figure 25: Singular locus in $\overline{I_1 I_6 I_2 I_3}$

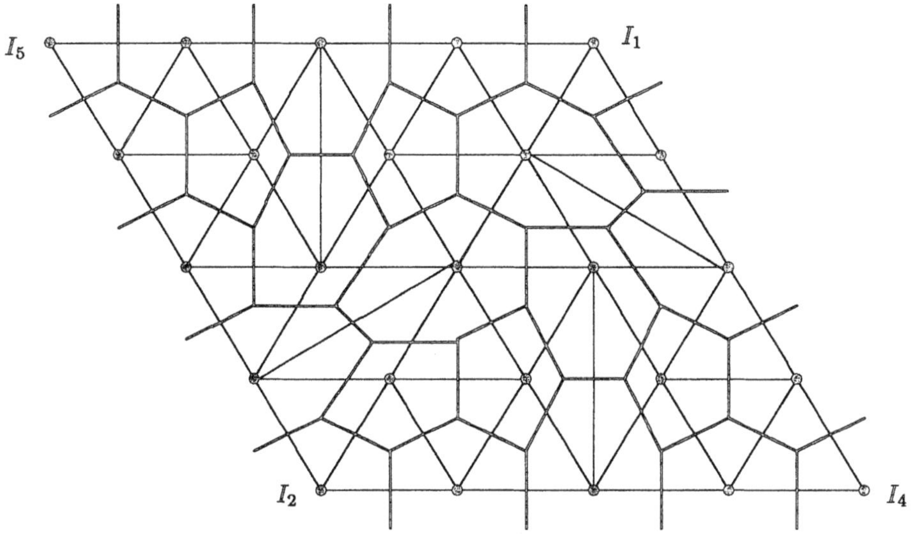

Figure 26: Singular locus in $\overline{I_1 I_5 I_2 I_4}$

One can easily observe similarity of these two graphs with the graph for $\overline{I_3 I_4 I_6 I_5}$. Totally, Γ consists of 3 square pieces Γ_{3465}, Γ_{1623}, Γ_{1524} and 8 triangle pieces Γ_{134}, Γ_{145}, Γ_{156}, Γ_{163}, Γ_{234}, Γ_{245}, Γ_{256}, Γ_{263}. This will give us the singular locus of the Lagrangian torus fibrations of Calabi-Yau hypersurfaces in the Grassmannian $\mathbf{Gr(2,4)}$. Notice that unlike the hypersurfaces in toric varieties, where the singular locus graph is completely unknotted, in the Grassmannian (and more generally the flag manifold) case, the singular locus graphs show slight knotting phenomenon. In fact such knotting phenomenon already show up in the case of complete intersections in toric varieties (in somewhat different fashion), which we will discuss in our forthcoming paper [34]. It is a rather interesting question to see if more complicated knotting is allowed for the singular locus graph of Lagrangian torus fibrations of more general Calabi-Yau manifolds.

In summary, the Calabi-Yau hypersurface in $\hat{P}(2,4)$ is transformed to Calabi-Yau hypersurface in the Grassmannian $\mathbf{Gr(2,4)}$ through conifold transition at 4 ordinary double points in the singular Calabi-Yau hypersurface in

$P(2,4)$. The corresponding singular locus graph locally undergoes the following basic graph change as mentioned in the introduction.

Figure 27: Graph degeneration corresponding to conifold transition

Singular locus graph and the singular fibres are modelled according to the local $T^2 \times \mathbb{R}$ fibration examples discussed in the previous section.

We can similarly discuss the Lagrangian torus fibrations in the mirror side. The mirror of the Calabi-Yau hypersurface in $\hat{P}(2,4)$ is transformed to the mirror of Calabi-Yau hypersurface in the Grassmannian $\mathbb{G}\mathrm{r}(2,4)$ also through conifold transition at 4 ordinary double points (of course in reverse direction). Singular locus graph and the singular fibres are modelled according to the local $S^1 \times \mathbb{R}^2$ fibration examples discussed in the previous section. Symplectic SYZ duality can also be proved in this situation. We hope to provide more details of these arguments in [35, 36].

Acknowledgement: I would like to thank Prof. S.T. Yau for constant encouragement, Prof. Kefeng Liu for suggesting the work of [5, 6] to me during the conference in Montreal, Prof. Yong-Geun Oh for pointing out the work of [27] to me during the KIAS conference and Prof. A. Libgober for helpful discussion. I would also like to thank Prof. Yong-Geun Oh for inviting me to the well organized conference on mirror symmetry held in KIAS. Major part of this work was done while I was in Columbia University and finished up in University of Illinois at Chicago. I am very grateful to both universities for excellent research environment. Thanks also go to Qin Jing for stimulating discussions and suggestions.

References

[1] Aspinwall, P.S., Greene, B.R., Morrison, D.R., "The Monomial-Divisor Mirror Map", *Inter. Math. Res. Notices* 12 (1993), 319-337.

[2] Aspinwall, P.S., Greene, B.R., Morrison, D.R., "Topological change in mirror symmetry", *Nucl. Phys.* B446 (1994), 414-480.

[3] Batyrev, V.V., "Dual polyhedra and mirror symmetry for Calabi-Yau hypersurfaces in toric varieties", *J. Alg. Geom.* 3 (1994), 493-535.

[4] Batyrev, V.V. and Borisov, L.A., "Dual cones and mirror symmetry for generalized Calabi-Yau manifolds", in *Essays on Mirror Symmetry II*, (edited by B. Greene, S.-T. Yau.) AMS and Int. Press (1997) 71-86.

[5] Batyrev, V.V., Ciocan-Fontanine, I, Kim, B, and van Straten, D., "Conifold transitions and mirror symmetry for Calabi-Yau complete intersections in Grassmannians", alg-geom 9710022.

[6] Batyrev, V.V., Ciocan-Fontanine, I, Kim, B, and van Straten, D., "Mirror symmetry and toric degenerations of Partial Flag Manifolds", alg-geom 9803108.

[7] Borisov, L.A., "Towards the Mirror Symmetry for Calabi-Yau Complete intersections in Gorenstein Toric Fano Varieties", alg-geom 9310001.

[8] Candelas, P., de la Ossa, X.C., Green, P., Parkes, L., "A Pair of Calabi-Yau Manifolds as an Exactly Soluble Superconformal Theory", in *Essays on Mirror Symmetry*, edited by S.-T. Yau.

[9] Chiang, T.-M., Greene, B.R., Gross, M, and Kanter, Y., "Black hole condensition and the web of Calabi-Yau manifolds", hep-th 9511204.

[10] Clemens, H., "Double solids", *Adv. Math.* 47 (1983), 107-230.

[11] Morrison, D., "Through the Looking Glass", alg-geom 9705028.

[12] Forsberg, M., Passare, M. and Tslkh, A., "Laurent determinants and arrangement of hyperplane amoebras, preprint, 1998.

[13] Gelfand, I. M., Kapranov, M. M. and Zelevinsky, A. V., *Discriminants, Resultants and Multidimensional Determinants*, Birkhauser Inc., Boston, MA, 1994.

[14] Goldstein, E., "Calibrated Fibrations", dg-ga 9911093.

[15] Goldstein, E., "Calibrated Fibrations on Complete Manifolds via Torus Action", dg-ga 0002097.

[16] Greene, B.R., Morrison, D.R., Strominger, A., "Black hole condensation and the unification of string vacua", *Nucl. Phys.* B451 (1995), 109, hep-th 9504145.

[17] Gross, M., "Special Lagrangian Fibration I: Topology", alg-geom 9710006.

[18] Gross, M., "Special Lagrangian Fibration II: Geometry", alg-geom 9809072.

[19] Gross, M., "Topological Mirror Symmetry", alg-geom 9909015.

[20] Gross, M., "Examples of Special Lagrangian Fibrations", alg-geom 0012002.

[21] Gross, M. and Wilson, P.M.H., "Mirror Symmetry via 3-torus for a class of Calabi-Yau Threefolds", to appear in *Math. Ann.*

[22] Harvey, R. and Lawson, H.B., "Calibrated Geometries", *Acta Math.* 148 (1982), 47-157.

[23] Hitchin, N., "The Moduli Space of Special Lagrangian Submanifolds", dg-ga 9711002

[24] Joyce, D., "Singularities of special Lagrangian fibrations and the SYZ Conjecture", dg-ga 0011179.

[25] Leung, N., Vafa, C., "Branes and Toric Geometry,", *Nucl. Phys.* B484 (1997), 562-582, hep-th 9711013.

[26] Lynker, M., Schimmrigk, R., "Conifold transitions and Mirror symmetries,", *Nucl. Phys.* B484 (1997), 562-582, hep-th 9511058.

[27] Mikhalkin, G., "Real algebraic curves, the moment map and amoebas", *Ann. of Math.* 151 (2000), 309-326.

[28] Reid, M., "The moduli space of 3-folds with $K=0$ may nevertheless be irreducible", Math. Ann. **278** (1987), 329–334.

[29] Ruan, W.-D., "Lagrangian torus fibration of quintic Calabi-Yau hypersurfaces I: Fermat type quintic case", dg-ga 9904012.

[30] Ruan, W.-D., "Lagrangian torus fibration of quintic Calabi-Yau hypersurfaces II: Technical results on gradient flow construction", (preprint).

[31] Ruan, W.-D., "Lagrangian torus fibration of quintic Calabi-Yau hypersurfaces III: Symplectic topological SYZ mirror construction for general quintics", dg-ga 9909126.

[32] Ruan, W.-D., "Lagrangian torus fibration and mirror symmetry of Calabi-Yau hypersurfaces in toric variety", dg-ga 0007028.

[33] Ruan, W.-D., "Newton polygon, string diagram and toric variety", dg-ga 0011012.

[34] Ruan, W.-D., "Lagrangian torus fibration and mirror symmetry of Calabi-Yau complete intersections in toric variety", (preliminary version).

[35] Ruan, W.-D., "Lagrangian fibration and mirror symmetry of conifold transition", (preliminary version).

[36] Ruan, W.-D., "Lagrangian torus fibration and mirror symmetry of Calabi-Yau complete intersections in flag manifold", (In preparation).

[37] Ruan, W.-D., "Monodromy near the large complex limit and horizontal sections of the Lagrangian torus fibration of Calabi-Yau hypersurface in toric variety", (In progress).

[38] Strominger, A.,Yau, S.-T. and Zaslow, E, "Mirror Symmetry is T-duality", *Nuclear Physics* B 479 (1996),243-259.

[39] Yau, S.-T., Essays on Mirror Manifolds, International Press, Hong Kong, 1992.

[40] Zharkov, I., "Torus Fibrations of Calabi-Yau Hypersurfaces in Toric Varieties and Mirror Symmetry", alg-geom 9806091

MORE ABOUT VANISHING CYCLES AND MUTATION

PAUL SEIDEL

1. Introduction

Not surprisingly, this is a sequel to "Vanishing cycles and mutation" [VC]. Notions from that paper will be used freely later on, but for starters a short recapitulation seems appropriate. Let $D \subset \mathbb{C}$ be the closed unit disc. An exact Morse fibration $\pi : E^{2n+2} \to D$ can be described roughly as a family of $2n$-dimensional exact symplectic manifolds parametrized by D, with singularities modelled on Morse-type critical points of holomorphic functions. Such objects are suitable for developing Picard-Lefschetz theory in a symplectic context. Suppose that (E,π) comes equipped with a relative Maslov map $\delta_{E/D}$, which is basically a trivialization of $(\Lambda^{n+1}_{\mathbb{C}} TE)^{\otimes 2}$. Then one can associate to it an algebraic invariant, the derived directed Fukaya category $D^b Lag^{\to}(\tilde{\Gamma})$, which encodes the symplectic geometry of the vanishing cycles of (E,π). It is constructed in three steps:

(i) Fix a base point $z_0 \in \partial D$, for definiteness $z_0 = -i$; the fibre $M = E_{z_0}$ is an exact symplectic manifold. Then make an admissible choice of paths [VC, Figure 2] from z_0 to the critical values of π. This determines a distinguished basis of vanishing cycles in M, which is an ordered family $\Gamma = (L_1, \dots, L_m)$ of embedded exact Lagrangian spheres $L_i \subset M$ (there is one more bit of information, so-called framings of the L_i, which we ignore for the moment). We will sometimes call such families Lagrangian configurations in M. Any other distinguished basis, arising from a different choice of paths, can be obtained from Γ through a sequence of Hurwitz moves.

(ii) $\delta_{E/D}$ induces a Maslov map δ_M on M, thus allowing one to speak of graded Lagrangian submanifolds in M. By choosing gradings in an arbitrary way, one lifts Γ to a graded Lagrangian configuration $\tilde{\Gamma} = (\tilde{L}_1, \dots, \tilde{L}_m)$. Next one introduces the directed Fukaya category $\mathcal{A} = Lag^{\to}(\tilde{\Gamma})$, which is an A_∞-category with objects $\tilde{L}_1, \dots, \tilde{L}_m$. The spaces of morphisms in $\mathcal{A}$ are essentially given by the natural cochain complexes underlying the Floer cohomology groups $HF^*(\widetilde{L}_i, \widetilde{L}_k)$, but the ordering of the objects also plays a role in the definition, which is what the word "directed" refers to.

(iii) The derived category $D^b(\mathcal{A})$ of $\mathcal{A}$ is no longer an A_∞-category but rather a category in the ordinary sense; in fact it is a triangulated category, linear over $\mathbb{Z}/2$ since we are using $\mathbb{Z}/2$-coefficients. Its objects are twisted complexes in $\mathcal{A}$, which are A_∞-analogues of chain complexes, and morphisms are chain homotopy classes of maps in a suitable sense. Unlike $\mathcal{A}$ itself, $D^b(\mathcal{A})$ is independent of the choice of Γ (and of $\widetilde{\Gamma}$), hence is an invariant of (E, π) and $\delta_{E/D}$. The reason for this is a relation between Hurwitz moves of Γ and the purely algebraic notion of mutations of $\mathcal{A}$.

All this is explained in more detail, although still without proofs, in [VC]. The present paper adds computations and some more far-reaching conjectures. It is divided into four more or less independent parts. The first part, formed by Sections 2–3, introduces some classes of examples. Section 2 concerns the toy model case when the fibre dimension is zero. Section 3 explains how to construct exact Morse fibrations from algebro-geometric objects, namely Lefschetz pencils or isolated hypersurface singularities. Also included is a discussion of the mirror manifold of $\mathbb{C}\mathrm{P}^2$.

The subject of the second part, spanning Sections 4–6, is Hochschild cohomology. The basic observation is that like $D^b(\mathcal{A})$, the Hochschild cohomology $HH^*(\mathcal{A}, \mathcal{A})$ is an invariant of (E, π) and $\delta_{E/D}$. One would expect this to have a more geometric meaning. Following a suggestion of Donaldson we propose a conjecture in this direction, which relates Hochschild cohomology to the global monodromy of the fibration. As a preliminary to this, Section 4 enlarges the topological quantum field theory framework from [VC, Section 3] to include the Floer cohomology of symplectic automorphisms.

The third part, Sections 7–8, concerns exact Morse fibrations with a real involution. In that situation, and under certain additional assumptions, the category $D^b(\mathcal{A})$ should have a description in terms of Morse theory on the real part. We state a precise conjecture, which is a theorem at least in the lowest nontrivial dimension $n = 1$.

Finally, Section 9 is a reflection on the aspect of dimensional reduction inherent in the theory. By this we mean that $\mathcal{A}$, which is defined in terms of pseudo-holomorphic curves in the $2n$-dimensional manifold M, produces invariants of the $(2n+2)$-dimensional object (E, π). In an attempt to push this further, we give a conjectural algorithm for determining the Floer cohomology of certain Lagrangian spheres in E through computations in $D^b(\mathcal{A})$. This, if correct, might be the first actual application of homological algebra methods to symplectic geometry.

Acknowledgments. This paper owes much to Donaldson; several of the ideas presented here arose during conversations with him. Givental and Kontsevich provided helpful explanations concerning mirror symmetry.

2. Branched covers

(2A) In [VC] it was tacitly assumed that all symplectic manifolds were of dimension > 0 (and that assumption will be resumed as soon as this section is finished). The zero-dimensional case is rather trivial, but it still requires some minor adaptations, which we now explain. Let's start with a basic dictionary of zero-dimensional symplectic geometry:

compact symplectic manifold	finite set M
Lagrangian sphere	subset $L \subset M$, $\lvert L\rvert = 2$
Floer cohomology	$HF(L_1, L_2) = (\mathbb{Z}/2)^{L_1 \cap L_2}$
Dehn twist	$\tau_L : M \to M$ swapping the two points of L, leaving the rest fixed
exact Morse fibration	generic branched cover $\pi : E \to S$

Here a **generic branched cover** is a proper map $\pi : E \to S$ between oriented surfaces with boundary, with $\pi^{-1}(\partial S) = \partial E$, having finitely many critical points, which are locally modelled on $\pi(z) = z^2$; and no two critical points may lie in the same fibre.

The TQFT framework deserves a brief mention, if only for its simplicity. Let S be a compact oriented surface with boundary, $\Sigma \subset \partial S$ a finite set of marked points, and (E, π) a generic branched cover over $S^* = S \setminus \Sigma$. Suppose moreover that we have a submanifold $Q \subset \pi^{-1}(\partial S^*)$ such that $\pi|Q : Q \to \partial S^*$ is a double cover; this corresponds to what in [VC] was called a Lagrangian boundary condition for (E, π). The behaviour of Q near a point $\zeta \in \Sigma$ can be described by a pair $(L_{\zeta,+}, L_{\zeta,-})$ of Lagrangian zero-spheres in some fibre of E close to ζ. A continuous section $u : S^* \to E$ of π which satisfies $u(\partial S^*) \subset Q$ singles out a point of $L_{\zeta,+} \cap L_{\zeta,-}$ for each ζ, and hence determines an element of $\bigotimes_\zeta HF(L_{\zeta,+}, L_{\zeta,-})$. Taking the sum over all sections defines an invariant

$$\Phi_{rel}(E, \pi, Q) \in \bigotimes_{\zeta \in \Sigma} HF(L_{\zeta,+}, L_{\zeta,-}). \tag{2.1}$$

For trivial reasons, these invariants satisfy the gluing law formulated in [VC, Section 3]. One can use them to find the correct zero-dimensional analogues of various familiar maps in Floer theory. For instance, the pair-of-pants product $HF(L_2, L_3) \otimes HF(L_1, L_2) \to HF(L_1, L_3)$ is derived from an invariant (2.1) with $S = D$, $\Sigma = \{\textit{three points}\}$ and $E = S^* \times M$, together with a boundary condition given by L_1, L_2, L_3 over the three connected components of ∂S^*. Working this out explicitly shows that the product takes a basis element $\bar{x} \otimes \bar{y} \in HF(L_2, L_3) \otimes HF(L_1, L_2)$, $x \in L_2 \cap L_3$ and $y \in L_1 \cap L_2$, to $\bar{x} \in HF(L_1, L_3)$ if $x = y$, and to zero otherwise. There is the same long exact sequence in Floer cohomology as in [VC, Theorem 3.3] and again, the more fundamental fact underlying it is a quasi-isomorphism

$$\mathrm{Cone}(a : HF(L, L_2) \otimes HF(L_1, L) \to HF(L_1, L_2)) \xrightarrow{(h,b)} HF(L_1, \tau_L(L_2)).$$

While a and b are as in higher dimensions, h is now the product $HF(L, L_2) \otimes HF(L_1, L) \cong HF(L, \tau_L(L_2)) \otimes HF(L_1, L) \to HF(L_1, \tau_L(L_2))$. Of course, both the result and proof are now elementary!

The next topic are gradings. We summarize their zero-dimensional theory in another dictionary:

Maslov map on a symplectic manifold	$\delta_M : M \to S^1$
graded Lagrangian sphere	$L \subset M$ with a map $\tilde{L} : L \to \mathbb{R}$ such that $\exp(2\pi i \tilde{L}) = \delta_M \vert L$
graded Floer cohomology	in $HF^*(\tilde{L}_1, \tilde{L}_2)$, the degree of $\bar{x}$ is $\tilde{L}_2(x) - \tilde{L}_1(x)$
shifting the grading	$\tilde{L}[\sigma] = \tilde{L} - \sigma$, $\sigma \in \mathbb{Z}$
graded symplectic automorphism	$\phi : M \to M$ with a map $\tilde{\phi} : M \to \mathbb{R}$ such that $\exp(2\pi i \phi(x)) = \delta_M(x)/\delta_M(\phi^{-1}(x))$
relative Maslov map	on $\pi : E \to S$, a trivialization $\delta_{E/S}$ of $TE^{\otimes 2} \otimes (\pi^* TS)^{\otimes -2}$

The action of a graded symplectic automorphism on graded Lagrangian submanifolds is defined by $(\tilde{\phi}\tilde{L})(x) = \tilde{L}(\phi^{-1}(x)) + \tilde{\phi}(x)$. A grading $\tilde{L}$ determines a preferred grading $\tilde{\tau}_{\tilde{L}}$ of the associated Dehn twist, characterized by

$$\tilde{\tau}_{\tilde{L}}(\tilde{L}) = \tilde{L}[1] \quad \text{and} \quad \tilde{\tau}_{\tilde{L}}(x) = 0 \text{ for } x \notin L;$$

compare [VC, Equation (4)] for the first property. Graded Lagrangian configurations $\tilde{\Gamma} = (\tilde{L}_1, \dots, \tilde{L}_m)$ are defined as one would expect. Their Hurwitz moves are

- $\tilde{\Gamma} \rightsquigarrow (\tilde{L}_1[\sigma_1], \dots, \tilde{L}_m[\sigma_m])$ for $\sigma_1, \dots, \sigma_m \in \mathbb{Z}$;

- $\widetilde{\Gamma} \rightsquigarrow c\widetilde{\Gamma} = (\tilde{\tau}_{\widetilde{L}_1}(\widetilde{L}_2), \dots, \tilde{\tau}_{\widetilde{L}_1}(\widetilde{L}_m), \widetilde{L}_1)$;
- $\widetilde{\Gamma} \rightsquigarrow r\widetilde{\Gamma} = (\widetilde{L}_1, \dots, \widetilde{L}_{m-2}, \tilde{\tau}_{\widetilde{L}_{m-1}}(\widetilde{L}_m), \widetilde{L}_{m-1})$.

Let (E, π) be a generic branched cover over D, with a relative Maslov map $\delta_{E/D}$. A drastically simplified version of the usual Picard-Lefschetz argument produces from it a graded (with respect to the induced Maslov map δ_M) Lagrangian configuration in $M = E_{z_0}$, unique up to Hurwitz equivalence.

Remark 2.1. *One can observe here two minor differences with respect to the positive-dimensional situation. Firstly, $\tilde{\tau}_{\widetilde{L}}$ depends on $\widetilde{L}$; secondly, the Hurwitz equivalence class of a graded configuration contains information that is lost if one forgets the gradings. Both phenomena go back to the obvious fact that there is a $\mathbb{Z}^2$ ambiguity in the choice of grading for a Lagrangian zero-sphere, as opposed to $\mathbb{Z}$ in higher dimensions.*

The directed Fukaya category $\mathcal{A} = Lag^{\rightarrow}(\widetilde{\Gamma})$ attached to a zero-dimensional graded configuration has objects $\widetilde{L}_i$. The morphisms $hom_{\mathcal{A}}(\widetilde{L}_i, \widetilde{L}_k)$ are $HF^*(\widetilde{L}_i, \widetilde{L}_k)$ for $i < k$, $\mathbb{Z}/2 \cdot \mathrm{id}_{\widetilde{L}_i}$ for $i = k$, and zero for $i > k$. The nontrivial compositions $HF^*(\widetilde{L}_j, \widetilde{L}_k) \otimes HF^*(\widetilde{L}_i, \widetilde{L}_j) \to HF^*(\widetilde{L}_i, \widetilde{L}_k)$, $i < j < k$, are given by the pair-of-pants product. Even though $\mathcal{A}$ is an ordinary $\mathbb{Z}/2$-linear and $\mathbb{Z}$-graded category, we prefer to regard it as a directed A_∞-category where all composition maps $\mu^d_{\mathcal{A}}$ of order $d \neq 2$ vanish. The details of the proof are slightly different than in the higher-dimensional case, but the main result of [VC] remains true, which is that Hurwitz moves of $\widetilde{\Gamma}$ give rise to mutations of $\mathcal{A}$; and one arrives at the familiar conclusion that $D^b Lag^{\rightarrow}(\widetilde{\Gamma})$ is an invariant of (E, π), $\delta_{E/D}$.

Before going on to concrete examples, we need to recall some algebraic terminology. Let T be a quiver (an oriented graph). The path category $\mathcal{P}T$ has one object for each vertex, and the space of morphisms between two objects is the $\mathbb{Z}/2$-vector space freely generated by all paths in T going from one vertex to the other. One can further divide the morphism spaces by some two-sided ideal, in order to kill certain designated morphisms. This is usually referred to as describing a category by a "quiver with relations". If the quiver is directed, which means that its vertices are numbered $1, \dots, m$ such that there are no paths from the i-th one to the k-th one unless $i < k$, $\mathcal{P}T$ and its quotients can again be seen as special cases of directed A_∞-categories, with morphisms only in degree zero and vanishing composition maps of order $\neq 2$. Note that in this situation our notion of derived category (defined through twisted complexes) agrees with the classical one (defined by considering the categories as algebras, and taking chain complexes of right modules over them).

(2B) Take the following quiver with m vertices, with the arrows oriented in an arbitrary way:

$$\bullet \longrightarrow \bullet \longleftarrow \bullet \longrightarrow \cdots \longleftarrow \bullet \tag{2.2}$$

and let $\mathcal{A}$ be its path category. It is a classical result that $D^b(\mathcal{A})$ is independent of the orientation of the arrows; what we will do is to explain this geometrically. First of all, one can find a configuration of Lagrangian zero-spheres $\Gamma = (L_1, \dots, L_m)$ in $M = \{1, \dots, m+1\}$ which, when lifted to a graded configuration in the trivial way ($\delta_M = 0$ and $\widetilde{L}_i = 0$), has $\mathcal{A}$ as its directed Fukaya category. The way to do that is best explained by an example:

$$\begin{array}{ccccccccccccc} \bullet & \longrightarrow & \bullet & \longrightarrow & \bullet & \longleftarrow & \bullet & \longleftarrow & \bullet & \longleftarrow & \bullet & \longrightarrow & \bullet \\ & \{1,2\} & & \{1,3\} & & \{1,4\} & & \{5,4\} & & \{6,4\} & & \{7,4\} & \{7,8\} \end{array}$$

$$\Gamma = (\{1,2\},\{1,3\},\{7,4\},\{6,4\},\{5,4\},\{1,4\},\{7,8\}). \tag{2.3}$$

Next one constructs an $(m+1)$-fold generic branched cover $\pi : E \to D$ such that Γ is one of its distinguished bases of vanishing cycles. Inspection of (2.3) and an Euler characteristic computation show that the total space E is connected and in fact a disc. It is a consequence of the classification of branched covers that there is just one such (E, π) up to isomorphism, and moreover it admits precisely one homotopy class of relative Maslov maps. Therefore all categories $D^b(\mathcal{A})$ arise from the same geometric situation, hence must be equivalent on grounds of the general theory.

(2C) Let $\pi : E \to D$ be a double cover branched along $2g+1 \geq 5$ points, so that the total space is a genus $g \geq 2$ surface with one boundary component. Write ι for the nontrivial covering transformation, which is a hyperelliptic involution of E. There is a unique homotopy class of relative Maslov maps which are invariant under ι; take $\delta_{E/D}$ in that class. All vanishing cycles of (E, π) are the same, and if we choose their gradings in the most obvious way, the resulting directed Fukaya category $\mathcal{A}_g$ can be described by the quiver with relations

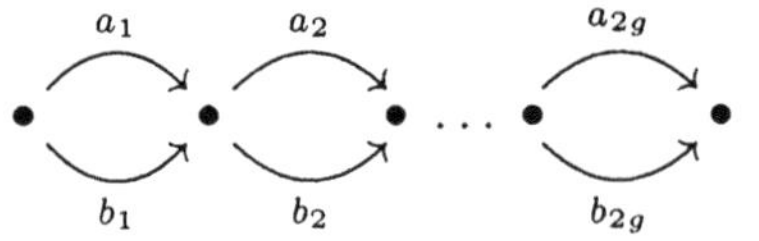

$$\begin{cases} b_{i+1} a_i = 0, \\ a_{i+1} b_i = 0. \end{cases}$$

We will now make some further remarks concerning $\mathcal{A}_g$ and its derived category. These may seem a bit unmotivated, but they will be at least partially put into context later on; see Remark 9.5. The following definition is taken from [23]:

Definition 2.2. *Let $\mathcal{C}$ be a triangulated category, linear over $\mathbb{Z}/2$ and such that the spaces $\mathrm{Hom}^*_{\mathcal{C}}(-,-)$ are finite-dimensional. $C \in \mathrm{Ob}\,\mathcal{C}$ is called spherical of* *dimension* $(n+1)$ *if* $\mathrm{Hom}^*_{\mathcal{C}}(C,C) \cong H^*(S^{n+1};\mathbb{Z}/2)$ *and the composition*

$$\mathrm{Hom}^*_{\mathcal{C}}(X,C) \otimes \mathrm{Hom}^{n+1-*}_{\mathcal{C}}(C,X) \to \mathrm{Hom}^{n+1}_{\mathcal{C}}(C,C) \cong \mathbb{Z}/2$$

is a nondegenerate pairing for any $X \in \mathrm{Ob}\,\mathcal{C}$.

Under some additional assumptions on $\mathcal{C}$, which are satisfied for derived categories of directed A_∞-categories, one can associate to any spherical object C an exact self-equivalence of $\mathcal{C}$, the twist functor T_C, which is well-defined up to isomorphism. If C_1, C_2 are spherical with $\mathrm{Hom}^*_{\mathcal{C}}(C_1,C_2) = 0$ then their twist functors commute; and if $\mathrm{Hom}^*_{\mathcal{C}}(C_1,C_2)$ is one-dimensional, one gets a braid relation $T_{C_1}T_{C_2}T_{C_1} \cong T_{C_2}T_{C_1}T_{C_2}$.

Denote the objects of $\mathcal{A}_g$ by $(X^1,\dots,X^{2g+1})$. The twisted complexes $C_i = (X^i[1] \oplus X^{i+1}, \left(\begin{smallmatrix} 0 & 0 \\ a_i+b_i & 0 \end{smallmatrix}\right))$, $1 \le i \le 2g$, are spherical objects of dimension one in $D^b(\mathcal{A}_g)$, and moreover

$$\dim \mathrm{Hom}^*_{D^b(\mathcal{A}_g)}(C_i,C_k) = \begin{cases} 0 & |i-k| \ge 2, \\ 1 & |i-k| = 1. \end{cases}$$

This implies that $T_{C_1},\dots,T_{C_{2g}}$ generate an action (in a weak sense) of the braid group B_{2g+1} on $D^b(\mathcal{A}_g)$. Now let $D^{per}(\mathcal{A}_g)$ be the category defined like $D^b(\mathcal{A}_g)$ but using $\mathbb{Z}/2$-graded twisted complexes. This is triangulated, and its two-fold shift functor [2] is isomorphic to the identity; we call it the periodic derived category of $\mathcal{A}_g$. One can adapt the theory of spherical objects and twist functors to such categories. In our case, this means that there are $C_i \in D^{per}(\mathcal{A}_g)$ defined as before, with the same properties. In addition there is now another spherical object

$$C_0 = (X^1 \oplus X^2[1] \oplus X^3 \oplus X^4[1], \begin{pmatrix} 0 & 0 & 0 & 0 \\ a_1 & 0 & 0 & 0 \\ 0 & b_2 & 0 & 0 \\ b_3b_2b_1 & 0 & a_3 & 0 \end{pmatrix})$$

The dimension of $\mathrm{Hom}^*_{D^{per}(\mathcal{A}_g)}(C_0,C_i)$ is one if $i=4$ and zero for all other $i>0$. The diagram representing this situation,

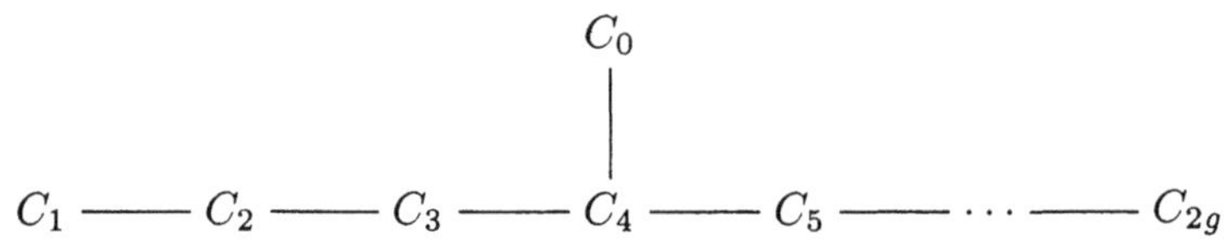

has the same structure as the configuration of curves in Wajnryb's presentation of $R_{g,1}$, the mapping class group of a genus g surface with one marked point [24]; and one can ask

Question 2.3. *Let $\mathcal{C} \subset D^{per}(\mathcal{A}_g)$ be the full triangulated subcategory generated by the C_i. Do $T_{C_0}, \dots, T_{C_{2g}}$ generate an action of $R_{g,1}$ on $\mathcal{C}$?*

Additional references. The classical paper about the quiver (2.2) is by Bernstein, Gelfand, and Ponomarev [7]. The category $D^b(\mathcal{A}_g)$ is closely related to those studied in [16], even though the quiver presentation looks different. However, since that paper does not consider periodic derived categories, it misses out on the additional object C_0.

3. Examples from algebraic geometry

(3A) Let X be a smooth projective variety, $\xi \to X$ an ample line bundle, and σ_0, σ_1 two holomorphic sections of ξ which generate a Lefschetz pencil of hypersurfaces $Y_z = \{x \in X \mid \sigma_0(x)/\sigma_1(x) = z\}$, $z \in \mathbb{CP}^1 = \mathbb{C} \cup \{\infty\}$. Suppose, in addition to the Lefschetz condition, that Y_∞ is smooth. What we want to look at is, in principle, the holomorphic Morse function

$$\sigma_0/\sigma_1 : X \setminus Y_\infty \longrightarrow \mathbb{C}.$$

To obtain additional symplectic data, one chooses a metric on ξ, with corresponding connection A, such that $\omega_X = (i/2\pi)F_A$ is a Kähler form. By a standard construction, A and σ_1 give rise to a one-form θ_X defined on $X \setminus Y_\infty$, such that $d\theta_X = \omega_X$. This still doesn't fit the definition of an exact Morse fibration, making some further modifications necessary.

Let $Z = Y_0 \cap Y_\infty$ be the base locus of the pencil, $G = \{(z, x) \in \mathbb{CP}^1 \times X \mid \sigma_0(x)/\sigma_1(x) = z\}$ its graph, and $p : G \to \mathbb{CP}^1$ the projection. (G, p) is a Lefschetz fibration with fibres Y_z, and contains a trivial subfibration $\mathbb{CP}^1 \times Z \subset G$. Let $\Omega_G \in \Omega^2(G)$ be the pullback of ω_X; it is closed, nondegenerate on the vertical tangent spaces $\ker(Dp) \subset TG$, and hence defines a symplectic connection away from the critical points of p. Moreover, the restriction of this connection to $\mathbb{CP}^1 \times Z \subset G$ is trivial. Similarly, pulling back θ_X gives a one-form Θ_G defined on the complement of $(\{\infty\} \times Y_\infty) \cup (\mathbb{CP}^1 \times Z) \subset G$, with $d\Theta_G = \Omega_G$.

After possibly rescaling σ_0, we may assume that Y_z is smooth for $|z| \geq 1$. Let $U \subset Y_0$ be a small open neighbourhood of Z. Using the symplectic parallel transport in radial directions one constructs an embedding Ξ,

$$\begin{array}{ccc} D \times U & \xrightarrow{\ \Xi\ } & p^{-1}(D) \subset G \\ & \searrow \quad \swarrow_{p} & \\ & D & \end{array}$$

with $\Xi \,|\, D \times Z$ and $\Xi \,|\, \{0\} \times U$ the obvious inclusions, which has the property that $\Xi^* \Omega_G - \omega_X | U$ vanishes on each diameter $[-z; z] \times U$, $z \in S^1$. Let $h \in C_c^\infty(U, [0; 1])$ be a function which is $\equiv 1$ on some smaller neighbourhood $U' \subset U$ of Z. Take the map $H : p^{-1}(D) \to p^{-1}(D)$ which is the identity outside $\mathrm{im}(\Xi)$ and satisfies $H\,\Xi(z, x) = \Xi((1 - h(x))z, x)$.

Set $E = p^{-1}(D) \setminus \Xi(D \times U'')$ for an even smaller open neighbourhood $U'' \subset U'$ of Z, which we choose such that $\partial U''$ is smooth. Together with $\pi = p|E$, $\Omega = (H^* \Omega_G | E) \in \Omega^2(E)$, $\Theta = (H^* \Theta_G | E) \in \Omega^1(E)$, and the given complex structure near the critical points of π, this forms an exact Morse fibration over D. Indeed, the required trivialization near $\partial_h E$ is given by Ξ itself; and the fact that H contracts along the same radial directions which are used to define Ξ ensures that Ω equals Ω_G on $\ker(D\pi) \subset TE$, hence is nondegenerate there.

Suppose now that the canonical bundle K_X satisfies

$$K_X^{-2} \cong \xi^a \text{ holomorphically, for some } a \in \mathbb{Z}. \tag{3.1}$$

Then σ_1^a gives a trivialization of K_X^{-2} over $X \setminus Y_\infty$, determined up to multiplication with a constant. Since the projection $G \to X$ identifies E with a compact subset of $X \setminus Y_\infty$, and Ω agrees with the pullback of ω_X except in a small neighbourhood of $\partial_h E$, one gets, at least up to homotopy, a preferred relative Maslov map $\delta_{E/D}$. The category $D^b Lag^{\to}(\widetilde{\Gamma})$ associated to (E, π) and $\delta_{E/D}$ actually depends only on X and on the Lefschetz pencil, and not on the other choices made during the construction of E. That is because different choices lead to exact Morse fibrations which are deformation equivalent in a suitable sense. For the same reason, if ξ is such that any two generic sections define a Lefschetz pencil, the category is an invariant of the pair (X, ξ).

As a concrete example take $X = \mathbb{CP}^2$, $\xi = \mathcal{O}(2)$. The fibre $M = E_{z_0}$ of (E, π) is, topologically, $\mathbb{CP}^1$ with four small discs removed. The relative Maslov map $\delta_{E/D}$ coming from (3.1) induces a Maslov map δ_M on M, which is characterized by having the same behaviour over each component of ∂M. Figure 1 shows a distinguished basis $\Gamma = (L_1, L_2, L_3)$ of vanishing cycles. One can read off directly that the directed Fukaya category $\mathcal{A} = Lag^{\to}(\widetilde{\Gamma})$, for a suitable choice of gradings, is described by the quiver with relations

(3.2)

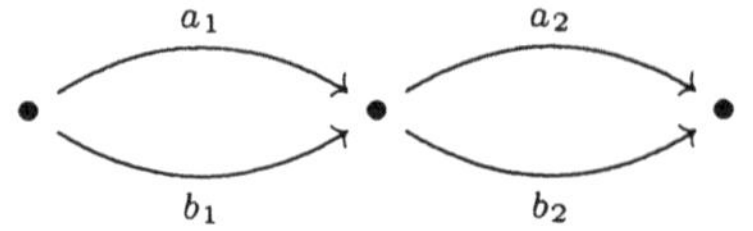

$$\begin{cases} a_2 a_1 = b_2 b_1, \\ b_2 a_1 = a_2 b_1. \end{cases}$$

Namely, the intersections $L_i \cap L_k$, $i < k$, consist of two points which are essential, in the sense that they cannot be removed by an isotopy in M. This implies that $hom_{\mathcal{A}}(\widetilde{L}_i, \widetilde{L}_k)$ is two-dimensional and $\mu^1_{\mathcal{A}} = 0$. And there are four triangles in M whose sides (in positive order) map to (L_1, L_2, L_3), giving rise to the nonzero products $\mu^2_{\mathcal{A}}(a_2, a_1)$, $\mu^2_{\mathcal{A}}(b_2, a_1)$, $\mu^2_{\mathcal{A}}(a_2, b_1)$, $\mu^2_{\mathcal{A}}(b_2, b_1)$.

Remark 3.1. *Suppose that* (3.1) *holds but with* $a \in \mathbb{Q}$, *which is to say* $K_X^{-2c} \cong \xi^b$ *for coprime* b, c. *Then* σ_1^b *defines a c-sheeted cyclic covering of* $X \backslash Y_\infty$, *hence also a covering of* E. *The total space of the latter can be made into an exact Morse fibration which then has a canonical relative Maslov map, so that our theory can be applied to it. The* $\mathbb{Z}/c$*-action on* E *by covering transformations induces a self-equivalence* σ *of* $D^b Lag^{\rightarrow}(\widetilde{\Gamma})$ *which satisfies* $\sigma^c \cong [1]$. *One can view this as a category with "fractional gradings", thinking of* $\mathrm{Hom}(X, \sigma^i Y)$ *as "the group* $\mathrm{Hom}^{i/c}(X, Y)$ *of morphisms of degree* i/c*".*

(3B) We need to extend the previous discussion to a degenerate situation. Let $X \subset \mathbb{CP}^N$ be a projective variety with singular set X^{sing}, and $\sigma_0, \sigma_1 \in (\mathbb{C}^{N+1})^*$ two linear forms which generate a pencil (Y_z) of hyperplane sections of X; so the line bundle we are looking at is $\xi = \mathcal{O}_X(1)$. Suppose that the following conditions are satisfied: $X^{\mathrm{sing}} \subset Y_\infty \backslash Y_0$; the base locus $Z = Y_0 \cap Y_\infty$ is smooth; and $(Y_z)_{z \neq \infty}$ satisfies the same nondegeneracy conditions as a Lefschetz pencil. Then the associated graph $p : G \to \mathbb{CP}^1$ is well-behaved except over $\infty \in \mathbb{CP}^1$.

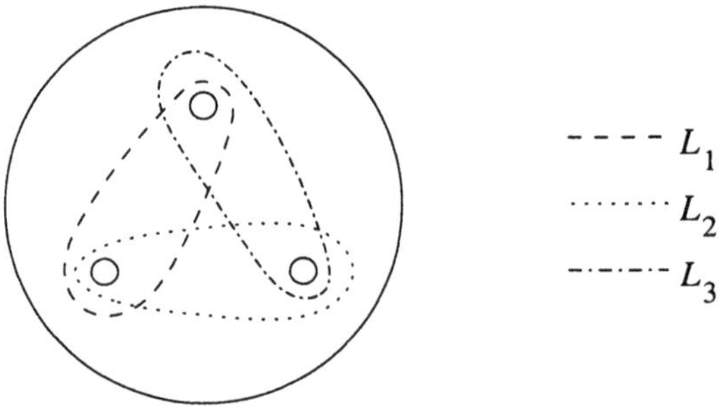

FIGURE 1.

The construction made above, in which the fibre at infinity plays no role, goes through without any changes, yielding an exact Morse fibration (E, π) over D. If we suppose moreover that X is locally a complete intersection, so that there is a well-defined canonical bundle K_X, then condition (3.1) again ensures the existence of preferred relative Maslov maps $\delta_{E/D}$.

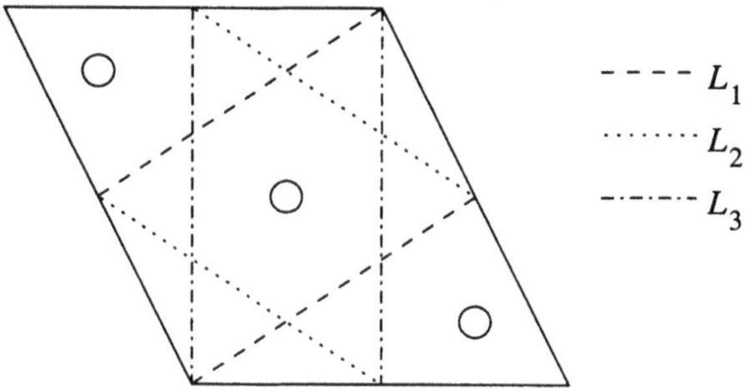

FIGURE 2.

The example which we have in mind is $X = \{x \in \mathbb{C}\mathrm{P}^3 \mid x_0^3 = x_1x_2x_3\}$, $\sigma_0(x) = x_1 + x_2 + x_3$, $\sigma_1(x) = x_0$. One can identify $X \setminus Y_\infty$ with the affine hypersurface $x_1x_2x_3 = 1$, and then σ_0/σ_1 becomes the function $x_1 + x_2 + x_3$, which appears in Givental's work [12, Theorem 5] as the object mirror dual to the projective plane. The next result exhibits the mirror phenomenon within our framework.

Proposition 3.2. *Let (E, π) be the exact Morse fibration constructed from (X, σ_0, σ_1), with relative Maslov map $\delta_{E/D}$ obtained from* (3.1). *Then for any graded distinguished basis of vanishing cycles ones has*

$$D^b Lag^{\rightarrow}(\tilde{\Gamma}) \cong D^b Coh(\mathbb{P}^2)$$

where $D^b Coh(\mathbb{P}^2)$ is the derived category, in the classical sense, of coherent sheaves on the projective plane over the field $\mathbb{Z}/2$.

The fibre $M = E_{z_0}$ is a torus with three small discs removed. Figure 2 shows a particular distinguished basis of vanishing cycles $\Gamma = (L_1, L_2, L_3)$. In that picture, the Maslov map δ_M is that given by the flat structure of M. One can therefore choose gradings such that the groups $CF^*(\tilde{L}_i, \tilde{L}_k)$, $i < k$, are concentrated in degree zero. Then $\mathcal{A} = Lag^{\rightarrow}(\tilde{\Gamma})$ is given by the quiver with

relations

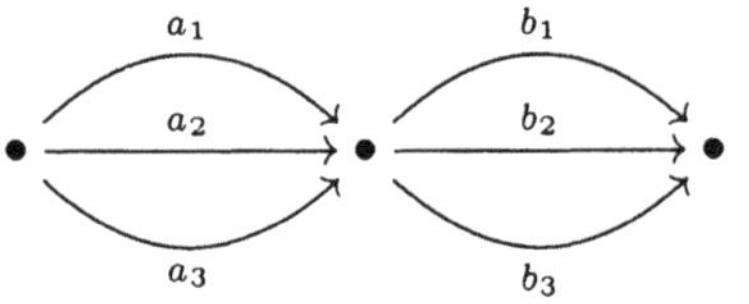

$$\begin{cases} b_i a_j = b_j a_i & i \neq j, \\ b_i a_i = 0. \end{cases}$$

A theorem of Beilinson [6] says that $D^b(\mathcal{A}) \cong D^b Coh(\mathbb{P}^2)$. Since $D^b(\mathcal{A})$ remains the same for all other distinguished bases, the Proposition is proved. The same approach can be used for the mirror partner of $\mathbb{C}P^1 \times \mathbb{C}P^1$, with analogous results; I have not checked any further cases.

Remarks 3.3. *(i) It is maybe helpful to mention one way of making drawings like Figure 2. By projecting to one coordinate, each fibre $(\sigma_0/\sigma_1)^{-1}(z)$ of $\sigma_0/\sigma_1 : X \setminus Y_\infty \to \mathbb{C}$ can be represented as a double covering of $\mathbb{C}^*$ branched over three points. The vanishing cycles are preimages of paths in $\mathbb{C}^*$ joining two branch points, and one finds them by looking at how the branch points come together as one varies z.*

(ii) One sees from Figure 2 that the Dehn twists τ_{L_i} can be represented by affine maps, at least away from a small neighbourhood of the missing discs. It follows that any Lagrangian configuration $\Gamma' = (L'_1, L'_2, L'_3)$ which is Hurwitz equivalent to Γ consists of curves L'_i isotopic in M to straight lines. An easy argument using the topology of E shows that no two L'_i can ever be isotopic to each other. Therefore $HF^*(\tilde{L}'_i, \tilde{L}'_k)$, $i < k$, is concentrated in a single degree, for any choice of gradings. This is related to the fact that mutations of the standard exceptional collection in $D^b Coh(\mathbb{P}^2)$ remain strongly exceptional; see [9] for much more about this subject.

(3c) Let $f \in \mathbb{C}[x_1, \dots, x_{n+1}]$ be a polynomial with an isolated critical point at the origin. A Morsification of f is a smooth family of polynomials $(f_t)_{0 \leq t < \rho}$ with $f_0 = f$ and such that the critical points of f_t, $t > 0$, which lie near the origin are nondegenerate. For sufficiently small $0 < \epsilon$, $0 < \delta \ll \epsilon$, and $0 < t \ll \delta$, one finds that

$$X = \{x \in \mathbb{C}^{n+1} \mid |x| \leq \epsilon, \ |f_t(x)| \leq \delta\}$$

is a manifold with corners, and f_t a holomorphic Morse function on it. One can equip X with the standard symplectic form on $\mathbb{C}^{n+1}$, and its canonical bundle has a standard trivialization obtained from the constant holomorphic volume form. After some modifications, this becomes an exact Morse fibration (E, π) over D together with a canonical homotopy class of relative Maslov maps $\delta_{E/D}$. We will not explain the details since the procedure is similar to that for Lefschetz pencils. A deformation equivalence argument shows that the resulting category $D^b Lag^{\to}(\tilde{\Gamma})$ is independent of all choices, and even of the Morsification, so that it is an invariant of f.

Proposition 3.4. *Let $f \in \mathbb{C}[x_1, x_2]$ be an ADE singularity. Then the category $D^b Lag^{\to}(\tilde{\Gamma})$ associated to it is equivalent to the derived path category of the Dynkin quiver of the same type, with an arbitrary orientation of the arrows.*

A'Campo [3, pp. 13–17] has constructed particularly nice Morsifications and distinguished bases $\Gamma = (L_1, \dots, L_m)$ for these singularities. These have the properties that (i) there is a bijection between the L_i and the vertices of the corresponding Dynkin diagram, such that $L_i \cap L_k$ consists of a single point if the i-th and k-th vertex are connected by a line, and is empty otherwise; (ii) there are no $i < k < l$ with $L_i \cap L_k \neq \emptyset$ and $L_k \cap L_l \neq \emptyset$. Figure 3 shows the E_6 case. One can choose gradings $\tilde{L}_i$ such that $HF^*(\tilde{L}_i, \tilde{L}_k)$ is concentrated in degree zero for all $i < k$. Then $Lag^{\to}(\tilde{\Gamma})$ is the path category of the Dynkin quiver for a particular orientation, which is such that each vertex is either a sink or a source. Changing the orientation does not affect the derived category of the path category, which completes the proof.

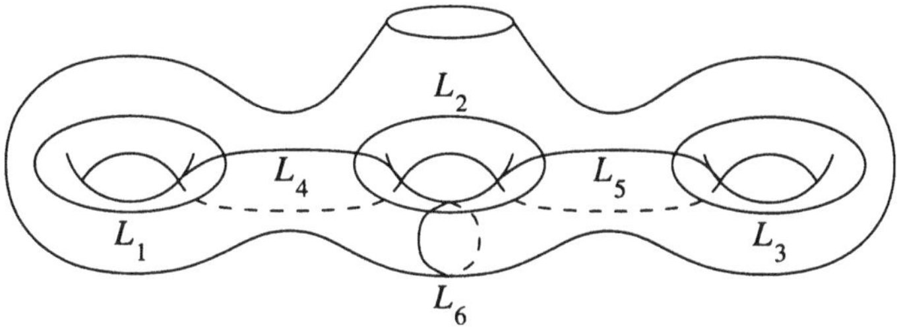

FIGURE 3.

We mention two more expected properties of the categories $D^b Lag^{\to}(\tilde{\Gamma})$ as invariants of singularities. Strictly speaking these are conjectures, but the proofs should not be difficult.

- (μ-invariance) They should remain the same when one deforms f inside its $\mu = const.$ stratum. Moreover, if f is adjacent to g then the category belonging to f should contain that of g as a full triangulated subcategory.
- (Stabilization) The categories associated to $f(x)$ and $\hat{f}(x, y) = f(x) + y_1^2 + \cdots + y_q^2$ should be equivalent, for any q.

For instance, the reader may have noticed that the categories of A_m-singularities in Proposition 3.4 are the same as those arising from certain branched covers, which were considered in Section 2b. This is an extreme case of the stabilization property, since the branched covers can be viewed as Morsifications of the one-dimensional singularities $f(x) = x^{m+1}$.

Additional references. Hori, Iqbal and Vafa [14] discuss mirror symmetry for Fano varieties from a physics point of view. The construction of exact Morse fibrations from singularities is similar to the definition of symplectic monodromy in [16, Section 6].

4. Floer cohomology for automorphisms

(4A) Let (M, ω, θ) be an exact symplectic manifold (which is, as always, assumed to have contact type boundary). Floer cohomology associates to each $\phi \in \mathrm{Symp}^e(M)$ a pair of vector spaces $HF(\phi, -)$ and $HF(\phi, +)$. These correspond to two ways of treating the fixed points near ∂M: let $H \in C^\infty(M, \mathbb{R})$ be a function with $H|\partial M \equiv 0$, whose Hamiltonian flow (ϕ_t^H) is equal to the Reeb flow on ∂M. To define $HF(\phi, -)$ and $HF(\phi, +)$ one perturbs ϕ to $\phi \circ \phi_t^H$ for some small $t < 0$, respectively $t > 0$. The difference is measured by a long exact sequence

$$HF(\phi, -) \longrightarrow HF(\phi, +) \longrightarrow H^*(\partial M; \mathbb{Z}/2) \longrightarrow HF(\phi,-) \tag{4.1}$$

For $\phi = \mathrm{id}_M$ this reduces to the usual cohomology exact sequence, with $HF(\mathrm{id}_M, -) \cong H^*(M, \partial M; \mathbb{Z}/2)$ and $HF(\mathrm{id}_M, +) \cong H^*(M; \mathbb{Z}/2)$. Other properties of $HF(\phi, \pm)$ are its invariance under isotopies within $\mathrm{Symp}^e(M)$; conjugation invariance, $HF(\phi_2\phi_1, \pm) \cong HF(\phi_1\phi_2, \pm)$; and Poincaré duality, $HF(\phi, \pm) \cong HF(\phi^{-1}, \mp)^\vee$. If M is equipped with a Maslov map δ_M, a grading $\tilde{\phi}$ determines a $\mathbb{Z}$-grading on Floer cohomology, denoted by $HF^*(\tilde{\phi}, \pm)$. This satisfies $HF^*(\tilde{\phi}[\sigma], \pm) \cong HF^{*-\sigma}(\tilde{\phi}, \pm)$ for $\sigma \in \mathbb{Z}$.

Let S be a closed oriented surface together with a finite set $\Sigma \subset S$ of marked points, and $(E, \pi) = (E, \pi, \Omega, \Theta, J_0, j_0)$ an exact Morse fibration over $S^* = S \setminus \Sigma$, with smooth fibres isomorphic to M. Suppose that around each $\zeta \in \Sigma$ we have oriented local coordinates $\psi_\zeta : D \to S$, such that there are commutative diagrams

$$\begin{array}{ccc} \mathbb{R}^{\leq 0} \times T(\phi_\zeta) & \xrightarrow{\Psi_\zeta} & E \\ \downarrow & & \downarrow \pi \\ D^* & \xrightarrow{\psi_\zeta | D^*} & S^*. \end{array} \tag{4.2}$$

Here $T(\phi_\zeta)$ is the mapping torus of $\phi_\zeta \in \mathrm{Symp}^e(M)$, meaning the quotient of $\mathbb{R} \times M$ by $(t, x) \sim (t-1, \phi_\zeta(x))$. The map from $\mathbb{R}^{\leq 0} \times T(\phi_\zeta)$ to $D^* = D \setminus \{0\}$ is $(s, [t, x]) \mapsto \exp 2\pi(s + it)$. The pullback of ω to $\mathbb{R} \times M$ is invariant under the $\mathbb{Z}$-action, hence descends to a two-form ω_{ϕ_ζ} on $T(\phi_\zeta)$; and a choice of function $K \in C_c^\infty(M \setminus \partial M, \mathbb{R})$ with $dK = \phi^*\theta - \theta$ yields a one-form θ_{ϕ_ζ} on the same space. Ψ_ζ is an isomorphism of differentiable fibre bundles between $\mathbb{R}^{\leq 0} \times T(\phi_\zeta)$ and $E|\psi_\zeta(D^*)$, satisfying $\Psi_\zeta^*\Omega = \omega_{\phi_\zeta}$, $\Psi_\zeta^*\Theta = \theta_{\phi_\zeta}$. By and large, this means that the behaviour of (E, π) around ζ is described by the exact symplectic automorphism ϕ_ζ. One then has a relative invariant

$$\Phi_{rel}(E, \pi) \in \bigotimes_{\zeta \in \Sigma} HF(\phi_\zeta, \pm), \tag{4.3}$$

where the signs $\pm$ can be chosen arbitrarily, subject to the restriction that there must be at least one ζ labeled $+$ in each connected component of S; in particular Σ may not be empty. Changing an existing $-$ to a $+$ gives a new relative invariant which is the image of the old one under the map $HF(\phi_\zeta, -) \to HF(\phi_\zeta, +)$ from (4.1). To define the relative invariant, one chooses a suitable complex structure on S^* and almost complex structure on E, and considers the moduli spaces of pseudo-holomorphic sections.

Remark 4.1. *The asymmetry between $+$ and $-$ can be explained as follows. By definition of an exact Morse fibration, one has a trivialization near ∂E. However, the almost complex structures J on E used here do not have the obvious product structure with respect to this trivialization; this is made necessary by the perturbation of ϕ_ζ near ∂M which defines $HF(\phi_\zeta, \pm)$. The convexity of ∂E with respect to J-holomorphic sections now becomes a more delicate matter, and it is there that the sign question appears.*

Suppose that $\zeta \neq \zeta'$ are points in Σ with $\phi_{\zeta'} = (\phi_\zeta)^{-1}$, and look at a relative invariant $\Phi_{rel}(E, \pi)$ in which ζ, ζ' have different signs. The surface $\overline{S}$ obtained by connect summing together ζ, ζ' comes with marked points $\overline{\Sigma}$ inherited from $\Sigma \setminus \{\zeta, \zeta'\}$, and with an exact Morse fibration over $\overline{S} \setminus \overline{\Sigma}$. The relative invariant of this new fibration (keeping the signs as before) can be computed, if it is well-defined, by applying the Poincaré duality pairing $HF(\phi_\zeta, \pm) \otimes HF(\phi_{\zeta'}, \mp) \to \mathbb{Z}/2$ to $\Phi_{rel}(E, \pi)$.

This setup can be generalized and unified with that in [VC, Section 3]. In the generalization one considers oriented compact surfaces S with boundary, with marked points $\Sigma \subset S$ which may lie on the boundary or in the interior, together with an exact Morse fibration (E, π) over $S^* = S \setminus \Sigma$ having a Lagrangian boundary condition Q. The structure of (E, π) near a marked point in the interior remains as before, while near boundary marked points (E, π, Q) is as in [VC]. One then gets relative invariants of a mixed kind,

$$(4.4)\qquad \Phi_{rel}(E,\pi,Q) \in \bigotimes_{\zeta\in\Sigma\cap\partial S} HF(L_{\zeta,+},L_{\zeta,-}) \otimes \bigotimes_{\zeta\in\Sigma\setminus\partial S} HF(\phi_\zeta,\pm).$$

The restriction that there must be at least one + now applies only to those connected components of S which have no boundary; and there are two different gluing formulas, for boundary and interior marked points respectively.

(4B) We can now explain several kinds of maps between Floer cohomology groups as special cases of the relative invariants (4.3) and (4.4).

- Take an exact symplectic manifold M and $\phi_1,\phi_2 \in \mathrm{Symp}^e(M)$. Let $S = S^2$ and $\Sigma = \{\mathit{3\ points}\}$. There is a unique exact Morse fibration (E,π) over S^* which has no critical points and is even flat (locally trivial), such that the symplectic monodromy around the three missing points is respectively, ϕ_1, ϕ_2 and $(\phi_1\phi_2)^{-1}$. By marking the points with suitable signs and using Poincaré duality, one obtains from $\Phi_{rel}(E,\pi)$ a map $HF(\phi_1,\pm)\otimes HF(\phi_2,+) \to HF(\phi_1\phi_2,\pm)$, which is the "pair-of-pants" product in the Floer cohomology of symplectic automorphisms.
- Let $L \subset M$ be an exact framed Lagrangian sphere. Then there is an exact Morse fibration (E,π) over D with a single critical point, such that the monodromy around ∂D is τ_L. If one extends this to $\mathbb{C} = S^2\setminus\{\infty\}$ in such a way that it is flat near ∞, one gets an invariant $\Phi_{rel}(E,\pi) \in HF(\tau_L^{-1},+)$.
- Take the pair-of-pants product with $\phi_1 = \phi\tau_L$ and $\phi_2 = \tau_L^{-1}$, and plug in the distinguished element of $HF(\tau_L^{-1},+)$ constructed above. This gives a canonical map $HF(\phi\tau_L,\pm) \to HF(\phi,\pm)$, which we call c. Thanks to the gluing formula, c can also be described as the relative invariant of a certain exact Morse fibration over $S^2 \setminus \{\mathit{2\ points}\}$ with a single critical point.
- For any $\phi \in \mathrm{Symp}^e(M)$, $\mathbb{R}^{\leq 0} \times T(\phi^{-1})$ is an exact Morse fibration over D^*. Denote by (E,π) its restriction to $D \setminus \{0,i\}$. Parallel transport along ∂D starting at the base point $z_0 = -i$ yields a trivialization $E\,|\,\partial D\setminus\{i\} \cong (\partial D\setminus\{i\})\times M$. Given an exact Lagrangian submanifold $L \subset M$, one can introduce the Lagrangian boundary condition which in this trivialization is $(\partial D \setminus \{i\}) \times L$. Because of the monodromy around 0, $\Phi_{rel}(E,\pi,Q)$ takes values in $HF(\phi^{-1},\pm) \otimes HF(\phi(L),L)$. Equivalently, one can see it as a map $d : HF(\phi,\pm) \to HF(\phi(L),L)$.

Theorem 4.2. *Let (M, ω, θ) be an exact symplectic manifold, $L \subset M$ an exact framed Lagrangian sphere, and $\phi \in \mathrm{Symp}^e(M)$ an exact automorphism. Suppose that $2c_1(M, L) \in H^2(M, L)$ is zero. Then there is a long exact sequence, with maps c and d as defined above,*

$$\begin{array}{ccc} HF(\phi\tau_L, \pm) & \xrightarrow{c} & HF(\phi, \pm) \\ & \nwarrow & \downarrow d \\ & & HF(\phi(L), L). \end{array} \tag{4.5}$$

If M has a Maslov map δ_M, and ϕ, L have gradings $\tilde{\phi}$, $\tilde{L}$ then, with respect to the natural gradings of the groups in (4.5), the maps c, d have degree zero while the remaining one has degree one. The proof of the exact sequence is similar to that of [VC, Theorem 3.3].

Example 4.3. *By taking $\phi = \mathrm{id}_M$ resp. $\phi = \tau_L^{-1}$ one gets exact sequences*

$$\begin{array}{ccc} HF(\tau_L, +) & \longrightarrow & H^*(M; \mathbb{Z}/2) \\ & \nwarrow & \downarrow \\ & & H^*(L; \mathbb{Z}/2), \end{array} \qquad \begin{array}{ccc} H^*(M; \mathbb{Z}/2) & \longrightarrow & HF(\tau_L^{-1}, +) \\ & \nwarrow & \downarrow \\ & & H^*(L; \mathbb{Z}/2). \end{array}$$

In the case when M is a surface, this fits in with the results of direct computation in [22], which are that $HF(\tau_L, +) \cong H^*(M, L; \mathbb{Z}/2)$ and $HF(\tau_L^{-1}, +) \cong H^*(M \setminus L; \mathbb{Z}/2)$. In fact, the same holds in all dimensions, as one can show by taking a closer look at the map d in the exact sequence.

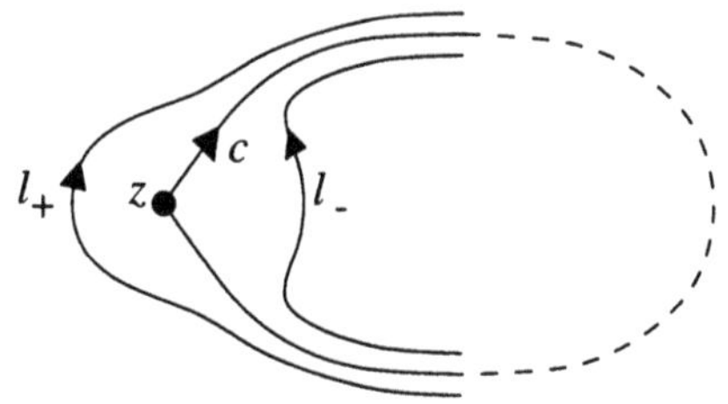

FIGURE 4.

Now take an exact Morse fibration (E, π) with arbitrary base S. To any loop $l : S^1 \to S \setminus S^{\mathrm{crit}}$ one can associate its symplectic monodromy ρ_l and hence Floer cohomology groups, which are invariant under homotopies of l that do not pass across S^{crit}. We write $HF(l, \pm)$ instead of $HF(\rho_l, \pm)$, in parallel with the notation $HF(c)$ adopted at the end of [VC, Section 3] for paths $c : [0;1] \to S$ with $c^{-1}(S^{\mathrm{crit}}) = \{0;1\}$. Theorem 4.2 implies that if one has loops l_-, l_+ and a path c as in Figure 4, there is a long exact sequence

$$HF(l_-) \longrightarrow HF(l_+) \longrightarrow HF(c).$$

The connection with the original formulation is made by the Picard-Lefschetz theorem: l_+ differs from l_- by a clockwise turn around $z \in S^{\mathrm{crit}}$, so that $\rho_{l_+}^{-1}\rho_{l_-}$ is the Dehn twist along the vanishing cycle arising from z.

5. Hochschild cohomology

(5A) The Hochschild cohomology $HH^*(\mathcal{A},\mathcal{A})$ of a (small) A_∞-category $\mathcal{A}$ is a graded vector space, defined via the following cochain complex $CC(\mathcal{A},\mathcal{A})$. A cochain of degree r is a sequence $h = (h^d)_{d\geq 0}$ of which each member h^d is a family of graded linear maps

$$\bigotimes_{i=1}^{d} hom_{\mathcal{A}}(X_i, X_{i+1}) \to hom_{\mathcal{A}}(X_1, X_{d+1})[r-d],$$

one for each $(d+1)$-tuple $X_1, \dots, X_{d+1} \in \mathrm{Ob}\,\mathcal{A}$, satisfying $h^d(a_d, \dots, a_1) = 0$ whenever some a_i is the identity map id_{X_i}. In particular, h^0 consists of an element in $hom^r_{\mathcal{A}}(X,X)$ for each X. The differential is

$$\begin{aligned}(\partial h)^d(a_d, a_{d-1}, \dots, a_1) &= \\ &= \sum_{i+j\leq d+1} \mu_{\mathcal{A}}^{d+1-j}(a_d, \dots, a_{i+j}, h^j(a_{i+j-1}, \dots, a_i), a_{i-1}, \dots, a_1) \\ &\quad + \sum_{i+j\leq d+1} h^{d+1-j}(a_d, \dots, a_{i+j}, \mu_{\mathcal{A}}^j(a_{i+j-1}, \dots, a_i), a_{i-1}, \dots, a_1).\end{aligned}$$

When $\mathcal{A}$ is directed, $CC(\mathcal{A},\mathcal{A})$ and hence $HH^*(\mathcal{A},\mathcal{A})$ are finite-dimensional.

Hochschild cohomology has two related but different interpretations. One is in terms of first order deformations of the maps $\mu_{\mathcal{A}}^d$, and $\partial h = 0$ appears there as the linearization of the structure equations of an A_∞-category, extended by an h^0 term. The other involves the A_∞-category $funct(\mathcal{A},\mathcal{A})$ of A_∞-functors from $\mathcal{A}$ to itself, and it says that

$$HH^*(\mathcal{A},\mathcal{A}) = H(hom_{funct(\mathcal{A},\mathcal{A})}(\mathrm{Id}_{\mathcal{A}}, \mathrm{Id}_{\mathcal{A}})) \tag{5.1}$$

is the space of A_∞-natural transformations from the identity functor to itself. In the case where $\mu_{\mathcal{A}}^d = 0$ for $d \geq 3$, which means $\mathcal{A}$ is a dg-category, there is yet another approach which is essentially equivalent to (5.1) but more in line with classical homological algebra. Let $\mathcal{A}-mod-\mathcal{A}$ be the dg-category of dg-functors from $\mathcal{A}^{opp} \times \mathcal{A}$ to chain complexes of vector spaces, and $D(\mathcal{A}-mod-\mathcal{A})$ its derived category, defined by inverting quasi-isomorphisms as in [8, Chapter 10]. Take the dg-functor $\Delta_{\mathcal{A}} \in \operatorname{Ob}\mathcal{A}-mod-\mathcal{A}$ which assigns to $(X,Y) \in \operatorname{Ob}(\mathcal{A}^{opp} \times \mathcal{A}) = \operatorname{Ob}\mathcal{A} \times \operatorname{Ob}\mathcal{A}$ the complex $hom_{\mathcal{A}}(X,Y)$ (to motivate the notation, we should say that if $\mathcal{A}$ had only a single object Z, objects of $\mathcal{A}-mod-\mathcal{A}$ would be just dg-bimodules over the dg-algebra $hom_{\mathcal{A}}(Z,Z)$, and $\Delta_{\mathcal{A}}$ would be $hom_{\mathcal{A}}(Z,Z)$ considered as a bimodule over itself). Then

$$HH^*(\mathcal{A},\mathcal{A}) = \operatorname{Hom}^*_{D(\mathcal{A}-mod-\mathcal{A})}(\Delta_{\mathcal{A}}, \Delta_{\mathcal{A}}). \tag{5.2}$$

The proof goes as follows: one observes that the Hochschild complex is isomorphic to $hom_{\mathcal{A}-mod-\mathcal{A}}(B_{\mathcal{A}}, \Delta_{\mathcal{A}})$, where $B_{\mathcal{A}}$ is the bar construction. $B_{\mathcal{A}}$ is quasi-isomorphic to $\Delta_{\mathcal{A}}$, and K-projective in the sense of [8, Definition 10.12.2.1].

(5B) An informal principle says that $HH^*(\mathcal{A},\mathcal{A})$ should be an invariant of $D^b(\mathcal{A})$. It is unknown whether this is rigorously true, but there is a weaker result which is sufficient for our purpose.

Proposition 5.1. *Mutation of a directed A_∞-category does not change its Hochschild cohomology.*

To prove this one goes through the list of moves [VC, Definition 5.2]. If $F : \mathcal{A} \to \mathcal{B}$ is a quasi-isomorphism, there are natural maps

$$HH^*(\mathcal{A},\mathcal{A}) \to H(hom_{funct(\mathcal{A},\mathcal{B})}(F,F)) \leftarrow HH^*(\mathcal{B},\mathcal{B}),$$

both of which can be shown to be isomorphisms by looking at the spectral sequence associated to the length filtration. Next, changing the gradings of the morphism groups $hom_{\mathcal{A}}(X^i, X^k)$ by some $\sigma_i - \sigma_k$ does not affect the Hochschild complex. Now consider $\mathcal{A} \rightsquigarrow c\mathcal{A}$. One can always find a directed dg-category $\mathcal{C}$ with a quasi-isomorphism $\mathcal{C} \to \mathcal{A}$, and then $c\mathcal{C}$ is quasi-isomorphic to $c\mathcal{A}$. Hence it is sufficient to prove $HH^*(c\mathcal{C}, c\mathcal{C}) \cong HH^*(\mathcal{C},\mathcal{C})$, which reduces the problem to the case of dg-categories. In that case there is a derived Morita equivalence between $\mathcal{C}$ and $c\mathcal{C}$, which means that there are objects $P \in \operatorname{Ob} D(c\mathcal{C}-mod-\mathcal{C})$, $Q \in \operatorname{Ob} D(\mathcal{C}-mod-c\mathcal{C})$ with

$$P \otimes^{\mathbf{L}}_{\mathcal{C}} Q \cong \Delta_{c\mathcal{C}}, \quad Q \otimes^{\mathbf{L}}_{c\mathcal{C}} P \cong \Delta_{\mathcal{C}};$$

for the definition of $\otimes^{\mathbf{L}}$ see again [8, Chapter 10]. The desired equality is a formal consequence of this and (5.2). The argument for $\mathcal{A} \rightsquigarrow r\mathcal{A}$ is identical.

Additional references. This section is essentially a review of known material. $HH^*(\mathcal{A},\mathcal{A})$ is defined in [20]; the A_∞-categories $funct(-,-)$ in [11]; and an account of derived Morita equivalences can be found in [17]. These are just sample references, and by no means the first ones historically.

6. Hochschild cohomology and global monodromy

(6A) What follows is my attempt to formulate an idea of Donaldson. Let (E, π) be an exact Morse fibration over D with a relative Maslov map $\delta_{E/D}$, M the fibre over the base point with its induced Maslov map δ_M, and $\mu \in \mathrm{Symp}^e(M)$ the global monodromy, defined to be the symplectic parallel transport around ∂D in positive direction. This comes with a canonical grading $\tilde{\mu}$, characterized by being zero near $\mathfrak{d}_M \times \partial M$. After making an admissible choice of paths $(c_1, \dots, c_m)$, with corresponding distinguished basis $\Gamma = (L_1, \dots, L_m)$, one has that $\tilde{\mu}$ is isotopic to $\tilde{\tau}_{L_1} \dots \tilde{\tau}_{L_m}$ within the group of graded exact symplectic automorphisms. Choose gradings $\widetilde{L}_i$ and let $\mathcal{A} = Lag^{\rightarrow}(\widetilde{\Gamma})$ be the directed Fukaya category. By Proposition 5.1, $HH^*(\mathcal{A}, \mathcal{A})$ is independent of the choice of Γ and of the gradings $\widetilde{\Gamma}$.

Conjecture 6.1. *There is a long exact sequence*

$$\begin{array}{ccc} HF^*(\tilde{\mu}, +) & \longrightarrow & H^*(E; \mathbb{Z}/2) \\ & \nwarrow & \downarrow \\ & & HH^*(\mathcal{A}, \mathcal{A}) \end{array} \tag{6.1}$$

with the $\nwarrow$ map having degree one, and the others degree zero.

It is elementary to check, using the classical Picard-Lefschetz formula, that the Euler characteristics add up in the right way. There is also an informal argument showing that, in a somewhat loose sense, $HF^*(\tilde{\mu}, +)$ is "constructed" from nearly the same pieces as $HH^*(\mathcal{A}, \mathcal{A})[-1]$, with $H^*(E; \mathbb{Z}/2)$ compensating for the small difference. We will now reproduce this argument, both because it seems to have been Donaldson's original motivation and because it uses nicely the two exact sequences in Floer cohomology. As in the discussion at the end of Section 4, it is convenient to use the notation $HF(l, \pm)$ and $HF(c)$ for Floer cohomology groups.

(i) Let $l_0, \dots, l_m$ be the loops in $D \setminus D^{\mathrm{crit}}$ shown in Figure 5. To each of them is associated a Floer group $HF(l_k, +)$. Our aim is to successively "decompose" $HF(l_m, +)$. To begin, by Theorem 4.2 one has long exact sequences

$$HF(l_k, +) \longrightarrow HF(l_{k-1}, +) \longrightarrow HF(d_{1,k}),$$

(with the third map going from $HF(d_{1,k})$ back to $HF(l_k, +)$), where the $d_{1,k}$ are as in Figure 6. We interpret this as saying that $HF(l_m, +)$ is "constructed" by putting together the pieces $HF(l_0, +)$, $HF(d_{1,1})$, ..., $HF(d_{1,m})$.

(ii) We next analyze $HF(d_{1,k})$ further. Take the paths $e_{i,k}$ and $e'_{i,k}$, $i < k$, from Figure 6. [VC, Theorem 3.3] yields long exact sequences

$$HF(d_{i,k}) \longrightarrow HF(d_{i+1,k}) \longrightarrow HF(e_{i,k}) \otimes HF(e'_{i,k}),$$

so that $HF(d_{1,k})$ is "built" from $HF(d_{k,k})$ and $HF(e_{i,k}) \otimes HF(e'_{i,k})$, $1 \leq i < k$.

(iii) A similar argument as in (ii) shows that $HF(e'_{i,k})$ can be "decomposed" into $HF(e_{i,k})^{\vee}$ together with $HF(e_{i,j})^{\vee} \otimes HF(e'_{j,k})$ for all $i < j < k$ (the dual vector space appears because the paths $e_{i,k}$, $e_{i,j}$ occur with reversed orientation). One can apply the same argument recursively to the second component $HF(e'_{j,k})$ of the tensor product. This ultimately yields a different "decomposition" of $HF(e'_{i,k})$, of which the pieces are

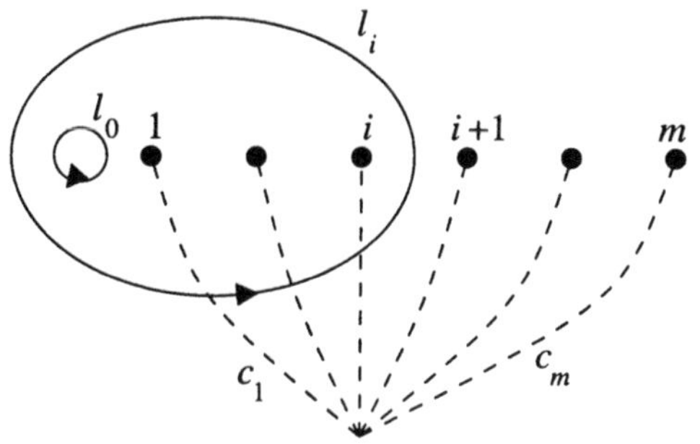

FIGURE 5.

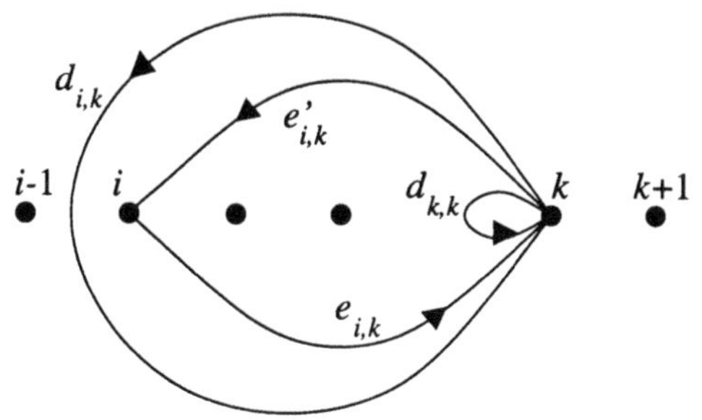

FIGURE 6.

$(HF(e_{i_d,i_{d+1}}) \otimes \cdots \otimes HF(e_{i_1,i_2}))^\vee$ ranging over all $d \geq 1$ and $i = i_1 < i_2 < \cdots < i_{d+1} = k$.

Putting all of this together yields a list of "elementary pieces" which make up $HF(l_m, +)$. At this point we revert to the more conventional notation, so that $HF(l_m,+) = HF(\mu,+)$, $HF(l_0,+) = HF(\mathrm{id}_M,+) \cong H^*(M;\mathbb{Z}/2)$, $HF(d_{k,k}) = HF(L_k,L_k) \cong H^*(L_k;\mathbb{Z}/2)$, and $HF(e_{i,k}) = HF(L_i,L_k)$. We also restore the gradings, which were omitted up to now for the sake of simplicity. Then the "elementary pieces" are:

$$\begin{cases} H^*(M;\mathbb{Z}/2); \\ (\mathbb{Z}/2)^m[-n-1]; \\ (\mathbb{Z}/2)^m[-1]; \text{ and} \\ HF^*(\widetilde{L}_{i_1},\widetilde{L}_{i_{d+1}})[-d-1] \otimes (HF^*(\widetilde{L}_{i_d},\widetilde{L}_{i_{d+1}}) \otimes \cdots \otimes HF^*(\widetilde{L}_{i_1},\widetilde{L}_{i_2}))^\vee \\ \qquad\qquad\qquad \text{for all } d \geq 1 \text{ and } 1 \leq i_1 < \cdots < i_{d+1} \leq m. \end{cases}$$

The second and third piece are the result of splitting each $H^*(L_k;\mathbb{Z}/2)[-1] = \mathbb{Z}/2[-1] \oplus \mathbb{Z}/2[-n-1]$. Now E is obtained up to homotopy equivalence by attaching m cells of dimension $(n+1)$ to M. Hence the first two pieces can be "put together" again, in the same informal sense as before, to form $H^*(E;\mathbb{Z}/2)$. On the other hand the spectral sequence induced by the length filtration, which converges to $HH^*(\mathcal{A},\mathcal{A})[-1]$, has E_1 term equal to the direct sum of the remaining pieces (that is, all except the first two). Of course this sort of bookkeeping falls far short of a proof of Conjecture 6.1, since it ignores all the maps in the long exact sequences, and equally the differentials in the spectral sequence.

(6B) For a better understanding of Conjecture 6.1 one would need to know what the maps in it are. We outline here a possible description of one of them, namely $H^*(E;\mathbb{Z}/2) \to HH^*(\mathcal{A},\mathcal{A})$. This is pure speculation, supported mainly by the fact that the construction, apart from technicalities, seems natural.

One may suppose that the L_i are in generic position. Choose Morse functions f on M and f_i on L_i, such that $f|L_i = f_i$ and the restriction of f to a tubular neighbourhood of each L_i is given by the sum of f_i and a positive definite quadratic form on the normal bundle to L_i. Moreover, $f|\partial M = const.$ and the gradient should point outwards there. After choosing suitable Riemannian metrics, one gets Morse cohomology complexes $C_{Morse}(M)$ and $C_{Morse}(L_i)$ with $\mathbb{Z}/2$ coefficients, together with natural restriction maps $r_i : C_{Morse}(M) \to C_{Morse}(L_i)$. Enlarge $C_{Morse}(L_i)$ to a complex $\tilde{C}_{Morse}(L_i)$ by adding one generator in degree -1 whose boundary is the only zero-dimensional cocycle (the sum of all local minima of f_i). Composing r_i and the inclusions

$C_{Morse}(L_i) \hookrightarrow \tilde{C}_{Morse}(L_i)$ gives a map $\tilde{r} : C_{Morse}(M) \to \bigoplus_i \tilde{C}_{Morse}(L_i)$, whose mapping cone satisfies

$$H^*(\text{Cone}(\tilde{r})) \cong H^*(E; \mathbb{Z}/2)[1]. \tag{6.2}$$

The point of using this rather strange model for the cohomology of E is that there are natural maps

$$\tilde{C}_{Morse}(L_i) \to CC(\mathcal{A}, \mathcal{A})[1], \quad C_{Morse}(M) \to CC(\mathcal{A}, \mathcal{A}). \tag{6.3}$$

The first of these is a chain homomorphism; the second isn't, but the sum of the two is a chain map $\text{Cone}(\tilde{r}) \to CC(\mathcal{A}, \mathcal{A})[1]$, which in combination with (6.2) gives the desired map $H^*(E; \mathbb{Z}/2) \to HH^*(\mathcal{A}, \mathcal{A})$.

Both maps in (6.3) are constructed by the method of "cutting down moduli spaces" which is familiar from the definition of the cap product on Floer cohomology. We will describe the first map in detail, and the second one only briefly. Fix $i \in \{1, \dots, m\}$ and let $Z \subset L_i$ be the unstable manifold, under the gradient flow, of some critical point of f_i of Morse index p. Suppose for the moment that $p > 0$, so that Z has positive codimension. Recall that the boundary operator $\mu^1_{\mathcal{A}}$ on $CF^*(\widetilde{L}_{i_1}, \widetilde{L}_{i_2})$, $i_1 < i_2$, is defined using pseudo-holomorphic maps $u : \mathbb{R} \times [0;1] \to M$ with boundary conditions $u(\mathbb{R} \times \{2 - \nu\}) \subset L_{i_\nu}$. If $i_\nu = i$ for some $\nu \in \{1, 2\}$, one can consider the subset of those maps u which satisfy $u(0, 2 - \nu) \in Z$. If one assumes that the almost complex structures have been chosen generically (similar assumptions will be made tacitly later on) then counting isolated points in that subset yields a map

$$h^1_Z : CF^*(\widetilde{L}_{i_1}, \widetilde{L}_{i_2}) \to CF^{*+p}(\widetilde{L}_{i_1}, \widetilde{L}_{i_2}). \tag{6.4}$$

Extend this to the other cochain groups in $\mathcal{A}$, that is to say to $CF^*(\widetilde{L}_{i_1}, \widetilde{L}_{i_2})$ with $i_1, i_2 \neq i$, by setting h_Z to be zero there. Now consider the composition maps of order $d \geq 2$, $\mu^d_{\mathcal{A}} : CF^*(\widetilde{L}_{i_d}, \widetilde{L}_{i_{d+1}}) \otimes \cdots \otimes CF^*(\widetilde{L}_{i_1}, \widetilde{L}_{i_2}) \to CF^{*+2-d}(\widetilde{L}_{i_1}, \widetilde{L}_{i_{d+1}})$ for $i_1 < \cdots < i_{d+1}$. Their definition uses moduli spaces of pairs (r, u), where $r \in \mathcal{R}^{d+1}$ is a point in the moduli space of discs with $d+1$ marked boundary point, $\mathcal{S}^{d+1,*}_r$ is the corresponding disc with the marked points removed, and $u : \mathcal{S}^{d+1,*}_r \to M$ is a map satisfying a pseudo-holomorphicity equation. $\partial \mathcal{S}^{d+1,*}_r$ consists of intervals $I^{d+1}_{r,\nu}$, $1 \leq \nu \leq d+1$, and the boundary conditions are

$$u(I^{d+1}_{r,\nu}) \subset L_{i_\nu}.$$

Assuming that $i_\mu = i$ for some μ, one considers the moduli space of triples (r, u, z), where z is a point in $I^{d+1}_{r,\mu}$ such that $u(z) \in Z$. Counting isolated points in this moduli space defines a map

$$h^d_Z : CF^*(\widetilde{L}_{i_d}, \widetilde{L}_{i_{d+1}}) \otimes \cdots \otimes CF^*(\widetilde{L}_{i_1}, \widetilde{L}_{i_2}) \to CF^{*+p+1-d}(\widetilde{L}_{i_1}, \widetilde{L}_{i_{d+1}}).$$

Again, this is extended trivially to the cases when $i_1, \dots, i_{d+1} \neq i$. The sequence $(h_Z^0 = 0, h_Z^1, h_Z^2, \dots)$ is a Hochschild cochain $h_Z \in CC^{p+1}(\mathcal{A}, \mathcal{A})$.

The case $p = 0$, which we had excluded above, is considerably easier: one takes (6.4) to be the projection to the subspace spanned by the points of $L_{i_1} \cap L_{i_2}$ which lie on Z, and all other h_Z^d to be zero. Finally there is a canonical element $\tilde{h} \in CC^0(\mathcal{A}, \mathcal{A})$, defined by setting $\tilde{h}^0$ to be the unit element in $hom^0_{\mathcal{A}}(\tilde{L}_i, \tilde{L}_i) \cong \mathbb{Z}/2$, with all other components equal to zero. The assignment $Z \mapsto h_Z$, extended to $\tilde{C}_{Morse}(L_i)$ by mapping the generator of degree -1 to $\tilde{h}$, gives the first map in (6.3).

The definition of the second map uses moduli spaces of triples (r, u, z) with a point z which can lie anywhere in $\mathbb{S}_r^{d+1,*}$, such that $u(z)$ is in some stable manifold of f. The failure of this to be a chain map reflects the fact that z can move to the boundary; in that case however $u(z)$ necessarily lies on some L_i, which provides the connection with the first map.

Remarks 6.2. *(i) This discussion may seem rather abstract, but it has concrete computational implications, most strikingly for* $n = 1$ *when* M *is a surface. In that case the* A_∞*-category* $\mathcal{A}$ *is constructed from* Γ *by a purely combinatorial count of immersed polygons, and the same is true for the maps* (6.3). *Suppose that Conjecture 6.1 is true and that our guess for the map* $H^*(E; \mathbb{Z}/2) \to HH^*(\mathcal{A}, \mathcal{A})$ *which occurs in it is correct. Then one gets a combinatorial algorithm for computing the Floer cohomology* $HF(\tau_{L_1} \dots \tau_{L_m}, +)$ *of an arbitrary product of Dehn twists in* M *(along exact curves* L_i *which admit gradings, to be precise).*

(ii) From a deformation theory point of view, what we have described is a first order infinitesimal deformation of $\mathcal{A}$ *parametrized by the graded vector space* $H^*(E; \mathbb{Z}/2)$. *This was done by inserting one marked point into the Riemann surfaces which define* $\mu^d_{\mathcal{A}}$. *It seems natural to expect that using more marked points will allow one to extend this to higher order. This is interesting because in principle there are obstructions to such an extension, which are expressed by a natural graded Lie bracket on* $HH^*(\mathcal{A}, \mathcal{A})[1]$. *If the extension to higher order is indeed possible, it would mean that the Lie bracket vanishes on the image of* $H^*(E; \mathbb{Z}/2) \to HH^*(\mathcal{A}, \mathcal{A})$.

(6c) To round off the discussion of Conjecture 6.1 we look at its implications in some specific cases. Suppose first that (E, π) and $\delta_{E/D}$ come from a Lefschetz pencil satisfying (3.1) for some $a \in \mathbb{Z}$. Then μ can be isotoped to the identity but only within a group of symplectic automorphisms which is larger than $\mathrm{Symp}^e(M)$, the isotopy being nontrivial along ∂M. Floer cohomology is not invariant under such isotopies. However, a more careful analysis shows that $HF(\mu, +) \cong HF(\mathrm{id}_M, -) \cong H^*(M, \partial M; \mathbb{Z}/2)$. More precisely, taking the canonical grading into account, this formula reads

$$HF^*(\tilde{\mu},+) \cong H^{*+a-4}(M,\partial M;\mathbb{Z}/2).$$

For $X = \mathbb{CP}^2$, $\xi = \mathcal{O}(2)$, $\mathcal{A}$ has been determined in Section 3a, and one finds by explicit computation that $HH^r(\mathcal{A},\mathcal{A}) \cong H^r(M,\partial M;\mathbb{Z}/2) \oplus H^r(E;\mathbb{Z}/2)$ for all r. This is compatible with an exact sequence (6.1) in which the horizontal arrow would be the zero map. There are reasons to believe that this arrow will in fact vanish for all Lefschetz pencils except the trivial one (the degree one pencil on $X = \mathbb{CP}^{n+1}$).

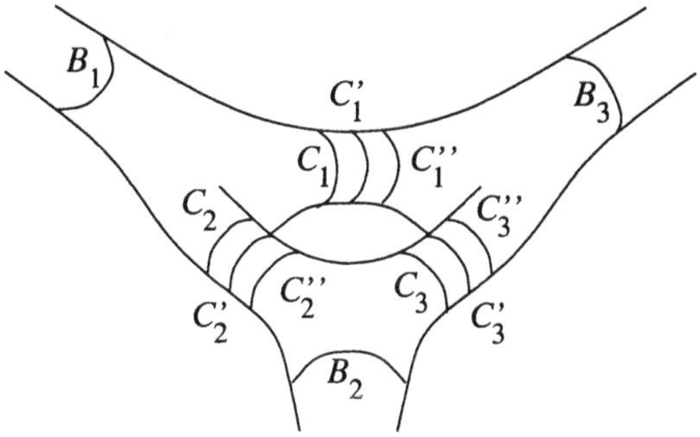

FIGURE 7.

Next consider the mirror dual of the projective plane, as in Section 3b. From the structure of the singular fibre Y_∞ one sees that the global monodromy can be written as a product of Dehn twists and their inverses along disjoint curves, namely

$$\mu = \prod_{i=1}^{3}(\tau_{B_i}\tau_{C_i}^{-1}\tau_{C_i'}^{-1}\tau_{C_i''}^{-1})$$

where the curves are as shown in Figure 7. Using [22] one computes that $HF^*(\tilde{\mu},+) = H^{*-2}(M \setminus \bigcup_{i=1}^{3} C_i \cup C_i' \cup C_i'', \partial M;\mathbb{Z}/2)$. Comparing dimensions shows that $HH^r(\mathcal{A},\mathcal{A}) \cong HF^{r+1}(\tilde{\mu},+) \oplus H^r(E;\mathbb{Z}/2)$, as in the previous example.

As a third and final case, suppose that (E,π) and $\delta_{E/D}$ come from a Morsification of an isolated hypersurface singularity. We conjecture that for all nontrivial singularities

$$HF(\mu) = 0; \tag{6.5}$$

this can be checked easily in many cases, e.g. for weighted homogeneous singularities. From (6.5) and Conjecture 6.1 it would follow that $HH^*(\mathcal{A},\mathcal{A}) \cong$

$H^*(E;\mathbb{Z}/2) = \mathbb{Z}/2$. Because of the role of Hochschild cohomology in deformation theory, this means that $\mathcal{A}$ is rigid (admits no nontrivial first order deformations of any degree). It might be interesting to have a more direct explanation of this phenomenon.

Additional references. For the cap product on Floer cohomology see [19], [21]. Equation (6.5), if true, improves on old topological results of A'Campo [2] and Lê [18].

7. Morse categories

(7A) Let N^{n+1} be a compact manifold with codimension two corners. More precisely, we assume that ∂N consists of three codimension one faces $\partial_- N$, $\partial_0 N$, $\partial_+ N$ (any of which may be empty) with $\partial_- N \cap \partial_+ N = \emptyset$, and where the corners are $(\partial_- N \cup \partial_+ N) \cap \partial_0 N$. Let $p : N \to [-1;1]$ be a Morse function satisfying $\partial_- N = p^{-1}(-1)$, $\partial_+ N = p^{-1}(1)$, and such that $p|\partial_0 N$, as a function on $\partial_0 N$, has no critical points. In addition we require that if $x \in N$ is a critical point of p with Morse index $i(x)$, then

$$\begin{cases} -1 < p(x) < 0 & \text{if } i(x) = 0, \\ p(x) = 0 & \text{if } 0 < i(x) < n+1, \\ 0 < p(x) < 1 & \text{if } i(x) = n+1. \end{cases} \tag{7.1}$$

For $n = 1$ this is more or less the classical notion of self-indexing Morse function, but in higher dimensions it is far more restrictive. To any function p with these properties one can associate a **Morse category** $Morse(N,p)$. To do that, choose a Riemannian metric such that ∇p is parallel to $\partial_0 N$. For two critical points x, y of p, let $\mathcal{G}(x,y)$ be the space of unparametrized gradient trajectories going from x to y. This is a smooth manifold; it is compact unless $i(x) = 0$ and $i(y) = n+1$, in which case there is a natural compactification to a manifold with boundary $\overline{\mathcal{G}}(x,y)$ such that

$$\partial\overline{\mathcal{G}}(x,y) = \coprod_{0<i(w)<n+1} \mathcal{G}(w,y) \times \mathcal{G}(x,w). \tag{7.2}$$

These are all standard facts from Morse theory. What is important is that there is no need to choose the metric generically, since (7.1) implies that ∇p is always Morse-Smale. A homotopy argument based on this fact shows that the topological type of the spaces $\mathcal{G}(x,y)$ and of their compactifications is independent of the metric.

The objects of $\mathcal{A} = Morse(N,p)$ are the critical points of p, and the morphisms are $hom_{\mathcal{A}}(x,y) = H_{-*}(\mathcal{G}(x,y);\mathbb{Z}/2)$; note the sign in front of the grading. The composition maps

$$H_{-*}(\mathcal{G}(w,y)) \otimes H_{-*}(\mathcal{G}(x,w)) \to H_{-*}(\mathcal{G}(x,y)) \tag{7.3}$$

are as follows. If $x = w$ then $[point] \in H_0(point;\mathbb{Z}/2) = hom^0_{\mathcal{A}}(x,w)$ acts as identity morphism, and similarly for $w = y$. If $i(x) = 0$, $0 < i(w) < n+1$, and $i(y) = n+1$, the composition comes from the inclusion of the boundary stratum (7.2), together with the fact that $\mathcal{G}(x,y) \hookrightarrow \overline{\mathcal{G}}(x,y)$ is a homotopy equivalence. In all remaining cases (7.3) is automatically zero, because the tensor product on the left hand side vanishes. Thus $\mathcal{A}$ is a $\mathbb{Z}/2$-linear and $\mathbb{Z}$-graded category. Ordering the critical points according to $p(x)$ shows that it can also be viewed as a directed A_∞-category with $\mu^d_{\mathcal{A}} = 0$ for $d \neq 2$.

Let $B(N,p) \subset N \setminus \partial N$ be the subspace of points whose gradient flow line exists for all time, which means that it converges to a critical point in both positive and negative direction. Define the fundamental object $[\![B(N,p)]\!] \in \mathrm{Ob}\, D^b(\mathcal{A})$ to be the twisted complex

$$(C,\delta_C) = \Big(\bigoplus_{dp(x)=0} x[-i(x)], (\delta_{xy})\Big),$$

with $\delta_{xy} \in hom^1_{\mathcal{A}}(x[-i(x)], y[-i(y)])$ the fundamental homology class $[\mathcal{G}(x,y)]$ if that space is compact and $x \neq y$, and zero otherwise. The generalized Maurer-Cartan equation reduces to $\mu^2_{\mathcal{A}}(\delta_C,\delta_C) = 0$, which is a consequence of (7.2). The notation $[\![B(N,p)]\!]$ is motivated by the fact that

$$\mathrm{Hom}^*_{D^b(\mathcal{A})}([\![B(N,p)]\!],[\![B(N,p)]\!]) \cong H^*(B(N,p);\mathbb{Z}/2). \tag{7.4}$$

To see this one identifies $hom_{\mathrm{Tw}\,\mathcal{A}}(C,C)$ with the E^1 term of the spectral sequence, converging to $H^*(B(N,p);\mathbb{Z}/2)$, which arises from a decomposition of $B(N,p)$ determined by the gradient flow. There is only one nonzero differential in this spectral sequence, and that coincides with the differential on $hom_{\mathrm{Tw},\mathcal{A}}(C,C)$, which completes the proof. If N is closed, so that $B(N,p) = N$, the fundamental object $[\![N]\!]$ has another property. Namely, for any $X \in \mathrm{Ob}\, D^b(\mathcal{A})$ the composition gives rise to a nondegenerate pairing

$$\mathrm{Hom}^*_{D^b(\mathcal{A})}(X,[\![N]\!]) \otimes \mathrm{Hom}^{n+1-*}_{D^b(\mathcal{A})}([\![N]\!],X) \to \mathrm{Hom}^{n+1}_{D^b(\mathcal{A})}([\![N]\!],[\![N]\!]) \cong \mathbb{Z}/2. \tag{7.5}$$

Remarks 7.1. *(i) Consider $B(N,p)$ as a space stratified by the stable manifolds of ∇p. It was pointed out to me by Khovanov that the bounded derived category of sheaves of $\mathbb{Z}/2$-vector spaces, constructible with respect to this stratification, is equivalent to $D^b(\mathcal{A})$; this appears to be a reformulation of a familiar result* [15]. *The objects $x[-i(x)]$ correspond to the constant sheaves along the* *strata, and* $[\![B(N,p)]\!]$ *to the constant sheaf on all of* $B(N,p)$. *This explains the properties listed above:* (7.4) *computes the cohomology of the constant sheaf, and* (7.5) *is Verdier duality.*

(ii) The Morse categories for p *and* $-p$ *are not equivalent. However, they are Koszul dual, so that their derived categories are equivalent.*

(7B) Assume now that $n = 1$, so that $p^{-1}(0)$ is a nodal curve on the surface N. Put a sign $\ominus, \oplus$ into each connected component of $N \setminus p^{-1}(0)$ that contains a local minimum respectively maximum of p. This diagram retains enough information about p to reconstruct $Morse(N,p)$.

FIGURE 8.

Let's say that two functions are related by a triple point move if the corresponding diagrams differ as shown in Figure 8. By a sequence of mutations that mimics the base changes in [5, Volume 2, I §4], one can prove

Proposition 7.2. *Up to equivalence, a triple point move does not change the derived Morse category* $D^b Morse(N,p)$.

It would be interesting to find a topological interpretation of $HH^*(\mathcal{A},\mathcal{A})$ for $\mathcal{A} = Morse(N,p)$. I have not succeeded in doing that, except for one special situation. Call (N,p) cellular if all spaces $\mathcal{G}(x,y)$ are contractible (we are still assuming that $n = 1$; anyway cellularity is impossible in higher dimensions, except for trivial cases).

Proposition 7.3. *If* (N,p) *is cellular,* $HH^*(\mathcal{A},\mathcal{A}) \cong H^*(B(N,p);\mathbb{Z}/2)$.

The first step in the proof is to note that the only possible nonzero groups are $HH^r(\mathcal{A},\mathcal{A})$ for $r = 0,1,2$. $HH^0(\mathcal{A},\mathcal{A}) \cong H^0(B(N,p);\mathbb{Z}/2)$ is always true. From the cellularity one derives that $HH^1(\mathcal{A},\mathcal{A}) \cong H^1(B(N,p);\mathbb{Z}/2)$, and also that the Euler characteristic of $HH^*(\mathcal{A},\mathcal{A})$ equals that of $B(N,p)$, which completes the argument.

Additional references. Objects similar to our Morse categories were introduced several years ago by Cohen, Jones and Segal [10] under the name flow categories. Proposition 7.3 was inspired by A'Campo's paper [1].

8. Real structures

(8A) For the duration of this section, we extend the definition of exact Morse fibration by allowing each fibre to contain several critical points. This does not affect the theory seriously. The notion of admissible choice of paths remains the same; each path may give rise to several disjoint vanishing cycles, which one can place in arbitrary order in the distinguished basis. The important thing is that such bases are still unique up to Hurwitz moves.

Let $(E, \pi) = (E, \pi, \Omega, \Theta, J_0, j_0)$ be an exact Morse fibration over D, with j_0 the standard complex structure on D, and $\delta_{E/D}$ a relative Maslov map. A real structure on it is an involution $\iota : E \to E$ such that

$$\pi(\iota(x)) = \overline{\pi(x)}, \quad \iota^*\Omega = -\Omega, \quad \iota^*\Theta = -\Theta, \quad \iota^* J_0 = -J_0.$$

There is always a $(j, J) \in \mathcal{J}_{E/D}$ with j the standard complex structure on D and such that $\iota^* J = -J$. Then the line bundle

$$\Delta_{E/D} \,|\, \{(j, J)\} \times E = \Lambda^{n+1}(TE, J)^{\otimes 2} \otimes \pi^*(TD, j)^{\otimes -2} \tag{8.1}$$

has a natural real involution $\iota_{(j,J)}$ which covers ι. Call ι compatible with the relative Maslov map if $\delta_{E/D} \,|\, \{(j, J)\} \times E$ is homotopic to a trivialization of (8.1) which takes $\iota_{(j,J)}$ to complex conjugation in $\mathbb{C}$. This condition is independent of the choice of (j, J).

Given a real structure, one can consider $N = E^\iota$ with the Morse function $p = \pi|N : N \to [-1; 1]$. N is a manifold with corners of the kind considered in the previous section, with $\partial_- N \cup \partial_+ N = (N \cap \partial_v E) = p^{-1}(\{\pm 1\})$ and $\partial_0 N = N \cap \partial_h E$; and $p|\partial_0 N$ has no critical points because the same holds for $\pi|\partial_h E$. We say that ι is a good real structure if the critical points of p satisfy (7.1), in which case there is a well-defined Morse category $Morse(N, p)$. We say that ι is complete if all critical points of π lie on N.

Conjecture 8.1. *Let (E, π) be an exact Morse fibration over D with a good and complete real structure ι, which is also compatible with a relative Maslov map $\delta_{E/D}$. Take an admissible choice of paths $(c_1, \dots, c_m)$, each of which lies in the lower half-disc $D \cap \{\mathrm{im}(z) \leq 0\}$, and let Γ be the corresponding distinguished basis. Then $Lag^{\to}(\tilde{\Gamma})$, for a certain choice of grading $\tilde{\Gamma}$, is quasi-isomorphic to $Morse(N, p)$.*

For $n = 1$ there is an elementary way of drawing the vanishing cycles which constitute Γ, starting from (N, p); see e.g. [4]. Using it I have proved Conjecture 8.1 in that dimension. There is also a general argument in support of it which, while falling short of a proof, is quite suggestive. Take an interval $[a; b] \subset [-1; 1] = D \cap \mathbb{R}$ such that $[a; b] \cap D^{\mathrm{crit}} = \{a; b\}$. The symplectic parallel transport along $(a; b)$ is ι-equivariant, and hence preserves the real part N; in fact its restriction to N is equal, up to reparametrization of the orbits, to the flow of ∇p with respect to some metric. Using property (7.1) one can show that intersection points of the vanishing cycles arising from the endpoints $\{a, b\}$ are in one-to-one correspondence with gradient flow lines joining the corresponding critical points of p.

Remark 8.2. *The combination of Conjecture 8.1, Proposition 7.3 and Conjecture 6.1 leads to intriguing relations between the global monodromy of an exact Morse fibration on one hand, and its possible good and complete real structures on the other hand.*

Real structures are not the most general context in which there is a relation between directed Fukaya categories and Morse categories. Take an exact Morse fibration (E, π) over D with relative Maslov map $\delta_{E/D}$. A pseudo-real submanifold is a pair of embeddings

$$\begin{array}{ccc} N & \xrightarrow{C} & E \\ {\scriptstyle p}\downarrow & & \downarrow{\scriptstyle \pi} \\ [-1;1] & \xrightarrow{c} & D. \end{array} \tag{8.2}$$

Here c is a curve in D with $c^{-1}(\partial D) = \{\pm 1\}$, meeting ∂D transversally. N is an $(n+1)$-dimensional manifold with corners as before. C is an embedding transverse to $\partial_h E$ and $\partial_v E$, such that $C^{-1}(\partial_v E) = \partial_- N \cup \partial_+ N$ and $C^{-1}(\partial_h E) = \partial_0 N$; we require that $C^*\Omega$ is zero and $C^*\Theta$ is an exact one-form. p is a Morse function such that $\partial_\pm N = p^{-1}(\pm 1)$. Furthermore, using the fact that $\mathrm{im}(C)$ is totally real with respect to J for any $(j, J) \in \mathfrak{J}_{E/D}$, one finds that there is a canonical homotopy class of sections of $S(\Delta_{E/D})$ over $\{(j, J)\} \times \mathrm{im}(C)$. If a section in that homotopy class can be chosen such that its composition with $\delta_{E/D}$ is a constant map $N \to S^1$, we say that our pseudo-real submanifold is compatible with the relative Maslov map. The notion of good pseudo-real submanifold is defined as before.

Conjecture 8.3. *Suppose that (N, p) is a good pseudo-real submanifold in (E, π), compatible with $\delta_{E/D}$. Then $D^b Lag^{\to}(\widetilde{\Gamma})$ contains $D^b Morse(N, p)$ as a full triangulated subcategory. In particular there is an object $[\![B(N,p)]\!] \in \mathrm{Ob}\, D^b Lag^{\to}(\widetilde{\Gamma})$ whose endomorphism ring is $H^*(B(N, p); \mathbb{Z}/2)$.*

Again, it is not difficult to verify this for $n = 1$.

(8B) As examples one can consider Lefschetz pencils as in Section 3a which are real, meaning that X and ξ have real involutions such that the composition of the two preserves σ_0, σ_1. Then the resulting exact Morse fibration has a natural real structure. To get the right kind of relative Maslov map, one must ask that the isomorphism (3.1) be compatible with the involutions on both sides. An example is $X = \mathbb{CP}^2$, $\xi = \mathcal{O}(2)$, $\sigma_0(x) = x_0^2 - x_2^2$, $\sigma_1(x) = x_0^2 + x_1^2 + x_2^2$, with complex conjugation. This gives rise to a complete and good real structure such that (N, p) is $\mathbb{RP}^2$ with its standard Morse function (having three critical points). One verifies easily that $Morse(N, p)$ is precisely given by the quiver with relations (3.2). Another possible application is to isolated hypersurface singularities which have real Morsifications. For instance, the Morsifications used in Section 3c for ADE plane curve singularities are real, and this provides another way of computing the associated categories.

Additional references. The basic relation between vanishing cycles and real structures comes from work of A'Campo [3] and Gusein-Zade [13].

9. Matching pairs and matching paths

(9a) Let $\mathcal{C}$ be a triangulated category, linear over $\mathbb{Z}/2$ and such that the spaces $\mathrm{Hom}^*_{\mathcal{C}}(X, Y)$ are finite-dimensional. Suppose that $(Y^1, \dots, Y^m)$ is a full exceptional collection in $\mathcal{C}$ such that for some $1 \leq i < m$, the following properties hold. Firstly

$$\mathrm{Hom}^*_{\mathcal{C}}(Y^i, Y^{i+1}) = H^*(S^n; \mathbb{Z}/2) \tag{9.1}$$

for some $n > 0$. Secondly, if a denotes the unique nontrivial morphism $Y^i \to Y^{i+1}$ of degree zero, the compositions

$$a \circ - : \mathrm{Hom}^*_{\mathcal{C}}(Y^k, Y^i) \longrightarrow \mathrm{Hom}^*_{\mathcal{C}}(Y^k, Y^{i+1}) \qquad \text{for } k < i, \text{ and}$$
$$- \circ a : \mathrm{Hom}^*_{\mathcal{C}}(Y^{i+1}, Y^l) \longrightarrow \mathrm{Hom}^*_{\mathcal{C}}(Y^i, Y^l) \qquad \text{for } l > i+1$$

are isomorphisms. We then call (Y^i, Y^{i+1}) a **matching pair** of dimension n. Let $C \in \mathrm{Ob}\,\mathcal{C}$ be the cone of a, which is well-defined up to isomorphism.

Lemma 9.1. *C is a spherical object of dimension $(n+1)$ in the sense of Definition 2.2.*

For future use we need to mention one more fact about C. Assume that $\mathcal{C} = D^b(\mathcal{A})$ is the derived category of some directed A_∞-category $\mathcal{A}$, with $\mathrm{Ob}\,\mathcal{A} = \{X^1, \dots, X^m\}$, and moreover that $(Y^1, \dots, Y^m)$ is obtained from $(X^1, \dots, X^m)$ by mutation, that is to say by applying the transformations [VC, Equations (6) and (7)] and their inverses, plus possibly shifting each object by some amount. In this situation the following holds:

Lemma 9.2. *If $F : \mathcal{A} \to \mathcal{A}$ is a quasi-isomorphism which preserves the numbering of the objects, then $(D^bF)(C) \cong C$.*

The proof is a general nonsense argument. Consider first a general triangulated category $\mathcal{C}$ as before with an exact self-equivalence G, and define $Inv(G) \subset \mathrm{Ob}\,\mathcal{C}$ to be the class of objects X such that $G(X) \cong X$. This has the following properties: (i) if X lies in $Inv(G)$ then so does $X[\sigma]$ for any $\sigma \in \mathbb{Z}$, as well as any object isomorphic to X; (ii) if $X, Y \in Inv(G)$ and there is a unique nonzero map $X \to Y$ of degree zero, then the cone over it lies in $Inv(G)$; (iii) if $X, Y \in Inv(G)$ then $T_X(Y)$, $T'_X(Y) \in Inv(G)$. Here $T_X(Y)$ is as in [VC, Section 5], and $T'_X(Y)$ is the object which fits into an exact triangle

$$T'_X(Y) \to Y \xrightarrow{ev^\vee} \mathrm{Hom}^*_{\mathcal{C}}(Y,X)^\vee \otimes X \to T'_X(Y)[1],$$

where $ev^\vee$ is the transpose of the evaluation $ev : \mathrm{Hom}^*_{\mathcal{C}}(Y,X) \otimes Y \to X$. Now specialize to $\mathcal{C} = D^b(\mathcal{A})$, $G = D^bF$. Then $X^i \in Inv(G)$, and by (iii) the same holds for the objects Y^i of any mutated exceptional collection. Applying (ii) shows that $C \in Inv(G)$.

(9B) Let $(E, \pi) = (E, \pi, \Omega, \Theta, J_0, j_0)$ be an exact Morse fibration over a compact base S, with $\dim E = 2n + 2$. As explained in [VC, Section 3], to any smooth path $c : [0; 1] \to S$ with $c^{-1}(S^{\mathrm{crit}}) = \{0; 1\}$ and $c'(0), c'(1) \neq 0$ one can associate two vanishing cycles $V_{c,0}, V_{c,1}$, which are exact framed Lagrangian spheres in, say, $E_{c(1/2)}$. Now suppose that c is an embedding, does not intersect ∂S, and that its vanishing cycles are isotopic, which means that there is a smooth family $I = (I_t)_{0 \le t \le 1}$ inside the space of all exact Lagrangian submanifolds of $E_{c(1/2)}$ joining $I_0 = V_{c,0}$ and $I_1 = V_{c,1}$. Pairs $(c, [I])$ of this kind, where $[I]$ denotes the homotopy class of I rel endpoints, are called **matching paths**. The interest of this notion lies in a construction which I learnt from Donaldson. Roughly speaking (see below for a precise formulation) this produces from a matching path a submanifold of the total space E, which is Lagrangian in a suitable sense. The idea is easy to see in the case where $V_{c,0} = V_{c,1}$ and I is the constant isotopy. By the definition of vanishing cycle, there are canonical $(n + 1)$-dimensional balls $D_{c,\nu} \subset E$, $\nu = 0, 1$, such that $\pi(D_{c,0}) = c([0; 1/2])$, $\pi(D_{c,1}) = c([1/2; 1])$, $\partial D_{c,\nu} = V_{c,\nu}$ and $\Omega | D_{c,\nu} = 0$. Under the assumption that $V_{c,0} = V_{c,1}$, the two balls fit together smoothly to give a closed submanifold of E. Fix a positive two-form β on S. Then $\Omega^{(r)} = \Omega + r \cdot \pi^*\beta$ is symplectic for large positive r, and $D_{c,0} \cup D_{c,1}$ is Lagrangian with respect to it.

In general, rather than choosing some r, it seems more natural to work with the entire family of forms $\Omega^{(r)}$. This requires a suitably adapted notion of Lagrangian submanifold. Let $\Lambda \subset \mathbb{R}^{\ge 0} \times (E \setminus \partial E)$ be an $(n + 2)$-dimensional submanifold such that the projection $\Lambda \to \mathbb{R}^{\ge 0}$ is a proper submersion; this is the same as a smooth family of closed $(n + 1)$-submanifolds $\Lambda^{(r)} \subset E$, $\Lambda^{(r)} = \Lambda \cap (\{r\} \times E)$. For simplicity assume that $H^1(\Lambda; \mathbb{R}) = 0$. Such a Λ

is called eventually Lagrangian if $\Omega^{(r)}|\Lambda^{(r)} = 0$ for $r \gg 0$. We can now state precisely the nature of the construction mentioned above: for any matching path $(c,[I])$ it gives an eventually Lagrangian submanifold $\Lambda_{(c,[I])}$, unique up to isotopy.

The first step is to trivialize E over c, away from its endpoints, using symplectic parallel transport. The trivialization is the unique embedding $\Psi : (0;1) \times E_{c(1/2)} \to E$ which satisfies $\Psi(\{t\} \times E_{c(1/2)}) = E_{c(t)}$, $\Psi \,|\, \{1/2\} \times E_{c(1/2)} = \mathrm{id}$, and $\Psi^*\Omega = \Omega|E_{c(1/2)}$. Because $D_{c,0}$ and $D_{c,1}$ are likewise defined by parallel transport, they satisfy $\Psi^{-1}(D_{c,0}) = (0;1/2] \times V_{c,0}$, $\Psi^{-1}(D_{c,1}) = [1/2;1) \times V_{c,1}$. One can assume that the isotopy (I_t) is constant near the endpoints. Then

$$A = \{x_0\} \cup \{x_1\} \cup \bigcup_{t\in(0;1)} \Psi(\{t\} \times I_t),$$

where $x_\nu \in E_{c(\nu)}$ are the critical points of π, is a closed submanifold of $E \setminus \partial E$ and contained in $\pi^{-1}c([0;1])$.

By dualizing the normal vector field $\partial I_t/\partial t$ one gets a function h_t (unique up to a constant) on each I_t. Extend the h_t to functions H_t on $E_{1/2}$ supported in a neighbourhood of I_t; these may be chosen such that $H_t = 0$ for t close to 0 or 1. One can then find an $\alpha \in \Omega^1(E)$ vanishing near ∂E, which satisfies $\alpha \,|\, \ker(D\pi) = 0$ and $\Psi^*\alpha = H_t\, dt$. With respect to the modified symplectic forms $\Omega^{(r)} + d\alpha$, $r \gg 0$, A is a Lagrangian submanifold. Moser's Lemma shows that $\Omega^{(r)} + d\alpha$ and $\Omega^{(r)}$ are diffeomorphic, and the diffeomorphism can be made to depend smoothly on r. Applying these diffeomorphisms to A yields a smooth family of submanifolds of E, parametrized by $r \gg 0$, which are Lagrangian with respect to $\Omega^{(r)}$. One extends this arbitrarily to small values of r to complete the definition of $\Lambda_{(c,[I])}$.

Remarks 9.3. *(i) We will often write c, Λ_c instead of $(c,[I])$, $\Lambda_{(c,[I])}$, for the sake of brevity. This is fully justified only when $n = 1$: in that case each component of the space of exact Lagrangian circles on the surface $E_{c(1/2)}$ is simply-connected, so that $[I]$ contains no information.*

(ii) By definition each $\Lambda_c^{(r)} \subset E$, $r \in \mathbb{R}^{\geq 0}$, is diffeomorphic to A. Since A is glued together from two balls, it is a homotopy $(n+1)$-sphere, but it may not be the standard one. To get standard spheres from the construction, one has to impose the additional condition that the composition

$$S^n \xrightarrow{f_0} V_{c,0} \longrightarrow V_{c,1} \xleftarrow{f_1} S^n$$

can be isotoped within Diff(S^n) *to an element of* $O(n+1)$. *Here* f_0, f_1 *are the framings of the vanishing cycles* $V_{c,0}, V_{c,1}$*; and the middle arrow is the diffeomorphism determined, up to isotopy, by* (I_t).

Assume now that no connected component of S is closed, so that $\beta \in \Omega^2(S)$ is exact. One then defines the Floer cohomology of two eventually Lagrangian submanifolds $\Lambda, \Lambda' \subset \mathbb{R}^{\geq 0} \times E$ by

$$HF(\Lambda, \Lambda') = HF(\Lambda^{(r)}, (\Lambda')^{(r)}) \tag{9.2}$$

where the right hand side is with respect to $\Omega^{(r)}$, for some $r \gg 0$. This makes sense because $\Omega^{(r)}$ is exact, and because there is a natural class of compatible almost complex structures on E such that pseudo-holomorphic curves do not reach out to ∂E. Moreover, one can show that the choice of r does not affect (9.2).

(9C) We will now explain the relation between the algebra and geometry introduced above. Let (E, π) and $\delta_{E/D}$ be an exact Morse fibration over D with a relative Maslov map; the notation M and δ_M is as usual. Let c be a matching path. One can find an admissible choice of paths $(c_1, \dots, c_m)$ such that c_i and c_{i+1}, for some $1 \leq i < m$, are as in Figure 9, and such that no other c_j intersects c. In the corresponding distinguished basis $\Gamma = (L_1, \dots, L_m)$, L_i and L_{i+1} are isotopic. Now choose gradings $\widetilde{\Gamma} = (\widetilde{L}_1, \dots, \widetilde{L}_m)$ such that $\widetilde{L}_i$ is isotopic to $\widetilde{L}_{i+1}$, and let $\mathcal{A} = Lag^{\rightarrow}(\widetilde{\Gamma})$ be the directed Fukaya category. Then $(\widetilde{L}_i, \widetilde{L}_{i+1})$ is a matching pair in $D^b(\mathcal{A})$. In fact one has

$$\begin{aligned} \mathrm{Hom}_{D^b(\mathcal{A})}(\widetilde{L}_i, \widetilde{L}_{i+1}) &= H(hom_{\mathcal{A}}(\widetilde{L}_i, \widetilde{L}_{i+1})) \\ &= HF^*(\widetilde{L}_i, \widetilde{L}_{i+1}) \cong H^*(L_i; \mathbb{Z}/2) \end{aligned}$$

which proves (9.1), and the other property is similarly easy. The informal idea is that the spherical object $C \in \mathrm{Ob}\, D^b(\mathcal{A})$ associated to this matching pair "represents", in some sense, the eventually Lagrangian submanifold Λ_c. We will now make a concrete conjecture based on this philosophy.

Suppose that c, c' are two matching paths. For each of them one can find an admissible choice of paths of the kind considered above, but in general these choices cannot be the same for both paths. Therefore one gets two distinguished bases Γ, Γ' of vanishing cycles, and two directed Fukaya categories $\mathcal{A}, \mathcal{A}'$ together with spherical objects $C \in \mathrm{Ob}\, D^b(\mathcal{A})$ and $C' \in \mathrm{Ob}\, D^b(\mathcal{A}')$, constructed from matching pairs of objects in $\mathcal{A}, \mathcal{A}'$ in the way which we have explained. We know of course that $D^b(\mathcal{A})$ and $D^b(\mathcal{A}')$ are equivalent, but for the present purpose we need to make this slightly more precise. There is a chain $\mathcal{A} = \mathcal{A}_0, \mathcal{A}_1, \dots, \mathcal{A}_r = \mathcal{A}'$ of directed A_∞-categories, of which each is related to the preceding one in one of four possible ways: either $\mathcal{A}_{i+1}$ is quasi-isomorphic to $\mathcal{A}_i$; or it is obtained from $\mathcal{A}_i$ by shifting the grading of the morphism spaces; or else $\mathcal{A}_{i+1} = c\mathcal{A}$ or $r\mathcal{A}$. In the three last cases there are canonical equivalences $D^b(\mathcal{A}_i) \cong D^b(\mathcal{A}_{i+1})$; in the first case we choose an equivalence induced by an arbitrary quasi-isomorphism. Composing all of them gives an equivalence $G : D^b(\mathcal{A}) \to D^b(\mathcal{A}')$.

Conjecture 9.4. *Let $\Lambda_c, \Lambda_{c'}$ be the eventually Lagrangian submanifolds associated to c, c'. Then*

$$HF(\Lambda_c, \Lambda_{c'}) \cong \bigoplus_{d \in \mathbb{Z}} \mathrm{Hom}^d_{D^b(\mathcal{A}')}(G(C), C'). \tag{9.3}$$

One can refine the statement by introducing suitable graded Floer cohomology groups for $(\Lambda_c, \Lambda_{c'})$ but we prefer to skip this, since it does not involve any really new ideas.

The two admissible choices of paths which lead to Γ, Γ' are not unique, and neither is the sequence of Hurwitz moves joining them; moreover, in the definition of G we admitted arbitrarily chosen quasi-isomorphisms. It is implicitly part of the conjecture that the right hand side of (9.3) is independent of all these choices. The most doubtful-looking bit is the one about quasi-isomorphisms. Imagine for instance that $\mathcal{A}_i$ and $\mathcal{A}_{i+1}$ are directed Fukaya categories arising from the same Lagrangian configuration, but with different choices of almost complex structures. As mentioned in [VC, Section 6], there is an explicit construction of quasi-isomorphisms between $\mathcal{A}_i$ and $\mathcal{A}_{i+1}$, using pseudo-holomorphic curves for a one-parameter family of choices of almost complex structures. One might think that for (9.3) to be correct, it would be necessary to choose the step between $D^b(\mathcal{A}_i)$ and $D^b(\mathcal{A}_{i+1})$ in the construction

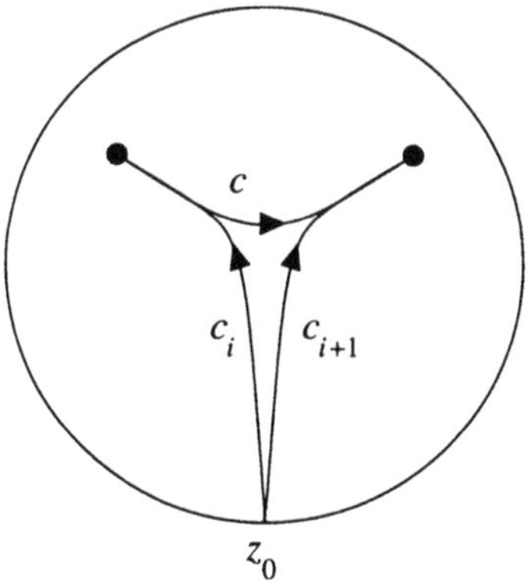

FIGURE 9.

of G to be such an analytically defined quasi-isomorphism. The answer to this objection is Lemma 9.2, which says that choices of quasi-isomorphisms can never affect how objects like C are mapped.

To see what Conjecture 9.4 does, consider the case $n = 1$. Then M is a surface, so that $\mathcal{A}$ and $\mathcal{A}'$ can be determined combinatorially, and so can $G : \mathcal{A} \to \mathcal{A}'$ (this is only true thanks to the freedom of choosing quasi-isomorphisms, which is why we were insisting on it). On the other hand, the left hand side of (9.3) is the Floer cohomology of Lagrangian two-spheres in the symplectic four-manifold $(E, \Omega^{(r)})$, which is difficult or even impossible to compute directly.

Remark 9.5. *With some modifications, the discussion can be adapted to the trivial dimension $n = 0$. Here we only want to point out one example, namely the double branched cover $\pi : E \to D$ from Section 2c. Because the smooth fibre consists of two points, any path is a matching path, so that there is a multitude of one-dimensional spherical objects. This explains the geometric meaning of the objects $C_1, \dots, C_{2g} \in \mathrm{Ob}\, D^b(\mathcal{A}_g)$ which we wrote down earlier. By pushing the argument further, one can show that the zero-dimensional version of Conjecture 9.4 implies the algebraic formulae for geometric intersection numbers on E which were derived, by a quite different argument, in* [16].

References

[VC] P. Seidel, *Vanishing cycles and mutation*, Proceedings of the 3rd European Congress of Mathematics (Barcelona, 2000), Birkhäuser, to appear.

[1] N. A'Campo, *A combinatorial property of generic immersions of curves*, Preprint.

[2] ———, *Le nombre de lefschetz d'une monodromie*, Indag. Math. **35** (1973), 113–118.

[3] ———, *Le groupe de monodromie du déploiement des singularités isolées de courbes planes I*, Math. Ann. **213** (1975), 1–32.

[4] ———, *Real deformations and complex topology of plane curve singularities*, Annales Fac. Sci. Toulouse **8** (1999), 5–23.

[5] V. I. Arnold, S. M. Gusein-Zade, and A. N. Varchenko, *Singularities of differentiable maps*, Birkhäuser, 1988.

[6] A. Beilinson, *Coherent sheaves on $\mathbb{P}^n$ and problems of linear algebra*, Funct. Anal. Appl. **12** (1978), 214–216.

[7] J. Bernstein, I. Gelfand, and V. Ponomarev, *Coxeter functors and Gabriel's theorem*, Russian Math. Surveys **28** (1973), 17–32.

[8] J. Bernstein and V. Lunts, *Equivariant sheaves and functors*, Lecture Notes in Mathematics, vol. 1578, Springer, 1994.

[9] A. Bondal and A. Polishchuk, *Homological properties of associative algebras: the method of helices*, Russian Math. Izvestiya **42** (1994), 219–260.

[10] R. L. Cohen, J. D. S. Jones, and G. B. Segal, *Floer's infinite-dimensional Morse theory and homotopy theory*, The Floer memorial volume (H. Hofer, C. Taubes, A. Weinstein, and E. Zehnder, eds.), Progress in Mathematics, vol. 133, Birkhäuser, 1995, pp. 297–325.

[11] K. Fukaya, *Floer homology for three-manifolds with boundary I*, Preprint, 1997.

[12] A. Givental, *Homological geometry and mirror symmetry*, Proceedings of the International Congress of Mathematics, Zürich, vol. 1, Birkhäuser, 1994, pp. 472–480.

[13] S. M. Gusein-Zade, *Dynkin diagrams for singularities of functions of two variables*, Functional Anal. Appl. **8** (1974), 295–300.

[14] K. Hori, A. Iqbal, and C. Vafa, *D-branes and mirror symmetry*, Preprint hep-th/0005247.

[15] M. M. Kapranov, *Mutations and Serre functors on constructive bundles*, Functional Anal. Appl. **24** (1990), 155–156.

[16] M. Khovanov and P. Seidel, *Quivers, Floer cohomology, and braid group actions*, Preprint Math.QA/0006056.

[17] S. Koenig and A. Zimmermann (eds.), *Derived equivalences for group rings*, Lecture notes in Math., vol. 1685, Springer, 1998.

[18] D. T. Lê, *La monodromie n'a pas de points fixes*, J. Fac. Sci. Univ. Tokyo Sect. IA **22** (1975), 409–427.

[19] H. V. Lê and K. Ono, *Cup-length estimates for symplectic fixed points*, Contact and symplectic geometry (Cambridge, 1994), Cambridge Univ. Press, 1996, pp. 268–295.

[20] M. Penkava and A. Schwarz, *A_∞-algebras and the cohomology of moduli spaces*, Amer. Math. Soc. Translations, vol. 169, pp. 91–108, American Mathematical Society, 1995.

[21] M. Schwarz, *A quantum cup-length estimate for symplectic fixed points*, Invent. Math. **133** (1998), 353–397.

[22] P. Seidel, *The symplectic Floer homology of a Dehn twist*, Math. Research Lett. **3** (1996), 829–834.

[23] P. Seidel and R. Thomas, *Braid group actions on derived categories of coherent sheaves*, Preprint math.AG/0001043. To appear in *Duke Math. J.*

[24] B. Wajnryb, *A simple presentation for the mapping class group of an orientable surface*, Israel J. Math. **45** (1983), 157–174.

Centre de Mathématiques, Ecole Polytechnique, U.R.M. 7640 du CNRS
E-mail address: seidel@math.polytechnique.fr

MOMENT MAPS, MONODROMY AND MIRROR MANIFOLDS

R. P. THOMAS
Department of Mathematics, Imperial College, Huxley Building, 180 Queen's Gate, London, SW7 2BZ. UK.
E-mail: richard.thomas@ic.ac.uk

Via considerations of symplectic reduction, monodromy, mirror symmetry and Chern-Simons functionals, a conjecture is proposed on the existence of special Lagrangians in the hamiltonian deformation class of a given Lagrangian submanifold of a Calabi-Yau manifold. It involves a stability condition for graded Lagrangians, and can be proved for the simple case of T^2.

1 Introduction

Just as explicit solutions of the Einstein and Hermitian-Yang-Mills equations exist only on spaces that are either low dimensional, non-compact and/or highly symmetric, so the equations for special Lagrangian (SLag) cycles, also important in physics, have the same properties. Physically there are also similarities in that we have two first order supersymmetric minimal energy equations (HYMs and SLag) implying the more standard second order equations (YMs and minimal volume equations).

There are powerful existence results of Calabi and Yau (and more recently Tian, Donaldson and others) for the Einstein equations, and of Donaldson-Uhlenbeck-Yau for the HYM equations, so long as we are on a Kähler (or projective) manifold; this often reduces an infinite dimensional problem in PDEs to a finite dimensional problem in linear algebra. Producing many Kähler-Einstein (e.g. Calabi-Yau) manifolds becomes trivial, and dealing with Hermitian-Yang-Mills connections requires only algebraic computations; in both cases the complicated role of the Kähler form and/or metric is almost removed. This can be thought of as possible because of the existence of some infinite dimensional geometry recasting the equations in terms of moment maps and symplectic reduction. A similar situation for SLags would therefore be highly desirable. In particular it might give a way of studying SLags using only Lagrangians and symplectic geometry, much as HYM connections are studied via stable bundles and algebraic geometry.

This paper explores the mirror symmetry of holomorphic bundles (on a Calabi-Yau 3-fold M, often referred to here as 'the complex side') and Lagrangians (on the mirror Calabi-Yau 3-fold W, 'the symplectic side', known as the Kähler side in the physics literature). Many people have worked and are still working on proving some kind of direct correspondence between such objects given an SYZ torus fibration [SYZ]; see for example [AP], [BBHM], [Ch], [Ful], [Gr], [LYZ], [PZ], [Ty], and see [MMM] for a review of this and many many more issues in mirror symmetry. Here, however, we work purely formally without reference to a particular pair of mirror manifolds, without worrying about what mirror symmetry might rigorously mean, and we will not try to give any explicit correspondence. Using mirror symmetry merely as motivation, we point out some similar structures on both sides of the mirror map. Under some conditions (in some 'large complex structure' or 'semi-classical' or somesuch limit) these structures might be genuinely dual; again it does not matter if they are not in general. For instance, physics [MMMS], [DFR] predicts that one should consider not the HYM equations and slope but some perturbation of them away from the large complex structure limit; however these equations also come from a moment map and, conjecturally, a stability condition (for a discussion of such matters see [Le] or [T3]). So while the slope and phase of Lagrangians discussed below might not be exactly mirror to slope of bundles, it should be mirror to something with analogous properties and significance.

Loosely, we would like to think of submanifolds in a fixed homology class as mirror to connections on a fixed topological complex bundle (with Chern classes mirror to the homology class); then Lagrangians should correspond to holomorphic connections (i.e. integrable connections; those with no $(0,2)$-curvature) and special Lagrangians to those with HYM curvature. These last two conditions should be stability conditions for the group actions of hamiltonian deformations and complex gauge transformations, respectively. The full picture is much more complicated, involving triangulated categories and so forth, as envisaged some six years ago in the seminal conjecture of Kontsevich [K]; we can ignore this in only using mirror symmetry as motivation. It could be noted, however, that the functionals defined below are additive under exact sequences of holomorphic vector bundles and sums of Lagrangians, so should extend to the derived category of coherent sheaves and the derived Fukaya category of Lagrangians respectively.

First note that while the connections side has a complex structure and a complex gauge group involved, the Lagrangian side needs complexifying. So motivated by Kontsevich [K] and by physics (e.g. [SYZ]) we add in connections on the submanifolds (which will later reduce to flat connections on Lagrangians). The dictionary we are aiming towards, much of which is already standard, is, in the 3-dimensional case, the table on the next page; all the terms used will be defined in due course.

Some notation: in the fourth line, $v(E)$ is the Mukai vector of E; in the last line, vol (L) is the cohomological volume measured with respect to $\mathrm{Re}\,\Omega$. A SLag cycle (of phase ϕ) in a Calabi-Yau is a Lagrangian with $\mathrm{Im}\,(e^{-i\phi}\Omega)|_L \equiv 0$ [HL]; then $\mathrm{Re}\,(e^{-i\phi}\Omega)|_L$ is the Riemannian volume form on L induced by the Ricci-flat metric on the Calabi-Yau. Obviously, rotating Ω by $e^{-i\phi}$ gives SLags in the more traditional sense of phase zero. The part of the theory to do with SLags will apply in all dimensions; it is only the functionals that are special to Calabi-Yau 3-folds.

We will partially justify the table, though the symplectic structure and moment map give problems that will appear in due course. However we can derive enough to arrive at a conjecture about Lagrangians and SLags for which evidence will be given by using monodromy and mirror symmetry from [ST] to interpret an example of Lawlor and Joyce [J].

Acknowledgements. The debt of any ex-student of Simon Donaldson writing a paper on moment maps should be clear. This work is also more immediately influenced by the papers [D], [J], [K]. In particular I was surprised to see the Lagrangian condition coming from a moment map in [D], [H], which does not fit into the scheme I always supposed was true. So the purpose of this paper, apart from trying to set a record for the number of m's in a title, is to expand on that scheme and to try to get the *special* condition from a moment map instead. This paper was finished in the summer of 2000 and reported on in [T3]; since then exciting new ideas have appeared in physics [Do] and mathematics [KS] better explaining mirror symmetry. I would like to thank S.-T. Yau, C. H. Taubes and Harvard University for support, and Yi Hu, Albrecht Klemm, Ivan Smith and Xiao Wei Wang for useful conversations. Communications with Mike Douglas, Dominic Joyce, Paul Seidel, S.-T. Yau and Eric Zaslow have been extremely influential.

Complex side M	Symplectic side W
$\Omega = \Omega_M \in H^{3,0}$	$\omega = \omega_W \in H^{1,1}$
H^{ev}	H^3
Connections A on a fixed C^∞ complex bundle E; $v(E) = ch(E)\sqrt{\mathrm{Td}\,X} \in H^{\mathrm{ev}}$	Submanifolds/cycles L in a fixed class $[L] \in H^3$, with a connection on $\mathbb{C} \times L$
$CS_{\mathbb{C}}\,(A = A_0 + a) = \frac{1}{4\pi^2}\int_M \mathrm{tr}\left(\bar{\partial}_{A_0} a \wedge a + \frac{2}{3} a^{\wedge 3}\right) \wedge \Omega$	$f_{\mathbb{C}}\,(A, L) = \int_{L_0}^{L} (F + \omega)^2 = \int_{L_0}^{L} (F^2 + \omega^2) + 2\int_{L_0}^{L} \omega \wedge F$
Critical points : $F_A^{0,2} = 0$ holomorphic bundles	Critical points : $\omega\vert_L = 0,\ F_A = 0$ Lagrangians + flat line bundles
Holomorphic Casson invariant [T1]	Counting SLags [J]
Gauge group	$U(1)$ gauge group on L
Complexified gauge group	Hamiltonian deformations
$\omega = \omega_M \in H^{1,1}$	$\Omega = \Omega_W \in H^{3,0}$
Moment map $F_A \wedge \omega^{n-1}$	Moment map $\mathrm{Im}\,\Omega\vert_L$
Stability, slope $\mu = \frac{1}{\mathrm{rk}\,E}\int \mathrm{tr}\,F_A \wedge \omega^{n-1}$	Stability, slope $\mu = \frac{1}{\mathrm{vol}(L)}\int_L \mathrm{Im}\,\Omega$

2 Chern-Simons-type functionals and critical points

Consider the space $\mathcal{A}$ of $(0,1)$-connections A on a fixed complex bundle on a Calabi-Yau 3-fold M. This infinite dimensional space has a natural complex structure, with respect to which it admits a holomorphic functional, Witten's holomorphic Chern-Simons functional [W1, DT],

$$CS_{\mathbb{C}}\,(A = A_0 + a) = \frac{1}{4\pi^2}\int_M \mathrm{tr}\left(\bar{\partial}_{A_0} a \wedge a + \frac{2}{3} a \wedge a \wedge a\right) \wedge \Omega,$$

where Ω is the holomorphic (3,0)-form. It is infinitesimally gauge-invariant (gauge transformations not homotopic to the identity can give periods to $CS_{\mathbb{C}}$) and its gradient is $F_A^{0,2}$, with zeros the integrable connections. That is, after dividing by gauge equivalence (under which $\operatorname{grad} CS_{\mathbb{C}}$ is invariant), the critical points of $CS_{\mathbb{C}}$ form the space of holomorphic bundles of the same topological type. As critical points of a functional, moduli of holomorphic bundles have virtual dimension zero, and one might try to make sense of counting them – a holomorphic Casson invariant [T1]. This is independent of deformations of the complex structure, but can have wall-crossing changes as the Kähler form varies. (This is because we count only stable bundles, and the notion of stability depends on a Kähler form.)

On the other hand, on a different Calabi-Yau 3-fold W (for instance the mirror, in some situation where this makes sense), Lagrangians are the critical points of a functional too, on the space of all 3-dimensional submanifolds (or cycles):

$$f_{\mathbb{R}}(L) = \int_{L_0}^{L} \omega \wedge \omega,$$

where ω is the symplectic form on W. Here L_0 is a fixed cycle in the same homology class, and we integrate over a 4-chain with boundary $L - L_0$; the functional $f_{\mathbb{R}}$ is invariant under the choice of different, homologous, 4-chains (picking non-homologous 4-chains can give periods to $f_{\mathbb{R}}$). It is invariant under deformations of L pulled back from hamiltonian deformations of W (deformations generated by vector fields v on W whose contraction with ω is exact $v \lrcorner \omega = dh$ at each point in time) as $\int_L \omega \wedge dh = 0$, and its gradient is $\omega|_L$. Thus its critical points are Lagrangian submanifolds. We would like to think of $f_{\mathbb{R}}$ as mirror to $CS_{\mathbb{C}}$, but to do so we must complexify it.

Thus we work on the space $\mathcal{A}$ of submanifolds L of W with $U(1)$ connections A on the trivial bundle $\mathbb{C} \times L$ on L. Notice these submanifolds are not parameterised by a map of a real 3-manifold into W; we are only interested in the image L. *From now on we shall restrict attention to smooth Lagrangian submanifolds.* Formally, we consider the tangent space to $\mathcal{A}$ at a point $(A, L \subset W)$ to be

$$\Omega^1(L; \mathbb{R}) \oplus \Omega^1(L; \mathbb{R}), \tag{2.1}$$

at least for those L with no J-invariant subspaces of its tangent spaces (J is the complex structure on W, and this is reasonable since we are looking for Lagrangian submanifolds after all). The first factor is the obvious tangent space to the connections on L; the second gives deformations of L via the vector fields produced by contracting with the Kähler form ω on W. That is, we use the metric on W to map $\Omega^1(L)$ to $\Omega^1(W)|_L$, then use the isomorphism provided by ω to get a vector field along L. Equivalently, using the metric on W, we may think of one-forms on L as tangent vectors to L, then apply the complex structure J on W to give W-vector fields on L. We denote this map from one-forms to normal vector fields by

$$\Omega^1(L) \to TW|_L, \qquad \sigma \mapsto \sigma \lrcorner \omega^{-1}. \tag{2.2}$$

Connections on L are carried along by the vector field to connections on nearby cycles, and we are identifying the space of $U(1)$ connections with $i\Omega^1(L;\mathbb{R})$.

There is a natural almost complex structure on $\mathcal{A}$, acting as

$$J = \begin{pmatrix} 0 & 1 \\ -1 & 0 \end{pmatrix},$$

with respect to the splitting (2.1) of the tangent spaces. With respect to this we claim to have the following holomorphic functional

$$f_{\mathbb{C}}(A, L) = \int_{L_0}^{L} (F+\omega)^2 = \int_{L_0}^{L} (F^2+\omega^2) \ + \ 2\int_{L_0}^{L} \omega \wedge F.$$

Here we have extended A to a connection on the trivial bundle on the whole of W (restricting to a fixed connection A_0 on L_0, and to A on L) and taken its curvature form F. We have again picked a 4-cycle bounding $L - L_0$; because F and ω are closed the resulting functional is independent of different homologous choices of the 4-cycle, and in general well defined up to the addition of some discrete periods. It is also (again) independent of hamiltonian isotopies of L. Notice that the $\int_{L_0}^{L} F^2$ term is just the *real* Chern-Simons functional $CS_{\mathbb{R}}$ of the connection A on L, whose critical points are well known to be flat connections. As pointed out to me by Eric Zaslow, the real and complex Chern-Simons functionals already appear in [W1] and [Va] as possible mirror partners (this is partially justified in [LYZ]), but without the terms in the symplectic form (and including instanton corrections from holomorphic disks which we are ignoring for our rough analogy). Asking for a real function to be equal to a complex one is possible when one restricts attention to a real slice such as the space of *Lagrangian* submanifolds in $\mathcal{A}$; deforming within this space the imaginary part of $f_{\mathbb{C}}$ remains constant and it reduces to $CS_{\mathbb{R}}$. But allowing the imaginary counterparts to these real deformations the right functional to consider is $f_{\mathbb{C}}$. Notice also that if $\omega/2\pi$ is integral, so that we can pick a connection B with curvature $-i\omega$, then the action functional can be written in the more familiar looking Chern-Simons form

$$f_{\mathbb{C}}(A, L) = \int_L (B+iA) \wedge d(B+iA) = \int_L C dC$$

for the 'complexified connection' $C = B + iA$ (a $\mathbb{C}^\times$-connection, instead of a $U(1)$-connection.) This makes more contact with the physics literature and allows one to extend the identification of $CS_{\mathbb{R}}$ and $CS_{\mathbb{C}}$ in [LYZ] to non Lagrangian sections, giving complex values. Tian has informed me that he and Chen have also considered the functional $f_{\mathbb{C}}(A, L)$ [Ch].

Mirror symmetry should relate Lagrangians not just to bundles but the whole derived category. For Riemann surfaces $C \subset M$, for instance, there is a functional in [DT], [W2] (rather like $f_{\mathbb{R}}$ above):

$$\int_{C_0}^{C} \Omega$$

is formally holomorphic and has as critical points the *holomorphic* curves C. Similarly for four-manifolds $S \subset M$ with connections on them the following functional (formally similar to $f_{\mathbb{C}}$)

$$\int_{S_0}^{S} \operatorname{tr} F \wedge \Omega$$

has critical points the holomorphic surfaces with flat connection on them. Alternatively, as $CS_{\mathbb{C}}$ is additive under extensions of bundles it does extend to the derived category. (Whether these two approaches are compatible; i.e. whether or not the functional associated to a curve or surface is the same as CS_C applied to a locally free resolution of its structure sheaf, up to a constant, seems to not have been worked out.)

That $f_{\mathbb{C}}$ is holomorphic follows from the computation that the derivative of $f_{\mathbb{C}}$ down $a \in \Omega^1(L) \oplus 0$ (that only changes the connection $A \mapsto A + \delta a$) is $\int_L 2F \wedge ia + 2\omega \wedge ia$, while the derivative down $-Ja \in 0 \oplus \Omega^1(L)$, i.e. down the vector field $a \lrcorner \omega^{-1}$, is $\int_L 2\omega \wedge a + 2a \wedge F$. The second expression is $-i$ times the first, so the derivative is complex linear and f is holomorphic. Equivalently we are saying that $df_{\mathbb{C}}$ is the 2-form

$$2i\,(F + \omega) \;\oplus\; 2\,(F + \omega),$$

which pairs with the tangent space (2.1) by integration over L to give a form of type (1,0) on (2.1).

Thus critical points of the functional are Lagrangian cycles with flat line bundles on them: exactly the basic building blocks of the objects proposed in [K] to be mirror dual to the holomorphic bundles that are the critical points of CS. So this ties in three well known moduli problems of virtual dimension zero (i.e. with deformation theories whose Euler characteristic vanishes) – flat bundles on 3-manifolds, holomorphic bundles on Calabi-Yau 3-folds, and Lagrangians (up to hamiltonian deformation) in symplectic 6-manifolds.

So as mirror to [T1] one would like to count Lagrangians (up to hamiltonian deformations) plus flat line bundles on them, and this is what Joyce's work [J] has begun to tackle (in the rigid case of L being a homology sphere). Mirroring precisely the behaviour of the holomorphic Casson invariant this count appears to be independent of deformations of the Kähler form and to have wall-crossing changes as the complex structure varies.

3 Gauge equivalence and moment maps

In fact what Joyce is proposing to count is *special* Lagrangian spheres with flat line bundles on them (hence the otherwise anomalous dependence on the complex structure), while [T1] counts *stable* bundles (i.e. by Donaldson-Uhlenbeck-Yau, modulo the technicalities of polystable and non-locally-free sheaves, we count *Hermitian-Yang-Mills* connections; hence the dependence on the Kähler form). (Tyurin [Ty] was perhaps the first to suggest that the holomorphic Casson invariant should be related by mirror symmetry to the real Casson invariant (here the $U(1)$ Casson invariant) of SLag submanifolds.)

The link should be, of course, that we want to consider holomorphic connections on one side, up to complex gauge equivalence, and Lagrangians on the other side, up to hamiltonian isotopy, and in both cases we try to do this by picking distinguished representatives of equivalence classes by the usual method of symplectic reduction. Bringing in a Kähler structure on the complex side, we get a moment map for the gauge group action, whose zeros give the HYM equations. Dually, we would like to bring in the holomorphic 3-form on the symplectic (Kähler) side, and get a complex group to act. So again complexify by adding flat line bundles: consider the critical points of the functional f of the last section, i.e. the space

$$\mathcal{Z} = \{(L, A) \,:\, L \subset W \text{ is Lagrangian, } A \text{ is a flat connection on } L\}$$

(*not* up to gauge equivalence). In fact consider this space on a Calabi-Yau manifold W of any dimension n. It has tangent space

$$T_{(L,A)}\mathcal{Z} = Z^1(L) \oplus Z^1(L)$$

($Z^1(L)$ denotes closed real one-forms on L), the first being tangent to the space of flat connections, the second giving normal vector fields (by contracting with ω^{-1}) preserving the Lagrangian condition. We have an obvious almost complex structure

$$J = \begin{pmatrix} 0 & 1 \\ -1 & 0 \end{pmatrix}. \tag{3.1}$$

Then the real group $C^\infty(L;\mathbb{R})/\mathbb{R}$ acts as the Lie algebra to the group of gauge transformations on the flat line bundles (taking d and adding to the connection) whose complexification $C^\infty(L;\mathbb{C})/\mathbb{C}$ acts complex linearly: the imaginary part $C^\infty(L;\mathbb{R})/\mathbb{R}$ acts by hamiltonian deformations through the normal vector field

$$h \mapsto dh \lrcorner \omega^{-1}.$$

Unfortunately, without using a metric this vector field is only defined up to the addition of tangent vector fields to L; the map (2.2) is really a map to $(TW|_L)/TL$ which we have lifted to $TW|_L$ using the metric. (Equivalently we can extend h to a first formal neighbourhood of L in different ways to get a different vector field.) How we pick this alters how we carry the flat connection along with L, and how the almost complex structure (3.1) acts. For instance suppose we are in the rather artificial case of L being transverse to an SYZ T^n-fibration. Then we can carry L and the flat connection up the fibres and identify the functions $C^\infty(L)$ from Lagrangian to Lagrangian using the projection. Thus the group remains constant as L moves (effectively what we are doing is extending functions from L to a neighbourhood of L in W by pulling up along the SYZ fibres). This does not work when L branches over the base of such a fibration. One can instead use the metric to define normal vector fields, but then identifying the Lie algebra $C^\infty(L)$ with a fixed $C^\infty(L_0)$ for all L becomes difficult.

This problem is perhaps not so surprising – the moment the Lagrangian has branching over the base of an SYZ fibration simple explicit correspondences between Lagrangians and vector bundles (such as [LYZ]) also break down due to our ignoring important holomorphic disk instanton corrections that appear in the physics. For instance recent work of Fukaya, Oh, Ohta and Ono [FO3], surveyed in [Fu1], show these provide the obstructions mirror to those of deformations of holomorphic bundles [T2] – one should not in general consider all (S)Lags (which are unobstructed) as mirror to holomorphic bundles, but only those whose Floer cohomology (whose definition involves holomorphic disks) is well defined.

However, what is clear is the totality of the group action, even if identifying individual elements causes problems, and this is all we really need. For instance in the $K3$ (or T^4) case one can get the same total group orbit, with a genuine fixed group acting, by hyperkähler rotating a construction due to Donaldson [D]. The end result is that one considers parametrised Lagrangian embeddings f from a Riemann surface L into the $K3$ such that the pullback of $\operatorname{Re}\Omega$ is a fixed symplectic form on L. Then the group of exact symplectomorphisms of

$$(L, f^*\mathrm{Re}\,\Omega)$$

provide a symmetry group of the space of maps f, which also carries a natural Kähler structure. Complexified orbits give hamiltonian deformations, and the moment map is $m(f) = f^*\,\mathrm{Im}\,\Omega$. The connection with our construction is that after fixing a line bundle η and connection with curvature $f^*\mathrm{Re}\,\Omega$, an infinitesimal symplectomorphism ϕ induces a flat connection, via parallel transport and pull back, on the bundle $\eta \otimes \phi^* \eta^*$. Globally the action is different (this action has non-zero Lie bracket, for instance, and a fixed group) but the total group orbit and the moment map (see below) are the same.

In general it is clear that the problem of identifying the group for different embeddings of L should be resolved by working with the space of maps from a fixed L_0 to W, and enlarging the group by including diffeomorphisms of L_0, giving a semi-direct product of $\mathrm{Diff}(L_0)$ and $U(1)$ gauge transformations on L_0. Then the moment map for the diffeomorphism part of the total group would be the Lagrangian condition as in [D], and the problems we are encountering would come from the fact that the group is a semi-direct product and not a product, so that we cannot separate the two out and divide by them separately, as in effect we have been trying to do. Unfortunately, I have not found the correct formulation of the problem, but it is not so important for follows.

So we shall not worry too much about whether the complex structure defined above is integrable, the group is fixed, or the symplectic structure below is closed. In 1 complex dimension it is trivial, in 2 we can use Donaldson's picture, and in 3 dimensions we could either try to use an abstract SYZ fibration to deform and identify Lagrangians transverse to it, or take everything in this section as motivation for finding the stability condition for Lagrangians of the next section.

Fix a homology class of Lagrangians and multiply Ω by a unit norm complex number so that $\int_L \mathrm{Im}\,\Omega = 0$. We induce a symplectic structure on $\mathcal{Z}$ from J and the following metric on the tangent space

$$\langle a, b \rangle = \int_L a \wedge ((b \lrcorner \omega^{-1}) \lrcorner \mathrm{Im}\,\Omega),$$

for a, b closed 1-forms. A computation in local coordinates shows this is symmetric in a and b; in fact it can be written as

$$\int_L a \wedge (\tilde{b} \lrcorner \mathrm{Re}\,\Omega) = \int_L \cos\theta\, (a \wedge *b), \tag{3.2}$$

where $\tilde{}$ is the isomorphism $T^*L \to TL$ set up by the induced metric on L, $\Omega|_L = e^{i\theta}\mathrm{vol}_L$, and vol_L the Riemannian volume form on L induced by the Ricci-flat metric. Thus for Lagrangians with $\theta \in (-\pi/2, \pi/2)$, i.e. those for which $\mathrm{Re}\,\Omega$ restricts to a nowhere vanishing volume form on L and so are not too far from being SLag ($\theta \equiv 0$), this gives a non-degenerate metric.

The symplectic form is invariant under the group action, and formally the moment map is indeed $m(L, A) = \mathrm{Im}\,\Omega$ in the dual $\Omega^n(L)_0$ of the Lie algebra (i.e. n-forms on L with integral zero). This follows from the computation

$$X \int_L h\,\mathrm{Im}\,\Omega = \int_L h\,d\,(X \lrcorner \mathrm{Im}\,\Omega) = \int_L dh \wedge ((\sigma \lrcorner \omega^{-1}) \lrcorner \mathrm{Im}\,\Omega),$$

where $X = \sigma \lrcorner \omega^{-1}$ is a normal vector field to the Lagrangian L down which we compute the derivative of the hamiltonian $\int_L h\,\mathrm{Im}\,\Omega = \langle m(L, A), h\rangle$ for the infinitesimal action of h. Here have extended h to a first-order neighbourhood of $L \subset W$ so that it is constant in the direction of $X = \sigma \lrcorner \omega^{-1}$. Then the right hand side of the above equation is the pairing using the symplectic form of dh and σ, as required.

Infinitesimally we can see the moment map interpretation very easily, and fitting naturally with the mirror bundle point of view. Deformations of holomorphic connections A modulo complex gauge equivalence are given by a $\ker \bar\partial_A / \mathrm{im}\,\bar\partial_A$ first cohomology group, related to deformations $\ker \bar\partial_A \cap \ker \bar\partial_A^*$ of the HYM equations (modulo *unitary* gauge transformations) via Hodge theory, with the moment map equation providing the $d^* = 0$ slice to the imaginary part of the linearised group action. Similarly, deformations of Lagrangians are given by closed 1-forms $\ker d : \Omega^1(L;\mathbb{R}) \to \Omega^2(L;\mathbb{R})$, so that dividing by hamiltonian deformations we get

$$H^1(L) = \ker d / \mathrm{im}\, d.$$

If instead of dividing we impose the *special* condition, we get a $\ker d^*$ slice

$$H^1(L) = \ker d \cap \ker d^*,$$

to the (imaginary) deformations (real deformations are given by changing the flat $U(1)$ connection that can be incorporated into this).

A symplectic example

To motivate a guess at the correct definition of stability for Lagrangians, we expand on an example of Lawlor and Joyce ([J] Sections 6 and 7, building on work of [Ha], [L]; see also a similar example in [SV] that is studied in [TY]), explaining its relevance to mirror symmetry, and giving a simple example in algebraic geometry that mirrors it.

First define the pointwise phase θ of a submanifold L: we may write

$$\Omega|_L = e^{i\theta}\,\mathrm{vol}$$

where vol is the Riemannian volume form on L induced by Yau's Ricci-flat metric [Y] on W. Thus vol provides a (local) orientation for L, and reversing its sign alters the phase θ by π. A SLag is a Lagrangian with *constant* phase θ.

At first sight θ is multiply-valued; we always choose it to be a fixed single-valued function to $\mathbb{R}$, lifting $e^{i\theta} : L \to S^1$ and thus providing the Lagrangian with a *grading* as introduced by Kontsevich [K], [S2]. Thus *we only consider Lagrangians of vanishing Maslov class* – for a Calabi-Yau this is the winding class $\pi_1(L) \to \pi_1(S^1)$ of the phase map

$$L \xrightarrow{e^{i\theta}} S^1,$$

which of course vanishes for a SLag. (The definition of grading in [K], [S2] is topological and uses the universal $\mathbb{Z}$-cover of the bundle of Lagrangian Grassmannians; here we first pass to the $\mathbb{Z}/2$ orientation cover of the Grassmannian, choosing an orientation of our Lagrangians, and then use a complex structure to pass to the universal $\mathbb{Z}$-cover of this. The two definitions are of course equivalent.)

Similarly we can define a kind of average phase $\phi = \phi(L)$ of a submanifold (or homology class) $L \subset W$ by

$$\int_L \Omega = A\, e^{i\phi(L)},$$

for some real number A; we then use $\mathrm{Re}\,(e^{-i\phi(L)}\Omega|_L)$ to orient L. Reversing the sign of A alters the phase by π and reverses the orientation. Again for a *graded* Lagrangian $L = (L,\theta)$, and *we will always implicitly assume a grading*, $\phi(L)$ is canonically a real number (rather than S^1-valued). Shifting the grading $[\,2n\,] : \theta \mapsto \theta + 2n\pi$ gives a similar shift to the phase $\phi(L)$.

The terminology comes from the fact that if there is a submanifold in the same homology class as L that is SLag with respect to some rotation of Ω, then it is with respect to $e^{-i\phi(L)}\Omega$. Slope, which we define as

$$\mu(L) := \tan(\phi(L)) = \frac{1}{\int_L \mathrm{Re}\,\Omega}\int_L \mathrm{Im}\,\Omega,$$

is defined independently of grading, is monotonic in ϕ in the range $(-\pi/2, \pi/2)$, and is invariant under change of orientation $\phi \mapsto \phi \pm \pi$. This agrees with the slope of a straight line SLag in the case of T^2, as featured in [PZ], and we think of it as mirror to the slope of a mirror sheaf, as is shown for tori in [PZ] (see [DFR] for corrections in higher dimensions away from the large complex structure limit).

Joyce describes examples of SLags which we interpret as follows. We have a family of Calabi-Yau 3-folds W^t as t ranges through (a small open subset of) the moduli space of complex structures on W *with fixed symplectic structure.* That is, the holomorphic 3-form Ω^t varies with t, but the Kähler form ω is fixed. We also have a family of SLag homology 3-spheres $L_1^t,\ L_2^t \subset W^t$ such that L_1^t and L_t^2 intersect at a point. If we choose a rotation of Ω^t such that L_2^t always has phase $\phi_2^t \equiv 0$ (this is possible locally at least; in the family described later it will have to be modified slightly), then we are interested as t varies only in the complex number

$$\int_{L_1^t} \Omega^t = R^t e^{\phi_1^t}$$

and its polar phase $\phi = \phi_1^t$; we plot this (i.e. the projection from the complex structure moduli space to $\mathbb{C}$ via this map) in Figure 1.

Then in Joyce's example, for $\phi < 0$ (and $R^t > 0$) there is a SLag L^t (of some phase ϕ^t) in the homology class $[L^t] = [L_1^t] + [L_2^t]$, such that as $\phi \uparrow 0$, this degenerates to a singular union of SLags of the same phase $L^t = L_1^t \cup L_2^t$ and then disappears for $\phi > 0$.

Most importantly, where L^t exists as a smooth SLag ($\phi < 0$) we have the slope (and phase) inequality

$$\mu_1^t < \mu_2^t, \qquad \text{i.e. } \phi = \phi_1^t < \phi_2^t \equiv 0; \tag{3.3}$$

at $t = 0$, L^t becomes the singular union of L_1^t and L_2^t, with

$$\mu_1^t = \mu_2^t \ (\phi_1^t = \phi_2^t);$$

then there is no SLag in L's homology class for

$$\mu_1^t > \mu_2^t \ (\phi_1^t > \phi_2^t),$$

though there is a Lagrangian, of course – the symplectic structure has not changed. Though we have been using slope μ in order to strengthen the analogy with the mirror (bundle) situation, from now on we shall use only the phase (lifted to $\mathbb{R}$ using the grading). While each is monotonic in the other for small phase (as $\tan\phi = \mu$), slope does not see orientation as phase does; reversing orientation adds $\pm\pi$ to the phase but leaves μ unchanged. This is related to the fact that we should really be working with complexes and so forth on the mirror side (the bundle analogy is too narrow) and changing orientation has no mirror analogue in terms of only stable bundles; it corresponds to shifting (complexes of) bundles by one place in the derived category.

While slopes of bundles cannot go past infinity (without moving degree in the derived category at least), for Lagrangians they certainly can, and phase ϕ continues monotonically upwards as its slope $\tan\phi$ becomes singular and then negative.

Importantly, we can think of the various SLags as independent of time when thought of as Lagrangians in the fixed symplectic manifold W^t:

Lemma 3.4 *For $t>0$ the SLags L^t are all in the same hamiltonian deformation class. Similarly for L_1^t, L_2^t, and for $t<0$.*

Proof Now choosing the phase of Ω^t such that $\phi(L^t)\equiv 0$,

$$\int_L \frac{d}{dt}(\operatorname{Im}\Omega^t) = \int_L \operatorname{Im}\dot{\Omega}^t = 0. \tag{3.5}$$

To show this deformation preserves the hamiltonian class of L, we need to find a corresponding first order hamiltonian deformation $dh\lrcorner\omega^{-1}$ under which the change in $\operatorname{Im}\Omega^t$,

$$\mathcal{L}_{dh\lrcorner\omega^{-1}}(\operatorname{Im}\Omega^t)|_L = d((dh\lrcorner\omega^{-1})\lrcorner\operatorname{Im}\Omega^t)|_L,$$

is $-\operatorname{Im}\dot{\Omega}^t|_L$. But as $\operatorname{Re}\Omega^t|_L$ is the induced Riemannian volume form vol^t on L, this means we want to solve

$$-\operatorname{Im}\dot{\Omega}^t|_L = d(J(dh\lrcorner\omega)\lrcorner\operatorname{Re}\Omega^t|_L) = d(\widetilde{dh}\lrcorner\mathrm{vol}^t) = d(*dh) = \Delta(*h),$$

where J is the complex structure and $\tilde{}$ is the isomorphism $T^*L\to TL$ set up by the induced metric on L. So the equation has a solution by the Fredholm alternative and (3.5). □

Thus for $\phi>0$ we consider the L^ts as the *same* as Lagrangian submanifolds (up to hamiltonian deformation) in the *fixed* symplectic manifold W^t; it is only the SLag representative that changes as Ω^t varies. We think of this as mirror to a *fixed* holomorphic bundle in a fixed complex structure, with varying HYM connection as the mirror Kähler form changes.

Lemma 3.6 *In the analogous 2-dimensional situation of SLags in a $K3$ or abelian surface, the obstruction does not occur.*

Proof Choose a real path of complex structures W^t, $t\in(-\epsilon,\epsilon)$ in complex structure moduli space such that there is a nodal SLag $L^0=L_1^0\cup L_2^0$ in W^0. Without loss of generality we can choose the phase of Ω^t so that both ω and $\operatorname{Im}\Omega^t$ pair to zero on the homology class of L^0. Now hyperkähler rotate the complex structures so that instead the new $\operatorname{Re}\Omega^t$ and $\operatorname{Im}\Omega^t$ pair to zero on the homology class of L^0 for all t. L^0 is now a nodal holomorphic curve C

in the central $K3$. We can understand deformations of C via deformations of the ideal sheaf $\mathcal{J}_C$, with obstructions in

$$\mathrm{Ext}^2(\mathcal{J}_C, \mathcal{J}_C) \to H^{0,2}(W) \cong \mathbb{C},$$

where the arrow is the trace map and is an isomorphism by Serre duality. Standard deformation theory shows the obstruction is purely cohomological – it is the derivative of the $H^{0,2}$-component of the class

$$[C] \in H^2(W) \cong H^{2,0}(W) \oplus H^{1,1}(W) \oplus H^{0,2}(W).$$

But we have fixed this to remain zero by the phase condition, so the curve deforms to all t (really we should assume the family is analytic in t here and extend to $t \in \mathbb{C}$, or just work with first order deformations). Hyperkähler unrotating gives back a family of SLags. □

There is a notion of connect summing Lagrangian submanifolds intersecting in a single point (probably due to Polterovich) – see for instance Appendix A of [S1] – which we claim gives the smoothings L^t of the singular $L^0 = L_1 \cup L_2$. This follows by comparing the local models [J], [S1] for the Lagrangians; see [TY] where it is studied in more detail for a related purpose, and our conventions (slightly different from those of [S2]) are described. While topologically we are just connect summing L_1 and L_2 by removing a small 3-ball containing the intersection point from each and gluing the resulting boundary S^3s (there are two ways, depending on orientation), symplectically the construction does not explicitly use orientations of the submanifolds. (Effectively we are using their relative orientation – the canonical orientation of the sum of the tangent spaces of L_1, L_2 at the intersection point given by the symplectic structure.)

Giving L_1 and L_2 in that order produces a Lagrangian, well defined up to hamiltonian isotopy (this will be shown in Section 4 in more generality; see (4.1)),

$$L_1 \# L_2,$$

with the singular union $L_1 \cup L_2$ a limit point in the hamiltonian isotopy class, which is *not* itself hamiltonian isotopic to $L_1 \# L_2$ (we have seen a family of hamiltonian deformations which has limit $L_1 \cup L_2$, but the deformations are singular at this limit).

There is also an obvious notion of *graded connect sum*, which is in fact what we shall always mean by $\#$. There is a unique grading on L_1 compatible with a fixed grading on L_2 such that we can give a (continuous) grading to the smoothing $L_1 \# L_2$. In the case of connect summing at multiple intersection points (Section 4) there is *at most* one such grading; in general the graded connect sum may not exist.

In n dimensions, if L_1 and L_2 are graded such that $L_1\#L_2$ exists, then on reversing the order of the L_i, the graded sum that exists is

$$L_2\#(L_1[2-n]) \quad \text{in the homology class} \quad [L_2]+(-1)^n[L_1]. \tag{3.7}$$

Here $L[m]$ means the graded Lagrangian L with its grading changed by adding $m\pi$ to θ, and the homology class of $L_1\#L_2$ is $[L_1]+[L_2]$ using the orientations on the L_is induced by the gradings.

This is closely related, as we shall see, to Joyce's obstruction, and the lack of it in dimension 2 (Lemma 3.6). In 2 dimensions, $L_1\#L_2$ and $L_2\#L_1$ are in the same homology class, though by a result of Seidel [S1] *not in general in the same hamiltonian isotopy class,*

$$L_1\#L_2 \not\approx L_2\#L_1,$$

importantly (we use $\approx$ to denote equivalence up to hamiltonian deformations). For $t>0$ in the above family L^t is in the constant hamiltonian deformation class of $L_1\#L_2$, for $t<0$ it is in the different class of $L_2\#L_1$, and at $t=0$ it is $L_1 \cup L_2$ – in neither class but in the closure of both. (For complex t the symplectic structure is no longer constant like it is for $t \in \mathbb{R}$, as one can see by following through the hyperkähler rotation; thus we do not get a contradiction to the above statement by going round $t=0$ in $\mathbb{C}$.) In 3 dimensions, however, the corresponding obvious choice for a SLag on the other side of the $\pi_1^t=0$ wall, $L_2\#(L_1[-1])$, is in the wrong homology class.

In the case that the L_i are Lagrangian spheres we can see this by going round the wall

$$\phi(L_1^t)=0 \simeq \phi(L_2^t), \tag{3.8}$$

and using monodromy. In the 2-dimensional $K3$ or T^4 case this works as follows.

Consider a disk in complex structure moduli space over which the family of Kähler $K3$ surfaces (with constant Kähler form) degenerates at the origin to a $K3$ with an ordinary double point (ODP) with the Lagrangian $L_1 \cong S^2$ as vanishing cycle. A local model is the standard Kähler structure on $x^2+y^2+z^2=u$, over the parameter u in the unit disk in $\mathbb{C}$. Now base-changing by pulling back to the double cover in u, $u \mapsto u^2$, we get the 3-fold

$$x^2+y^2+z^2=u^2,$$

with a 3-fold ODP which has a small resolution at the origin putting in a holomorphic sphere resolving the central $K3$ fibre $u=0$. Choosing a nowhere-zero holomorphic section Ω_u of the fibrewise $(2,0)$-forms (using the fact that

$$\mathbb{C} = \left\{ \int_{L_1} \Omega = R e^{i\phi(L_1)} \right\}$$

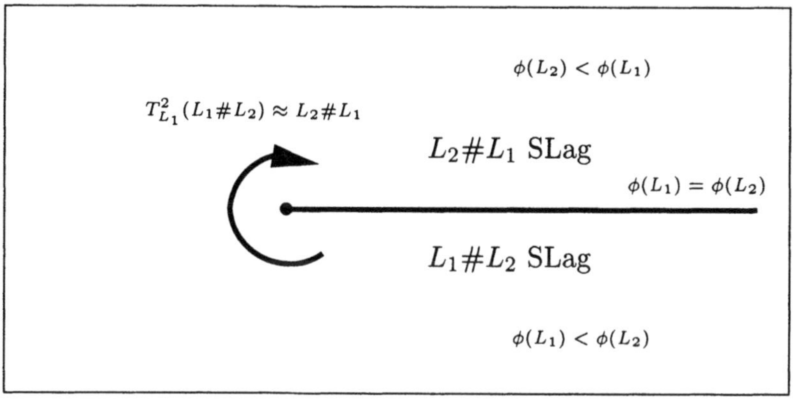

Figure 1. $\left(\int_{L_1} \Omega\right)$-space, as Ω on $K3$ varies, with polar coordinates $(R,\ \phi(L_1))$

the relative canonical bundle of either family is trivial), this restricts to zero on the exceptional $\mathbb{P}^1$ (which is homologous to the vanishing cycle L_1). Therefore the function

$$\int_{L_1} \Omega_u \tag{3.9}$$

has a simple zero at $u = 0$, i.e. it vanishes to order 1 in u. (The same expression vanished only as $\sqrt{u}$ in the original family with the singular fibre, and as such its sign was not well defined; this is because the class $[L_1]$ was defined globally only up to the monodromy $T_{L_1}[L_1] = -[L_1]$, i.e. up to a sign. In our new family the monodromy action $T^2_{L_1}$ is trivial on homology so it makes sense to talk about the homology class $[L_1]$ in any fibre, and (3.9) is single valued.)

Then our loop of complex structures is given by taking the loop $u = e^{it}$ and setting $\Omega^t = \Omega_{e^{it}}$. Pulling back the Kähler form from the original family, we get a locally trivial fibre bundle of symplectic manifolds over the circle whose monodromy is the Dehn twist $T^2_{L_1}$ (since the monodromy round the un-base-changed loop is T_{L_1} [S1]). As the family no longer has a singular fibre this monodromy is trivial as a diffeomorphism, but it is a result of [S1], [S2] that as a symplectic automorphism *it is non-trivial.* This is possible because although the family is a locally trivial bundle of symplectic manifolds over the punctured disk, over $u = 0$ the symplectic form becomes degenerate since it was pulled back via the resolution map.

Measuring $[L_1]$ against Ω_u as in (3.9) we see a principle familiar in physics (in issues of 'marginal stability', and taught to me by Eric Zaslow) – we detect a monodromy, like the degree 1 map $S^1 \to \mathbb{C}^\times$ given by $t \mapsto \int_{L_1} \Omega^t$ in this example, by counting wall crossing where a certain real part (here $\int_{L_1} \operatorname{Im} \Omega^t$, or the phase ϕ_1^t) hits $0 \simeq \phi_2^t$.

(Here we can no longer choose the phase of Ω such that $\phi_2^t = \phi(L_2^t) \equiv 0$ in the whole family, as the homology class of L_2 is not preserved in the family:

$$[T_{L_1}^2 L_2] = [L_2] + 2[L_1].$$

However, for a sufficiently small loop about the ODP, i.e. for $\left|\int_{L_1} \Omega\right|$ sufficiently small, this will not affect us much and we can write $\phi_2^t \simeq 0$: we are only interested in topological information like winding numbers and ϕ_1^t crossing the wall at $\phi_2^t \simeq 0$, which are unaffected by small perturbations.)

So instead of going through the $\phi(L_1^t) = \phi(L_2^t) \simeq 0$ wall we can go round it. If the loop is sufficiently small we do not encounter any more walls where the homology class $[L_1] + [L_2]$ can be split into classes of the same phase to possibly make the SLag a singular union of distinct SLags of equal phase. For instance the wall at phase 0 does not extend past $u = 0$ to phase $\phi_1^t = \pi$ (even though there $\mu_1^t = 0$) – the phase of L_1 is not zero but π, and is only zero for L_1 with the opposite orientation, so it does not exist as a SLag (e.g. in the hyperkähler rotated situation, we are saying there is no complex curve in L_1's homology class to possibly make L the nodal union of L_1 and something else, there is only an anti-complex curve). So we really can go round the wall; it ends at $u = 0$.

So this monodromy description shows that on the other $t \uparrow 2\pi$ side of the wall the SLag deforming $L_2 \cup L_1$ is in the hamiltonian deformation class

$$T_{L_1}^2 L = T_{L_1}^2 (L_1 \# L_2) = T_{L_1}^2 (T_{L_1}^{-1} L_2) \approx T_{L_1} L_2 \approx L_2 \# L_1, \tag{3.10}$$

as claimed (for the above equalities see [S1], [S2]).

Notice that the alternative connect sum description of the above Lagrangian

$$L_2 \# L_1 = T_{L_1}^2 (L_1 \# L_2) \approx T_{L_1}^2 (L_1) \# T_{L_1}^2 (L_2) \approx L_1[-2] \# T_{L_1}^2 (L_2), \tag{3.11}$$

does not violate the phase inequality to (3.3), as

$$-2\pi + \epsilon \approx \phi(L_1[-2]) < \phi(T_{L_1}^2 (L_2)) \simeq 0.$$

This is why it is important here to keep track of gradings – assigning the phase ϵ to $\phi(T^2_{L_1}(L_1))$ would give the opposite inequality, but one would not be able to form the above graded connect sum without also shifting the phase of $T^2_{L_1}(L_2)$ by -2π.

$$\mathbb{C} = \left\{ \textstyle\int_{L_1} \Omega = Re^{i\phi(L_1)} \right\}$$

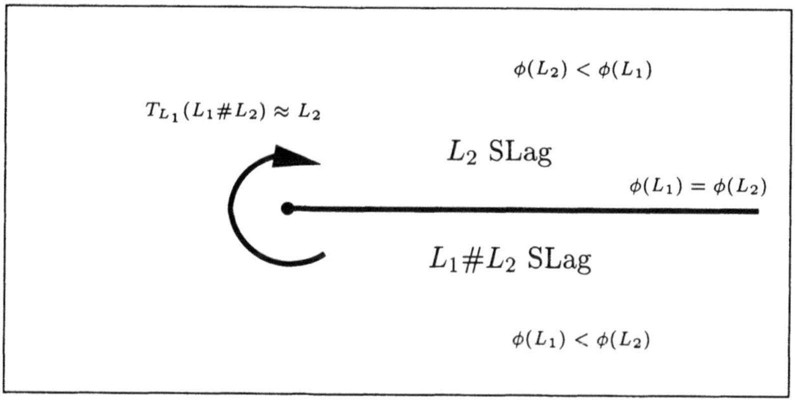

Figure 2. $\left(\int_{L_1} \Omega\right)$-space, as Ω on a 3-fold varies, with polar coordinates $(R,\ \phi(L_1))$

The 3-fold case (which Dominic Joyce has also, independently, considered) is slightly different; we need only take a single Dehn twist T_{L_1} corresponding to the local family

$$x^2 + y^2 + z^2 = u,$$

over $u \in \mathbb{C}$ to get a winding number one loop in the phase of L_1. This is because

$$T_{L_1} L_1 \approx L_1[1 - n]$$

in dimension n, so in 3 dimensions the homology class $[L_1]$ is preserved instead of being reversed. The corresponding picture is displayed in Figure 2.

Again there is a SLag on the other side of the $\phi = 0$ wall, but it is in the wrong homology class $[L_2]$:

$$T_{L_1} L \approx L_2. \tag{3.12}$$

Analogously to (3.11) this has a number of decompositions as connect sums induced by monodromy,

$$T_{L_1}(L_1\#L_2) \approx L_1[-2]\#(L_2\#(L_1[-1])) \approx L_2 \approx (L_1\#L_2)\#(L_1[1]),$$

none of which violate the phase inequality (3.3). The only other obvious choice for a (S)Lag on the other side of the $\phi = 0$ wall (given the $K3$ result) is $T_{L_1}^2(L_1\#L_2) \approx L_2\#(L_1[-1])$; this however is also in the wrong homology class, and in any case does violate (3.3) and so, by Joyce's analysis, should not be represented by a SLag. Thinking of $T_{L_1}^2$ as rotating through -4π in Figure 2, it is at roughly -3π that the phase inequality (3.3) gets violated, and the $-\pi$ rotation of L_2 splits as a SLag into the union of the $-\pi$ rotations of $(L_1\#L_2)$ and $L_1[1]$: these both have phase approximately zero.

A holomorphic bundle example

These phenomena are similar to wall-crossing in bundle theory on the complex side – in a real one-parameter family of Kähler forms, for fixed complex structure, stable holomorphic bundles for $t > 0$ can become semistable at $t = 0$ and unstable for $t < 0$.

An example that mirrors Joyce's is the following. Suppose we have two stable bundles (or coherent sheaves) E_1 and E_2 with

$$\mathrm{Ext}^1(E_2, E_1) \cong \mathbb{C}.$$

This is $H^1(E_1 \otimes E_2^*)$ in the case of bundles and is the mirror ^{K} of the one dimensional Floer cohomology $HF^*(L_2, L_1) \cong \mathbb{C}$ that is defined by the single intersection point of L_1 and L_2 (see Section 4 for more details of this, and an explanation of why we are dealing with Ext^1 and HF^1 here). We then form E from this extension class

$$0 \to E_1 \to E \to E_2 \to 0. \tag{3.13}$$

Take a family of Kähler forms ω^t such that $\mu^t(E_2) - \mu^t(E_1)$ is the same sign as t (here $\mu^t(F) = c_1(F).(\omega^t)^{n-1}/\mathrm{rk}(F)$ is the slope of F with respect to ω^t). Supposing that the E_i are stable for all $t \in (-\epsilon, \epsilon)$, we claim that E is stable for sufficiently small $t > 0$, while it is destabilised by E_1 for $t \leq 0$. Without loss of generality take $\mu^t(E_2) = \mu$ fixed, and $\mu^t(E_1) = \mu - t$. As E_2 is stable, for t sufficiently small there are no subsheaves of E_2 of slope greater than $\mu - t$, so for any stable destabilising subsheaf F of E, the composition

$$F \hookrightarrow E \to E_2$$

cannot be an injection (unless it is an isomorphism, but (3.13) does not split. So $F \cap E_1 \neq 0$, and the quotient $Q = F/(F \cap E_1)$ has slope $\mu(Q) > \mu(F) > \mu - t$

by the stability of F and instability of E. But Q injects into E_2, which we know is impossible.

In the 2-dimensional case, by Serre duality $\mathrm{Ext}^1(E_1, E_2) \cong \mathrm{Ext}^1(E_2, E_1)^* \cong \mathbb{C}$ on $K3$ or T^4, so for $t < 0$ we can instead form an extension

$$0 \to E_2 \to E' \to E_1 \to 0, \tag{3.14}$$

to give a new bundle E' which is also stable, and has the same Mukai vector

$$v(E') = v(E_1) + v(E_2);$$

compare (3.7). At $t = 0$ we take the (polystable) bundle

$$E_1 \oplus E_2.$$

This is because the semistable extension (3.13) no longer admits a Hermitian-Yang-Mills metric, but $E_1 \oplus E_2$ does. Also, the algebraic geometry of the moduli problem shows that while a semistable bundle gets identified in the moduli space with the other ("S-equivalent") sheaves in the closure of its gauge group orbit, there is a distinguished representative of its equivalence class – the polystable direct sum (of the Jordan-Hölder filtration, which here is $E_1 \oplus E_2$).

Thus, while the HYM connections vary, the bundle has only 3 different holomorphic structures – for $t > 0$, $t = 0$, and $t < 0$. Put another way (to spell out the analogy with the Lagrangians L^t, L_1, L_2) as ω_t varies with $t > 0$ we take different points in a fixed complexified gauge group orbit, and at $t = 0$ we take as limit point something in a different orbit that is nonetheless in the closure of the $t > 0$ (and $t < 0$) orbit. The *stable* deformations of the polystable $E_1 \oplus E_2$ (which we are thinking of as the mirror of the singular union $L_1 \cup L_2$, of course) are precisely (3.13) for $t > 0$ and (3.14) for $t < 0$.

In the 3-fold case, however, Serre duality gives $\mathrm{Ext}^2(E_1, E_2) \cong \mathrm{Ext}^1(E_2, E_1)^* \cong \mathbb{C}$ instead, and so no stable extension (3.14). In fact one would expect there to be no stable bundle with the right Chern classes; instead the one dimensional Ext^2 gives us a complex E' in the derived category $D^b(M)$ fitting into an exact sequence of complexes

$$0 \to E_2 \to E' \to E_1[-1] \to 0,$$

where $E_1[-1]$ is E_1 shifted in degree by one place to the right as a complex. This has Mukai vector

$$v(E') = v(E_2) - v(E_1),$$

compare (3.7). Thus, just as in the case of SLags, as we pass through $t = 0$ there is no natural stable object on the other side in the same homology class in 3 dimensions (though there is in 2 dimensions) and so an element of the appropriate moduli space disappears.

In fact, as in the Lagrangian example, the natural stable object on the other side of the wall is E_2 if we consider monodromy. The mirror of the symplectic Dehn twists of above are described in [ST] (in the case that the bundles E_i are *spherical* in the sense of [ST]: $\mathrm{Ext}^k(E_i, E_i) \cong H^k(S^n; \mathbb{C})$; this is the natural mirror analogue of the L_is being spheres). These are the twists T_{E_1} of [ST] on the derived category of the Calabi-Yau that act on the extension bundle E of (3.13) to give precisely the extension (3.14),

$$T_{E_1}^2 E = E'$$

(compare (3.10)), as a short calculation using [ST] shows. Similarly

$$T_{E_1} E = E_2,$$

the analogue of (3.12). (In both of these calculations it is important to calculate this monodromy in the derived category; in the $K3$ case the action of $T_{E_1}^2$ is trivial on K-theory and cohomology, and we cannot distinguish between (3.13) and (3.14), but they are very different as holomorphic bundles and as elements of the derived category.)

The mirror wall crossing, with a SLag splitting into two and then disappearing, is interpreted in [DFR] (and in [SV] in a different case) as the state it represents decaying as we reach a point of 'marginal stability'. Despite this dealing with only SLags (and so with only a priori *stable* Lagrangians in our mathematical sense of stability), this suggestive language does in fact have something to say about the stability, in our sense of group actions, of (non-special) Lagrangians, by considering the nodal limit $L_1 \cup L_2$ to be a *semistable* Lagrangian.

Thus the Lagrangian $L_1 \# L_2$ (which *always exists as a Lagrangian* as the complex structure varies with fixed Kähler form) becomes semistable at $t = 0$ and is represented by something in a different orbit of the hamiltonian deformation symmetry group (but in the closure of the original orbit), and is unstable for $t < 0$ so exists there only as a Lagrangian and *not* as a SLag. This, and the bundle analogue described above, leads us to think of the Lagrangian L_1 as destabilising $L = L_1 \# L_2$ when $\phi(L_1) \geq \phi(L_2)$. This motivates the now obvious definition of stability in Section 5; first we explain more about the connections to mirror symmetry, and generalisations to connect sums at more intersection points.

4 Relationship to Kontsevich's mirror conjecture

The inspiration behind most of this paper is of course Kontsevich's mirror conjecture [K]. In particular, Kontsevich proposes that the graded vector spaces Ext* and HF^* should be isomorphic for mirror choices of bundles E_i and graded Lagrangians L_i (or more exotic objects in their derived categories)

$$HF^*(L_2, L_1) \cong \mathrm{Ext}^*(E_2, E_1);$$

this corresponds to the equality of (graded) morphisms on both sides. Here HF^* is Floer cohomology [Fl] – a symplectic refinement of the intersection number of L_1 and L_2 – which can be $\mathbb{Z}$-graded for graded Lagrangians [S2], whenever it is defined [FO3], [Fu1]. (More precisely it is the cohomology of a chain complex built out of the free vector space generated by the intersection points, with the differential defined by counting holomorphic disks with boundary in the Lagrangians running from one intersection point to another.) In mirror symmetry, and so in this paper, one should only really consider those Lagrangians whose Floer cohomology is well defined [Fu1].

Thus the point of intersection of the L_1 and L_2 of the last section define the Floer cohomology $HF^*(L_2, L_1) \cong \mathbb{C}$, and the grading of [S2] is designed specifically so that $L_1 \# L_2$ can be graded precisely when the relative gradings of the L_i force the Floer cohomology to be concentrated in degree 1; $HF^*(L_2, L_1) = HF^1(L_2, L_1)$. We then think of the connect sum $L_1 \# L_2$ as being mirror to the extension (3.13) defined by $\mathrm{Ext}^1(E_2, E_1) \cong \mathbb{C}$. Fukaya, Seidel, and perhaps others have also proposed that Lagrangian connect sum should be mirror to extensions [Fu2], [S3].

We also consider connect sums of Lagrangians intersecting at n points p_i. Then the connect sum is not unique up to hamiltonian deformation: H^1 is added to the Lagrangian as loops between the intersection points, giving additional deformations of its hamiltonian isotopy class. The upshot is that there is a scaling of the neck of the connect sum at each intersection point; we denote any such resulting Lagrangian by $L_1 \# L_2$. Since we insist on all intersection points having Floer (Maslov) index one (so that the connect sum can be graded), the Floer differential vanishes in this case, and these scalings define a class in $HF^1(L_2, L_1)$.

Deformations (up to those which are hamiltonian) as such a connect sum are given by the elements of

$$H^1(L_1 \# L_2) \cong H_{n-1}(L_1 \# L_2)$$

spanned by the S^{n-1} vanishing cycles S_i at the points of intersection $p_i \in L_1 \cap L_2$. Given a particular connect sum, the deformation represented by

$\sum_i a_i S_i$ simply scales the local gluing parameter in a Darboux chart around each p_i by a factor $(1 + a_i)$ (here a_i is considered to be infinitesimal). Since the sum of these spheres separates $L_1 \# L_2$ into $L_1 \backslash \cup \{p_i\}$ and $L_2 \backslash \cup \{p_i\}$ and so is zero in homology

$$\sum_i [S_i] = \pm\partial[L_1 \backslash \cup \{p_i\}] = \mp\partial[L_2 \backslash \cup \{p_i\}] = 0 \in H_{n-1}(L_1 \# L_2),$$

the infinitesimal deformation represented by $\sum_i S_i$ is zero (it is pure hamiltonian) and dividing out gives the projectivisation

$$\mathbb{P}(\oplus_i \mathbb{R}_{p_i}). \tag{4.1}$$

(Replace $\mathbb{R}$ by $\mathbb{C}$ when including flat bundles and their gluing parameters at the p_is.) This explains the earlier claim that connect sums at one intersection point are uniquely defined up to hamiltonian deformations. More precisely, when holomorphic disks are taken into account and we consider only those Lagrangians whose Floer cohomology is defined [FO3], [Ful], hamiltonian deformation classes of connect sums *whose Floer cohomology can be defined* should be parameterised by $\mathbb{P}(HF^1(L_2, L_1))$. (On the mirror side *isomorphism classes* of extensions of E_2 by E_1 are parametrised by $\mathbb{P}\,\mathrm{Ext}^1(E_2, E_1)$.)

We would then expect that the resulting connect sum has a canonical homomorphism from L_1; that is there should be a canonical element

$$\mathrm{id}_{L_1} \in HF^0(L_1, L_1 \# L_2)$$

for any graded Lagrangians L_i for which the graded connect sum exists. While a local model suggests this is true (see for instance [TY]), a complete proof is still not available. This homomorphism we think of as expressing L_1 as a *subobject* of $L_1 \# L_2$; i.e. as giving an injection. It should be emphasised that subobject does not make sense in a triangulated category such as the derived Fukaya category of Lagrangians; in the context of the derived category of sheaves, subobject only makes sense for an abelian category such as that of the sheaves themselves (i.e. complexes with cohomology in degree zero only). What we are proposing is that it also makes sense in the category of (complexes of sheaves mirror to) graded Lagrangians, and is vital to make definitions of stability (which involve such subobjects). While there are now more Homs to consider, in particular those of higher order (i.e. Homs to Lagrangians shifted in phase by some $2\pi n$), the targets of these Homs have higher phase and so do not disturb the definition of stability below – this is seemingly a huge piece of luck that means we can extend the stability condition for bundles to all Lagrangians. For similar reasons, the many connect sum decompositions of the L_is given in the last section also do not destabilise them.

There are other operations, however, which can also be thought of as Ext^1-type extensions. For instance, taking the product of a single Lagrangian curve L_1 in T^2 with a (graded) connect sum $L_2 \# L_3$ in another T^2, we get a Lagrangian $L_1 \times (L_2 \# L_3)$ in T^4 which is some kind of extension of the Lagrangians $L_1 \times L_2$ and $L_1 \times L_3$ in T^4. Supposing that the L_is are mirror to some (complexes of) sheaves E_i, and that the connect sum $L_2 \# L_3$ is mirror to an extension represented by an element $e \in \mathrm{Ext}^1(E_3, E_2)$. Then by the Künneth formula for sheaf cohomology, we see that $L_1 \times (L_2 \# L_3)$ is indeed mirror to an extension

$$\mathrm{id} \otimes e \in \mathrm{Hom}\,(E_1, E_1) \otimes \mathrm{Ext}^1(E_3, E_2) = \mathrm{Ext}^1(E_1 \boxtimes E_3, E_1 \boxtimes E_2),$$

and so this sort of relative connect sum (which is not $\#$ on T^4: $L_1 \times L_2$ and $L_1 \times L_3$ do not intersect transversely) should also be considered.

So we consider Lagrangians L_1, L_2 intersecting *cleanly* (see e.g. [S1] Definition 2.1), that is $N = L_1 \cap L_2$ is a smooth submanifold, and $TN = TL_1|_N \cap TL_2|_N$. Basic results of Weinstein allow us to identify a neighbourhood of N with a neighbourhood of the zero section N in $T^*N \oplus E$, where the total space of T^*N has its canonical symplectic structure, and

$$E \equiv (TL_1|_N)/TN \oplus (TL_2|_N)/TN$$

is the annihilator, under the symplectic form, of $TN \subset TX|_N$ (to which the symplectic form therefore descends, making E a symplectic bundle).

Choosing a metric on E, compatible with its symplectic structure, such that its transverse subbundles $(TL_1|_N)/TN$, $(TL_2|_N)/TN$ are orthogonal, we can now perform the family connect sum of these, over the base N, since the local model in [S1] is $O(n)$ invariant. As before we insist that this can be compatibly graded again denote it by $\#$; given a grading on L_1 there will be at most one grading on L_2 such that this graded relative connect sum exists.

It should be noted that although such a clean intersection could be hamiltonian isotoped to be transverse, the resulting intersection points would not necessarily all be of Floer/Maslov index one, and so the pointwise graded connect sum could not be formed at every point; we would end up with an immersed Lagrangian. Studying which immersed Lagrangians should be included in the Fukaya category, and which embedded Lagrangians they should be considered equivalent to, is an important part of mirror symmetry and will need to be better understood to refine our conjecture. For instance forming extensions of bundles which also have nonzero homorphisms between them would appear to be mirror to forming connect sums between graded Lagrangians at index one intersection points, leaving the index zero intersection points immersed. In general one would like to consider two objects of the Fukaya category to be equivalent if their Floer cohomologies with any other objects are the same. This would include hamiltonian deformation equivalence, but also more exotic equivalences for immersed Lagrangians (thanks to Paul Seidel for pointing this out to me). A start in understanding the Floer cohomology of immersed Lagrangians is [Ak]; in the present paper we are largely ignoring singularities.

5 Stability

Definition 5.1 *Take graded Lagrangians (L_1, θ_1) and (L_2, θ_2), hamiltonian isotoped to intersect cleanly, and such that the graded (relative) Lagrangian connect sums $(L_1 \# L_2, \theta_1 \# \theta_2)$ exist as above. Then a Lagrangian L of Maslov class zero is said to be destabilised by the L_i if it is hamiltonian isotopic to such an $L_1 \# L_2$, and the phases (*real *numbers, induced by the gradings) satisfy*

$$\phi(L_1) \geq \phi(L_2).$$

If L is not destabilised by any such L_i then it is called stable.

Remarks

- There is an obvious notion of a flux homomorphism for isotopies of smooth Lagrangians, taking a deformation to an element of $H^1(L; \mathbb{R})$ (and linearising to give the usual deformation theory of Lagrangians). Namely, take a deformation $\Phi_t(L)$ through a vector field X_t, $t \in [0, 1]$ to the one form

$$\int_0^1 (X_t \lrcorner \omega) dt \in H^1(L; \mathbb{R}).$$

  Alternatively, the homomorphism takes a loop $\gamma \subset L$, tracing out the 2-cycle $f(\gamma \times [0, 1])$ in W under the isotopy, to the real number $\int_{\gamma \times [0,1]} \omega$. (See Chapter 10 of [MS] for the analogous map for symplectomorphisms.) If the isotopy Φ_t is hamiltonian, the flux is zero; the converse is also easily proved using the methods of ([MS] Theorem 10.12): we may assume without loss of generality that the 1-form $\int_0^1 X_t \lrcorner \omega$ is identically zero in $\Omega^1(L)$. [To see this, write the 1-form as $d\phi$, and compose the deformation with the time one map of the hamiltonian flow with vector field $d\phi \lrcorner \omega^{-1}$; this does not alter the flux in $H^1(L; \mathbb{R})$ or the property of being hamiltonian.] Then let Σ^s be the closed 1-form on L defined by

$$\Sigma^s = \int_0^s X_t \lrcorner \omega dt,$$

and let Ψ_t^s be the corresponding flow through time t. Then the flow $\phi_t = \Psi_t^t \circ \Phi_t$ is the corresponding *hamiltonian* flow from $\Phi_0(L)$ to $\Phi_1(L)$; see [MS].

Thus it is not too hard to check if two Lagrangians are hamiltonian deformations of each other, at least through smooth Lagrangians, if we know they are deformations of each other as Lagrangians. This second condition, however, is harder to test, as the results of [S1] demonstrate.

- As mentioned in the last section, holomorphic disks are crucial in both mirror symmetry and Floer cohomology; thus one should perhaps restrict attention in the above definition to those Lagrangians whose Floer cohomology is defined [FO3],

- As pointed out to me by Conan Leung, this definition and the resulting conjecture below may only be reasonable close to the large complex structure limit point where the mirror symmetric arguments used to motivate the conjecture are most valid.

Conjecture 5.2 *A Lagrangian of Maslov class zero has a special Lagrangian in its hamiltonian deformation class if and only if it is stable, and this SLag representative is unique.*

Again, we have been vague about singularities: which we allow, and what hamiltonian deformation equivalence would mean for them. We should also restrict attention to those Lagrangians whose Floer cohomology exists [FO3], and whose Oh spectral sequence $H^*(L) \Rightarrow HF^*(L,L)$ [Oh] degenerates; this will be discussed more in [TY]. We might also want to restrict to Lagrangians whose phase function varies only by a certain bounded amount; in the example worked out in [TY] this is required. In [TY] it is shown there that the gradient of the norm-squared $|m|^2$ of the moment map can be taken to be the mean curvature vector of the Lagrangian, so mean curvature flow (which is hamiltonian for Maslov class zero) should converge to this SLag representative if the Lagrangian is stable and the phase satisfies certain bounds.

6 The 2-torus

Everything works rather simply on T^2; Grayson [G], building on work of Gage, Hamilton and others (e.g. [GH]), has shown that mean curvature flow for curves (of Maslov class zero) converges to straight lines and so we get the mirror symmetric analogue of Atiyah's classification [At] of sheaves on an elliptic curve – they are basically all sums of stable sheaves. The only exceptions are the non-trivial extensions of certain sheaves by themselves; these correspond to thickenings of the corresponding special Lagrangian (giving fat SLags, as they are known in Britain, or multiply-wrapped cycles in physics speak).

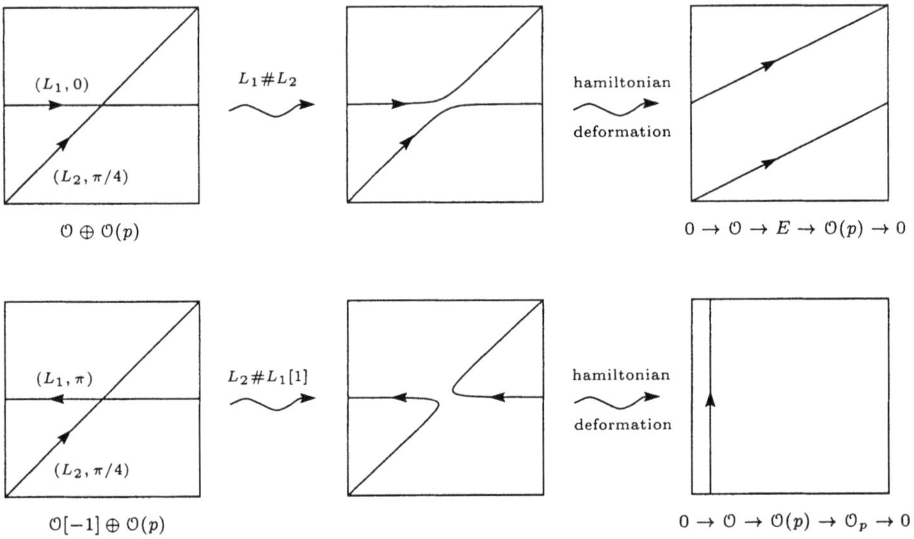

Figure 3. $L_1\#L_2$ and $L_2\#(L_1[1])$, equivalent SLags, and their mirror sheaves

We give an example to demonstrate why one cannot form smooth *unstable* Lagrangians on T^2 in Figure 3. First, giving L_1 and L_2 the gradings such that their phases are 0 and $\pi/4$, we expect $L_1\#L_2$ to be stable, and indeed we see it is hamiltonian deformation equivalent to the slope 1/2 SLag mirror to the stable extension E of $\mathcal{O}$ by $\mathcal{O}(p)$ (where p is a basepoint of T^2 with corresponding line bundle mirror to the diagonal SLag drawn).

If one then tries to form an unstable SLag $L_2\#L_1$, the *graded* connect sum does not exist – the phase would become discontinuous. To form $L_2\#L_1$ we see from the diagram that we have to take the phase of L_1 to be π, thus reversing its orientation, and in fact forming $L_2\#(L_1[1])$. Then the stability inequality (3.3) is not violated, and in fact this Lagrangian is stable and hamiltonian deformation equivalent to the SLag in T^2 represented by the vertical edge of the square (and so drawn with a little artistic license in Figure 3). Under the mirror map this corresponds to replacing the extension Ext^1 class by a Hom (as we have shifted complexes of sheaves by one place) and taking the cone of this in the derived category; this is the cokernel $\mathcal{O}_p$ of Figure 3.

As pointed out to me by Markarian and Polishchuk, one can play with lots of pictures of connect sums on tori to recover descriptions of certain moduli of sheaves, their special cycles (for instance where one connect-sum neck parameter goes to zero), and so forth, giving results similar to some of those in [FO].

This example can be extended to show that we cannot form the graded connect sum $L_1\#L_2$ of any two Lagrangians (via a class in $HF^*(L_1, L_2)$) if $\phi(L_1) > \phi(L_2)$. Namely, replace L_1 and L_2 by their hamiltonian deformation equivalent SLag representatives, which are straight lines of constant phase $\theta_i = \phi(L_i)$. As Figure 3 shows, $L_1\#L_2$ can be compatibly graded about an intersection point if and only if we have the local inequalities

$$\theta_2 > \theta_1 > \theta_2 - \pi.$$

Thus we require $\phi(L_2) > \phi(L_1)$. (We will explain this kind of phenomenon more generally in [TY] in terms of the grading on Floer cohomology.) Each intersection point is Floer coclosed since the Floer grading is the same as the relative orientation of the Lagrangians, mod 2, and this is the same at each intersection point of the straight lines. So each possible connect sum of the SLags defines a class in HF^*, and any other connect sum, defined on hamiltonian deformations of L_1 and L_2 by a class in HF^*, will be hamiltonian deformation equivalent to the appropriate connect sum of the SLags, and so satisfy the same phase inequality.

If two smooth Lagrangians have the same phase then their representative SLags will either be the same or disjoint parallel SLags. Either way there are no connect sums (though as mentioned above to account for the mirror symmetry of bundles one should also include non-trivial thickenings of SLags in the Fukaya category).

So unstable Lagrangians do not exist, and by the result of [G] mentioned earlier, the conjecture is true on T^2.

Thus complex dimension 1 is too simple – in trying to make the phase of one Lagrangian become larger than the phase of another, the two must cross, thus reversing their relative orientations and changing the order of the connect sum. Far more complicated phenomena arise in 2 and 3 dimensions, however.

References

[Ak] M. Akaho, *An intersection theory of Lagrangian immersions*, preprint.

[At] M. F. Atiyah, *Vector bundles over an elliptic curve*, Proc. London Math. Soc. **7** (1957), 414–452.

[AP] D. Arinkin and A. Polishchuk, *Fukaya category and Fourier transform*, preprint math.AG/9811023.

[BBHM] C. Bartocci, U. Bruzzo, D. Hernandez Ruiperez and J. Munoz Porras, *Mirror symmetry on $K3$ surfaces via Fourier-Mukai transform*, Comm. Math. Phys. **195** (1998), 79–93.

[Ch] J. Chen, *Lagrangian sections and holomorphic $U(1)$-connections*, preprint.

[D] S. K. Donaldson, *Moment maps and diffeomorphisms*, Asian Jour. Math. **3** (1999), 1–16.

[DT] S. K. Donaldson and R. P. Thomas, *Gauge theory in higher dimensions*. In: The Geometric Universe: Science, Geometry and the work of Roger Penrose, S. A. Huggett et al (eds), Oxford University Press, 1998.

[Do] M. R. Douglas, *D-branes, Categories and N=1 Supersymmetry*, preprint hep-th/0011017.

[DFR] M. R. Douglas, B. Fiol, and C. Römelsberger, *Stability and BPS branes*, preprint hep-th/0002037.

[FO] B. Feigin and A. Odesskij, *Vector bundles on an elliptic curve and Sklyanin algebras*, q-alg/9509021. In: Topics in quantum groups and finite-type invariants, B. Feigin et al (eds), AMS Transl. 185 (38), 65–84, 1998.

[Fl] A. Floer, *Morse theory for Lagrangian intersections*, Jour. Diff. Geom. **28** (1988), 513–547.

[Fu1] K. Fukaya, *Floer homology and Mirror symmetry I*, preprint (1999) http://www.kusm.kyoto-u.ac.jp/~fukaya/fukaya.html.

[Fu2] K. Fukaya, *Mirror symmetry of Abelian variety and multi theta functions*, preprint (2000) http://www.kusm.kyoto-u.ac.jp/~fukaya/fukaya.html.

[FO3] K. Fukaya, Y.-G. Oh, H. Ohta, and K. Ono, *Lagrangian intersection Floer theory - anomoly and obstruction*, book to appear in International Press.

[GH] M. Gage and R. Hamilton, *The shrinking of convex plane curves by the heat equation*, Jour. Diff. Geom. **23** (1986), 69–96.

[G] M. Grayson, *Shortening embedded curves*, Ann. of Math. **129** (1989), 71–111.

[Gr] M. Gross, *Special Lagrangian Fibrations I: Topology*. In: Integrable systems and algebraic geometry (Kobe/Kyoto, 1997), 156–193, World Sci. Publishing, 1998.

[Ha] F. R. Harvey, *Spinors and Calibrations*. Academic Press, San Diego, 1990.

[HL] F. R. Harvey and H. B. Lawson, *Calibrated Geometries*, Acta Math. **148**

(1982), 47–157.

[H] N. Hitchin, *Lectures on Special Lagrangian Submanifolds*, preprint math.DG/9907034.

[J] D. Joyce, *On counting special Lagrangian homology 3-spheres*, preprint hep-th/9907013.

[K] M. Kontsevich, *Homological Algebra of Mirror Symmetry*, International Congress of Mathematicians, Zürich 1994. Birkhäuser, 1995.

[KS] M. Kontsevich and Y. Soibelman, *Homological mirror symmetry and torus fibrations*, preprint math.SG/0011041.

[L] G. Lawlor, *The angle criterion*, Inv. Math. **95** (1989), 437–446.

[Le] N. C. Leung, *Mirror symmetry without corrections*, preprint math.DG/0009235.

[LYZ] N. C. Leung, S.-T. Yau and E. Zaslow, *From Special Lagrangian to Hermitian-Yang-Mills via Fourier-Mukai Transform*, preprint math.DG/0005118.

[MS] D. McDuff and D. Salamon, *Introduction to symplectic topology*, 2nd edition, Oxford University Press, 1998.

[MMM] Y. Manin, *Moduli, Motives, Mirrors*, preprint math.AG/0005144.

[MMMS] M. Marino, R. Minasian, G. Moore and A. Strominger, *Nonlinear Instantons from Supersymmetric p-Branes*, preprint hep-th/9911206.

[Oh] Y.-G. Oh, *Floer cohomology, spectral sequence and the Maslov class of Lagrangian embeddings*, Internat. Math. Res. Notices **7** (1996), 305–346.

[PZ] A. Polishchuk and E. Zaslow, *Categorical mirror symmetry: the elliptic curve*, Adv. Theor. Math. Phys. **2** (1998), 443–470.

[S1] P. Seidel, *Lagrangian two-spheres can be symplectically knotted*, Jour. Diff. Geom. **52** (1999), 145–171.

[S2] P. Seidel, *Graded Lagrangian submanifolds*, Bull. Soc. Math. France. **128** (2000), 103–146.

[S3] P. Seidel, private communication.

[ST] P. Seidel and R. P. Thomas, *Braid group actions on derived categories of sheaves*, preprint math.AG/0001043.

[SV] A. Shapere and C. Vafa, *BPS structure of Argyres-Douglas superconformal theories*, preprint hep-th/9910182.

[SYZ] A. Strominger, S.-T. Yau and E. Zaslow, *Mirror Symmetry is T-Duality*, Nucl. Phys. **B479** (1996), 243–259.

[T1] R. P. Thomas, *A holomorphic Casson invariant for Calabi-Yau 3-folds, and bundles on K3 fibrations*, preprint math.AG/9806111.

[T2] R. P. Thomas, *An obstructed bundle on a Calabi-Yau 3-fold*, Adv. Theor. Math. Phys. **3** (1999).

[T3] R. P. Thomas, *D-branes: Lagrangians, bundles and derived categories*, to appear as a chapter in the book *"Mirror Symmetry"*, ed. C. Vafa and E. Zaslow. Available at http://geometry.ma.ic.ac.uk/~rpwt.

[TY] R. P. Thomas and S.-T. Yau, *More moment maps, monodromy and mirror manifolds*, preprint.

[Ty] A. N. Tyurin, *Non-abelian analogues of Abel's theorem*, ICTP preprint (1997).

[Va] C. Vafa, *Extending Mirror Conjecture to Calabi-Yau with Bundles*, Comm. Contemp. Math. **1** (1999), 65–70.

[W1] E. Witten, *Chern-Simons gauge theory as a string theory*, The Floer memorial volume, 637–678. Progr. Math., 133, Birkhäuser, Basel, 1995.

[W2] E. Witten, *Branes And The Dynamics Of QCD*, Nucl. Phys. **B**507 (1997), 658–690.

[Y] S.-T. Yau, *On the Ricci curvature of a compact Kähler manifold and the complex Monge-Ampere equation*, Comm. Pure Appl. Math. **31** (1978), 339–411.

www.ingramcontent.com/pod-product-compliance
Ingram Content Group UK Ltd.
Pitfield, Milton Keynes, MK11 3LW, UK
UKHW020731190926
8624IPUK00035B/11

9 789810 247140